DYNAMIC RESEARCH, INC.

Pregnant Occupant Biomechanics - Advances in Automobile Safety Research

Other SAE International books of interest:

Forensic Biomechanics
By Jeffrey A. Pike
(Product Code R-379)

Child Anthropometry for Improved Vehicle Occupant Safety
Edited by Matthew Reed and Kathleen Klinich
(Product Code PT-142)

Vehicle Accident Analysis and Reconstruction Methods, Second Edition
By Raymond M. Brach and Matthew Brach
(Product Code R-397)

Pregnant Occupant Biomechanics -
Advances in Automobile Safety Research

Edited by

Stefan M. Duma

Chapter Authors:

Greg G. Duma

Stefan M. Duma

Melissa C. Hulse

Kathleen D. Klinich

Sarah J. Manoogian

Jonathan D. Rupp

Lawrence W. Schneider

Joel D. Stitzel

Published by
SAE International
400 Commonwealth Drive
Warrendale, PA 15096-0001 USA

Phone: (724) 776-4841
Fax: (724) 776-5760
www.sae.org
April 2011

PT-150

SAE *International*

400 Commonwealth Drive
Warrendale, PA 15096-0001 USA

E-mail: CustomerService@sae.org
Phone: 877-606-7323 *(inside USA and Canada)*
724-776-4970 *(outside USA)*
Fax: 724-776-1615

ISBN 978-0-7680-4849-0
Library of Congress Catalog Number 2011922588
SAE Order No. PT-150

Applicable to SAE International technical papers: The Engineering Meetings Board has approved these papers for publication. They have successfully completed SAE's peer review process under the supervision of the session organizer. This process requires a minimum of three (3) reviews by industry experts. Positions and opinions advanced in these papers are those of the author(s) and not necessarily those of SAE. The author is solely responsible for the content of the paper.

To purchase bulk quantities, please contact:
SAE Customer Service
E-mail: CustomerService@sae.org
Phone: 877-606-7323 *(inside USA and Canada)*
724-776-4970 *(outside USA)*
Fax: 724-776-1615

Visit the SAE Bookstore at
http://store.sae.org

Dedication

To *Chrissy, Lauren, and Brock*
For putting me on the pregnant occupant research path

TABLE OF CONTENTS

PREFACE

This book provides substantial new analysis and summary data for the field of pregnant occupant biomechanics. In addition, full-length manuscripts of selected publications are included to augment the new materials. The overall goal of this book is to provide the reader with a complete resource for issues relating to the pregnant occupant. The target audience is anyone in the broad field of automobile safety including engineers, physicians, and others who are dedicated to advancing automobile safety.

The book is divided into six chapters that were each authored separately by individual topic experts. The first three chapters set the foundation for the pregnant occupant problem, as the female anatomy, injury incidence, and case study are presented. The next three chapters present possible solutions in the form of pregnant dummies, tissue testing, and computational models. Each chapter begins with a summary and analysis of the research for that topic. Next, to provide depth and additional materials, a selection of references is followed by selected full-length paper reproductions.

This book will be a critical asset to anyone in the field of automobile safety. It will be of specific interest to safety design engineers, trauma physicians, OEM engineers, accident reconstruction specialists, occupant restraint suppliers, law enforcement officials, impact biomechanics researchers, and human factors specialists. Many government agencies will find this book useful, including the Department of Transportation (National Highway Traffic Safety Administration), the Department of Defense, and the Federal Aviation Administration.

The authors of each chapter selected the papers to include based on the significance and long-term reference value of each paper. Once the papers were selected, permission was obtained to reproduce each manuscript. In a few cases with older papers, full reproductions were not included due to inability to find high-quality copies of the original materials.

ACKNOWLEDGMENTS

First, I want to thank Martha Swiss and SAE International for their invaluable assistance and for providing the forum for this important publication. Second, I thank all of the chapter authors for taking the time to write, and assemble and review, their respective chapters. Finally, I thank Melissa Hulse for her effort in tracking down all of the copyright permissions as well as handling the review process with the separate chapter authors. It took nearly two years to complete, but this book will serve our field for decades to come.

I also want to acknowledge the Virginia Tech – Wake Forest University School of Biomedical Engineering and Sciences for the financial support of this book. In addition, I thank the organizations that granted permission to reproduce their papers: Association for the Advancement of Automotive Medicine, American Journal of Obstetrics and Gynecology, Biomedical Sciences Instrumentation, Journal of Biomechanics, Journal of Biomechanical Engineering, Journal of Trauma, Stapp Car Crash Journal, and SAE International.

CHAPTER 1

FEMALE ANATOMY

CHAPTER 1: FEMALE ANATOMY

Greg G. Duma, Sarah J. Manoogian, Kathryn L. Loftis, Joel D. Stitzel, and Stefan M. Duma
Virginia Tech – Wake Forest University, Center for Injury Biomechanics

Kathleen D. Klinich, Carl A. Miller, Jingwen Hu, Julia Samorezov, Jesse Buehler, Mark D. Pearlman, Lawrence W. Schneider, and Jonathan D. Rupp
University of Michigan Transportation Research Institute

The pelvic anatomy of the female has been studied since recorded history. Female anatomy has developed many accommodations to facilitate child bearing. Not only are there the obvious differences in the external anatomy, but even changes in the bony pelvic structure all aid in the birthing process. The pelvic anatomy of a female is somewhat complex and becomes more so in the state of pregnancy. This chapter is designed to be an overview for the student of engineering and is not meant to be a surgical atlas.

NONPREGNANT ANATOMY

Bony Pelvis

The bony pelvis is similar, although not exactly the same, in males and females. The anatomy of the bones is slightly different to adapt for childbearing. The pelvic bones are made up of the ilium, ischium, pubis, sacrum, and coccyx. In the adult female, the hip bones are formed from the ilium, ischium, and pubic bones which are fused together with cartilage. The hip bones are fused anteriorly in the midline at the symphysis pubis and connected posteriorly via the sacrum and coccyx. The ischium supports the weight of the body while in the seated position with the weight resting on the ischial tuberosities.

On the medial surface of the ilium is a structure known as the arcus tendineus. The muscles of the pelvic floor attach to the hip bones at this point. The sacrum and coccyx are extensions of the vertebra which have been fused. The most anterior portion of the sacrum is termed the sacral promontory. This landmark, palpable on a physical exam, is useful clinically as it represents the level where the internal iliacs divide, and is a point used in obstetrics to clinically assess the adequacy of the pelvis for labor. The ischial spines extend from the ischium, and the distance between the two represents the shortest diameter of the pelvis. This distance can be clinically measured to help assess risk of dystocia - difficult passage of the baby - during childbirth.

The typical female has a gynecoid-shaped pelvis with the anterior and posterior portions being fairly equal and forming a round pelvic diameter. This pelvic shape facilitates the birth process. The wider subpubic angle and increased distance from the symphysis to the acetabulum, or hip bone fossa, allows for easier delivery of the fetus.

Vagina and Cervix

The vagina, or birth canal, is the tubular structure connecting to the cervix. While in the standing position, the vagina lies in a horizontal plane. The cervix, attached to the uterus, extends into the vagina. Directly posterior to the cervix in the vagina, the tissue is thin. A needle can easily be passed through this thin tissue to enter into the abdominal or peritoneal cavity. The peritoneal cavity contains the internal structures such as the uterus and ovaries, as well as the loops of intestine, etc. This area of the abdominal cavity is known as the posterior cul de sac, as it is a sac structure posterior to the uterus. Likewise, an incision made in the vagina at this point will allow a larger access into the abdominal cavity.

Internal Organs

The uterus is located in the lower pelvis in a nonpregnant female. It is situated between the bladder anteriorly, and the rectum posteriorly. The intraperitoneal space in between is known as the anterior and posterior cul-de-sac, respectively. The posterior cul-de-sac is clinically important in early pregnancy, because a needle can be passed transvaginally into this space to obtain some of the intra-abdominal fluid. The uterus is divided into two parts, the upper body and the lower cervix. The cervix projects into the vaginal cavity. The portion of the uterus between the cervix and body is known as the isthmus. During late pregnancy, this area thins to become the lower uterine segment. A caesarean section incision is made in this segment. The dome-shaped area of the uterine body, in between the fallopian tubes, is known as the fundus. The uterus varies greatly in size and shape depending on the age of the female. Before puberty, the uterus will be 2.5-3.5 cm in length. An adult uterus is similar in shape to an inverted pear. A nulliparous adult uterus will be closer to 6-8 cm in length, 5 cm width, and 70 g weight. A multiparous uterus will be about 1-2 cm larger in each direction and weigh an additional 20 g. The uterine body is made up of smooth muscle in three distinct layers. During pregnancy, this muscle hypertrophies, and the weight of the uterus can increase to 1100 g with a volume of about 5 L.

The inner cavity of the uterus is a flat triangular-shaped area known as the endometrium. This area is lined with endometrial glands. The corners of the triangle end at the cervical opening and the two lateral fallopian tube openings. Only the outer layer of the endometrial glands responds to the changing hormones of the menstrual cycle and pregnancy. The majority of fluid excreted during the menstrual cycle is comprised of endometrial glands, with true blood making up only a small portion of a normal cycle.

The cervix consists primarily of fibrous material. The central canal running the length of the cervix is lined with endocervical glands. This canal connects the endometrial cavity of the uterus to the vagina (Fig. 1). In the adult female, the cervical length is approximately 3-4 cm. The cervical opening is normally open only enough to let out blood during the menses. The fibrous connective tissue in the body of the cervix will slowly stretch to open the canal to a diameter of 10+ centimeters during labor to allow a baby to pass through.

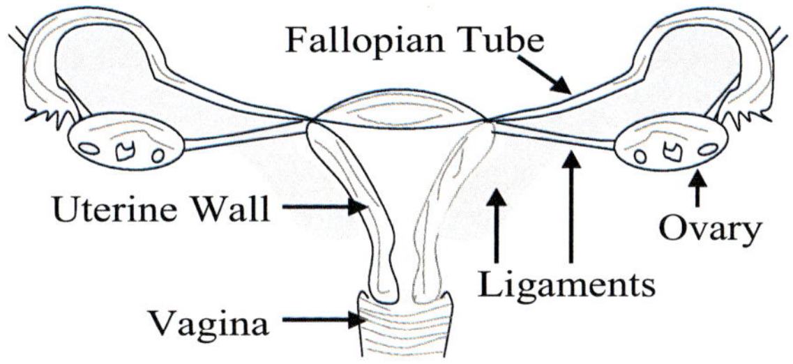

Figure 1: Diagram of the vagina and surrounding structures.

The fallopian tubes, or oviducts, emerge laterally from the uterus near the superior aspect. The peritoneal covering of the fallopian tubes extends to the pelvic sidewall. This structure is known as the broad ligament. The fallopian tubes connect the inside of the uterine cavity to the inside of the peritoneal cavity. Externally, they are 10-14 cm in length and about 1 cm wide. The internal canal is 1-3 mm in diameter. The portion extending from the uterus is known as the isthmus. The central portion is the ampulla, while the distal fimbriated end is the infundibulum. This fimbriated end has fan-like fingers projecting into the abdomen. These lie near the ovary and pick up the monthly ovulated egg. Typically, fertilization takes place in the ampullary portion of the tube. The fertilized egg then continues into the endometrial cavity to implant. If this egg is fertilized and remains in the fallopian tube, an ectopic pregnancy results. A tubal ectopic pregnancy is by far the most common type of ectopic, but by definition, an ectopic pregnancy is any pregnancy outside of the uterus.

The ovaries, which are about 5 cm in length, 2.5 cm width and height, lie on the posterior side of the broad ligament. They are attached by peritoneum as well as the ovarian and utero-ovarian blood vessels. At birth, the ovaries contain 1-2 million eggs, or oocytes. Less than 10,000 of these oocytes will start to develop as ovarian follicles, and over the reproductive years, only approximately 400 eggs will actually be released. The ovaries are mobile, but held in proximity to the fallopian tubes by the broad ligament, the ovarian ligament, and the infundibulopelvic ligament. The ovarian ligaments are connective tissue joining the uterus to the ovary. The infundibulopelvic ligaments are the names given to the structures containing the ovarian vessels arising from the pelvic sidewall.

Pelvic Ligaments

The broad ligament is comprised of two layers of peritoneum which are continuous with the pelvic sidewall. The two layers drape over the fallopian tubes and uterus. Running inside of the broad ligaments bilaterally are the uterine and ovarian arteries, the ureters, and fallopian tubes. The broad ligament is very mobile and provides little true support to the uterus. At the base of the uterus where it joins the cervix, the broad ligament condenses to form the cardinal ligament. The uterine arteries enter the uterus at this point. The cardinal ligaments are a thickened condensation of fibrous tissue and fascia that connect the cervix laterally to the pelvic diaphragm. These ligaments hold the cervix in place in the upper vagina and provide the main support for the uterus and cervix.

The round ligaments extend from the lateral fundus, just anterior to the fallopian tubes. They are comprised of smooth muscle and fibrous tissue, covered in peritoneum, and continuous with the broad ligament. The round ligaments pass anteriorly through the inguinal canal and terminate in the labia majora. They also supply very little support to the pelvic structures.

The uterosacral ligaments are one of the main supports for the nonpregnant uterus. They are comprised of smooth muscle and connective tissue. They leave the uterus bilaterally at the lower posterior surface and pass on either side of the rectum to insert into the sacral fascia.

Although all of the ligaments mentioned attach the uterus to the pelvis, the only fixed attachment is at the cervix. This allows the uterus to remain mobile within the abdominal cavity. As the uterus grows in pregnancy, it is confined in movement only by the abdominal wall.

Blood Supply

The blood supply to the female reproductive organs is all bilateral, paired, and enters the organs laterally only to mix with the vessels from the opposing side near the midline. This extensive collateral system allows for the large flow volume of blood during pregnancy. This also allows for normal pregnancies even after ligation of both uterine arteries.

The uterus receives its main blood supply from the uterine and ovarian arteries. The uterine artery arises from the hypogastric artery. The hypogastric artery itself has an anterior and posterior branch. The posterior branch gives off the iliolumbar, lateral sacral, and superior gluteal arteries. The anterior branch has wide variation from person to person, but typically gives off the uterine and vaginal arteries, the obturator and internal pudendal arteries, the superior and middle vesicle arteries, and the middle rectal and hemorrhoidal arteries. The uterine artery enters the uterus at the level of the uterine cervical junction in the cardinal ligament. This is a clinically important location, as the ureter passes directly underneath. The uterine artery then passes up the lateral side of the uterus in the broad ligament to anastomose (join) with the ovarian artery.

The ovarian artery is a direct branch of the aorta. It leaves the pelvic sidewall in the infundibulopelvic ligament. It interconnects with the ovary in the broad ligament and then continues on to the uterus. The main branches of both the uterine artery and ovarian branches communicate with each other on the lateral sides of the uterus. During pregnancy, the uterus receives over 500 ml of blood per minute. This is clinically relevant, as any injury to these vessels or the uterus can quickly cause massive hemorrhage. The great extent of communication between these vessels allows a female who has previously had both of the uterine vessels ligated to carry a normal pregnancy. In addition, in cases of severe hemorrhage in the female pelvis, it can be necessary to ligate the hypogastric artery. Because of the extensive collateral system in the pelvis, ligation can stop the hemorrhage while not resulting in hypoxia of the pelvic organs.

The venous supply of the pelvis typically follows the accompanying arteries. They then drain into the inferior vena cava. The exception is the left ovarian vein, which drains into the left renal vein.

FEMALE ANATOMY DURING PREGNANCY

The female genital tract and overall anthropometry undergoes considerable changes during pregnancy (Acar 2006 and 2009). The hormones estrogen and progesterone are responsible for many of these changes early, while stretching of the uterus and dislocations of pelvic organs from the growth of the fetus are responsible for later changes (Fig. 2).

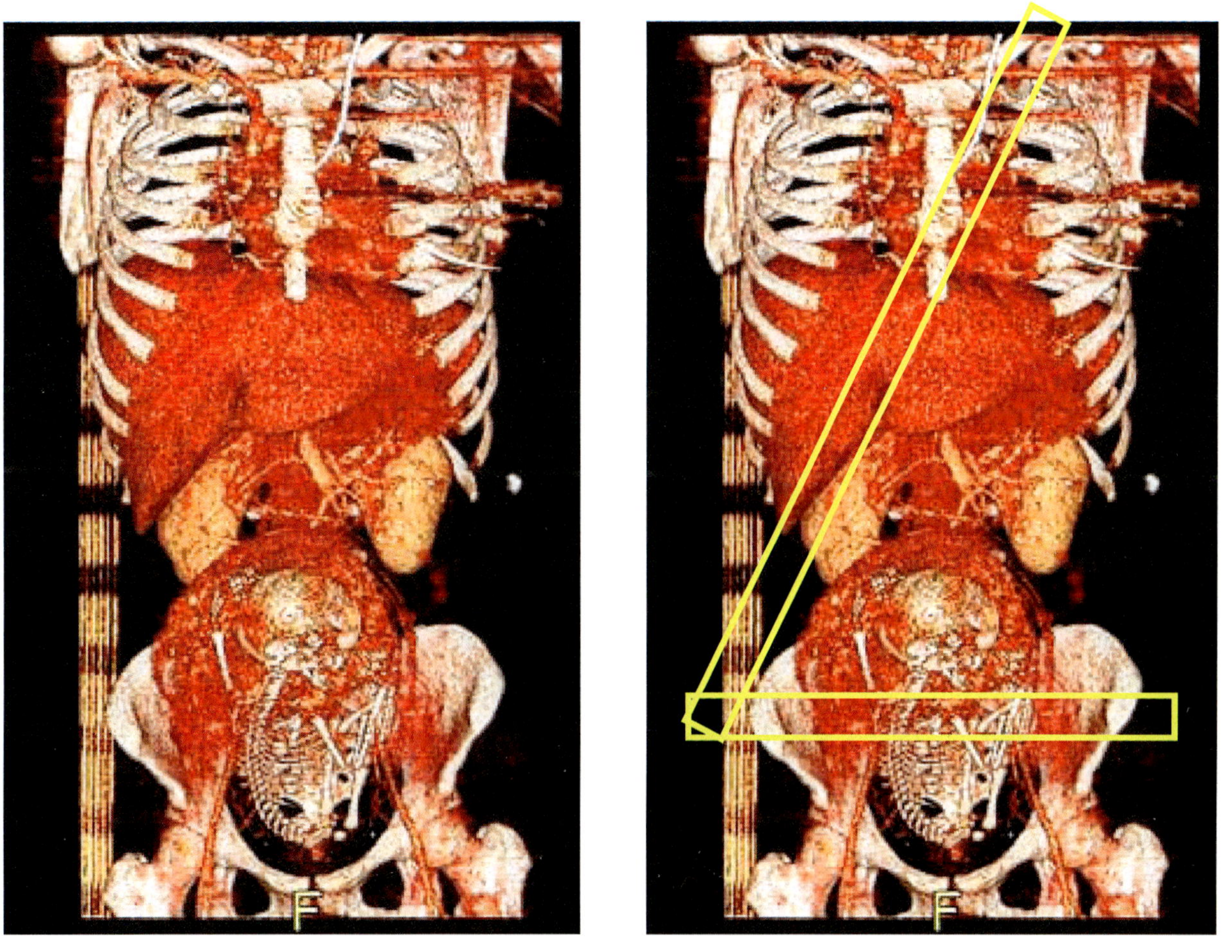

Figure 2: CT Scan of a pregnant occupant involved in a motor vehicle crash. Notice the correct placement of the seatbelt superimposed over the scan on the right side.

The uterus at full term (40 weeks) will be much greater in size and can have a volume of up to 5 L. The term fetus and amniotic fluid will lie within the uterus. The uterus will have undergone physiological changes to sustain fetal life and prepare for delivery of the fetus and fetal membranes. The anatomy discussed in this section will pertain to the uterus and the placenta (Fig. 3).

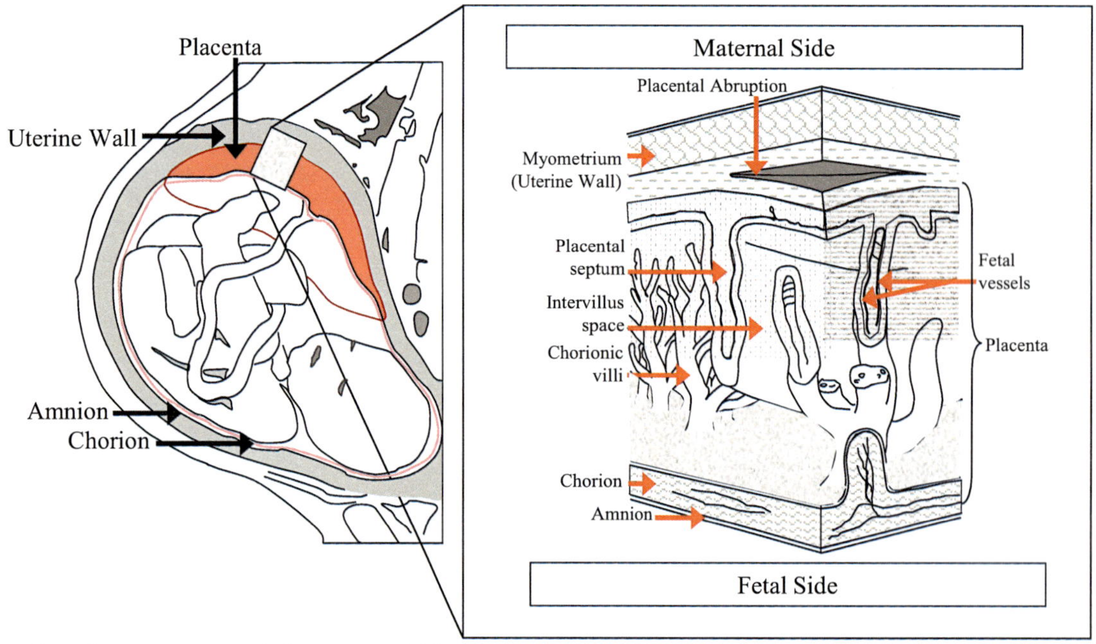

Figure 3: The pregnant abdomen at full term normally has the placenta located at the fundus of the uterus.

The uterus expands to include a term fetus, the placenta, and amniotic fluid. A nonpregnant uterus weighs approximately 50-70 g. During pregnancy, hypertrophy and hyperplasia of the smooth muscle occurs. This is due to hormonal changes as well as the growing fetus. The uterine body at term weighs approximately 800-1200 g. The uterine body has a 10-ml internal volume in the nonpregnant state, which increases to around 5 L at term but can be as high as 20 L. As the size of the uterine muscle cells increase, there is also an increase in the fibrous tissue which adds strength to the uterine wall. Early in pregnancy, there is a thickening of the uterine wall, but by around 12 weeks, as the fetus becomes the main distending factor, the uterine wall thins. The thickness for the 32 week gestation uterus varies from superior to anterior location (Fig. 4) (Loftis 2008). By term, the uterine wall is very thin due to the anatomic distension from the fetus, placenta, and amniotic fluid. A term uterus will have a thickness of about 1.5 cm at the fundus and only a few millimeters in the lower uterine segment.

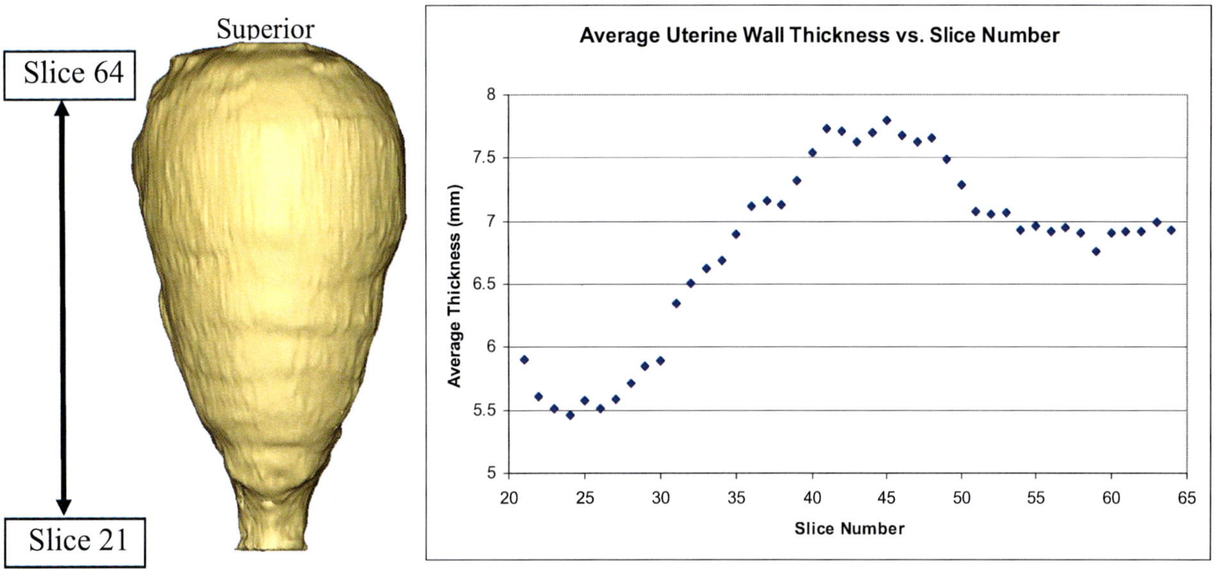

Figure 4: Uterine wall thickness at different locations for 32 week gestation.

The uterus assumes various shapes during pregnancy (Fig. 5). Prior to 6 weeks it has a pyramid shape. At 12 weeks, when the uterus leaves the pelvis, it is spherical in shape. As previously mentioned, the isthmic portion of the uterus becomes remarkably stretched during pregnancy to become the lower uterine segment, which can be only a few millimeters thick. After 12 weeks, as the uterus fills the abdominal cavity, the internal organs of the abdomen are displaced upwards (Fig. 6). The appendix, which typically lies in the right lower quadrant, will be at the level of the umbilicus at 20 weeks gestation. The enlargement of the uterus is not symmetrical, with the majority occurring in the fundus. The uterus remains mobile during pregnancy and is limited in movement only by the abdominal cavity. The uterus is held in place in the standing position by the abdominal wall. The rectus muscles in the abdominal wall meet in the midline. During pregnancy these can separate to such a degree that the uterus protrudes anteriorly and is only covered by the peritoneum and skin.

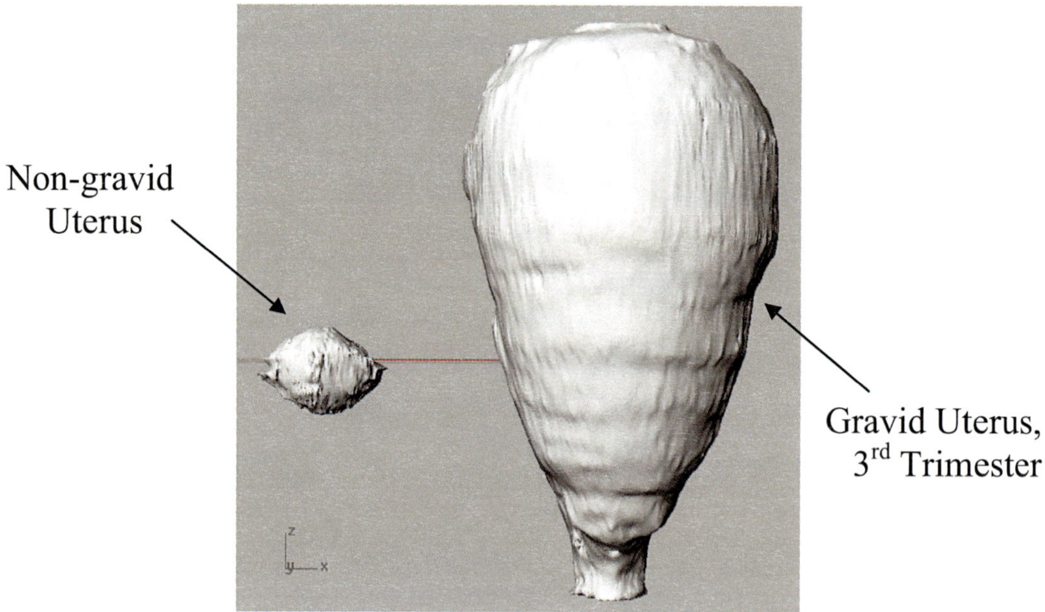

Figure 5: Uterus shown when non-gravid and into the 3rd trimester.

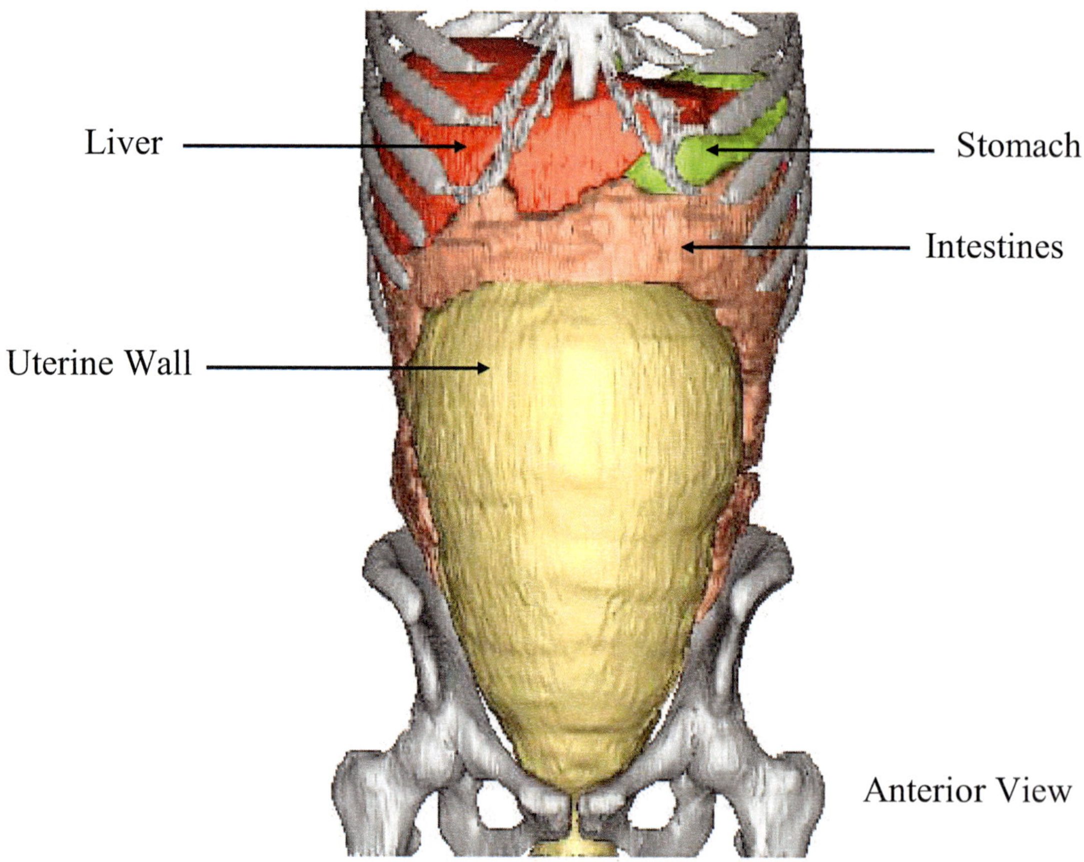

Figure 6: 3D rendering of a prone 5th percentile pregnant female in early 3rd trimester.

The cervix changes in position during pregnancy from typical anteversion early, to a more vertical position as the uterus leaves the pelvis. The cervix becomes markedly softer as the proliferation of the glandular tissue occupies about one half of the cervix.

There is a dramatic increase in blood flow to the uterus in pregnancy, which corresponds to the increase in size. While the nonpregnant baseline flow is 30-40 ml/min, this increases to 500 ml/min at term. The blood volume during pregnancy increases up to 45% over the nonpregnant state. An increase in cardiac output to approximately 5 L/min, along with a decrease in uterine vascular resistance, account for this increase in uterine blood flow.

In the uterus, not at the location of the placenta, there are three layers between the fetus and the myometrium. The myometrium is the smooth muscle layer of the uterine wall. Inside the smooth muscle layer, the uterus transitions to a decidua layer. The decidua layer is a result of hormones activating a transformation in the innermost surface of the endometrium referred to as the decidua vera. The next two layers, the chorion and amnion, are important to molecular transfer and metabolic activity. The chorion is a thin membrane, less than 1 mm thick, that is between the amnion and decidua layers. The innermost fetal membrane is the amnion, which is tough and pliable and defines the amniotic sac. The body of the placenta lies under the chorionic membrane. The chorionic membrane is directly adherent and under the amniotic membrane. The amniotic membrane forms the amniotic sac in which the amniotic fluid and fetus lie. Inside the amniotic sac, the fetus is surrounded by amniotic fluid.

The placenta is normally located at the fundus of the uterus, although it can be located in any position in the uterus (Figs. 7 and 8). The fetal connection to the placenta is through the umbilical cord, which contains two arteries and one vein that spiral together from the fetus to the placenta. The fetal surface of the placenta is covered by the amnion and then contains the chorionic plate where the fetal vessels from the umbilical cord branch out to the edges of the placenta (Fig. 9). The blood flow reaches the lobes and the smaller villi of the placenta for fetal maternal transfer. As the placenta grows, the larger villi continuously branch into smaller villi. The original stem villi and all of its smaller branches form a single cotyledon, or lobe. Each cotyledon has a single artery and vein from the mother. There are multiple shapes and sizes of placentas. The placenta at term weighs 508 g and has an average diameter of 18.5 cm with a variable thickness (Fig. 10). The maternal surface of the placenta consists of these lobes separated by grooves referred to as placenta septa. These grooves are filled in on the maternal side of the placenta with the decidua layer. At the placental bed site, the decidua layer is the decidua basalis. The decidua layer is then attached to the myometrium of the uterine wall. The fetus is supplied with oxygen and nutrients via the placenta. Maternal blood flowing to the uterus reaches the placenta by endometrial arteries. The blood actually leaves the maternal vessels and bathes the villi of the placenta. The fetal blood remains in the intervillous capillaries of the placenta. Under normal conditions, the fetal blood and maternal blood never come into contact with one another. This is relevant clinically for a pregnant occupant who has been involved in a motor vehicle crash (MVC) to detect possible traumatic injury. After an MVC or any type of maternal trauma, a blood test is done on the mother in an attempt to detect any fetal blood cells. Fetal blood cells can act as foreign bodies and cause the mother to mount an immune response against the pregnancy.

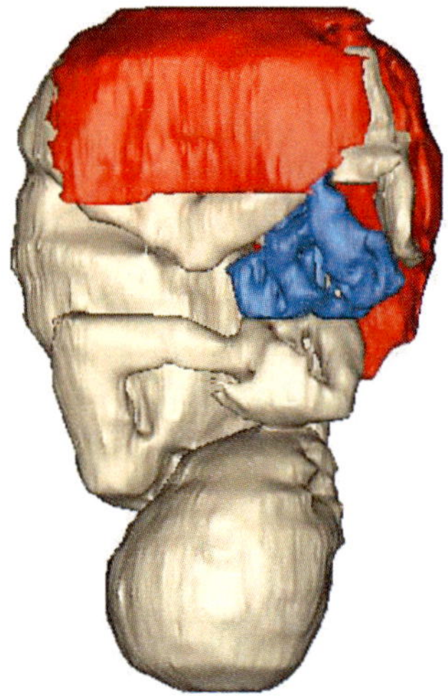

Figure 7: 3-D representation of a fetus with umbilical cord and fundal placenta (posterior view).

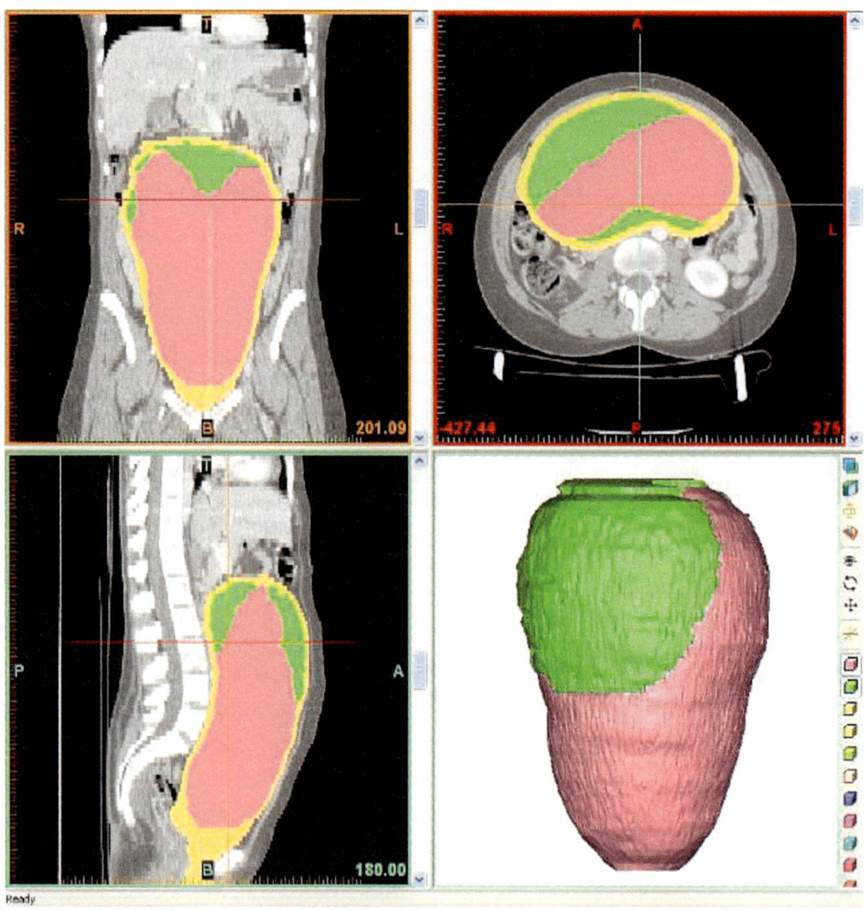

Figure 8: Different CT scan views of a placenta.

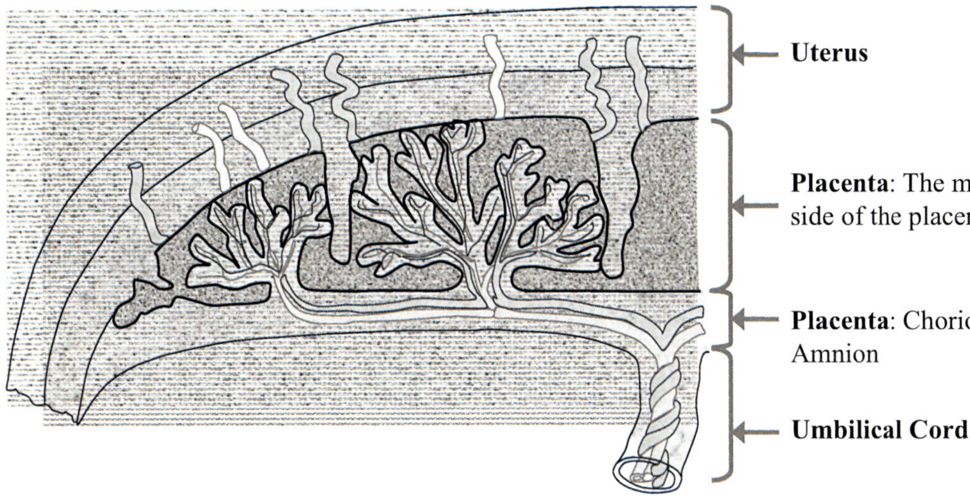

Figure 9: The structure of the placenta varies from the maternal to the fetal side.

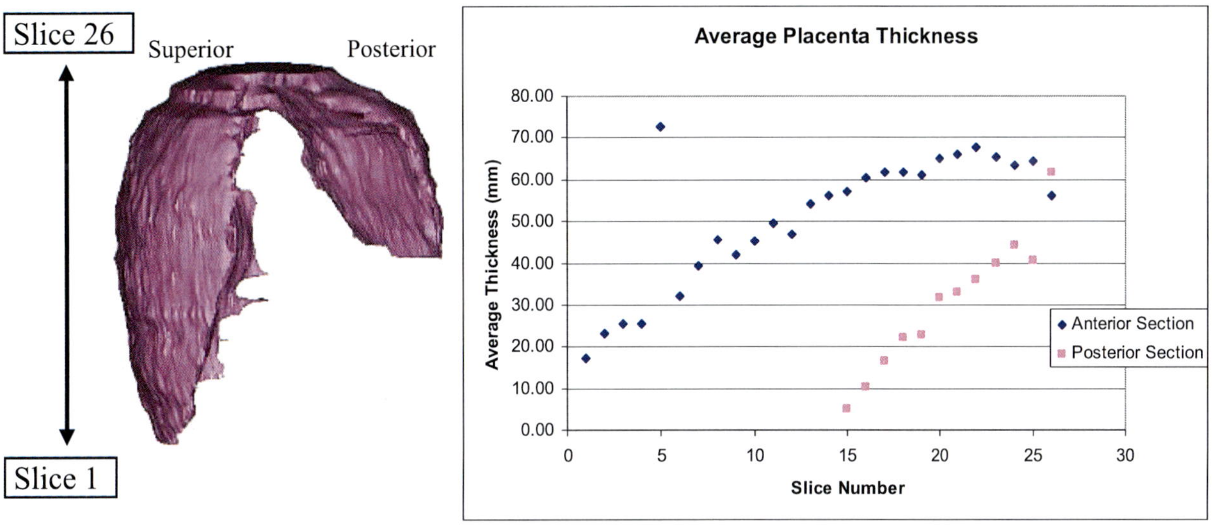

Figure 10: CT scan of an isolated placenta with the average placenta thickness by slice.

In natural birth, the fetus is delivered and the fetal membranes are delivered in the afterbirth, separating from the uterine wall at the decidua layer. The contraction of the uterus aids in the delivery of these membranes.

Fetal positioning in the abdomen is described in terms of lie, presentation, and position. The position refers to the relationship between a presenting part and the right and left side of the birth canal. The presentation is the part of the fetus lowest in the birth canal, such as the head in a cephalic presentation or the buttock in a breech presentation. The lie of the fetus relates the long axis of the fetus to that of the mother. It is either longitudinal, transverse, or oblique.

Fetal Development

The development of the fetus is described in terms of gestational age. The beginning point is the first day of the last menstrual cycle. This date is actually around two weeks before ovulation, and 4 weeks before implantation. Thus, when a pregnancy of 4 weeks gestational age is described, the mother has actually only been pregnant for 2 weeks. This still remains the terminology for dating pregnancies. The pregnancy is divided into trimesters, each lasting 14 weeks. The "due date" is given as the date of 40 weeks gestational age.

During the first 8 to 10 weeks of gestational age, the pregnancy is referred to as an embryo. The first embryonic cell multiplication takes place approximately 3 days after fertilization. The developing embryo – or blastocyst – will implant within the uterine cavity between 21-28 days: 4 weeks gestation. The human blastocyst differs from many other species in that it is entirely embedded in the endometrium after implantation. During this embryonic period, the heart is completely formed. Fetal cardiac activity will begin at about 5 weeks gestational age and can be detected on transvaginal ultrasound between 6 and 6.5 weeks. Once fetal cardiac activity is demonstrated, it is unusual to have a fetal loss (Goldstein and Snyder). The neural plate, which forms the central nervous system of the fetus, begins to form at 4 weeks gestation. This will become a neural tube and close by 6 weeks. Once this neural tube has closed, no teratogens can reopen it to cause defects. After completion of 8 weeks, the embryo will be approximately 2 cm long. By the end of 10 weeks, the arms, legs, fingers, and toes are also present. At the end of 10 weeks, the embryo is called a fetus.

At 12 weeks, the fetus will show spontaneous movements. At this point, the uterus has enlarged enough to be palpable during an abdominal exam. By careful examination, the gender of the fetus can usually be determined by the end of the first trimester – 14 weeks. The digits on each extremity will be visible at 12 weeks. During this time the head is relatively big compared to the rest of the body. At 24 weeks gestation, the lungs will start to form respiratory sacs and release surfactant. Surfactant reduces the surface tension in the alveoli and enables them to remain open for oxygen exchange after birth. At this time the developing fetus will attempt to breath. This is the gestational age where a fetus born will have a small chance of normal survival. At 26 weeks, the fetus will be able to hear sounds. By the end of the second trimester, the fetal eyes are sensitive to light.

Ultrasound is used to determine the gestational age of a fetus. In early gestation, prior to 8 weeks, the crown rump length is used for the calculation. This will have an accuracy of ±3 days. As the fetus matures, the accuracy of the ultrasound decreases. A general rule of thumb states that in the second trimester, an ultrasound will have an error of ±2 weeks, and ±3 weeks in the third trimester. Different parts of the fetus can be measured. The average is taken to determine gestational age (Fig. 11).

Approximate Gestational Age*		
Measurement	**Data (mm)**	**Age (weeks)**
Crown to Heel	360.91	29.5
Avg. Foot Length	58.13	30.1
Avg. Femur Length	63.87	32.5
Avg. Humerus Length	58.94	33.9
Occipitofrontal Diameter	101.84	32.0
Biparietal Diameter	87.83	34.9
Head Circumference	297.78	32.7
Avg. Approximate Gestational Age		32.2

*Clinicians estimate that all gestational age calculations are +-3 weeks accuracy at this stage of gestation.

Arneson, S., et al., J Obstet Gynecol Neonatal Nurs, 1986. 15(4): p. 339-44.
Oyelese, Y. and C.V. Ananth, Obstet Gynecol, 2006. 108(4): p. 1005-16.
ACOG educational bulletin. Int J Gynaecol Obstet, 1999. 64(1): p. 87-94.

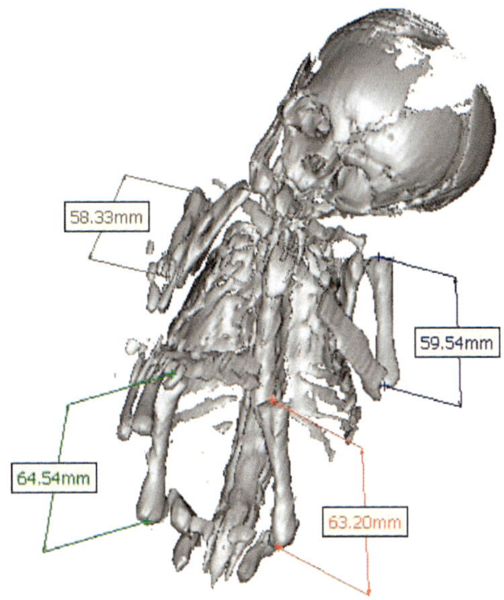

Figure 11: Linear skeletal measurements from 3-D rendering from masks.

INTERSPECIES COMPARISON OF THE PLACENTA, UTERUS, AND UTEROPLACENTAL INTERFACE

The characteristics of the uteroplacental interface (UPI) in humans have not been studied in vivo because the placenta naturally separates from the uterus and is delivered after birth. Researchers have used a variety of animal models to study the UPI, but choosing an appropriate animal model to use as a surrogate for human placenta and its interface is challenging, as the placenta is the organ with the greatest variation in structure and function among all species (Carter 2007, Ramsey 1975, Leiser and Kaufmann 1994). The structure of the uterus also varies among species. Placentas are classified partly by how they are formed, so a simplified explanation of placenta development is described below (Johnson and Everitt 2000, Benirschke and Kaufmann 2000).

Before a fertilized egg implants into the uterine wall, it is called a blastocyst. The outer wall of the blastocyst is called the trophoblast. Within the blastocyst, some cells will become the embryo, some become the yolk sac, which provides nutrition, and some become the allantois, or embryonic bladder, which handles gas and waste exchange. In most mammals, the placenta is formed by the trophoblast, and the fetal vasculature develops from the allantois; this is called an allantochorial placenta. If the fetal vasculature develops from the yolk sac, it results in a choriovitelline placenta.

Table 1 summarizes the key anatomical characteristics and functions of the uterus and placenta of humans compared to other species (Barry and Anthony 2008, Bernirschke and Kaufmann 2000, Carter 2007, DiGiacomo et al. 1978, Kaufmann and Davidoff 1977, Khong and Pearce 1987, Leiser and Kaufmann 1994, Leiser et al. 1997, Moll and Kunzel 1973, Moll et al. 1973, Perry 1981, Ramsey 1975, Suzuki et al. 1997). The table is limited to species that have been used experimentally and that have allantochorial placentas like the human.

Table 1: Characteristics of placenta and uterus of different species with allantochorial placentas.

Species	Gestational age (weeks)	Placental Shape	Fetomaternal Inter-digitation	Flow Pattern	Placental Barrier	Maternal Attachment	Efficiency	Uterus Shape	Maternal Blood Pressure at Placenta
Human	38	Discoid	Villous	Multivillous	Hemomono-chorial	Deciduate	6	Simplex	
Rhesus monkey	23	Bi-discoid	Villous	Multivillous	Hemomono-chorial	Deciduate		Simplex	
Guinea pig	10	Discoid	Labyrinthine	Countercurrent	Hemomono-chorial	Deciduate	20	Bicornuate	12
Rabbit	5	Discoid	Labyrinthine	Countercurrent	Hemodi-chorial		6	Duplex	8
Rat	3	Discoid	Labyrinthine	Countercurrent	Hemotri-chorial	Deciduate	10	Bicornuate	14
Mouse	3	Discoid	Labyrinthine	Countercurrent	Hemotri-chorial				
Cat	9	Zonary	Lamellar	Crosscurrent	Endothelio-chorial		8		
Pig	16	Diffuse	Folded	Double crosscurrent	Epithelio-chorial		9		
Sheep	21	Cotyledonary	Villous	Multivillous/ crosscurrent	Epithelio-chorial	Nondeciduate	10	Bicornuate	85

For the species listed, gestation durations range from 3 weeks in smaller rodents to 38 weeks for humans. After the human, the sheep and Rhesus monkey have the longest gestations, while the guinea pig has the longest gestation among rodents.

There are five main placental shapes. Diffuse placentas are attached along the entire interior uterine wall. Cotyledonary placentas consist of multiple regions of attachment which appear to be multiple small placentas (up to 100). In ruminants such as the sheep, each placental portion forms within a pocket in the uterus called a caruncle and the combination is called a placentome. Zonary placentas attach to the uterus in a cylinder around the fetus. Bidiscoid placentas have two main disc-shaped placentas, while discoid placentas have a single disc-shaped placenta (human).

To increase the maternofetal contact area, different species have developed various forms of interdigitation, or interlocking connective structure, between the fetal and maternal tissues. These interdigitations are classified as folded, lamellar, villous, or labyrinthine. Diagrams of each type are shown in Fig. 12. A folded placenta consists of "grooves" of fetal tissue that fit into "grooves" of uterine tissue. Lamellar placentas are similar, except the "grooves" are more complex, branching into multiple parallel layers. With a villous placenta, the fetal tissue grows as a tree-like structure into blood-filled pockets on the maternal side (both human and sheep). The tree-like vessels have stems that branch off into intermediate and terminating villi. For a labyrinthine placenta, the trophoblast forms a block of tissue that contains the fetal capillaries and maternal blood.

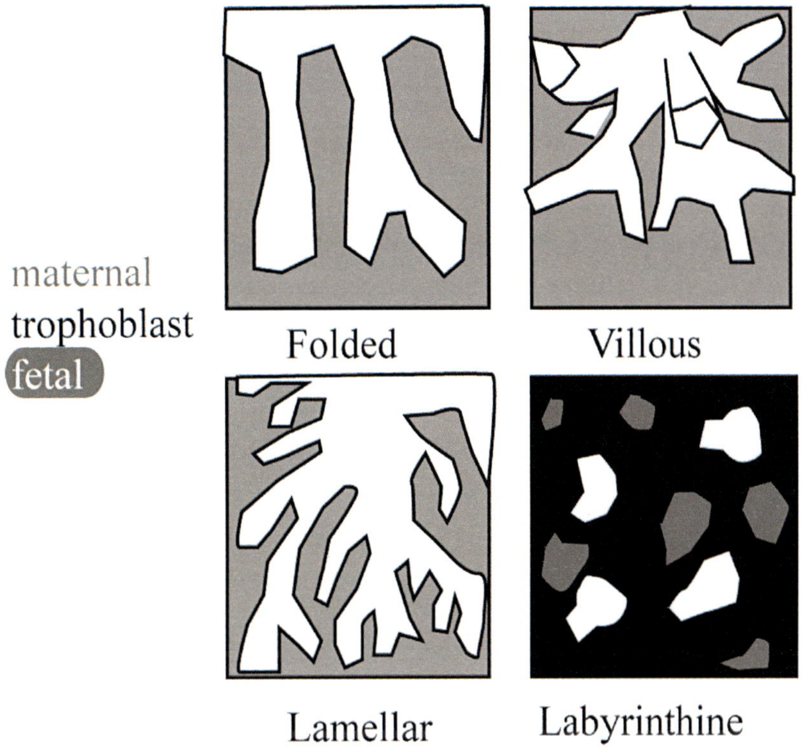

Figure 12: Comparison of villous, labyrinthine, lamellar, and folded fetal-maternal interdigitation (based on Bernirschke and Kaufmann 2000). Trophoblast is black, maternal tissue is gray, and fetal tissue is white.

Another characteristic of the fetal-maternal interface is the way blood flows in the fetal and maternal vasculature. Possible flow patterns are described as multivillous, crosscurrent, or countercurrent. In multivillous systems (human), the maternal blood collects in the spaces between villi. Transfer of nutrients to the fetus occurs through the villous fetal blood vessels that are immersed in these blood-filled spaces. With countercurrent flow patterns, the maternal and fetal vessels are adjacent, with the maternal and fetal blood flowing in opposite directions. This allows the most efficient nutrient and waste transfer between the vessels. For crosscurrent flow, the maternal and fetal vessels are perpendicular to each other, and a similar transfer process occurs. With some species, such as sheep, goats, and cows, some of the smaller maternal vessels penetrate the villi to intertwine with the fetal vessels, resulting in a combination of villous and crosscurrent flow.

Different types of tissue separate the maternal and fetal systems; this tissue is called the maternofetal barrier or membrane. Up to six layers of cells can form the membrane: trophoblast, connective tissue, and endothelium on the fetal side, and epithelium, connective tissue, and capillary tissue on the maternal side. With an epitheliochorial placenta, all of the layers remain intact throughout the pregnancy. This is the most complete barrier between the uterus and placenta, because the blastocyst is attached to but does not penetrate the endometrium. With an endotheliochoral placenta, the maternal epithelium and connective tissues are destroyed by the invasion of the trophoblast. With a hemochorial placenta (human), these two tissues, as well as the maternal capillary tissue, are destroyed, such that the trophoblast tissue directly contacts the maternal blood. The classification into hemotrichorial, hemodichorial, and hemomonochorial refers to the number of trophoblast epithelial layers.

Related to the maternofetal barrier is the classification of placentas according to the attachment to the maternal side as deciduate or nondeciduate. In deciduate species, such as humans, the fetal trophoblast fuses into the uterine endometrium, where some of the trophoblast cells develop into the placenta. After delivery, part of this maternal endometrium remains attached to the placenta and is delivered with the placenta. In nondeciduate species, such as sheep, fusion of the placenta and the uterine wall does not occur and separation of the placenta from the uterus is accomplished by a simple peeling off process between the fetal and maternal tissue layers.

One method for determining the suitability of one species as a surrogate for another in placental studies is the efficiency of placental nutrient transfer, because nutrient transfer depends on the type of connection between maternal and fetal tissues. In nature, humans have one of the least efficient placentas, with only 6 g of fetus produced for every gram of placenta. Mice are one of the most efficient at 30:1. Sheep are closer to humans at 10:1.

The shape of the uterus is classified as simplex, duplex, or bicornuate. A simplex uterus has a single cavity and is only found in humans and higher primates. A bicornuate uterus appears heart shaped, with two horns of the uterus extending upwards from the cervix. A duplex uterus has two completely separate horns.

Another factor to consider in evaluating the suitability of an animal model to study human placenta is the blood pressure of the vessels entering the placenta from the maternal side. Of the animals for which this measure is available, sheep are closest to humans.

This review indicates that no animal placenta matches all characteristics under consideration for studying human placenta and its interface to the uterus. Placental studies in the literature indicate that pregnant sheep are the most commonly used surrogate for pregnant humans because of many similarities in placental structure and function (Barry and Anthony 2008, Carter 2007, Leiser and Kaufman 1994, Leiser et al. 1997). However, most previous studies have focused on maternal-fetal nutrient transfer mechanisms.

The Rhesus monkey has one of the most closely related placental structures to humans, sharing a discoid shape, villous interdigitation, and deciduate nature. While the Rhesus monkey shares many characteristics with humans, its availability, expense, and ethical considerations have led to its limited use in research studies.

Rodents may offer another good UPI surrogate for humans. Like humans, the rat and guinea pig have decidual, discoid placentas. However, rats and guinea pigs possess a labyrinthine, rather than a villous fetal-maternal interdigitization, and have crosscurrent blood flow. Compared to sheep, rodents are a better match regarding the deciduate nature of the UPI, but the labyrinthine structure may be a more intertwined connection than the villous connection of sheep and humans. Guinea pigs are also different from humans in their higher level of nutritional efficiency (20 vs. 6).

REFERENCES

Acar, B. S. and A. M. Weekes. 2006. "Measurements for Pregnant Occupant Safety." *International Journal of Vehicle Design* 42(1-2):101-118.

Acar, B. S., D. van Lopik, and A. M.Weekes. 2009. "'Expecting': Occupant Model Incorporating Anthropometric Details of Pregnant Women." *International Journal of Vehicle Design* 51(3-4):374-385.

Barry, J. S. and R. V. Anthony. 2008. "The Pregnant Sheep as a Model for Human Pregnancy." *Theriogenology* 69(1):55-67.

Benirschke, K. B. and P. K. Kaufmann. 2000. *Pathology of the Human Placenta*, 4th Ed. Berlin: Springer.

Carter, A. M. 2007. "Animal Models of Human Placentation - A Review." *Placenta* 28(A):S41-S47.

DiGiacomo, R. F., P. W. Shaughnessy, and S. L. Tomlin. 1978. "Fetal-Placental Weight Relationships in the Rhesus (Macaca mulatta)." *Biology of Reproduction* 17:749-753.

Finn, C. A. and D. G. Porter. 1975. *The uterus.* London: Elek.

Kaufmann, P. and M. Davidoff. 1977. *The Guinea-Pig Placenta.* Berlin: Springer-Verlag.

Khong, T. Y. and J. M. Pearce. 1987. "Development and Investigation of the Placenta and Its Blood Supply." In *The Human Placenta: Clinical Perspectives.* Aspen Publishers, Rockville, MD. 25-45.

Leiser, R. and P. Kaufmann. 1994. „Placental Structure: In a Comparative Aspect." *Experimental and Clinical Endocrinology* 102:122-134.

Leiser, R., C. Krebs, B. Ebert, and V. Dantzer. 1997. "Placental Vascular Corrosion Cast Studies: A Comparison Between Ruminants and Humans." *Microscopy Research and Technique* 38:76-87.

Loftis, K., M. Halsey, E. Anthony, S. M. Duma, and J. D. Stitzel. 2008. "Pregnant female anthropometry from ct scans for finite element model development." *Biomedical Sciences Instrumentation* 44:355-360.

Marrable, A. W. 1971. *The Embryonic Pig: a Chronological Account.* Pitman Medical: London.

Mizrahi, J. and Z. Karni. 1975. "A Mechanical Model for Uterine Muscle Activity during Labor and Delivery." *Israel Journal of Technology* 13:185-191.

Moll, W. and W. Kunzel. 1973. "The Blood Pressure in Arteries Entering the Placentae of Guinea Pigs, Rats, Rabbits, and Sheep." *Pflugers Arch* 338:125-131.

Moll, W., W. Kunzel, L. A. M. Stolte, J. Kleinhout, P. A. de Jong, and A. F. L. Veth. 1973. "The Blood Pressure in the Decidual Part of the Uteroplacental Arteries (Spiral Arteries) of the Rhesus Monkey." *European Journal of Physiology* 346(4):291-297.

Perry, J. S. 1981. "The Mammalian Fetal Membranes." *Journal of Reproduction and Fertility* 62:321-335.

Ramsey, E. M. 1975. *The Placenta of Laboratory Animals and Man.* New York: Holt, Rinehart, and Winston.

Suzuki, K., M. Kobayashi, K. Kobayashi, Y. Shiraishi, S. Goto, and T. Hashino. 1997. "Structural and Functional Change of the Blood Vessel Labyrinth in Maturing Placenta of Mice." *Trophoblast Research* 9:155-164.

Weiss, S., T. Jaermann, P. Schmid, P. Staempfli, P. Boesiger, P. Niederer, R. Caduff, and M. Bajka. 2006. "Three-dimensional fiber architecture of the nonpregnant human uterus determined ex vivo using magnetic resonance diffusion tensor imaging." *The Anatomical Record, Part A* 288(1):84-90.

Young, R. C. and R. O. Hession. 1999. "Three-dimensional structure of the smooth muscle in the term-pregnant human uterus." *Obstetrics and Gynecology* 93(1):94-99.

The following paper is included in its entirety to augment this chapter:

Loftis, K., M. Halsey, E. Anthony, S. M. Duma, and J. D. Stitzel. 2008. "Pregnant female anthropometry from ct scans for finite element model development." *Biomedical Sciences Instrumentation* 44:355-360.

PREGNANT FEMALE ANTHROPOMETRY FROM CT SCANS FOR FINITE ELEMENT MODEL DEVELOPMENT

Kathryn Loftis[1], Michael Halsey[1], Evelyn Anthony[2], Stefan M. Duma[1], Joel Stitzel[1]
[1]Virginia Tech – Wake Forest University Center for Injury Biomechanics
[2]Wake Forest University School of Medicine, Medical Center Blvd. Winston-Salem, NC 27157

ABSTRACT

In this study, anthropometry data is collected from a CT scan of a pregnant abdomen at 32 weeks gestation. Over 1500 fetal losses occur each year in the United States due to motor vehicles crashes. Pregnant occupants involved in motor vehicle crashes are at risk for pregnancy-specific injuries. Masks of the fetus, uterus, placenta, and each of the abdominal organs are created by segmentation of the CT slices and three-dimensional volume renderings are formed. The volume and Hounsfield unit ranges for the masks of each abdominal organ are calculated. The total volume of the uterus in the 3rd trimester is measured as 3378 cm^3. By measuring the length of bones on the fetal skeleton from CT slices and the 3D rendering, the gestational age of the fetus is estimated to be 32 weeks by comparison with literature values. Measurements of each of the abdominal organs are also obtained from the 3D rendering to create a blueprint of the pregnant anatomy. The masks developed and the anthropometric measurements taken will be used to develop a more accurate FE model of the pregnant female for use in the research and development in academia, industry, and government.

Keywords: pregnant, anthropometry, finite element, modeling, computed tomography

INTRODUCTION

Motor vehicle crashes (MVCs) caused 42,642 deaths in the United States in 2006 [1]. MVCs involving pregnant mothers are reported as causing an estimated 1500-5000 fetal losses each year in the United States [2]. These demonstrate the need for an accurate crash test model of the pregnant female at different stages of pregnancy. This paper introduces methods and data for creating an anthropometrically correct pregnant female to incorporate into a finite element model of the female human for use with virtual crash tests.

In the latter stages of pregnancy, the gravid fetus causes the mother's abdomen to protrude from her body, and this makes the abdomen susceptible to trauma during a MVC. Common injuries to a pregnant female after a MVC include: excessive bleeding, placental abruption, uterine rupture, and premature delivery [3]. Even when the mother does not receive serious injury during a MVC, the fetus or placenta are injured such that the outcome is fatal for the fetus [4-6]. This is because the protruding abdomen absorbs some of the force of the crash, and this force is transmitted to the uterus, placenta, and fetus. The most common fatal injury for the fetus when the mother survives blunt force trauma, such as that sometimes experienced in a MVC, is abruptio placentae [7]. This is caused by a shearing effect between the placenta and the uterus during rapid impact and deceleration. The uterus can also be torn from the placenta due to the rebound of the uterus after the impact has occurred [7]. Often, this injury is not evident until it is too late to save the fetus [8].

An accurate pregnant model could be used to research the protective effect of current restraint systems and could also be used to develop new restraint system advancements to protect mother and child more successfully. This model would be helpful in determining the results of airbag deployment in MVCs involving pregnant females. It has been suggested in the literature that pregnant females should turn off their airbags during the late stages of pregnancy. The inflation of the airbag has been hypothesized to result in higher abdominal pressures because the gravid uterus protrudes out to the steering wheel. [9] An accurate pregnant female model could test this theory in order to make public health recommendations on the best methods for travelling safely in a vehicle while pregnant.

Changes during pregnancy drastically alter the abdominal anatomy of the pregnant female. In this study, a CT scan from an early 3rd trimester pregnant female is segmented using medical imaging software, and 3D models of the pregnant abdomen are created. This study investigates the anatomy of a pregnant female and seeks to determine gestational age by measuring 3D renderings of the fetus from CT scan data.

METHODS

A scan most representative of a fifth percentile female in the early 3rd trimester with a healthy fetus and no abdominal injury was selected for analysis from available abdominal computed tomography (CT) scans taken over the past ten years. The selected scan was then imported into a medical image analysis program (Mimics, Materialise Software, Ann Arbor, MI) for thresholding. The software was used to segment the CT scan into masks and create 3D models of the uterus, placenta, volume inside the uterus, fetus, fetal skeleton, maternal skeleton, and maternal internal organs.

The masks of the pregnant subject were created using a combination of automatic functions available in the software and manual editing with the aid of a touch screen monitor. Manual error was reduced by viewing different threshold levels for each type of soft and hard tissue within the CT scan. After the fetal skeletal mask was created, the skeletal measurements of the fetus were taken using a 3D measurement tool provided by the software. The fetal skeleton 3D rendering was used to measure the crown-to-heel length, foot length, femur length, and humerus length. A combination of the 3D rendering and fetal brain CT slices were used to measure the occipitofrontal and biparietal skull measurements. For accuracy, the crown-to-rump length measurement and rump-to-heel length measurement were added together to get the total crown-to-heel length measurement. The foot length was measured from toe to heel on each foot, and then these measurements were averaged. Each femur, humerus, and foot length was measured for each side of the body and then they were averaged. The measurements were then compared with literature values to estimate gestational age [10-12].

The placental mask was created differently than the other masks to ensure an exact fit. The placental surface adjoining the uterine wall mask was defined using a boolean operation to subtract the uterine wall from the placental mask. This allowed for a uniform assessment of the uterus-placental attachment. A mask was next created to define the volume inside the uterus and placenta. This mask was created using the cavity fill function. A single layer cavity fill was performed on each slice of the uterine wall and placenta mask so that a new mask defining the inner volume of the uterus was created. A boolean operation was then performed to subtract the placenta from the volume mask. The resulting mask was the volume interior to the uterine wall and placenta.

From the volume renderings, measurements were taken for each region of the pregnant female using the 3D measurement tool. These measurements were exported to a mathematical program where the length, height, width, and depth in each plane were found for each region.

RESULTS

The following results were found by creating masks of a pregnant female abdomen and all internal structures from transverse slices of a CT scan. All of the results are from the same individual in her 3rd trimester without any anatomical irregularities.

By measuring the height of T8-T12 on the CT scan and comparing it to literature data, the female is considered to be 5th percentile with an estimated height of 4' 11" and estimated weight of 104 lbs, as shown in Table 1 [13]. The T8-T12 distance of 101.51 mm is in the range listed for a female of this size. Her size had to be estimated using T8-T12 height from the CT scan because her height and weight were not available in medical records. She appeared to be of normal to low weight as evidenced by the amount of subcutaneous fat in her abdomen.

Table 1. Female T8-T12 size, based on ranges for a 5th percentile female.

Small Female Measurement	
Description	**T8-T12 Measurement (mm)**
Pregnant subject	101.5
Literature mean	128.9
Literature low	94.8
Literature high	135.4

Figure 1 shows how the uterus is positioned inside the abdomen. This also shows how crowded the abdomen has become in this stage of pregnancy. The intestines and other abdominal organs have been displaced superiorly due to the enlarged uterus. Figure 2 shows how crowded the fetus is within the uterus and also how the placenta is wrapped around the fundus.

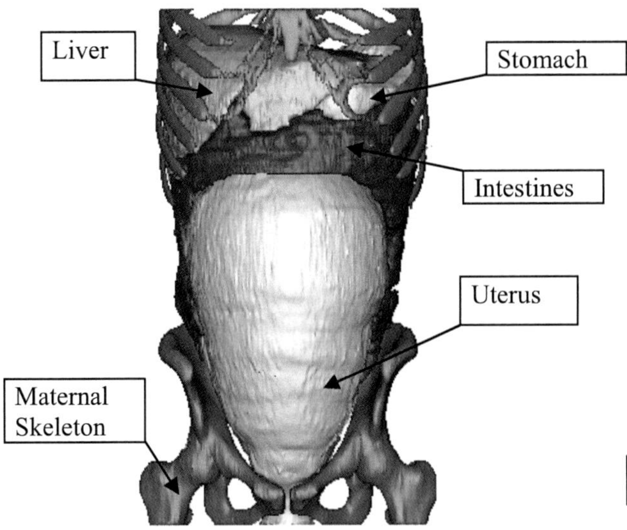

Figure 1. Anterior view of pregnant female masks with fat, skin, and muscles removed

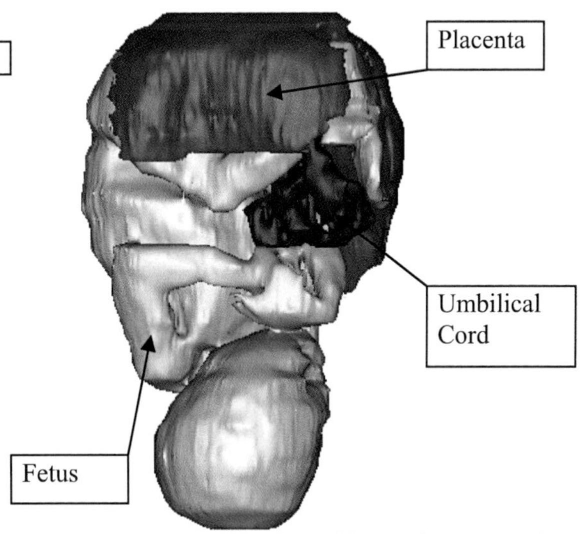

Figure 2. Posterior view of fetus, placenta, and umbilical cord

Table 2 contains the mask data collected. The computer program provided the volume, CT Hounsfield unit range, and average Hounsfield units for each mask created. Using these volume measurements, the 3rd trimester uterus and all its contents have a volume of 3377 cm^3.

Table 2: Mask data

Mask Name	Volume (cm^3)	Final Thresholds (HU)			Details
		Low	High	Average	
Uterine Volume	2579.0	-150	1177	42	Volume interior to uterine wall and placenta
Skin	329.2	-1024	60	-444	Rough mask of skin
Pancreas	84.6	-134	222	113	Pancreas mask
Uterine Wall	111.0	-1024	578	62	Uterus mask
Fat	6082.0	-199	141	-96	Rough mask of body fat under skin
Maternal Skeleton	1953.8	-328	1582	382	Maternal skeleton
Gallbladder	36.2	-134	174	42	Gallbladder mask
Stomach	455.8	-1024	417	-35	Stomach and duodenum mask
Placenta	687.7	-69	271	102	Placenta mask
Abdominal Muscles	1018.9	-150	174	29	Abdominal muscles mask
Fetal Skeleton	197.1	109	1177	250	Fetal skeleton mask
Liver	1769.7	-555	433	129	Liver mask
Intestines	1800.6	-1024	595	-50	Intestines mask
Fetus	1697.7	-150	3071	59	Fetal mask
Kidneys	402.0	-101	287	169	Kidney mask
Spleen	180.7	-134	287	143	Spleen mask
Umbilical Cord	95.0	-69	3071	33	Umbilical cord mask

Bone length measurements of the fetus were taken from the 3D rendering and CT scan and were compared to literature data to estimate the gestational age of the fetus. These results can be found in Table 3.

Table 3. Table of measurements used in determination of gestational age

Approximate Gestational Age		
Measurement	**Data (mm)**	**Age (weeks)**
Crown to Heel [12]	360.91	29.5
Avg. Foot Length [11]	58.13	30.1
Avg. Femur Length [10]	63.87	32.5
Avg. Humerus Length	58.94	33.9
Occipitofrontal Diameter	101.84	32.0
Biparietal Diameter	87.83	34.9
Head Circumference	297.78	32.7
Avg. Approximate Gestational Age		32.2

DISCUSSION

This study has been successful in generating quantitative data on the geometry of the pregnant female and fetus and has resulted in masks of the 3rd trimester pregnant abdomen. In estimating the pregnant female's size it was determined that she was approximately the same size as current 5th percentile female crash test dummies and this makes her measurements applicable to many current dummy and computational models. The masks of the fetus, uterus, placenta, liver, kidneys, gallbladder, pancreas, stomach, and spleen are representative volume calculations for this individual in her 3rd trimester. These measurements provide a starting point for creating an accurate Finite Element model of the pregnant female for use with crash tests for the automotive industry and academia.

Gestational age estimations are usually made from the brain and skull anatomy using ultrasound [14]. Using CT, it was difficult to locate the specific brain landmarks where these skull measurements were to be taken because of slice spacing and resolution of scans. The measurements were compared with data found in the literature values in Table 3, and these measurements estimated the fetus to be approximately 32 weeks gestation. The emergency department reported medical record of the pregnant female listed the gestation age to be 32 weeks. This report agreed with the gestational age as determined by the measurements. At this stage of gestation, clinicians estimate that all gestational approximations are ±3 weeks accuracy. Upon measurement, the biparietal diameter was slightly large and the occipitofrontal diameter was small. The head circumference calculation averaged the two measurements and allowed for a measurement independent of the shape of the head [15]. The approximate gestational age predicted by crown-to-heel measurement could be underestimated, because at this gestational stage it is difficult to perform this measurement due to the size and position of the fetus. This is why many measurements of the fetus were taken as a means of confirming the gestational age and also as a means to record the dimensions of a fetus in the proper stage of development.

While the skin, fat, and maternal skeletal volumes and measurements will differ for all females, the organ size and uterus, placenta, and fetus sizes are relatively consistent for pregnant females. The size of the uterus in the 3rd trimester is determined by the size of the fetus, not the size of the mother. This makes these organ volumes applicable for primarily all pregnant females, independent of body size. There are, however, some limitations to this study because it only utilized one pregnant female. While it is common for the placental attachment point to be at the fundus, other placental locations and individual anatomical differences will create different results. Additional studies at different gestational times would also be useful in creating multiple pregnant models for testing, but the 3rd trimester pregnant female was chosen to obtain geometry in a more distended state.

In the future, further studies using a larger study group could be used to validate the results of this study. The average uterus area for each slice could be calculated and used to create an accurate representation of the amount of amniotic fluid that is present. This could then be incorporated into the pregnant female model to predict how the amount of incompressible fluid within the uterus affects the fetus during blunt force trauma. After creating an accurate model of the uterus-placenta attachment, it may be possible to predict where abruption will occur by looking at thin points in the uterine wall and thick areas of the placenta. This prediction technique would also be helpful in formulating fetal injury criteria independent of maternal injuries.

CONCLUSION

Masks and 3D volume renderings have been created for each maternal abdominal organ from a CT scan of a pregnant female. Organ volume, fetal gestational age, and anthropomorphic measurements have been calculated from these masks and renderings. The uterine wall thickness was found to differ depending on the slice location within the CT scan of the pregnant female. The placenta thickness also varied in relation to slice thickness and was thicker at the fundus. It was possible to correctly predict fetal gestational age from CT scan by comparing fetal skeleton measurements to literature values. Lastly, anthropometric measurements were collected for the pregnant female and fetus, so that there is now a blueprint for creating a pregnant female finite element model. These organ renderings and measurements will aid in creating a more correct anatomical model of the pregnant female and fetus.

ACKNOWLEDGMENTS

We would like to thank the Ford University Research Program for sponsorship.

REFERENCES

[1] United States Department of Transportation, National Highway Traffic Safety Administration, *Traffic Safety Annual Assessment 2006*. DOT HS 810 791, Washington, DC, 2007.

[2] M. Hitosugi, Y. Motozawa, M. Kido, T. Yokoyama, H. Kawato, K. Kuroda, and S. Tokudome, "Traffic injuries of the pregnant women and fetal or neonatal outcomes," *Forensic Sci Int,* vol. 159, pp. 51-4, May 25 2006.

[3] L. K. Hyde, L. J. Cook, L. M. Olson, H. B. Weiss, and J. M. Dean, "Effect of motor vehicle crashes on adverse fetal outcomes," *Obstet Gynecol,* vol. 102, pp. 279-86, Aug 2003.

[4] Y. Bunai, A. Nagai, I. Nakamura, and I. Ohya, "Fetal death from abruptio placentae associated with incorrect use of a seatbelt," *Am J Forensic Med Pathol,* vol. 21, pp. 207-9, Sep 2000.

[5] P. F. Agran, D. E. Dunkle, D. G. Winn, and D. Kent, "Fetal death in motor vehicle accidents," *Ann Emerg Med,* vol. 16, pp. 1355-8, Dec 1987.

[6] D. L. Farmer, N. S. Adzick, W. R. Crombleholme, T. M. Crombleholme, M. T. Longaker, and M. R. Harrison, "Fetal trauma: relation to maternal injury," *J Pediatr Surg,* vol. 25, pp. 711-4, Jul 1990.

[7] W. M. Crosby and J. P. Costiloe, "Safety of lap-belt restraint for pregnant victims of automobile collisions," *N Engl J Med,* vol. 284, pp. 632-6, Mar 25 1971.

[8] A. M. Connolly, V. L. Katz, K. L. Bash, M. J. McMahon, and W. F. Hansen, "Trauma and pregnancy," *Am J Perinatol,* vol. 14, pp. 331-6, Jul 1997.

[9] S. Aschkenazi, J. Kovanda, J. Stingl, J. Karl, and A. Schoenfeld, "Airbags: benefits and risks in the pregnant population," *International Journal of Risk & Safety in Medicine,* vol. 11, p. 85, 1998.

[10] "Fetus Growth Charts," in *Ultrasound Examination Measurements*. www.Baby2see.com, 2007.

[11] M. Chatterjee, L. Izquierdo, B. Nevils, G. Gilson, C. Barada, "Fetal foot: evaluation of gestational age." Albuquerque, NM: TheFetus.net, 2007.

[12] "The Medical Algorithms Project," Release 19.1, Houston, Texas: Institute for Algorithmic Medicine, 2006-2007.

[13] L. Schneider, D. Robbins, M. Pflug, R. Snyder, "Development of Anthropometrically Based Design Specifications for an Advanced Adult Anthropomorphic Dummy Family." vol. 1, The University of Michigan Transportation Research Institute, Ed.: U.S. Department of Transportation, p.426, 1983.

[14] J. Woo, "Obstetric Ultrasound." vol. 2007 Hong Kong. http://www.ob-ultrasound.net/xbpd.html.

[15] "Diagnostic and Interventional Radiology Department," in *Ultrasound of Fetal Biometrics and Growth.* Omaha, Nebraska: Creighton University Medical Center. http://radiology.creighton.edu/fetalbio.htm#section2, 2007.

CHAPTER 2

MATERNAL AND FETAL INJURY INCIDENCE

CHAPTER 2: MATERNAL AND FETAL INJURY INCIDENCE

Melissa C. Hulse, Sarah J. Manoogian, and Stefan M. Duma
Virginia Tech – Wake Forest University, Center for Injury Biomechanics

The overall objective of this chapter is to explore maternal trauma and fetal death and injury rates as they relate to motor vehicle crashes (MVCs). The chapter is divided into two major sections, maternal injury and fetal death rates. The maternal injury section examines previous studies which address how many pregnant women are affected by trauma, and more specifically, MVCs. It also addresses the severity of the trauma from MVCs, from no trauma to death of the pregnant woman.

While the United States does not record all fetal deaths as they relate to MVCs, research has been conducted to estimate this number. The second section of this chapter looks at the different methods used to establish fetal death rates.

MATERNAL INJURY FROM TRAUMA

In 2005, there were an estimated 6.41 million pregnancies in the United States. Of those pregnancies, 4.14 million were live births, 1.21 million were terminated, and 1.06 million were fetal losses (Ventura et al, 2009). While 4.14 million represents the majority of pregnancy outcomes, it also includes a percentage of women who suffered some sort of trauma. This trauma can be minor and have no short- or long-term effect, or it can be life changing to the pregnant woman and/or fetus.

Overall Maternal Related Traumas

Peckham and King (1963) suggest that 6.9% of women who are pregnant will suffer some sort of trauma while pregnant. The authors reviewed the cases of 3072 pregnant women from three months before their last menstrual period (LMP) up to the termination of the pregnancy (Table 1). The women were all participants in a prepaid health plan in Oakland, California. Their findings revealed that there were 85 traumas suffered by 83 women before their LMP and an additional 222 traumas suffered by 214 pregnant women after their LMP. So of the 3072 women who were pregnant, 214 suffered some type of trauma (6.9%).

Table 1: The numbers and types of traumas suffered by pregnant women as reported by Peckham and King (1963).

	Trimester of pregnancy			
Condition	1st	2nd	3rd	Total
Motor vehicle related crashes	19	22	18	59
Drug or food poisoning	1	2	1	4
Accidental falls	12	26	22	60
Accident caused by cutting or piercing instrument	13	9	5	27
Accident purposely inflicted by another person	5	1	3	9
Other accident or injury	20	17	26	63
Total	70	77	75	222

In a study conducted in North Carolina by Connolly et al. (1997), they examined the medical records of 476 women whose pregnancies were complicated by trauma from July 1987 through October 1993. Motor vehicle crashes (MVCs) caused 54.6% of all traumas (Fig. 1). Twenty-two point three percent were due to domestic violence. Falls accounted for 21.8% of traumas, and the remaining 1.3% were due to other traumas (burns, puncture wounds, or animal bites). Connolly et al. (1997) found that the MVCs tended to occur throughout the entire pregnancy, with fewer crashes after 36 weeks. Domestic abuse-related traumas (32%) mainly occurred before 18 weeks, and falls (33.1%) were more common between 25 and 30 weeks of pregnancy.

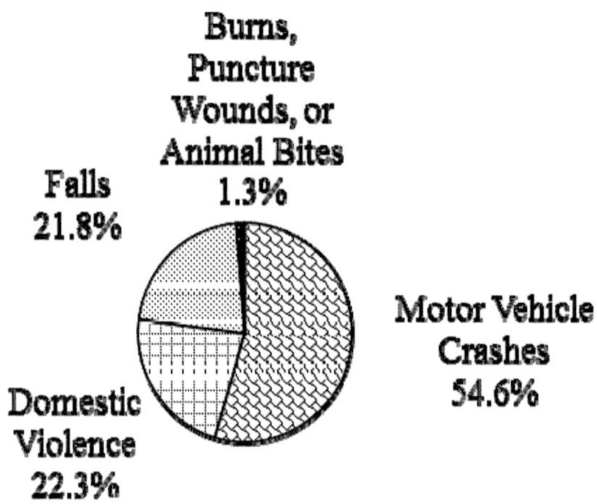

Figure 1: Breakdown of pregnancy-related traumas as reported by Connolly et al. (1997).

Pearlman et al. (1989) found similar results for a study looking at the records of 85 women who suffered trauma during pregnancy: MVCs accounted for 60% of traumas, falls were 25.9%, and direct blows to the abdomen were 14.1% (Fig. 2). The results also showed that of the 85 women in the study, 88.2% appeared to have minor or no injuries when they presented at the hospital, 9.4% had moderate injuries, and 2.4% of the injuries were critical. The study further reported that 9.4% of the pregnant women suffered severe complications (placental abruption) as a result of their traumas, but those traumas were not predictable based on their injury severity being minor or major traumas. In fact, Pearlman et al. (1989) report that placental abruption occurred in 1-5% of minor injury cases and 20-50% in major traumas for pregnant women.

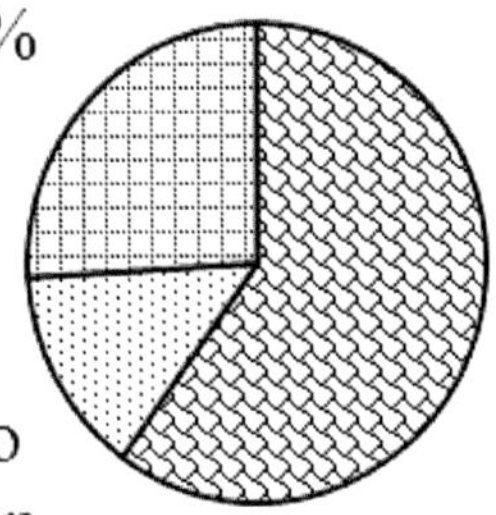

Figure 2: Breakdown of pregnancy-related traumas as reported by Pearlman et al. (1990).

Since the study by Peckham and King (1963), the number of miles that women drive has increased. This may help to explain why later studies (Connolly et al. [1997] and Pearlman et al. [1989]) find that women have a higher percentage of traumas caused from vehicular crashes than they do from falls or domestic abuse/violence.

Overall Maternal Vehicular Trauma

Reviewing pregnant occupants in the National Automotive Sampling System (NASS) from 1999 to 2008, 98.1% had moderate, minor or no injuries and 3.2% had critical, major or serious, major injuries, based on the maximum abbreviated injury scale (MAIS) and adjusted for actual fatalities that occurred (Table 2). The MAIS represents the highest abbreviated injury scale (AIS) for any body region. These findings are somewhat similar to those of Pearlman et al. (1989), even though these findings are only for traumas related to pregnancy and MVCs, not all traumas. Unweighted data show a total of 946 tow-away MVCs between 1999 and 2008; therefore, an average of 94.6 tow-away MVCs involving pregnant women per year.

Table 2: Weighted percentage of pregnant women who are involved in tow-away crashes and their corresponding MAIS score (NASS CDS 1999 to 2008).

MAIS score	No injury (0)	Minor (1)	Moderate (2)	Serious (3)	Severe (4)	Critical (5)	Unsurvivable (6)
Unweighted number of pregnant women	213	560	77	55	10	10	21
Weighted number of pregnant women	102738	198505	14880	3990	331	409	1365
Average number per year	10274	19850	1488	399	33	41	136
Percent of pregnant occupants in tow-away crashes	31.9%	61.6%	4.6%	1.2%	1.0%	1.0%	0.4%

Klinich et al. (1999) applied census data (census.gov) of the number of females between the ages of 15 and 44 to the number of births from the National Center for Health Statistics (cdc.gov) to help to derive the percentage of women who were over 20 weeks pregnant in 1996 (# births/U.S. population=% pregnant) (Table 3). The authors then took the percentage of pregnant women per age group and multiplied it by the number of women killed, injured, or uninjured in 1996 via MVCs per age group (which was obtained from Traffic Safety Facts [http://www-nrd.nhtsa.dot.gov/Cats/listpublications.aspx?Id=E&ShowBy =DocType]) and multiplied that figure by 20/52 to represent only those pregnant women over 20 weeks gestation. For example, they took the number of women killed multiplied by the percent pregnant multiplied by 20/52. Those calculations resulted in an estimate of the number of pregnant women who were killed, injured, or uninjured in MVCs. They estimated that a total of 128,255 pregnant women were involved in vehicular damage crashes, with 29,923 of those women being injured and 164 being killed.

Table 3: An estimate of the number of pregnant women who were involved in MVCs and their injury-related outcomes from Klinich et al. (1999).

	Age Groups (years)*				
	15-19 or 16-20	20-24 or 21-24	25-34	35-44	Total
U.S. Population 11/98 (census.gov)	9,478,000	8,666,000	19,232,000	22,395,000	59,771,000
1996 Births (cdc.gov)	494,272	951,247	1,982,740	472,473	3,900,732
Percentage pregnant	5.20%	11.00%	10.30%	2.10%	6.50%
From 1996 Traffic Safety Facts					
women killed (AIS 6)	1,725	919	1,916	1,713	6,273
women injured (AIS 1-5)	3,010,000	164,000	367,000	297,000	3,838,300
women uninjured** (AIS 0)	956,228	548,243	1,193,475	1,049,804	3,747,750
total women in vehicle-damage crashes	1,258,953	713,162	1,562,391	1,348,517	4,883,023
Number of pregnant women greater than 20 weeks gestational age (percentage pregnant * 20/52)					
pregnant women killed (AIS 0)	35	39	76	14	164
pregnant women injured (AIS 1-5)	6,037	6,924	14,552	2,410	29,923
pregnant women uninjured (AIS 0)	19,180	23,146	47,324	8,518	98,168
pregnant women in vehicle-damage crashes	25,251	30,109	61,952	10,942	128,255

*first age grouping used by census and CDC, second age grouping used in Traffic Safety Facts

**number of drivers reported explicitly; number of passengers estimated based on ratios of drivers and passengers killed or injured.

Motor Vehicle Crashes and Seatbelt Use

Pearlman and Phillips (1996) surveyed 350 pregnant women at their first prenatal visit and again at their 28-32 weeks' gestational visit concerning use and placement of seat belts. Their findings showed that nearly 20% reported rarely or never using their seatbelts during pregnancy. They also found that when given information of seat belt use from their prenatal provider pregnant women were more likely to wear their seatbelts (83% vs. 65%; $p < .001$) and wear them properly (77% vs. 57%; $p<.001$) than those who did not receive information.

In the same paper, the authors suggested that to find the number of MVC traumas during pregnancy one would take the number of pregnant women who delivered, and multiply it by the number of pregnant women who suffer trauma complications (6.5%). Then to separate out only MVCs from other forms of trauma, one would multiple that figure by 50%, which Pearlman and Phillips (1996) used as the percentage of women who suffer

MVC traumas (# of deliveries x 6.5% trauma occurrences x 50% trauma from MVCs = # of pregnant women who suffer trauma from MVCs).

Hyde et al. (2003) studied whether maternal involvement in MVCs affected the pregnancies themselves and whether seatbelt use played a part in those outcomes. The population they studied was pregnant women drivers in Utah for the years of 1992-1999. They looked at MVC records and birth and fetal death certificate databases. The authors only included singleton births. Eight thousand, nine hundred thirty eight of those who gave birth, or 2.8%, were involved in MVCs during pregnancy. The findings also showed that women who were pregnant and involved in a MVC were not significantly more at risk than pregnant women who were not in a MVC. With regard to seatbelt use, unbelted pregnant women were twice as likely to suffer excessive maternal bleeding as their counterparts who wore seatbelts. Additionally, pregnant women who did not wear seatbelts and were involved in MVCs were 2.8 times more likely to suffer the loss of the fetus than those pregnant women who wore seatbelts.

Weiss and Strotmeyer (2002) found after looking at weighted results for NASS/CDS for the years 1995-1999, of all the 12,399 women aged 15 to 39 years old who were involved in MVCs, 2.8% or 427 were pregnant. They found that pregnant women were similar to nonpregnant with regard to restraint use (no belt use for 14% of pregnant drivers vs. 13% for non-pregnant drivers), and driver status (70% of pregnant women were the drivers in the MVCs vs. 71% for nonpregnant). Pregnant women were more likely to be transported to the hospital and hospitalized even though their mean severity score was lower than those women who were not pregnant.

Summary of Vehicular Traumas in Pregnant Occupants

Figure 3 combines all of the information above to give the reader an overall summary of injury findings for pregnant women. One can see that of the 6.1 million pregnancies, 4.1 million resulted in live births. Of those live births, 6.9% or 282,900 suffer some form of trauma. Fifty-four point six percent to 60% of those traumas are from MVCs, which averages out to be 162,102 pregnancies. Ninety-seven point six percent to 98.1% of those pregnant women suffer no injuries to moderate injuries (averaging 158,617). The remaining 2.4-3.2% (averaging 4134) of pregnant women who suffer trauma from MVCs suffer serious to critical trauma. The findings show that the number of deaths varies from 136 to 164.

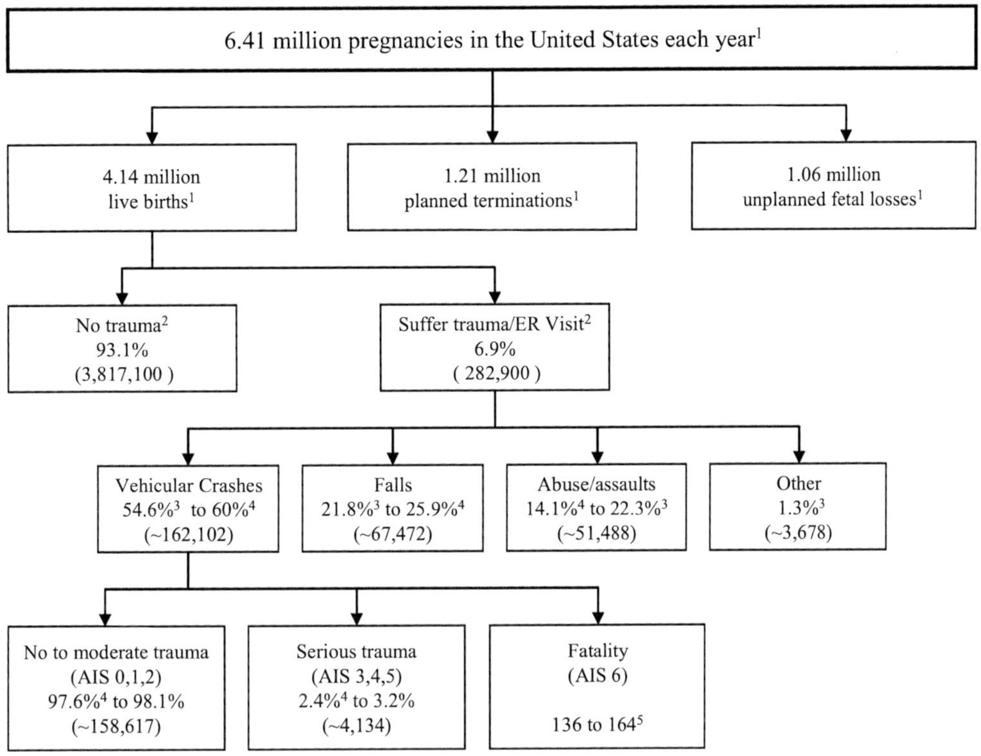

1 – Ventura et al. (2009), 2 – Peckham and King (1963); 3 – Connolly et al (1997); 4 – Pearlman et al (1990); 5 – Klinich et al (1999)

Figure 3: A summary of annual U.S. findings and how they relate to noninjury, injury, and death percentages for pregnant women.

International Research

In France, a study conducted by Barre et al. (2006) looked at abdominal trauma suffered by pregnant women during a three-year period to see if current methods of evaluating their injuries would determine whether hospitalization is warranted. They found that of the 95 patients who presented with abdominal trauma, 49 were from MVCs (51%), 39 were from falls (41%), and seven were from high-energy trauma (8%). They also found that the only women who presented fetal complications (three patients) were from MVCs.

Scandinavian countries have a national healthcare and road accident register where researchers can gather data on the exact number of injuries and fatalities related to MVCs. Kvarnstrand et al. (2008) found that between the years of 1991-2001 there were a total of 1,094,559 pregnant women in Sweden; 2270 of whom were involved in MVCs. Of those 2270 women, 15 died, 251 had major injuries, 1455 had minor injuries, and 549 were uninjured. Therefore, the results found that 88.3% of pregnant women involved in MVCs in Sweden had minor or no injuries related to the MVCs, and 11.01% had major injuries. The study further found that MVCs were responsible for 30% of maternal deaths. The authors note that it might be difficult to compare the data in Sweden to other countries due to the differences in the populations, number of live births, MVC fatality rates, and the number of vehicles.

While MVCs are shown to be the leading cause of injuries to pregnant women, Japan does not have a mandated seatbelt law for pregnant women. Research conducted by

Ichikawa et al. (2003) found that while 70-80% of women surveyed wore seatbelts before becoming pregnant, by the 30th week that percentage was down to 37.9%. They also found that those pregnant women who knew wearing a seatbelt was beneficial continued to do so, while those who knew wearing a seatbelt was not mandated while pregnant decreased their seatbelt use. Ichikawa et al. (2003) further note that these findings differ from percentages in the United States where seatbelt use is mandatory during pregnancy. Seatbelt wearing percentages increase for pregnant women in California from 70% before pregnancy to 86% during pregnancy and in New Mexico from 27% before pregnancy to 42% while pregnant. Hitosugi et al. (2007) states that only 1/3 of pregnant Japanese women wear their seatbelts after 30 weeks gestation vs. 74.6% in the United Kingdom and 83.8% in the United States.

FETAL DEATH RATES

According to the Center for Disease Control, 25,600 fetal deaths occurred in the United States in 2002 from all causes. The exact incidence of fetal deaths that occur from trauma during MVCs is not well documented. Current literature estimates come from various methods that indicate the number of fetal deaths per year from motor vehicle trauma can be as low as 92 or as high as 4800. The authors analyzed all previous estimates of fetal loss rates from motor vehicle trauma and tried to determine a more accurate range that represents all of the various methods in the literature.

A review of the available literature for pregnant occupant data and fetal loss rates was completed. All papers which included estimations of the annual fetal loss data in the United States from motor vehicle trauma were included in the study. The underlying data used by the individual studies come from individual National Automotive Sampling System (NASS) studies, Center for Disease Control (CDC) live birth and fetal mortality records, Traffic and Safety Facts, and previous literature regarding fetal loss. The minimum and maximum values for estimated fetal deaths from each study were used to calculate a more accurate range.

The results include a description of the calculations as well as the resulting maximum and minimum estimates of fetal losses per year in the U.S. from MVCs for the ten methods found in the literature. Weiss (2001) used the percentage of fetal deaths that have trauma as calculated from Pennsylvania death certificates (0.0065%) and the percentage of fetal injuries that are a result of MVC (80%) to calculate the number of fetal deaths each year equal to 208. Similarly, Weiss (1999) used the same calculation with a percentage of fetal deaths as a result of MVC, 0.0023%, determined from death certificates in 16 states during 1995-1997 to approximate 92 fetal deaths from MVC each year.

Jernigan (2002) used the NASS database (1995-2000) to determine the number of pregnant women who died directly from MVC (509) and the number of severe placental injuries that resulted in fetal loss over a six-year period (434) to acquire a minimum estimate of fetal losses per year equal to 157.

Klinich et al. (1999) estimated the number of pregnant women injured in crashes with vehicle damage to be 29,923. This value in conjunction with the percentage of pregnancies that have trauma with subsequent fetal loss (1-3%) plus the number of pregnant women killed (163) estimates the average number of fetal deaths each year to be 462-1061. This estimate does not include fetal losses from accidents in which there was no vehicle damage or the woman was uninjured and is based on an approximation of pregnant women injured in crashes.

Klinich et al. (1999) also used the percentage of women with an ISS score greater than or less than 20 and the respective risk of fetal loss based on the severity to approximate the number of fetal deaths from pregnant women injured in vehicle damage crashes. Those approximations, in addition to the number of pregnant women killed, estimates 496 fetal losses each year.

Jernigan et al. (2002) utilized the NASS database to calculate the number of crashes involving pregnant women for three different crash severities. The groups included crash severities with a change in velocity greater than 30 mph, between 15 and 30 mph, and those less than 15 mph. The incidence by crash severity was combined with the risk curve previously developed by Klinich (1999) relating the risk of adverse fetal outcome with crash severity. The associated risk of adverse fetal outcome and the percentage of that number that was fetal death were used accordingly to determine an average number of fetal losses. Each of these estimates was averaged over the six-year period and combined to get an average annual fetal death rate from MVC ranging from 743 to 1858.

Pearlman and Phillips (1996) used the national average number of births, the percentage of pregnancies that have trauma (6.5%), the percentage of trauma that is MVC (50%), and the percentage of MVC that result in death (1-3%) to obtain a range of fetal deaths per year of 1300-3900. Klinich et al. (1999) used the same equation assuming the percentage of pregnancies that have trauma as 6% and the percentage of trauma that is MVC as 67%, which changed the estimated fetal deaths per year to 1600-4800.

Klinich et al. (1999) used the total number of pregnant women in crashes with vehicle damage in which the occupant was killed, injured, or uninjured (128,255) and multiplied that by the percentage of pregnancies that have trauma and subsequent fetal loss (1-3%). The result was an average number of fetal deaths from MVC trauma equal to 1283-3848 each year.

Moreover, Klinich et al. considered the total number of women injured in crashes with vehicle damage and multiplied that number by the percentage of women at childbearing age in the NASS database that experienced crashes with a change in velocity greater than 30 mph, between 15 and 30 mph, and less than 15 mph (Klinich et al. 1999). The associated risk of adverse fetal outcome from the risk curve relating adverse fetal outcome to crash severity and the percentage of that number that was direct fetal deaths were used to determine an average number of fetal losses. This value plus the number of women killed estimates 1653-3887 fetal deaths each year.

A summary range of all motor vehicle trauma fetal death rates for fetal ages equal to or greater than 20 weeks gestation is calculated from averaging the minimum and maximum approximations from each of the ten methods (Table 4). The range of rates is narrowed to a low of 865 fetal deaths per year and a high of 2795 fetal deaths per year. There are limitations to these estimates, including variation in selection of gestational range.

Table 4: Estimations of annual fetal death rate from motor vehicle crashes in the United States.

Author	*Brief Description*	*Minimum*	*Maximum*
Weiss	Pennsylvania Death Certificates	n/a	208
Weiss	US Death Certificates	92	n/a
Jernigan	NASS	157	n/a
Klinich	1-3% Fetal Loss of Pregnant Women Injured plus Pregnant Women Killed	462	1061
Klinich	Fetal Loss Adjusted for Crash Severity plus Pregnant Women Killed	496	n/a
Jernigan	Fetal Loss Calculated by Delta V	743	1858
Pearlman	6.5% Trauma, 50% MVC, 1-3% Fetal Loss	1300	3900
Klinich	6.0% Trauma, 67% MVC, 1-3% Fetal Loss	1600	4800
Klinich	1-3% Fetal Loss of Pregnant Women in MVC with Damage	1283	3848
Klinich	Fetal Loss Calculated by Delta V plus Pregnant Women Killed	1653	3887
	Average	**865**	**2795**

REFERENCES

Barre, M., N. Winer, Y. Caroit, G. Boog, and H. J. Philippe. 2006. Traumatisme au cours de la grossesse: pertinence des elements de surveillance dans l"evaluation des suites obstetricales: A propos d'une serie de 98 patientes." *Le Revue Sage-Femme* 5(6): 312-216.

Connolly, A., V. Katz, K. Bash, M. McMahon, and W. Hansen. 1997. Trauma and Pregnancy." *American Journal of Perinatology* 14(6):331-336.

Hitosugi, M., Y. Motozawa, K. Masahito, M. Maegawa, T. Nagui, and S. Tokudome. 2007. Letter to the Editor: The benefits of seatbelt use in pregnant women drivers. *Forensic Science International* 169:274-275.

Hyde, L. K., L. H. Cook, L. M. Olson, H. B. Weiss, and J. M. Dean. 2003. "Effect of Motor Vehicle Crashes on Adverse Fetal Outcomes." *Obstetrics and Gynecology* 102(2):279-86.

Ichikawa, M., T. Nakahara, T. Okubo, and S. Wakai. 2003. "Car Seatbelt Use During Pregnancy In Japan: Determinants And Policy Implications." *Injury Prevention* 9(2):169-172.

Jernigan, M. V. 2002. "Statistical analysis and computational modeling of injuries in automobile crashes." M.S. thesis, Virginia Tech, Blacksburg.

Klinich, K. D., L. W. Schneider, J. L. Moore, and M. D. Pearlman. 1999. "Investigations of Crashes Involving Pregnant Occupants." UMTRI-99-29. University of Michigan Transportation Research Institute.

Kvarnstrand, L., I. Milsom, T. Lekander, H. Druid, and B. Jacobsson. 2008. "Maternal fatalities, fetal and neonatal deaths related to motor vehicle crashes during pregnancy: A national population-based study." *Acta Obstetricia et Gynecologica* 87:946-952.

Pearlman, M. D. and M. E. Phillips. 1996. "Safety Belt Use During Pregnancy." *Obstetrics and Gynecology* 88(6):1026-1029.

Pearlman, M. D., J. E. Tintinalli, and R. P. Lorenz. 1989. "A prospective controlled study of outcome after trauma during pregnancy." *American Journal of Obstetrics and Gynecology* 162(6):1502-1510.

Peckham, C. H. and R. W. King. 1963. "A study of intercurrent conditions observed during pregnancy." *American Journal of Obstetrics and Gynecology* 87(5):609-624.

Weiss, H. B. 1999. "The Epidemiology of Perinatal Traumatic Fetal Injury Mortality." PhD dissertation, University of Pittsburgh, Pittsburgh.

Weiss, H. B. 2001. "The epidemiology of traumatic injury-related fetal mortality in Pennsylvania, 1995-1997: the role of motor vehicle crashes." *Accident Analysis and Prevention* 33(4):449-54.

Weiss, H. B. and S. Strotmeyer. 2002. "Characteristics of pregnant women in motor vehicle crashes." *Injury Prevention* 8(3):207-213.

Ventura, S. J., J. C. Abma, W. D. Mosher, and S. K. Henshaw. 2009. "Estimated pregnancy rates for the United States, 1990-2005: An update." *National vital statistics reports* 58(4).

The following papers are included in their entirety to augment this chapter:

Pearlman, M. D., J. E. Tintinalli, and R. P. Lorenz. 1989. "A prospective controlled study of outcome after trauma during pregnancy." *American Journal of Obstetrics and Gynecology* 162(6):1502-1510.

Peckham, C. H. and R. W. King. 1963. "A study of intercurrent conditions observed during pregnancy." *American Journal of Obstetrics and Gynecology* 87(5):609-624.

A prospective controlled study of outcome after trauma during pregnancy

Mark D. Pearlman, MD,[a] Judith E. Tintinalli, MD,[b] and Robert P. Lorenz, MD[a, c]
Royal Oak and Detroit, Michigan

In a prospective cohort study, 85 women who suffered varying degrees of trauma during pregnancy (12 to 41 weeks) were compared with a control group of pregnant women matched for gestational age. Fetomaternal transfusion occurred significantly more frequently in the study group (30.6% vs 8.2%, $p < 0.05$). Study subjects whose placentas were anteriorly placed were at increased risk for fetomaternal transfusion on comparison with other placental positions (47% vs 23.5%, $p < 0.05$). Immediate adverse outcomes including abruptio placentae occurred frequently in the study group (9.4%) and were *not* predictable on the basis of injury severity. When immediate adverse outcomes were excluded, there was no difference in pregnancy outcome. Four hours of cardiotocographic monitoring used as a screening tool was found to be an extremely sensitive but nonspecific indicator of immediate adverse outcomes. On the basis of these findings, it is recommended that routine screening for fetomaternal transfusion occur in all pregnant women who suffer trauma during pregnancy beyond 11 weeks' gestation and that a minimum of 4 hours of cardiotocographic monitoring occur in women >20 weeks' gestation. (AM J OBSTET GYNECOL 1990;162:1502-10.)

Key words: Abruptio placentae, pregnancy, injury, pregnancy outcome, fetomaternal hemorrhage

Trauma during pregnancy is associated with an increased risk of spontaneous abortion, preterm labor, abruptio placentae, fetomaternal transfusion, and stillbirth.[1-9] The frequency and time of onset of these complications are not clear and may be related to the type of injury, location of injury, gestational age, or severity of injury. This uncertainty has prompted a variety of recommendations for the management of the pregnant patient who has suffered trauma.[10-13]

This cohort study was designed to prospectively evaluate pregnant women who have suffered trauma during pregnancy, with uninjured pregnant women serving as controls. Acute complications (i.e., abruptio placentae, fetal death, rupture of membranes, onset of labor, preterm delivery), and the incidence and volume of fetomaternal transfusion, as well as pregnancy outcome, are compared.

From the Departments of Obstetrics and Gynecology[a] and Emergency Medicine,[b] William Beaumont Hospital, and the Department of Obstetrics and Gynecology, Wayne State University.[c]
Supported by a grant from the William Beaumont Hospital Research Institute.
Annual Prize Award, presented at the Fifty-seventh Annual Meeting of the Central Association of Obstetricians and Gynecologists, Scottsdale, Arizona, October 12-14, 1989.
Reprint requests: Mark D. Pearlman, MD, Department of Obstetrics and Gynecology, Baylor College of Medicine, 6550 Fannin, Suite 701C, Houston, TX 77030.
6/6/20126

Material and methods

From November 1985 to August 1988, 85 women presenting to William Beaumont Hospital with a complaint of trauma during pregnancy were enrolled. Subjects were included if they sustained any severity of trauma at or beyond 12 weeks' gestation. Patients with isolated distal extremity injuries were excluded. Each subject was first evaluated in the Emergency Department, and after stabilization, those patients at or beyond 20 weeks' gestation were transferred to the labor and delivery suite for a minimum of 4 hours of cardiotocographic monitoring (model 8040A; Hewlett-Packard, Palo Alto). In many cases monitoring was begun in the Emergency Department. Subjects at or beyond 20 weeks' gestation also had a complete ultrasonographic examination performed (biparietal diameter, abdominal circumference, femur length, amniotic fluid volume, placental location, cerebral ventricular diameter, and calculated estimated fetal weight). Patients <20 weeks' gestation with absent fetal heart tones, abdominal pain, or vaginal bleeding underwent a similar ultrasonographic examination. The retroplacental area was carefully evaluated in all patients for findings suggestive of abruptio placentae.

Historical information included the type of injury: motor vehicle accident, fall to horizontal, fall from height, or direct blow to the abdomen (kick, punch). If the subject was involved in a motor vehicle accident, the restraint system used (if any), speed of the vehicle, damage to the vehicle, and patient position in the vehicle were recorded if possible. The time from injury to presentation was noted as were any symptoms or complaints on admission. A thorough physical examination and appropriate roentgenography or other diagnostic studies (e.g., peritoneal lavage) were performed as indicated.

Cardiotocographic monitoring of all study patients >20 weeks' gestation was performed for a minimum of 4 hours. The 4-hour monitoring record was evaluated for uterine contractions, baseline fetal heart rate, periodic changes, and overall assessment of fetal well-being by a single investigator who was blinded to outcome (R. P. L.). Subjects were categorized into one of five groups on the basis of the frequency of uterine contractions. Those not having contractions were identified as group 0, those with fewer than one contraction every 15 minutes were placed in group I, patients with contractions every 10 to 15 minutes were placed in group II, those with contractions every 5 to 9 minutes were put in group III, and those with contractions every 2 to 5 minutes were placed in group IV.

A control group matched for gestational age (±1 week) was selected from a large group of patients presenting to the Department of Fetal Imaging for obstetric ultrasonography. Subjects were also selected on the basis of the likelihood of obtaining complete follow-up (i.e., delivery at William Beaumont Hospital). The only exclusion criterion was a history of recent vaginal bleeding (within 1 month).

Both groups had an acid elution (Kleihauer-Betke) assay performed at the time of entry into the study. The method used has been previously described.[14] A minimum of 1000 cells were counted. The estimated volume of fetomaternal transfusion was based on the number of fetal cells seen on the smear and an estimate of maternal blood volume. Apgar score, birth weight, gestational age at delivery, and length of neonatal stay were recorded. Subjects who were delivered outside of the institution were contacted by telephone and their records reviewed for outcome.

All data were analyzed with the SAS package. Discrete variables were compared by χ^2 analysis. The two-tailed Student t test was used for continuous variables, and the Fisher exact test was used when appropriate. A p value ≤ 0.05 was considered to be statistically significant.

The study was approved by the Human Investigation Committee at William Beaumont Hospital, and appropriate consent forms were signed by each subject. Eight trauma subjects who were eligible refused admission into the study.

Table I. Demographics

	Trauma (n = *85*)	*Control* (n = *85*)
Age, yr (mean, range)	27.1 (17-41)	28.4 (19-39)
Gravidity (mean, range)	2.5 (1-7)	2.3 (1-6)
Parity (mean, range)	1.0 (0-6)	1.1 (0-5)
Gestational age at entry, wk (mean, range)	29.3 (12-42)	29.1 (12-41)

There was no significant difference between the two groups.

Results

A total of 170 women were enrolled in the study (85 trauma subjects and 85 controls). Outcomes were available for all pregnancies. The age, gravidity, parity, and gestational age at the time of entry are listed in Table I. Prior preterm deliveries were also noted. There was no significant difference between the two groups for any of these variables.

Injury type and severity. Motor vehicle accidents were the most frequent cause of trauma (n = 51; 60%), followed by falls (n = 22; 25.9%) and direct blow to the abdomen (n = 12; 14.1%). Falls were categorized as either fall to horizontal (n = 12) or fall from height (n = 10). Thirty-nine patients involved in motor vehicle accidents were restrained (76.5%), with one exception all subjects were in three-point restraint. Thirty-seven women were drivers of the vehicle, 12 were front-seat passengers, and two were rear-seat passengers. The severity of the motor vehicle accident ranged from minor car damage with no demonstrable maternal injury to extensive vehicle damage and critical illness of the victim at presentation (e.g., hemorrhagic shock, comatose from traumatic brain injury, multiple organ system involvement). Of all trauma victims, 75 (88.2%) had minor or no injuries on clinical examination (minor bruising, lacerations, contusions), 8 (9.4%) had moderate injuries (broken long bones, fractured ribs, extensive bruising), and 2 (2.4%) were critically ill at presentation.

Immediate adverse outcomes. Immediate adverse outcomes were prospectively defined as either abruptio placentae within 72 hours of injury, rupture of membranes within 4 hours of injury, onset of labor within 4 hours of injury that resulted in delivery during the index hospitalization, or fetal death within 7 days of the traumatic event. Table II lists all immediate adverse outcomes along with injury severity. Although both critically injured women suffered abruptio placentae, immediate adverse outcomes occurred more frequently among those less severely injured, as three of the five subjects who suffered abruptio placentae occurred in

Table II. Immediate adverse outcome

Injury severity	*n*	*Abruptio placentae*	*Ruptured membranes*	*Fetal death*	*Delivered (total)*
Minor	75	3	1*	0	5†
Moderate	8	0	1	0	1
Severe	2	2	2*	1‡	2
TOTAL	85	5 (5.9%)	4 (4.7%)	1 (1.2%)	8 (9.4%)

*Rupture of membranes occurred in addition to abruptio placentae.

†Includes two patients who were delivered without evidence of abruptio placentae.

‡Fetal death resulted from complete abruptio placentae.

Table III. Outcome of pregnancies complicated by abruptio placentae (all delivered during index hospitalization)

Patient No.	*Fetal monitor tracing*	*Apgar score*		*Venous cord pH*
		1 min	*5 min*	
1	Persistent type II decelerations	3	6	7.03
2	Tachycardia	5	8	7.16
3	Persistent type II decelerations; tachycardia	4	8	7.13
4	Deep variable decelerations; tachycardia	8	9	7.26
5	No fetal heart tones	0	0	Not performed

the minor injury group. Two of these women had entirely normal results at physical examination on admission.

Of the pregnancies that were complicated by abruptio placentae, all ended with delivery during the index hospitalization. One of the fetuses was dead on presentation as a result of a complete abruptio placentae, and the remaining four were delivered urgently by cesarean section because of abnormal fetal heart tracings. The outcomes of all pregnancies complicated by abruptio placentae are shown in Table III.

Three patients without evidence of abruptio placentae developed labor during the index hospitalization that resulted in delivery. These pregnancies were at 36, 38, and 41 weeks' gestation. No neonatal morbidity or mortality resulted from these pregnancies.

Fetomaternal transfusion. The rate of fetomaternal transfusion was 26 of 85 in the trauma group compared with 7 of 85 in the control group (30.6% vs 8.2%, $p = 0.0001$). The mean estimated volume of transfusion was 12.2 ml (range 5 to 69 ml) in the study group compared with 3.2 ml (range 2 to 8 ml) in the control group ($p = 0.02$).

In comparison with other placental locations in the trauma group anterior placental site was more frequently associated with fetomaternal transfusion (47% vs 23.5%, $p < 0.05$). There was no similar association in the control group. The physical finding of uterine tenderness also was more frequently associated with fetomaternal transfusion in comparison with absence of this finding (65% vs 32%, $p < 0.05$). No other physical finding was associated with fetomaternal transfusion. Trauma subjects who were restrained during motor vehicle accidents had a tendency to fetomaternal transfusion on comparison with those not restrained, but this difference did not reach statistical significance ($p = 0.06$). There was no definite relationship between uterine activity, fetal heart tracing abnormalities, and the risk of fetomaternal transfusion. None of the subjects demonstrated a sinusoidal fetal heart rate pattern.

The pregnancy complicated by the largest fetomaternal transfusion (69 ml) was terminated shortly after the accident because of fetal distress associated with abruptio placentae. The cord hematocrit was 36% and the neonate required transfusion after delivery. No other complications directly attributable to fetomaternal transfusion were noted. All Rho (D)–negative women who had fetomaternal transfusion were treated with Rh immune globulin in the appropriate dosage.

Pregnancy outcome. Birth weight, 1- and 5-minute Apgar scores, gestational age at delivery, type of delivery, and neonatal length of stay in both groups, excluding those subjects with immediate adverse outcomes, are shown in Table IV. When immediate adverse outcomes and their controls were excluded, there was no difference in any of the variables between the two groups.

Cardiotocographic monitoring and ultrasonography. All subjects >20 weeks' gestation were monitored for a minimum of 4 hours after the traumatic event ($n = 60$). Only four of the 60 subjects had essentially no uterine activity during the 4-hour monitoring period (6.7%). Fig. 1 illustrates the uterine activity among all contracting patients. Nearly two thirds of the subjects were having contractions at a frequency of every 2 to 5 minutes during the first hour of monitoring (group IV). The size of this group decreased progressively over the 4 hours, at which time only 28.6% of women were having contractions at that frequency. Conversely, 19.6% of women were having contractions at an interval greater than once in 15 minutes (group I) during the first hour, and this group increased in size over time to include a total of 37.5% of all subjects at the 4-hour mark. Groups II (contractions every 6 to 10 minutes) and III (contractions every 11 to 15 minutes) increased slightly over time at the expense of group IV.

All women who had immediate adverse outcomes (except one patient with only rupture of the membranes) were in group IV at some point during the 4 hours of monitoring. One woman who had a clinical apparent abruptio placentae 6 hours after the traumatic event began in group I at hours 1 and 2 and progressed to group IV at hours 3 to 6, at which time she had vaginal bleeding and eventually revealed persistent late decelerations of the fetus necessitating urgent operative delivery.

The use of tocolytic drugs was generally discouraged, but one patient was treated with parenteral ritodrine; she was seen at 28 weeks' gestation with contractions every 2 to 5 minutes after a motor vehicle accident. The frequency of uterine activity decreased over the next 45 minutes. The medication was discontinued after 90 minutes as chest pain developed. No other tocolytic therapy was begun, and contractions stopped spontaneously over the next 180 minutes. She was admitted for a total of 44 hours and was discharged without medication. She was delivered of a healthy 3.42 kg male infant at 39 weeks.

All subjects who were in groups II, III, or IV at any time during the 4-hour monitoring period were admitted beyond 4 hours (for a minimum of 24 hours). Subjects in group 0 or I during the entire 4-hour monitoring period were discharged at the end of the 4 hours unless uterine tenderness, a nonreassuring fetal monitor strip, vaginal bleeding, rupture of the membranes, or any serious maternal injury was present. Of the 60 women eligible for admission (>20 weeks' gestation), 42 were admitted beyond the 4-hour period (70%). All patients discharged at the end of the 4 hours or at 24 hours had a live birth. Gestational age at delivery, Apgar score, and birth weight were similar to those of the matched control group in all discharged patients. In patients admitted beyond the 24 hours, there was one fetal death (2.4%) and seven other deliveries (16.7%).

Ultrasonography predicted abruptio placentae in two of five patients (40%), but both of these patients had a clinically evolving abruptio placentae (vaginal bleeding and frequent uterine activity) at the time of the ultrasound. In three patients who did suffer abruptio placentae, the admitting sonogram did not suggest the diagnosis.

Comment

The management of trauma during pregnancy has evolved from a variety of protocols developed in nonpregnant trauma victims. Although these principals generally translate in very serious injuries, the less seriously injured pregnant woman is more problematic. The protocol developed herein was able to predict the major fetal morbidity experienced in this study. Interestingly, in the women who did not suffer immediate adverse outcomes, the accident appears to have had a minimal, if any, impact on pregnancy outcome in terms of birth weight, gestational age at delivery, or Apgar score. This finding is important to allay the fears of both the obstetrician and the pregnant woman after she is discharged home following the monitoring period. Nonetheless, it is important to instruct the patient to return if any signs or symptoms of preterm labor or abruptio placentae do occur after discharge, as the apparently rare episode of delayed abruptio placentae has been described.[15]

Several studies have noted that the incidence of abruptio placentae after major automobile accidents varied between 6% and 35%, but most of these series involve women who have suffered major injuries.[2, 3, 12] In this series, which involved mostly minor trauma, the incidence of abruptio placentae was 5.9% (4.0% in the minor injury group). Importantly, it appears that even relatively minor deformation forces are sufficient to shear the placental attachments away from the decidua basalis. It is clear that neither physical findings nor the severity of injury is able to rule out immediate adverse outcomes. Routine cardiotocographic monitoring for a minimum of 4 hours appears to be very sensitive for predicting immediate adverse outcomes including abruptio placentae. Ultrasonography does not appear to be particularly useful in predicting abruptio placentae after trauma (sensitivity = 40%). Nonetheless, ultrasonography may still be useful in these patients for other reasons (estimating fetal weight and gestational age, biophysical profile, amniotic fluid volume).

The current protocol recommended that tocolysis not be attempted, although the ultimate decision was left to the attending physician. Of the 35 women who were having frequent contractions during the monitoring period (group IV, 2 to 5 minutes apart), five subsequently suffered an abruptio placentae requiring delivery, and three others were delivered of their infants during the index hospitalization (at 36, 38, and 41 weeks). Tocolysis was discouraged as it was felt that regular uterine activity after trauma was a result of abruptio placentae until proved otherwise. This hypothesis was mostly borne out during the study because five of the eight patients who were delivered did suffer an abruptio placentae. In those women with frequent contractions, contractions stopped in 90% without tocolysis. Of those patients who were delivered during the index hospitalization without evidence of abruptio placentae, all were at or near term, and no serious fetal morbidity resulted from these deliveries. Uterine activity after trauma may result from a uterine "contusion" resulting in extravasation of blood from myometrial capillaries and subsequent irritability. As this extravasated blood is resorbed, the uterine irritability diminishes. We feel that the initial presentation of

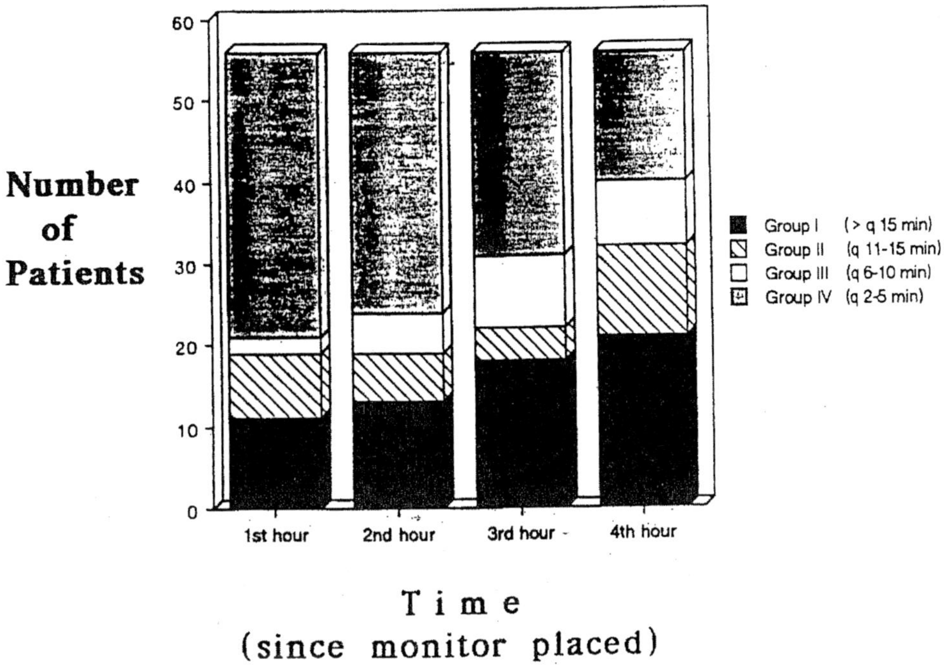

Fig. 1. Number of patients with uterine contractions at various frequencies during the first 4 hours of monitoring ($N = 56$).

Table IV. Pregnancy outcome (excluding immediate adverse outcomes)

	Trauma	*Study*
Birth weight (kg, mean ± SD)	3.44 ± 0.58	3.53 ± 0.57
Gestational age at delivery (wk, mean ± SD)	38.9 ± 4.5	39.8 ± 2.2
Apgar score		
1 min	7.5	7.7
5 min	8.8	8.9
Delivery type (% cesarean section)	21	24
Neonatal length of stay (days ± SD)	4.1 ± 4.0	4.1 ± 6.8

There was no significant difference for any variable between the two groups.

abruptio placentae and "contusion" are indistinguishable and tocolysis should not be attempted.

Pregnant patients who suffer trauma are clearly at risk for fetomaternal transfusion; this finding confirms an earlier report by Rose et al.[5] Neither the severity of injury nor the presence of uterine activity was predictive of fetomaternal transfusion, although an anteriorly positioned placenta and the presence of uterine tenderness did place patients at increased risk. With one exception, patients with fetomaternal transfusion experienced no difference in pregnancy outcome on comparison with controls. This contrasts with results of Rose et al.,[5] where three of nine fetuses who suffered fetomaternal transfusion had some demonstrable abnormality. Fetuses who suffer fetomaternal transfusion appear to be at some risk for adverse outcome, although most will not experience problems. This group of patients is at risk for Rh sensitization in Rh-negative mothers and should be treated with adequate prophylaxis. The risk for fetomaternal transfusion is high (30.6%) after trauma during pregnancy, and if testing is not readily available to assess fetomaternal transfusion, routine administration of Rh immune globulin is recommended in addition to continuous fetal monitoring.

In conclusion, this study suggests that all patients >20 weeks' gestation who suffer trauma during pregnancy should be monitored for a minimum of 4 hours as the risk for abruptio placentae is significant and associated with frequent uterine activity. In addition, all patients >12 weeks' gestation should have Kleihauer-Betke analysis to detect fetomaternal transfusion with Rh immune globulin administration to appropriate candidates. Although only one fetus appeared to suffer any serious morbidity on the basis of fetomaternal transfusion in our series, serial testing may be indicated to substantiate that further bleeding does not occur after the initial test. If serial testing is indicated, it is not clear whether testing should occur in all patients or only those with fetomaternal transfusion present initially.

We express our appreciation to Drs. Christine Comstock, Robert Bree, and S. S. Yang for their significant contributions to this study.

REFERENCES

1. Agran PF, Dunkle DE, Winn DG, Kent D. Fetal death in motor vehicle accidents. Ann Emerg Med 1987;16:1355-8.
2. Crosby WM, Costiloe JP. Safety of lapbelt restraint for pregnant victims of automobile collisions. N Engl J Med 1971;284:6326-9.
3. Rothenberger D, Quattlebaum FW, Perry JF, Zabel J, Fisher RP. Blunt maternal trauma: a review of 103 cases. J Trauma 1978;18:173-9.
4. Lizan G, Bordenave C, Sellam, et al. Posttraumatic feto-maternal hemorrhage. Apropos of a case. J Gynecol Obstet Biol Reprod 1988;17:917-21.
5. Rose PG, Strohm PL, Zuspan FP. Fetomaternal hemorrhage following trauma. AM J OBSTET GYNECOL 1985; 153:844-7.
6. Fort AT, Harlin RS. Pregnancy outcome after noncatastrophic maternal trauma during pregnancy. Obstet Gynecol 1970;35:912-5.
7. Peckham CH, King RW. A study of intercurrent conditions observed during pregnancy. AM J OBSTET GYNECOL 1963;87:609-24.
8. Dyer I, Barclay DL. Accidental trauma complicating pregnancy and delivery. AM J OBSTET GYNECOL 1961;83:907-29.
9. Poulson A, Gabert HA. Fetal death secondary to nonpenetrating trauma to the gravid uterus. AM J OBSTET GYNECOL 1973;116:580-2.
10. Chescheir NC. Trauma in pregnancy. Clin Obstet Gynecol 1988;1:11-2.
11. Trauma in pregnancy. In: Advanced trauma life support. Chicago: Committee on Trauma, 1985:303-7.
12. O'Keeffe DF. When the accident victim is pregnant. Contemp Ob/Gyn July 1985:148-63.
13. Smith CV, Phelan JP. Trauma during pregnancy. In: Clark SL, Phalen JP, Cotton DB, eds. Critical care obstetrics. Oradell, New Jersey: Medical Economics Books, 1987:362.
14. Virgilio LA, Simon NV. Measurement of fetal cells in the maternal circulation. Obstet Gynecol 1977;50:364-6.
15. Higgins SD, Garite TJ. Late abruptio placentae in trauma patients: implications for monitoring. Obstet Gynecol 1984;63:10s-2s.

Discussion

DR. WARREN M. CROSBY, Oklahoma City, Oklahoma. This is important work. The design of the study is superb and so we can rely on the lessons learned. The authors confined their study to the group of pregnant trauma victims in which we should have the most interest: those mothers that do not have life-threatening injuries. Here the obstetrician and emergency physician can learn which maternal findings and treatments correlate best with fetal outcome. Such a study is sorely needed to provide scientific evidence on which to rely when one is advising not only the pregnant trauma victims themselves as to the likely outcome of their pregnancy but their lawyers as well, who seem bent on establishing trauma as the certain cause of a bad perinatal outcome, no matter how unlikely that conclusion seems to be.

I wish to emphasize two areas of importance:

1. I am delighted to see that 76.5% of the motor vehicle accident victims were wearing restraints. This is a much higher percentage of women drivers wearing seat belts than national surveys have found. There is little doubt that the three-point restraint, available in all automobiles today, is very effective at reducing injury. This is particularly true of the pregnant driver because belts are very successful in preventing impaction of the mother's pregnant abdomen and chest into the steering wheel. Indeed, the overrepresentation of belted victims in this study attests to this fact. Nonbelted victims that are injured at all are likely to have more serious internal injuries, injuries severe enough to prevent their having been included in this study.

2. The most important finding is that if the trauma did not produce an adverse outcome within the first day or two after the traumatic event, the trauma did not affect the pregnancy at all.

This carefully documented observation brings up the question of the so-called "delayed abruptio" following trauma. To my knowledge there have been only two cases of "delayed abruptio" reported. Both were unbelted drivers in the early third trimester.[1,2] Both hit the steering wheel and had noncatastrophic injuries. In one case the nonstress was nonreactive on admission, but clear-cut fetal compromise was not seen until the membranes ruptured 5 days later. The fetus was thought to be uncompromised in the second case for 5 days, when bleeding and contractions began. The patient was transferred to a tertiary care center and continuous fetal monitoring was not remarkable until the next day when vaginal bleeding and uterine tenderness developed along with increased uterine contractions. Late decelerations were seen and a cesarean delivery was done. Both of these fetuses were small for dates, and both lived. In both cases extensive placental separation was found.

When trauma results in disruption of the placental attachments to the decidua, the bleeding that occurs at that time is usually self-limited. When the bleeding is not self-limited, the typical signs of severe placental separation occur, beginning within the first few hours after the traumatic event. When the amount of separation is less, the blood is released into the decidua clots promptly. The decidual cells involved in the process release prostaglandins, which cause almost all trauma victims in late pregnancy to have uterine contractions. Only 7% of the victims in this study did not have uterine contractions within the first few hours after admission. Again, the process usually is self-limited. Uterine contractions stop within a few hours as they did in 91% of the victims reported here. Labor may indeed progress in some, probably related to the amount of decidual damage, the release of decidual prostaglandins, and the length of the pregnancy. In those pregnant trauma victims that are not delivered within the first 48 hours or so, the clot organization process is underway. The area of separation ultimately becomes an infarct. The process bcomes stable in virtually all cases within 24 to 48 hours of the initial event. In my opinion, the two cases of "delayed abruptio" were not "delayed" at all. It was the diagnosis that was delayed. The fetus was just so compromised that it didn't quite die initially, and diagnosis was not made until increased uterine activity or better observation a few days later called attention to its plight. In both cases delivery occurred within 6 days of the traumatic event, and in both cases the fetus lived.

I am aware of two prospective studies of pregnant trauma victims so far unpublished, in addition to the one presented today and the one my colleagues and I reported in 1970.[3] These four prospective studies account for 797 pregnant victims of trauma. There were 17 cases of abruptio placentae among the 797 cases, and each was identified within the first few hours after the traumatic event. In short, I don't believe that there is such a thing as "delayed abruptio." The placental separation occurs at the time of the trauma and either is severe enough to cause labor, fetal death, maternal bleeding, and the like or is self-limited, with the initial symptoms of uterine contractions or spotting stopping within 48 hours in virtually every case.

Thus I agree with the authors, as well as Higgins and Garite,[2] that the fetal monitor should be applied as soon as appropriate after immediate emergency measures are taken when the pregnant trauma victim arrives at the hospital. If the fetus is far enough along to be considered viable, fetal monitoring should probably be continued for a period of a few hours to begin with. If the patient is totally asymptomatic and no evidence of injury or fetal distress can be found, it is a rare patient indeed that would return later with a dead fetus or bleeding. For patients with symptoms or signs that indicate the possibility of more severe injury, more prolonged observation in the hospital will be appropriate. I suggest that such patients have fetal monitoring during the time that they are in the hospital, because, during observation of the mother for serious injury, the fetus may very well be ignored by physicians and nurses that don't normally care for pregnant women.

But when do you stop the monitoring? When is it safe to let the patient go home? I think the best advice is to use the information presented today and tell the patient that "there are a few cases in which things turned bad several days after a traumatic event, but this is rare. It occurs much less frequently than one in a thousand cases." And then add: "But, Mrs. Smith, if you insist upon it, we'll keep you on the monitor for a full week. Since no case of trauma-related abruptio has ever been reported longer than a week later, you can go home with confidence then. This doesn't guarantee you a healthy baby. It only removes the possibility of the fetus dying from a placental separation that we haven't been able to discover." This strategy ought to allow you to let most people go home in a reasonable time. It also transfers the responsibility for going home at that time from you to the patient.

REFERENCES

1. Lavin JP, Miodovnik M. Delayed abruption after maternal trauma as a result of an automobile accident. J Reprod Med 1981;26:621-3.
2. Higgins SD, Garite TJ. Late abruptio placentae in trauma patients: implications for monitoring. Obstet Gynecol 1984;63:105-25.
3. Crosby WM, Costiloe JP. Safety of lap belt restraint for pregnant victims of automobile collisions. N Engl J Med 1971;281:25632-6.

Dr. David B. Cotton, Houston, Texas. Dr. Pearlman and his colleagues have studied 85 women suffering varying degrees of trauma during pregnancy and compared them prospectively with a control group matched for gestational age. In this study, which represents the experience of a large, sophisticated community hospital, motor vehicle accidents were the most frequent cause of trauma. Overall 97.6% of trauma cases were classified as mild to moderate and only 2.4% as critical. However, three of the five cases of abruptio placentae seen in the study, as well as two cases of preterm delivery, occurred in the minor trauma group. These five cases combine to give a 6.7% incidence of immediate adverse outcome in minor trauma cases, which are defined as minor bruising, lacerations, or contusions. This information emphasizes strongly that even minor trauma during pregnancy must be taken seriously by the obstetrician and a consistent and thorough evaluation implemented.

The incidence of fetomaternal transfusion in the trauma group was 30.6% as compared with 8.2% in the control group. The mean volume of fetomaternal transfusion in the trauma group (12.2 ml) was three times that of the control group (3.2 ml). Additionally, an anterior placental location site appeared to be much more susceptible to trauma as fetomaternal transfusion was twice as frequent in anterior placentas than other placental locations.

Importantly, for those evaluating pregnant trauma victims, uterine tenderness was associated with a fetomaternal transfusion 65% of the time. However, 32% or roughly one third of patients with a demonstrable fetomaternal transfusion had no significant uterine tenderness.

Additionally, in this group undergoing fetal monitoring for a minimum of 4 hours there was no relationship between uterine activity or fetal heart rate tracing abnormalities and the risk of fetomaternal transfusion. It would appear from these data that there are no clinically reliable signs or symptoms that predict fetomaternal transfusion in trauma cases. In cases of even minor trauma involving Rh-negative patients a Kleihauer-Betke test must be considered mandatory.

Ninety-three percent of patients in this study had some uterine activity following trauma. In this protocol all patients with uterine contractions more frequent than one every 10 minutes were admitted for a 24-hour observation period. All patients experiencing immediate adverse outcomes had contractions every 2 to 5 minutes (group IV) at some time during the initial 4-hour monitoring period. The data from this study suggest that a simple 4-hour fetal monitoring period combined with a Kleihauer-Betke screen will identify virtually all the fetal morbidity associated with trauma in pregnancy.

Last, and for the patient perhaps most important, this study documents that when pregnant patients undergo trauma but do not have an immediate adverse outcome there are no subsequent effects on maternal, fetal, or neonatal outcome.

Questions that this study raise include the following:

1. Is William Beaumont Hospital a level I referral center for trauma? If not, perhaps this explains the relatively low incidence of critical trauma.

2. Was there a gestational age below which immediate adverse outcomes such as abruptio placentae or preterm delivery did not occur in the minor trauma group? Perhaps minor trauma before 20 weeks' gestation rarely results in immediate adverse outcome.

3. In those patients with minor trauma and abruptio placentae or preterm labor did each patient experience a blow specifically to the abdomen?

4. Is the 8.2% incidence of fetomaternal transfusion in the control group somewhat higher than one would normally expect?

5. What was the incidence of fetomaternal transfusion in the minor versus moderate trauma group and were these mostly concentrated in the late third trimester?

6. Last, although this study has defined the incidence and degree of fetomaternal transfusion with trauma during pregnancy, what is the purpose of documenting fetomaternal transfusion in patients who are Rh-positive if management primarily hinges on fetal heart rate monitoring?

The authors are to be congratulated for a well-designed, clinically relevant study.

Dr. Charles A. Stedman, New Orleans, Louisiana. There have been several studies of ≥1000 consecutive Rh-negative patients who after the ultimate trauma, the delivery process, have been found to have significant fetal-maternal hemorrhage. Only three to ten patients per thousand have been noted under those circumstances. How do the authors explain the fetal-maternal hemorrhage incidence of 80 per thousand in their control group? This suggests some sort of method error in the laboratory.

The second question is do they get their Kleihauer-Betke results back within 4 hours? At most institutions a pathologist is required to give those results. In most of the hospitals in which I've worked, the pathologist is unavailable at nighttime, and we have to wait until the next morning for results.

Dr. Sharon L. Dooley, Chicago, Illinois. I think this is a wonderful study that tremendously advances our knowledge. I have a comment about terminology. Dr. Pearlman used the term *severe fetal acidosis* to show the cord artery pH values for the fetuses that required urgent delivery. For a long time, and for reasons that are not clear, we used a pH of 7.20 as the lower limit of normal for cord artery pH. There are now considerable data published that indicate that the lower limit of normal is closer to 7.10 or 7.12 and that a pH really has to be less than 7.00 before we call that a severe fetal acidemia.

I think it's time for us all to stop "digging our own graves" for the lawyers and use proper terminology.

Dr. James R. Scott, Salt Lake City, Utah. I wonder if you could give us details about the management of the cases of abruptio placentae and specifically what happened with the one stillborn infant. The reason I ask is that many of these fetuses are premature when being monitored, and we consider it to be ominous if there are any abnormalities. I can recall at least one patient who was being monitored and the fetal heart rate suddenly dropped to zero.

Dr. Daniel A. Rightmire, Springfield, Illinois. Do the authors have information about biochemical assays, such as maternal serum α-fetoprotein or β-human chorionic gonadotropin, which might be markers of trophoblast injury or death available in a hospital that does not have reliable Kleihauer-Betke techniques?

Dr. Pearlman (Closing). Dr. Cotton, Beaumont Hospital's emergency center is a level I trauma referral center and sees approximately 75,000 patients per year. However, because of its suburban location, we do not see the same frequency of severe trauma as an inner city institution.

The earliest gestational age at which we saw an immediate adverse outcome was 26 weeks—an abruptio placentae. However, I would not discount the possibility of abruptio placentae before that gestational age as it most definitely occurs. Prior studies have also shown that the more advanced a pregnancy is, the more susceptible the fetus and placenta are to injury.[1]

The issue of whether patients experienced a direct blow to the abdomen is particularly problematic. When patients who have been involved in motor vehicle accidents are asked if they've been struck directly in the abdomen, the answer is not always reliable. Many patients that we saw who stated they were not struck in the abdomen had seat belt bruises or other evidence of injury to the abdomen, so I don't think you can rely on history alone. In terms of immediate adverse outcome, all our patients, either historically or by examination, had evidence of a direct blow to the abdomen.

The incidence of fetomaternal transfusion in our control group was approximately 8%, which is similar to the 6% reported by Rose et al.[2] in their prospective trauma study's control group. In addition, there are two additional studies that evaluated the incidence of fetomaternal transfusion in normal pregnancies, revealing some evidence of fetomaternal transfusion in at least 5% to 10% during the third trimester, although they did count 5×10^4 maternal cells.[3,4] I think that our results are similar to these studies.

With regard to the relationship between fetomaternal transfusion and gestational age, most of these injuries occurred during the third trimester, so a majority of transfusions were found in this group. However, we saw evidence of fetomaternal bleeding in a pregnancy of 15 weeks. So I think that the 12-week limit for testing is reasonable (i.e., when the uterus becomes an abdominal organ and is susceptible to injury).

As Dr. Cotton stated, the major purpose of documenting fetomaternal transfusion is to decrease the incidence of Rh isoimmunization in the Rh-negative patient. However, one of the fetuses in this study suffered a 69 ml transfusion and on continued monitoring had an aminous fetal heart tracing. The fetus was delivered and was severely anemic at birth, requiring blood transfusion. That was our only complication related directly to fetomaternal transfusion. The only other study that prospectively evaluated fetomaternal transfusion after trauma exhibited even a higher frequency of complications related to the fetomaternal transfusion.[2] So I think that Kleihauer-Betke testing in all patients, whether Rh negative or Rh positive, is justified.

Regarding the question about the use of the term *fetal acidosis,* I agree that the nomenclature is evolving with regard to defining fetal acisosis; however, I think it is clear that three of the four venous pH values are lower than what we generally consider to be normal.

Dr. Scott asked about the history of the stillborn infant. That case report actually is already published.[5] Briefly, this was a woman at 32 weeks' gestation whose car was struck by a cement truck, and she was trapped inside the car for approximately 45 minutes. Her fetus was dead on presentation to the emergency center. The patient also had multiple severe intraabdominal injuries and underwent laparotomy. Cesarean section was done at the time of laparotomy because she also had an unstable pelvic fracture. The placenta was anterior and completely separated from the uterine wall, which explains the fetal loss. There was no evidence of direct fetal injury on examination of x-ray film.

We do not have any information on β-human chorionic gonadotropin or maternal serum α-fetoprotein in these patients. I would like to mention two things regarding maternal serum α-fetoprotein for assessing the need for isoimmunization prophylaxis. First, you cannot accurately determine the volume of the transfusion that prohibits the reliable administration of Rh immune globulin. Second, you would need a preinjury maternal serum α-fetoprotein level to determine the presence of fetomaternal bleeding, and this may or may not be available.

REFERENCES

1. Crosby WM, Snyder RG, Snow CC, Hansen PG. Impact injuries in pregnancy. I. Experimental studies. AM J OBSTET GYNECOL 1968;101:100-10.
2. Rose PG, Strohm PL, Zuspan FP. Fetomaternal hemorrhage following trauma. AM J OBSTET GYNECOL 1985; 153:844-7.
3. Schroeder J. Transplacental passage of blood. J Med Genet 1975;12:230-42.
4. Jorgensen J. Feto-maternal bleeding during pregnancy and at delivery. Acta Obstet Gynecol Scand 1977;56:487-90.
5. Pearlman MD, Tintinalli JE. Trauma in pregnancy (clinical conference). Ann Emerg Med 1988;17:829-34.

A study of intercurrent conditions observed during pregnancy

C. H. PECKHAM, M.D.
R. W. KING, M.D.
Oakland, California

RECENT obstetric literature is fairly replete with studies concerning the effect of certain individual diseases on the course and outcome of pregnancy and also the counter effect of the pregnancy on the disease. This is particularly true with reference to the more common chronic conditions, such as tuberculosis, cardiac disease, and diabetes. A few acute disease conditions such as pneumonia have been similarly studied. Information is lacking, however, on the broader subject of accidental complications of pregnancy, in general, although it is obvious that any condition which may attack the nonpregnant woman during her childbearing years is equally as likely to affect her during gestation.

An important reason for the difficulty encountered in studies of this sort has been that most groups available for investigation have not been representative of the childbearing population as a whole. Furthermore, a complete picture of their medical as contrasted with their obstetrical care has been lacking.

From The Cooperative Study of Child Development, the Division of Biostatistics of the University of California in Berkeley, and The Kaiser Foundation Research Institute.

Supported by Grant No. 02640 from the National Institutes of Health.

Presented at the Eighty-sixth Annual Meeting at the American Gynecological Society, New Orleans, Louisiana, May 13-15, 1963.

Clinic patients, those most available for investigation, are notoriously migratory; they frequently fail to seek care when needed for financial reasons, and what attention they do get may come from a number of different facilities, rendering a composite picture difficult to obtain. Those series obtained from private practice do not represent a true cross section of society and, although their medical care is, in general, more than adequate, it is likely to come from a number of specialized sources also difficult to assemble.

A study of intercurrent conditions occurring during puregnancy can best be done on a population whose entire medical and obstetrical care is obtained from a single facility and recorded in one unit record. Also the cost per clinic visit should be sufficiently low that the patient will seek advice concerning the most minor (to her) medical conditions. Such a population can probably only be found among the subscribers to a prepaid health plan like those in this study. The Cooperative Study of Child Development is an independent research project operating under the direction of the Division of Biostatistics at the University of California in Berkeley and the Kaiser Foundation Research Institute, and supported by a grant from the National Institutes of Health. The study has been in progress for about 4 years. During this time, it has followed to termination over 10,000 pregnancies. The patients included were women who are members of the Kaiser Foundation Health Plan and who have attended a Permanente Clinic in either Oakland, California, or one of its environs. These women, in general, have obtained their entire care from one of the above-mentioned facilities. As members of the Health Plan, they have had to pay only one

Table I. Interval in weeks between last menstrual period and contact with Obstetrics-Gynecology Clinic (based on 3,072 patients)

Weeks	*Frequency*	*Cumulative frequency*	*%*	*Cumulative %*
1 to 8	946	946	30.9	30.9
9 to 12	957	1,903	31.2	62.1
13 to 16	454	2,357	14.8	76.9
17 to 20	267	2,624	8.7	85.6
21 to 28	279	2,903	9.1	94.7
29 to 36	139	3,042	4.5	99.2
37 to 40	20	3,062	0.7	99.9
41 +	1	3,063	—	—
Unknown	9	3,072	0.1	100.0

Table II. Number of prenatal visits per patient

*No. of visits**	*No. of patients*
0	133†
1	169
2	121
3	94
4	105
5	127
6	170
7	206
8	242
9	289
10	337
11	341
12	270
13	168
14	98
15	66
16	34
17	16
18	14
19 and over	7
Number of visits unknown	65
Total	3,072

*Mean number of visits per patient is 8.3.

†Includes 75 patients eventually determined to be "not pregnant."

dollar per clinic visit and, hence, have sought care for many trivial as well as significant problems.

For purposes of this investigation, it was decided to examine the case records of all those Health Plan patients who became part of the Child Development Study (CDS) and whose last menstrual period (LMP) began after April 1, 1960, through March 31, 1961. It should be stated that this CDS series is comprised of all those patients under observation by the Permanente Clinic except a few individuals who did not care to participate in the study and unmarried minors in which cases it was felt that follow-up of the children would not be productive.

It is felt that, with a few exceptions, this series is comprised of a fairly representative cross section of the childbearing population. To determine this, the cohort was examined according to the race or ethnic background of the patient, age, parity, and education. Just over two thirds of the patients were classified as white and of these about one-half were predominantly of Western European extraction. A significant number of women claimed some American Indian background while about one-fifth were Negroes. The mean age of the patients in the series was about a year over that found for this Birth Registration Area of the United States in 1959. About a quarter of the total women were primigravidas. From a socioeconomic standpoint, the group was fairly representative of the community and included the wives of government employees, university faculty and students, members of various trade unions, as well as individuals of varied types of occupation. In only two respects does this series deviate from the average. First, it contains only a few indigent individuals and second, the educational level is high with 40 per cent of the patients having had at least some college education. With these exceptions, however, it is felt that the claim that this series is representative is, in the main, justified.

For purpose of study of intercurrent conditions occurring during pregnancy, it is important that the patient be under observation from an early date. In this series, generally, the date of contact was early. From Table I it will be seen that by the time a second period had been missed, 30.9 per cent were under observation; at the end of 12 weeks this figure had doubled, while 85.6 per cent of the women in the series were seen by the end of the twentieth week of gestation. In addition, it should be stated that 40.7 per cent of the patients had been seen in some clinic during the 3 calendar months prior to the LMP. The majority of the remaining three-fifths had been health plan members for varying periods of time, often since childhood; and, hence, would probably have been seen had an illness occurred.

Since the majority of patients in the group came under observation early in pregnancy, it was to be expected that there was adequate prenatal care in terms of multiple clinic visits (Table II). The figures shown here become somewhat more impressive when it is stated that 75 of the 133 patients in the "no prenatal visit" group were placed here because, following a period of amenorrhea, the eventual diagnosis was as not pregnancy. Furthermore, an additional 182 pregnancies terminated in abortion and in this group, obviously, the number of prenatal visits was very small.

The analysis of conditions occurring before and during pregnancy in the 3,072 patients conveniently may be divided into four categories; namely, intercurrent medical conditions, complications of pregnancy, accidents and injuries, and a supplementary list of examinations, most not diagnosed as conditions. Also, the above categories may be divided appropriately into four time periods according to whether the clinic visit occurred during the 3 months prior to the LMP or in the first, second, or third trimester of pregnancy, respectively. The following tables are constructed on this basis and will be discussed separately.

Intercurrent conditions

Table IIIA lists in summary form the intercurrent conditions occurring in the series divided into groups according to the International Classification of Diseases. It will be noted that in the time period "Before LMP" the conditions most frequently found were those of the genitourinary system with 318 listed. Over two thirds of these were in the groups of menstrual disorders, cervicitis, and vaginitis-vulvitis. There were 217 instances of respiratory disease with over half of these cases of acute pharyngitis. The next three most common categories were diseases of the nervous system and sense organs (159, with 70 of these refractive errors); diseases of the skin and cellular tissues (147, with 67 being various types of dermatoses); and allergic, endocrine, metabolic, and nutritional diseases (131, with 48 instances of exogenous obesity). Among the five most frequently occurring groups of a list entitled "Symptoms," totaling 200, in which these symptoms were sufficient to cause a clinic visit but not definite diagnosis, has been omitted. Conditions occurring before the LMP totaled 1,499 and were present in 1,177 or 40.7 per cent of the total series. Over three fifths of the entire group under discussion had had no illness sufficient to warrant a clinic visit in the 3 months prior to conception.

First trimester. An almost identical number of conditions (1,417) was observed during the first trimester of pregnancy. Here again, diseases of the genitourinary system were most frequent (375), although vaginitis and cervicitis were the two conditions most commonly observed. Respiratory conditions were the group second most frequently noted, with the common cold leading this list. Numbers 3, 4, and 5 were as before, but now the frequency was altered to allergies, endocrine, etc.; skin and cellular tissues; and nervous system and sense organs, respectively.

Second trimester. When intercurrent medical conditions were divided according to these time periods, those occurring in the second trimester of pregnancy were noted with the greatest frequency (1,597). Both at this time and in the third trimester symptoms without a definitive diagnosis were the group most often encountered; abdominal pain, headaches, and low back pain being often of

Table IIIA. Summary of conditions or visits from 3 months before LMP to termination of pregnancy (Excluding complications of pregnancy, accidents and injuries, and supplementary classification)

Condition	Time of onset: Before LMP	Trimester of pregnancy: 1st	2nd	3rd	Total	Total
Infective and parasitic	26	24	40	26	90	116
Neoplasms	64	70	66	32	168	232
Allergic, endocrine, metabolic, and nutritional diseases	131	101	123	111	335	466
Diseases of the blood and blood-forming organs	7	5	1	6	12	19
Mental, psychoneurotic, and personality disorders	62	56	37	13	106	168
Diseases of the nervous system and sense organs	159	85	88	55	228	387
Diseases of the circulatory system	30	47	105	142	294	324
Diseases of the respiratory system	217	216	295	245	756	973
Diseases of the digestive system	76	77	72	83	232	308
Diseases of the genitourinary system	318	375	288	123	786	1,104
Diseases of the skin and cellular tissue	147	96	109	84	289	436
Diseases of the bones and organs of movement	61	45	29	8	82	143
Congenital malformations	1	7	4	5	16	17
Symptoms	200	213	340	290	843	1,043
Total	1,499	1,417	1,597	1,223	4,237	5,736

a degree sufficient to cause the patient to seek medical aid. Of conditions definitely diagnosed, those of the respiratory system lead the list in the second trimester, with the common cold accounting for more than half of the total (295) in this group. Diseases of the genitourinary system fell to second place, with half of these being instances of vaginitis. Allergic and endocrine ailments remained in third place with skin and cellular tissue manifestations in fourth. Fifth in degree of frequency were diseases of the circulatory system with varicose veins and hemorrhoids affording the most common complaints.

Third trimester. Accidental complications of the third trimester of pregnancy numbered less than in any of the preceding time periods, 1,223. Respiratory conditions were still found to be the most frequent, but now being followed by circulatory conditions with diseases of the genitourinary system, third, and allergic, endocrine, etc., fourth. Diseases of the skin and cellular tissue were now observed fifth in order of frequency.

Table IIIA affords a summary of the intercurrent conditions noted by system involved and divided according to the time of observation. Total conditions observed in the entire study were 5,736 of which 4,237 occurred during the pregnancy, itself, indicating 1.36 medical conditions per patient. However, no condition at all was found in 1,841 of the group, so that the remaining 1,231 averaged 3.44 conditions each.

Table IIIB gives in detail many of the conditions previously summarized and is considered the basic part of the study. Comment on the individual diseases found here does not seem warranted but a study of this table will afford recognition of the multiplicity of the morbid processes encountered.

Complications of pregnancy

Table IV indicates the frequency of certain complications resulting from the pregnancy, itself, that were found in the 3,072 patients in the series. Vomiting, severe enough to be termed hyperemesis gravidarum, was noted 62 times or in 2 per cent of total pregnancies. Bleeding, diagnosed as threatened abortion, occurred in 221 patients during the study, an incidence of 7.2 per cent. The diagnosis of threatened abortion was made in 21 instances before the LMP and abortion, itself, was noted 40 times, in each instance denoting a pregnancy

Table IIIB. Conditions noted in 3,072 patients (from 3 months before last menstrual period through termination of pregnancy)

Code No.	Condition	Time of onset					Total
		Before LMP	Trimester of pregnancy				
			1st	2nd	3rd	Total	
Infective and parasitic							
001	Tuberculosis	1	1	0	0	1	2
002	Syphilis	3	2	9	3	14	17
003	Gonorrhoea	1	0	0	1	1	2
004	Other venereal	1	4	5	5	14	15
005	Typhoid	1	0	0	0	0	1
006	Dysentery	1	1	0	0	1	2
007	Food poisoning	0	0	1	0	1	1
013	Streptococcus infection of throat	7	2	2	0	4	11
019	Other bacterial	3	2	1	1	4	7
022	Measles	0	0	0	1	1	1
023	Rubella	0	1	2	2	5	5
025	Mumps	0	1	1	3	5	5
026	Infectious hepatitis	1	2	1	0	3	4
027	Exposure to rubella	1	2	12	4	18	19
028	Exposure to hepatitis	0	1	2	1	4	4
029	Other viral or parasitic disease	6	5	4	5	14	20
	Total	26	24	40	26	90	116
Neoplasms							
52	Malignancy of cervix (carcinoma in situ)	2	0	0	0	0	2
59	Other malignant disease	1	3	3	0	6	7
60	Neoplasm, benign, breast	5	3	13	3	19	24
61	Neoplasm, benign, uterus	20	35	24	5	64	84
62	Neoplasm, benign, ovary	10	6	3	0	9	19
69	Neoplasm, benign, other (moles, polyps, cysts, etc.)	25	21	19	21	61	86
70	Suspicious Papanicolaou smear	1	2	4	3	9	10
	Total	64	70	66	32	168	232
Allergic, endocrine, metabolic, and nutritional diseases							
100	Hay fever	2	5	11	4	20	22
101	Asthma	4	4	11	3	18	22
102	Allergic rhinitis	17	17	28	13	58	75
104	Urticaria	6	3	3	3	9	15
105	Drug allergies	8	11	18	14	43	51
106	Food allergies	1	1	1	0	2	3
109	Other allergic (includes poison oak, etc.)	18	12	14	5	31	49
110	Simple goiter	6	1	0	0	1	7
111	Nontoxic nodular goiter	1	2	0	1	3	4
112	Hyperthyroidism	9	1	0	0	1	10
113	Hypothyroidism	7	0	0	0	0	7
120	Diabetes mellitus	1	2	3	2	7	8
129	Other endocrine disease	2	1	0	0	1	3
132	Obesity	48	40	34	66	140	188
139	Other metabolic disease	1	1	0	0	1	2
	Total	131	101	123	111	335	466
Diseases of the blood and blood-forming organs							
151	Iron deficiency anemia (not anemia of pregnancy)	6	3	1	5	9	15
154	Sickle cell anemia	0	1	0	1	2	2
159	Other	1	1	0	0	1	2
	Total	7	5	1	6	12	19
Mental, psychoneurotic, and personality disorders							
200	Psychoses	2	4	4	0	8	10
201	Anxiety reaction (somatic symptoms not mentioned)	27	24	22	7	53	80
202	Psychoneurosis (symptoms circulatory)	0	1	0	0	1	1

Table IIIB—Cont'd

Code No.	Condition	Time of onset: Before LMP	Trimester of pregnancy: 1st	2nd	3rd	Total	Total
203	Psychoneurosis (symptoms digestive)	6	4	1	1	6	12
204	Psychoneurosis (other symptoms)	15	5	3	1	9	24
205	Alcoholism	1	0	0	0	0	1
207	Mental deficiency	0	0	1	0	1	1
209	Other behavior and intelligence disorders	11	18	6	4	28	39
	Total	62	56	37	13	106	168
Diseases of the nervous system and sense organs							
252	Epilepsy	3	3	1	3	7	10
253	Migraine	4	2	5	2	9	13
259	Other diseases of the nervous system	9	8	7	10	25	34
260	Conjunctivitis	14	2	5	4	11	25
261	Other inflammatory diseases of the eye	8	5	2	0	7	15
262	Refractive errors	70	36	34	10	80	150
263	Blindness	0	0	1	1	2	2
269	Other conditions of the eye	11	7	8	8	23	34
270	Otitis	19	9	9	5	23	42
271	Deafness	4	1	2	0	3	7
272	Deaf mutism	1	0	0	0	0	1
279	Other diseases of ear (otitis externa and earache)	16	12	14	12	38	54
	Total	159	85	88	55	228	387
Diseases of the circulatory system							
300	Rheumatic fever	1	1	1	1	3	4
301	Rheumatic heart disease	2	3	2	3	8	10
304	Benign hypertensive heart disease	0	0	3	0	3	3
305	Malignant hypertensive heart disease	0	0	1	0	1	1
306	Functional heart murmurs	3	8	13	6	27	30
307	Hypertension antedating pregnancy*	2	2	5	1	8	10
310	Varicose veins	5	11	38	39	88	93
311	Hemorrhoids	10	12	26	70	108	118
312	Phlebitis and thrombophlebitis	0	3	3	5	11	11
319	Other circulatory disorders (epistaxis)	7	7	13	17	37	44
	Total	30	47	105	142	294	324
Diseases of the respiratory system							
350	Common cold	111	97	167	141	405	516
351	Acute sinusitis	21	14	23	12	49	70
352	Acute tonsillitis	21	30	16	13	59	80
353	Influenza	31	34	32	24	90	121
354	Influenzal pneumonia	1	0	0	0	0	1
355	Viral disease with digestive manifestations	4	11	15	14	40	44
356	Pneumonia	1	4	0	2	6	7
357	Bronchitis	15	13	15	16	44	59
360	Peritonsillar abscess	1	0	1	0	1	2
361	Pleurisy	4	2	2	1	5	9
369	Other respiratory conditions	7	11	24	22	57	64
	Total	217	216	295	245	756	973
Diseases of the digestive system							
400	Dental disorders	12	9	13	7	29	41
401	Gastric and duodenal ulcers	6	2	1	1	4	10
402	Appendicitis	2	0	0	0	0	2
403	Hernia	2	6	6	8	20	22
404	Gastroenteritis and colitis	26	24	18	18	60	86
405	Anal fissure and fistula	5	2	0	2	4	9
406	Rectal abscess	0	0	2	2	4	4
408	Constipation	12	22	23	20	65	77
409	Other diseases of the intestinal tract	7	12	8	19	39	46
410	Cholelithiasis	2	0	0	1	1	3

*Not otherwise stated.

Table IIIB—Cont'd

Code No.	Condition	Time of onset: Before LMP	Trimester of pregnancy: 1st	2nd	3rd	Total	Total
411	Cholecystitis	2	0	0	2	2	4
419	Other liver and gall bladder disease	0	0	1	3	4	4
	Total	76	77	72	83	232	308
	Diseases of the genitourinary system						
450	Acute nephritis	1	0	0	0	0	1
451	Cystitis	28	33	29	13	75	103
452	Pyelitis (before LMP)	11	2	1	1	4	15
453	Renal calculi	0	3	2	0	5	5
459	Other diseases of the urinary system	8	13	15	8	36	44
460	Chronic cystic disease of breast	2	3	1	2	6	8
461	Breast abscess	4	0	1	2	3	7
470	Acute pelvic inflammatory disease	4	7	4	1	12	16
471	Chronic pelvic inflammatory disease	8	11	0	0	11	19
472	Vaginitis and vulvitis	53	103	136	64	303	356
473	Rectocele, cystocele, descensus, prolapse	6	10	30	9	49	55
474	Menstrual disorders (amenorrhoea)	105	70	22	5	97	202
475	Cervicitis	56	74	31	9	114	170
479	Other conditions of the generative system	32	46	16	9	71	103
	Total	318	375	288	123	786	1,104
	Diseases of the skin and cellular tissue						
600	Boil or carbuncle	11	5	8	8	21	32
601	Cellulitis of finger or toe	6	7	6	3	16	22
602	Other cellulitis	7	7	12	5	24	31
603	Adenitis, acute	10	8	5	0	13	23
609	Other local infections	8	4	7	8	19	27
610	Warts	10	9	14	10	33	43
611	Eczema	14	8	3	5	16	30
612	Acne	14	6	6	1	13	27
619	Other conditions of skin (dermatoses)	67	42	48	44	134	201
	Total	147	96	109	84	289	436
	Diseases of the bones and organs of movement						
650	Arthritis	7	1	2	3	6	13
651	Muscular rheumatism	2	1	0	0	1	3
652	Sacroiliac joint, affection of	5	1	0	0	1	6
653	Bursitis, synovitis, ganglion	16	15	5	1	21	37
654	Myositis	11	12	6	1	19	30
655	Displacement of intervertebral disk	2	0	2	0	2	4
657	Hallux valgus	1	1	0	0	1	2
659	Other	17	14	14	3	31	48
	Total	61	45	29	8	82	143
	Congenital malformations of mother						
675	Total malformations	1	7	4	5	16	17
	Symptoms						
700	Tachycardia and palpitation	3	2	4	3	9	12
701	Syncope	5	6	22	10	38	43
702	Breast pain*	3	3	0	1	4	7
703	Abdominal pain*	35	52	83	49	184	219
704	Pain, other*	28	27	46	47	120	148
705	Dysuria	7	7	6	12	25	32
7[illegible]	Hematuria	1	0	0	1	1	2
7[illegible]	Fever of unknown origin	4	4	3	3	10	14
7[illegible]	Headache	43	40	51	31	122	165
7[illegible]	Fatigue	21	25	31	15	71	92
7[illegible]	Nervousness*	15	12	16	23	51	66
7[illegible]	Low back pain*	20	20	39	37	96	116
7[illegible]	Other symptoms	15	15	39	58	112	127
	Total	200	213	340	290	843	1,043

[illegible] otherwise stated.

Table IV. Complications of pregnancy and childbirth*

Code No.	Condition	Time of onset: Before LMP	Trimester of pregnancy: 1st	2nd	3rd	Total	Total
500	Pyelitis of pregnancy	0	5	24	14	43	43
501	Hyperemesis gravidarum	0	35	16	11	62	62
502	Threatened abortion	21	132	83	6	221	242
503	Abortion	40	64	60	1	125	165
504	Ectopic pregnancy	2	0	1	1	2	4
505	Antepartum hemorrhage	0	1	27	57	85	85
506	Anemia of pregnancy	16	63	400	384	847	863
507	False labor	5	1	19	141	161	166
508	Hydramnios	2	0	2	13	15	17
509	Hypertension	5	6	69	305	380	385
510	Proteinuria (alone)	4	7	21	44	72	76
511	Toxemia of pregnancy	1	4	94	132	230	231
512	Excess weight gain (over 25 pounds)	14	0	9	953	962	976
513	Premature rupture of membranes	2	0	5	75	80	82
514	Multiple pregnancy	1	0	3	24	27	28
515	Prematurity (Less than 36 weeks)	0	1	17	75	93	93
516	Postmaturity (More than 42 weeks)	1	0	0	65	65	66
518	Hydatidiform mole	0	0	1	0	1	1
519	Other noteworthy symptoms	1	3	15	44	62	63
520	Nausea and vomiting, not hyperemesis	2	66	21	12	99	101
521	Other minor symptoms of pregnancy	6	22	79	139	240	246
	Total	123	410	966	2,496	3,872	3,995

*From 3 months before last menstrual period to termination of pregnancy.

Table V. Accidents and injuries from 3 months before LMP to termination of pregnancy

Code No.	Condition	Time of occurrence: Before LMP	Trimester of pregnancy: 1st	2nd	3rd	Total	Total
Accidents, external cause							
800	Motor vehicle accident, driver or passenger	10	14	18	18	50	60
801	Motor vehicle accident, other	1	4	2	0	6	7
802	Other vehicular accident	1	1	2	0	3	4
803	Drug or food poisoning	0	1	2	1	4	4
805	Accidental falls	15	12	26	22	60	75
806	Accident caused by cutting or piercing instrument	12	13	9	5	27	39
807	Accident purposely inflicted by another person	7	5	1	3	9	16
809	Other accident or injury	39	20	17	26	63	102
	Total	85	70	77	75	222	307
Accidents, nature of injury							
850	Fracture, skull, spine or trunk	2	1	1	0	2	4
851	Fracture, extremity	6	6	5	6	17	23
852	Dislocation	2	0	1	0	1	3
853	Sprains and strains, excluding sacroiliac	24	20	16	18	54	78
854	Sacroiliac sprain or strain	2	6	2	1	9	11
855	Lacerations and open wounds	16	12	14	15	41	57
856	Superficial injury	9	13	9	8	30	39
857	Contusions	10	3	9	5	17	27
858	Burns	4	1	1	3	5	9
859	Drug poisoning or food poisoning	0	1	2	2	5	5
860	Internal injury	0	0	0	1	1	1
869	Other injuries	12	7	10	7	24	36
879	No injury from accident	5	9	12	19	40	45
	Total	92	79	82	85	246	338

terminating within 3 months prior to the onset of the pregnancy included in the study. Threatened abortions were noted in the second trimester in 83 cases and in the third trimester, 6 times. When diagnosed after the twentieth week of gestation these are instances of so-called "missed" abortions. Abortion, complete, incomplete, or inevitable was noted on 125 occasions while 56 additional patients with this condition were admitted directly to the hospital without a clinic examination. Anemia of pregnancy denotes a hemoglobin of under 11 Gm. or a hematocrit below 35. Such a condition was noted 847 times, an incidence of 27.6 per cent. Hypertension is understood to be one or more rises of blood pressure to a systolic level of 140 or above, or a diastolic reading of at least 90, without the diagnosis of toxemia of pregnancy being made by the clinician. Hypertension was noted 380 times (12.4 per cent) while toxemia was diagnosed in an additional 230 instances (7.5 per cent).

Weight gain was termed excessive if before delivery the patient had gained more than 25 pounds. This occurred in 962 women, the incidence of 31.3 per cent practically disqualifying it from the list of abnormalities or complications. Premature rupture of the membranes was diagnosed if more than 24 hours elapsed between its occurrence and delivery. There were 28 instances of multiple pregnancy or 1 in 110. The incidence of prematurity, as determined by termination of pregnancy beyond the abortion stage but less than 252 days after the LMP, was low, 3.0 per cent. On the other hand, postmaturity, as determined by a gestation of over 294 days, was noted 65 times, an incidence of 2.1 per cent. Total pregnancy complications according to the above list were noted 3,872 times during the gestation under consideration, an average of 1.3 per patient. However, 919 of the pregnancies were without any complication here listed so that the remaining 2,153 averaged 1.8 conditions each.

Table VI. Number of different clinics other than Obstetrics-Gynecology visited by patients from 3 months before LMP through termination of pregnancy

Type of clinic	*Total No. of patients*
Medicine	1,237
Surgery	216
Orthopedic	150
Eye, ear, nose and throat	359
Urology	27
Dermatology	242
Dental	4
Psychiatry and neurology	23
Emergency room	746
2 or more clinics, not Obstetrics-Gynecology	86
Unknown	18
Total	3,108*

*Included are 152 instances where patient visited another clinic and also Obstetrics-Gynecology for the same condition.

Accidents and injuries

During the 3 months prior to the LMP, 83 women in the series had a total of 85 accidents, while during the pregnancy, itself, an

Table VII. Frequency distribution of x-rays taken according to a broad anatomical classification (from 3 months before LMP, through termination of pregnancy)

Type of x-ray	*Before LMP*	*Trimester of pregnancy* 1st	2nd	3rd	Total	Total
Head	13	9	5	8	22	35
Extremity	54	31	30	18	79	133
Spine	16	14	8	3	25	41
Chest survey	171	131	111	41	283	454
Upper gastrointestinal or gall bladder	14	4	0	0	4	18
Lower gastrointestinal or barium enema	3	2	0	0	2	5
Other abdominal including kidney, ureter, bladder	7	11	3	1	15	22
Pelvis and uterus, diagnostic	9	1	4	88	93	102
Pelvimetry	2	1	0	41	42	44
Total	289	204	161	200	565	854

additional 214 patients had 222. Thus the accident rate during the entire pregnancy was 6.9 per cent. The total number of accidents was about the same for each trimester of pregnancy. Whether an individual becomes more accident prone during pregnancy is not clear, but is not indicated by Table V, since the total number of mishaps suffered during the 3 months prior to LMP is greater than that for any trimester of pregnancy. Table V indicates, also, that the external cause of the accident and the nature of the injury occasioned by it.

Supplementary classification

Additional clinic visits on the part of the 3,072 patients in the series were noted and totaled 1,689. Included in this list are routine examinations, premarital examinations, visits where symptoms were found but no diagnosis ever made although several possibilities existed, and a heterogeneous group including postpartum checks, postoperative visits, weight reduction visits, etc.

The use made of clinics other than Obstetrics and Gynecology in the care of patients in the series is shown in Table VI. On most occasions, one clinic, alone, treated a given condition during one or more visits. However, in 152 instances of a single disease, the patient was seen in another clinic and also in Obstetrics-Gynecology, the referral on certain occasions being from Gynecology to the other clinic, while in others, the reverse was the case. In a total of 3,108 instances, a nonobstetrical condition was observed in another clinic, indicating the use the patient or obstetrician made of consultant help. It will be recalled that about one fourth of the total conditions under study occurred prior to the LMP, which would leave approximately 2,300 to have occurred in pregnancy when the patient presumably was under the care of the Prenatal Clinic. As would be expected, the Medical Clinic and its subdivisions and the Emergency Room were the agencies most frequently employed. However, a rather surprising use was made of the Ophthalmology, Laryngology, and Dermatology Clinics. In 86 instances, two or more outside clinics were employed in the care of a single condition without the patient being observed in Obstetrics-Gynecology at all.

Table VIII. Frequency distribution of operations performed according to a broad anatomical classification*

Type of operation	Before LMP	Trimester of pregnancy: 1st	2nd	3rd	Total	Total
Dental extraction	0	1	3	2	6	6
Tonsillectomy and adenoidectomy	0	1	0	0	1	1
Other operations, head, face, mouth	1	4	1	1	6	7
Reduction of fracture	0	0	4	0	4	4
Excision benign tumor, breast	3	1	2	0	3	6
Hemorrhoidectomy	0	0	0	1	1	1
Appendicectomy	2	1	0	1	2	4
Cholecystectomy	3	1	0	0	1	4
Herniorrhaphy	0	1	0	0	1	1
Procto- and sigmoidoscopy	4	3	1	0	4	8
Cystoscopy	4	6	2	1	9	13
Nephrotomy	1	0	2	0	2	3
Uterus, suspension	1	0	0	0	0	1
Uterus, myomectomy	1	1	1	0	2	3
Shirodkar operation	1	0	1	1	2	3
Ovarian cystectomy	2	0	1	0	1	3
Other abdominal	1	0	1	0	1	2
Other, nonabdominal, major	1	2	0	1	1	4
Other, nonabdominal, minor	71	28	42	23	93	164
Total	96	50	61	31	142	238

*From 3 months before LMP, through termination of pregnancy.

X-rays and operations

A total of 854 x-rays were obtained on the patients in the series, 289 before the LMP and 665 during the pregnancy, itself. Table VIII shows the type of x-ray obtained and when it was taken. Over half of the total number were chest surveys, while pelvimetry was done on only 44 patients, a rate of 1.4 per cent. During the pregnancy, itself, 16.2 per cent of total patients were x-rayed, and in only 156, or 5.1 per cent, was the film abdominal in character.

It will be seen from Table IX that a total of 238 operations were performed on the women in the group, and 96 of these were done during the 3 months prior to LMP. One hundred and sixty-four or 68.9 per cent of the total were listed as nonabdominal, minor. No major operations were performed in the last trimester of pregnancy and only 5 during the second trimester. A low rate of 3.8 per cent of total patients had any type of operative procedure done during the pregnancy, itself.

Drugs

Although the clinics giving medical and obstetric care to the patients in the series did not make excessive use of drug therapy, it was rather surprising to note the number of medications ordered for these women during the period of observation (Table IX).

Two hundred and forty-four or 7.9 per cent of the 3,072 patients in the series went through their entire period of observation with no drugs being ordered at all, and in an additional 563 (18.3 per cent), medications were limited to one each. On the other hand, more than 5 drugs were used in 617 individuals (20.8 per cent), while 121 were given 10 or more (3.9 per cent). The average number of drugs employed per patient was 3.6.

Table IX shows the frequency of drug administration to the patients in the series according to type of substance used. No attempt was made to group these drugs according to chemical composition. It is realized that many of the medicines listed here contain more than one ingredient and may be expected to have multiple effects. They are, however, treated according to condition-relief sought and are standardized only in that each individual substance is coded similarly throughout.

Specific therapy in the form of antibiotics or sulfa drugs used to combat infection was ordered in 632 instances. Antihistamines, mostly used for the relief of symptoms of upper respiratory infections, were employed

Table IX. Frequency distribution of drugs prescribed according to therapeutic indication

Type of drug	*No. of patients*
Antibiotics, oral and parenteral	388
Sulfa drugs	244
Urinary antiseptics	68
Antihistamines	533
Cold remedies (neither antibiotic nor antihistamine)	123
Antinauseants	454
Antiobesity and appetite depressants	443
Estrogens	13
Progestins	90
Cortisone	51
Insulin	4
Thyroid compounds	80
Digestive aids and antacids	384
Laxatives	158
Diuretics and antihypertensives	987
Headache remedies	36
Sedatives and hypnotics, barbiturates	605
Tranquilizers and antianxiety	370
Narcotics (includes compounds)	430
Analgesics	716
Stimulants and antidepressants	14
Prenatal capsules	2,306
Vitamins, alone	88
Iron compounds	375
Antivaginitis therapy	443
Dermatologic lotions and creams	426
Nose drops, eye and ear drops, sprays, etc.	142
Other unlisted drugs	422
Undecipherable	131
Inoculation or vaccination:	
Diphtheria	5
Influenza	133
Pertussis	1
Poliomyelitis	181
Smallpox	20
Tetanus	65
Typhoid	2
Gamma globulin	12
Total	10,923

533 times. In 454 cases (14.8 per cent of total patients), vomiting of pregnancy was sufficiently annoying to warrant the use of an antinauseant drug. Weight gain was such in 443 instances (14.4 per cent), that an appetite depressant was ordered. Progestins were used in an effort to prevent abortion in 90 patients.

Heartburn, indigestion, and other similar symptoms necessitated the use of an antacid or digestive aid on 384 occasions. Hypertension, edema, and similar symptoms were sufficient to suggest the use of some form of diuretic or antihypertensive drug on 987 occasions. Here it should be emphasized that the number of these drugs listed does not represent the number of patients involved, for frequently 3 or 4 of them were used in the same case before one was found to produce the desired effect. The nervous element, so frequently accompanying pregnancy, was well emphasized by the employment of 975 sedatives, hypnotics, and tranquilizers. Narcotics were noted in 430 cases, these usually forming an ingredient of some cough medicine or cold remedy. The presence of any narcotic drug in a compound was sufficient to cause the substance to be listed in this category. Analgesic drugs were employed to relieve various types of discomfort not necessitating narcotics in 716 instances.

Prenatal capsules were ordered 2,306 times, representing three quarters of the total patients, while additional iron therapy was used in 375 more.

The foregoing drugs were administered orally or parenterally. Local applications were frequent, with antivaginitis therapy being used on 443 occasions, dermatological lotions and creams on 426, and drops, sprays, etc., in 142. A total of 422 medications were ordered which did not fit into any of the above categories. Finally on 131 occasions, the handwriting of the clinician did not permit recognition of the drug which was used.

Table IX also lists the various types of inoculation or vaccination employed in the group during their period of observation and these together totaled 407. Finally, 12 women were given gamma globulin during pregnancy because of exposure to rubella or infectious hepatitis. The above list of drugs used in the 3,072 patients in the study had accordingly a grand total of 10,923.

Summary and conclusions

The case records of 3,072 young women having their last menstrual period during one calendar year and obtaining their entire medical and obstetrical care from a single facility have been abstracted, coded and analyzed by the punch-card method. From 3 months before LMP to the termination of pregnancy, information has been obtained on prenatal care, intercurrent medical conditions, complications of pregnancy, x-rays, operations, and drugs prescribed. It is felt that the patients involved represent fairly well a cross section of the childbearing population. Emphasis has been placed upon the large number of intercurrent and accidental complications encountered involving multiple facets of the medical discipline. We wish to make clear that at present there is no indication that the pregnant woman encounters more of these illnesses than her nonpregnant sister during a similar period of time, although it is planned to investigate this interesting possibility in the near future the role the obstetrician should play in the intercurrent conditions, while the patient is under his care.

Should the internist, surgeon, laryngologist, etc., assume full responsibility for the treatment of conditions within his sphere while the pregnancy exists, or should he be used as a consultant, only, with the obstetrician directing the entire care of the patient, rather than merely the pregnancy itself?

We wish to acknowledge our indebtedness to the entire staff of the Child Development Study for their aid in abstracting the case records involved and also for the preparation of the statistical tables herein presented.

3783 Piedmont Avenue
Oakland 11, California

Discussion

Dr. R. Gordon Douglas, New York, New York. A section of a large series of 10,000 pregnancies in a child development study has been selected for study of intercurrent diseases, namely, those patients whose last menstrual period occurred between April 1, 1960, and March 31, 1961, comprising 3,072 individuals who are members of the Kaiser Foundation Health Plan, have attended a Permanente Clinic and have, in general, obtained their entire care from such a facility. Clinic visits, hospitalization, diagnosis, and so on, from 3 months prior to the last menstrual period to termination of pregnancy were included in the study.

This series was considered by the authors to be a representative cross section of childbearing women. In respect to ethnic background, 67.3 per cent were white, 19.4 per cent Negro, 8.6 per cent other races, and 4.6 per cent unknown. It would be of value to have the proportion of each in the population in the area for comparison. At The New York Lying-In Hospital in 1962, 78.2 per cent of total births were white, 9.2 per cent Puerto Rican, and 12.6 per cent nonwhite. Among live births in the City of New York for 1961, 63.6 per cent were white, 14.7 per cent Puerto Rican, and 21.7 per cent nonwhite. Thus, the patients on our service are not representative of our community.

One could question the implication that the patients in the authors' series represent a cross section of the childbearing population as a whole because (1) only slightly more than two-thirds were white; (2) unwed minors were excluded, and (3) indigent patients were not included.

In a total of 5,300 pregnancies (4,768 deliveries and 532 abortions) terminating at The New York Lying-In Hospital in the year 1962, in which 99 per cent were registered patients, antepartum and concurrent conditions routinely recorded in our Annual Reports were tabulated to correspond as nearly as possible to the arrangement in Dr. Peckham's study. Figures were thus obtained from our service to compare with the data in the authors' paper, much of which was not presented to you this morning. In many instances there were striking similarities. In some instances, discrepancies were explainable while in others they were not. For instance, in Table I the incidence of maternal congenital anomalies in the authors' material was 0.5 per cent, while we had more than four times as many. Ours included circulatory, 34, digestive, 2, organs of movement, 16, generative tract, 43, and urinary tract, 15. It is hardly reasonable to suppose that there is such a difference in the incidence of anomalies between the West Coast and the East Coast. This and other recorded facts, such as lack of information on basic data such as age in 6.5 per cent and ethnic background in 4.6 per cent, suggest that there may be some minor deficiencies in the way of omissions in the authors' series.

In Table I, the number and per cent incidence of these conditions at The New York Lying-In Hospital summarized by broad classifications are compared with those encountered in the authors' study. The column headed "New York Lying-In" represents conditions in all pregnant patients who were delivered or who had abortions at The New York Lying-In Hospital in 1962. These figures are compared to the nearest to comparable figures in the authors' material, i.e., those conditions encountered during pregnancy, itself, in 3,072 patients, some of whom eventually proved to be not pregnant. In the category of infective and parasitic diseases, although the percentages appear fairly comparable, the recording of individual diseases is different. One case of, presumably, active pulmonary tuberculosis was included by the essayist whereas 4 active and 60 inactive or currently arrested cases were included in our figures. On the other hand, 4 streptococcus infections of the throat, 4 other bacterial, and 14 "other venereal" infections are in this category in the authors' study which would be assigned to the anatomic system affected in our figures.

Among the pregnancies in 1962, we diagnosed 2 new cases of carcinoma in situ of the cervix and recorded 6 with a previous history of this diagnosis, whereas 2 were diagnosed prior to pregnancy in Dr. Peckham's study and none during pregnancy, itself. Our 0.2 per cent of malignant disease other than gynecological is the same as in this study. Our recording of neoplasms of the breast and uterus is quite comparable to that of this study but for ovary and other benign neoplasms our recording is considerably higher.

In recent years we have been recording both a history of asthma and diagnosed asthma as potentially significant; our rate of incidence is over twice that of this study. Inclusion in our figures of 546 patients with a history of drug sensitivity accounts for more than one half of this broad category. The incidence of hay fever is similar in both studies, but allergic rhinitis is much higher in Peckham's study. His rate of diabetes is about one third of ours. Skin rash and skin reaction due to drugs or questionably due to drugs are classified under "skin" in our figures.

At The New York Lying-In Hospital, iron deficiency anemia was recorded in 580 patients based on the specific stated diagnosis, a hematocrit of 35 or less, or a hemoglobin of 11 Gm. or less which improved on iron therapy. An additional 1,586 patients had low readings at some time during the antepartum course without a specific diagnosis of anemia having been made. This latter group is not included here. The number with blood dyscrasias, exclusive of iron deficiency anemia, was 68 or 1.3 per cent. In Peckham's study, 27.6 per cent were classified as having anemia of pregnancy compared to 41 per cent on our service.

Psychosis was present in 0.5 per cent of our patients and in 0.3 per cent of Peckham's, including 2 cases diagnosed prior to pregnancy. An incidence of 69 cases or 2.2 per cent of psychoneurotic disorders was noted during pregnancy, itself, a larger rate than was recorded in our patients, 0.4 per cent.

Diseases of the eye and ear were much more frequently reported in this study than in our figures. In our statistics, refractive errors are not included but, on excluding these from Peckham's study, there are still more of these conditions recorded.

Epilepsy, migraine, and other conditions of the nervous system comprise 1.3 per cent in Peckham's study, 1.2 per cent of ours.

Rheumatic heart disease and hypertension are considerably less frequently recorded than at The New York Lying-In Hospital and one wonders whether there is a definite regional difference. We recorded more varicose veins, 5.3 per cent in comparison with 3 per cent, and fewer cases of hemorrhoids. Antepartum thrombophlebitis had an incidence of 0.2 per cent in 1962, compared to 0.4 per cent in Peckham's study. A large number of cases of edema is recorded in our figures, accounting for some of the difference.

The differences are greatest in the respiratory category, partly because we do not record the common cold as an intercurrent condition unless it is an associated condition on admission to the hospital or just prior to termination of pregnancy. During pregnancy, itself, 13 per cent were recorded in this study, contrasted with 2.2 per cent in our institution. Tonsillitis, sinusitis, pneumonia, and influenza were also low by comparison, although we would record the latter too if they occurred at any time during pregnancy.

The much greater incidence of gastroenteritis and colitis, constipation, and other conditions of the gastrointestinal tract accounts for most of the difference here.

Cystitis was recorded in 75 or 2.4 per cent of the study patients during pregnancy, itself, but in only 0.3 per cent of ours. Only 4 cases of pyelitis were diagnosed during pregnancy, compared to 24 or 0.5 per cent at The New York Lying-In Hospital. There are more included later in the "complications" table. Chronic renal disease, if it occurred, must be among other diseases of the urinary system in Peckham's study.

Among gynecological diseases, pelvic inflammatory disease has about the same incidence in both series; vaginitis and vulvitis, an incidence of 9.8 per cent in the study group in contrast to our 1.8 per cent. Rectocele, cystocele, descensus, and prolapse were recorded more frequently and cervicitis less frequently at The New York Lying-In Hospital.

Boils, cellulitis, and other local infections were diagnosed in 2.6 per cent of the patients during pregnancy, itself, compared with 0.3 per cent in our series. Dermatoses were reported in approximately 6.4 per cent compared with 1.1 per cent at our institution.

It is of interest that, in spite of many intrinsic differences, there were 1.1 per cent diagnoses per patient in both series when symptoms and conditions before pregnancy were excluded for comparison.

The essayist has presented us with a mass of information, and one must be impressed with the amount and frequency with which his patients employ medication and the number that have respiratory conditions. As no conclusions are included, we are left to draw our own. I am sure we all would like to know the impact of these many conditions, if any, on the maternal and fetal organism. Our pregnancy wastage from abortion was 10 per cent as compared with his 9.3 per cent. His perinatal mortality of 3.4 per cent compared with 2.3 per cent in our material. Such comparisons may be meaningless for many obvious reasons.

The study does emphasize certain responsibilities of the obstetrician. As I understand the operation of the facility where the authors' patients were cared for, a principle of referral for nonobstetrical conditions is the rule. This

Table I. Intercurrent conditions

Condition	N.Y.L.I. (5,300)		Peckham (3,072)	
	No.	%	No.	%
Infective	129	2.4		2.9
Neoplasms	424	8.0		5.5
Allergic, etc.	1,009	19.0		10.9
Blood forming	648	12.2		0.4
Mental, etc.	53	1.0		3.5
Nervous, sense	125	2.4		7.4
Circulatory	998	18.8		9.6
Respiratory	276	5.2		24.6
Digestive	200	3.8		7.6
Genitourinary	1,823	34.4		25.6
Skin	113	1.9		9.4
Bones, etc.	98	1.6		2.7
Malformations	110	2.1		0.5
Symptoms			843	27.4
Total	6,016*		4,237*	

*1.1 diagnoses per patient exclusive of symptoms.

practice is widely employed throughout the country. This I do not believe to be best for our patients or our discipline. The preferred principle, I believe, is the one in which the obstetrician remains at the helm, seeks consultation freely but remains in charge, and directs the over-all care of the patient.

The authors are to be congratulated for the accumulation of these data, representing a portion of a larger study. We will look forward to additional reports, detailing the impact of these conditions on the newborn.

Dr. William C. Keettel, Iowa City, Iowa. Dr. Peckham and his associates have presented a comprehensive study of intercurrent conditions observed during pregnancy. Comparable data are difficult to obtain, since most available statistics are from hospitalized patients. However, in certain states limited information concerning pregnancy complications are available from the birth certificate. Using this source of information, the Iowa figures for parity are very similar. In Peckham's series, 25 per cent were pregnant for the first time, in Iowa, 24.3 per cent. The number of women having three or more children varies depending upon the year selected; in 1952, 23.4 per cent of the women in Iowa had three or more children but by 1961 there were 35.5 per cent.

The number of intercurrent problems encountered immediately prior to pregnancy is surprising; perhaps young women have more health problems than commonly realized. In clinical practice one is impressed that the more serious complications of pregnancy are commonly seen in the first and third trimesters. This study indicates that, when minor complications are included, more of them are encountered in the second trimester than closer to term, which is difficult to explain. Our experience has indicated the frequency and severity of respiratory infection during pregnancy. Dr. Peckham has indicated this is a frequent problem even in California.

The incidence of carcinoma in situ and suspicious Papanicolaou smears seems low. In a recent study of 6,804 pregnancies in our institution, 55 women had suspicious or positive smears, carcinoma in situ was found in 25 patients, and invasive carcinoma in 5. Thus, 1 in every 225 patients had a premalignant lesion of the cervix during pregnancy.

The number of x-ray studies performed in the first trimester of pregnancy is higher than one would expect. One wonders about the necessity for gastrointestinal, kidney, and pelvimetry studies at this time.

One is concerned with the variety of drugs employed during the course of a pregnancy; one fifth of the patients received 5 or more drugs during gestation. With the variety and number of medicines employed, this would make a study of undesirable fetal effects almost impossible. Is the pregnant patient really in need of such a variety of medication for a physiologic process?

I would like to ask how reliable he considers the reporting of minor complications? Certain physicians might consider varicose veins, hemorrhoids, or backache as unimportant. Despite such limitation, studies of this type are of importance to gain insight into prenatal care and office practice. Teachers will find this material of value in student and resident training.

Dr. Peckham has posed a very important question to our specialty concerning the management of obstetrical patients with complications. Unless our future specialists are well trained and interested in the management of the entire care of the pregnant patient, we may find ourselves truly obstetricial technicians.

Dr. Arthur T. Hertig, Boston, Massachusetts. At the risk of seeming somewhat facetious after a very thorough discussion of this most important problem, I would like to remind the membership that the late and beloved Dr. Frederick C. Irving once said that, "the pregnant patient can have any disease that any other woman can have except sterility."

Dr. Peckham (Closing). In line with Dr. Hertig's remarks, we originally noted but corrected 2 instances of patients listed as having sterility investigations during the third trimester of pregnancy.

Dr. Douglas said that we had far fewer malformations than were noted in his series. Many of our patients had minor anomalies quite compatible with normal existence and we have reported only the major ones, such as congenital heart disease, polycystic kidneys, and so on. In our series there were no major operations performed during the third trimester of pregnancy and only 5 during the second trimester.

The large number of minor complications of pregnancy may seem unusual. Some of those listed were noted during a routine prenatal visit and were not such that a special visit was made because of them. It should be emphasized that in a clinic of this sort, many patients seek medical aid who would probably not do so if it were more difficult or expensive to obtain. Many women went through the entire period without any complications being recorded or drugs given, but there were others who had prescribed for them more than 20 drugs and one woman had 28 separate conditions listed, the last diagnosis being acute anxiety neurosis.

CHAPTER 3

INVESTIGATION AND ANALYSIS OF CRASHES INVOLVING PREGNANT OCCUPANTS

CHAPTER 3: INVESTIGATION AND ANALYSIS OF CRASHES INVOLVING PREGNANT OCCUPANTS

Kathleen D. Klinich, Carol A. C. Flannagan, Jonathan D. Rupp, Lawrence W. Schneider
University of Michigan, University of Michigan Transportation Research Institute

The leading cause of fetal loss in motor-vehicle crashes is placental abruption, which is premature separation of the placenta from the uterus (Pearlman et al. 1990). Little is known about the mechanisms or tolerance of trauma-induced placental abruption, either in motor-vehicle crashes or in other situations such as falls. However, detailed investigations of motor-vehicle crashes involving pregnant occupants can provide insight on the possible mechanisms of placental abruption and other adverse fetal outcomes, such as uterine laceration or direct fetal injury. Detailed crash investigations provide estimates of the crash severity, confirm occupant restraint use, and locate occupant contact points with the vehicle interior, all of which help to identify how the pregnant occupant and fetus were injured in a particular crash.

CRASH INVESTIGATION METHODS

A key part of crash investigations is the inspection, measurement, and photo documentation of the damaged vehicle exterior using standard reconstruction techniques (Fig. 1). For most types of crashes, crush measurements of the vehicle exterior can be used to estimate the crash severity, expressed either as delta V* or the Equivalent Barrier Speed (EBS)**. Crush distances measured at fixed intervals along the damaged vehicle plane are entered into established computer reconstruction programs (such as WinSmash), along with values for vehicle mass and stiffness. These programs use basic equations of momentum and energy conservation to estimate the severity of the crash for the case vehicle. If vehicle damage cannot be measured, expert investigators are often able to estimate the severity of the crash from photographs of the vehicle damage. However, using speed limits or estimated travel speed to estimate crash severity generally does not provide reliable estimates of calculated crash severity, because they do not account for factors such as preimpact braking, the size and stiffness of the struck object, and the direction of impact.

*Delta V is the change in speed of the vehicle at the time of impact loading by another vehicle or object.
** Equivalent Barrier Speed is the impact speed a vehicle would need to be moving into a rigid flat barrier to cause the same extent of exterior damage.

Figure 1: Photo of standard crush measurement used for estimated crash severity.

The vehicle interior is also inspected, photographed, and measured for damage, deformation, and contact marks caused by occupant interactions with the interior components. The extent of intrusion of vehicle components, such as steering wheels, knee bolsters, and roof, into the occupant space is measured using standard methods. Evidence of occupant contact with the interior, such as scuff marks, clothing transfers, deformations, or bloodstains, is also documented. An important part of the interior inspection is identifying evidence of seat belt use or nonuse by vehicle occupants. Evidence of belt use includes loading marks on the belt webbing, D-ring, or continuous loop of the buckle latchplate (Fig. 2) that confirm if the occupant was using her belt restraint. The seat belt being cut by rescue personnel or blood on an extended portion of the seatbelt webbing are also evidence of belt use. Other evidence of belt use is deployment of a belt pretensioner, extension or tearing of a seatbelt load-limiting element, or the retractor locked with the seat belt in an extended mode.

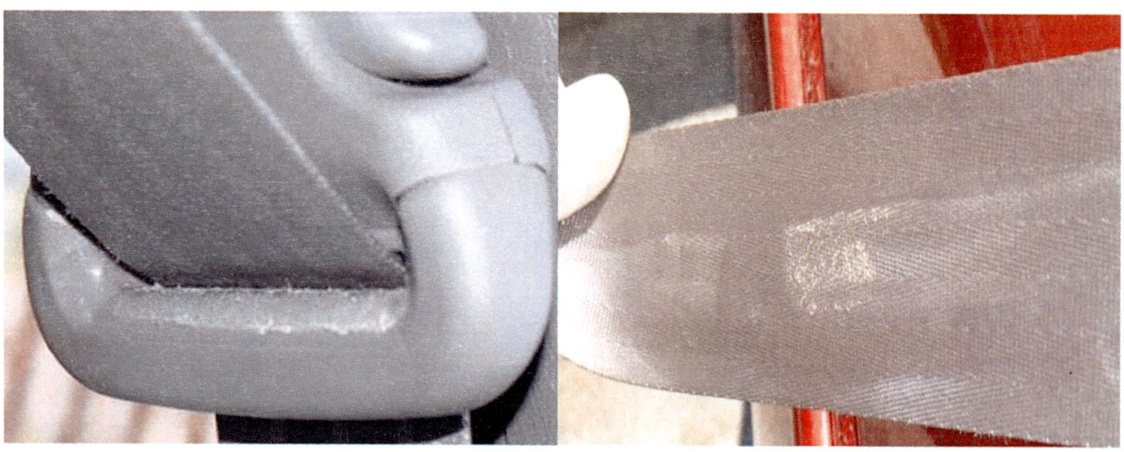

Figure 2: Webbing loading marks on the D-ring (left) or belt webbing (right) indicate that the seat belts were in use during the crashes.

In addition to the case vehicle inspection, the accident site, other involved vehicles or object struck (e.g., tree), and an exemplar vehicle are also usually inspected, photographed, and measured. This information is used to create a scene diagram (Fig. 3) that illustrates the crash scenario before, during, and after the crash. Scene diagrams can include the positions of the involved vehicles throughout the crash, the roadway configuration, and locations of road signs and traffic signals.

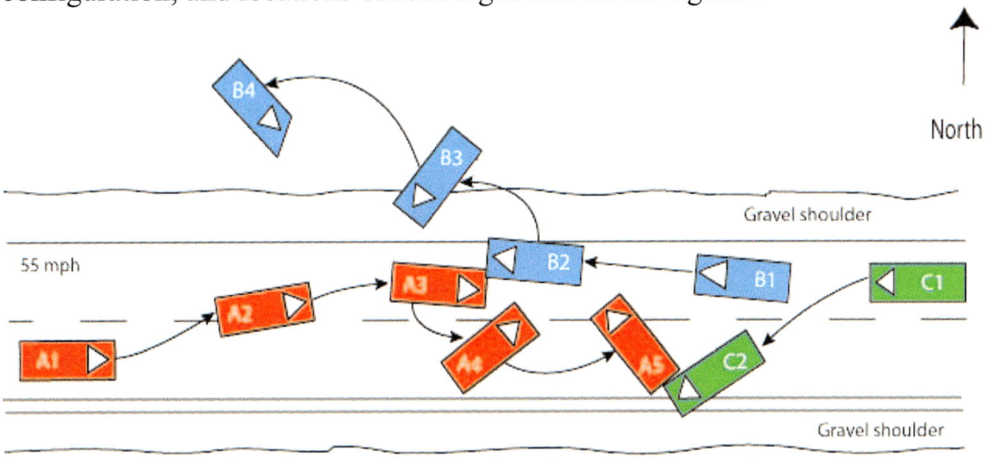

Figure 3: Example of scene diagram showing crash scenario.

When possible, occupants of the involved case vehicle are interviewed for their recollection of the crash scenario and whether they were using belt restraints. However, restraint use must be confirmed by interior evidence described above. When appropriate, written consent is obtained to retrieve information on injuries sustained from medical and hospital records. Occupant injuries are coded using the abbreviated injury scale (Baker 1974), and the suspected cause of each injury is documented, along with a confidence level for the source. In some instances, injuries may be attributed to more than one possible source or as unknown.

For crash investigations of pregnant occupants, it is possible that the fetus may have sustained injury in utero that cannot be detected prior to the birth. For this reason, it is important to follow up with the pregnant subject after the birth of the fetus to confirm that there were no adverse outcomes that may have resulted from trauma sustained during the crash.

DATABASES WITH CRASHES INVOLVING PREGNANT OCCUPANTS

The National Highway Traffic Safety Administration's National Automotive Sampling System (NASS) is a crash investigation program that collects data on a probability sample of tow-away crashes from throughout the United States. NASS began adding the pregnancy status of case occupants to the dataset in 1995, indicating the trimester of pregnancy. In addition, they added a variable in 2006 to indicate fetal death. However, a limitation of crashes with pregnant occupants investigated by NASS is that there is no explicit way to document other fetal outcomes after a crash. In addition, fetal death cannot be assumed prior to 2006 if placental abruption and uterine rupture are coded for the pregnant occupant because it is possible for a fetus to survive these injuries. If a fetal loss or direct fetal injury occurs immediately after a crash, it is likely (but not guaranteed) to be noted in the case documentation. However, if the pregnant occupant gives birth some weeks or months after the crash, the NASS investigation does not follow up to determine if any adverse outcomes may have occurred as a result of the crash. In addition, while the NASS record contains the trimester of pregnancy, some adverse fetal outcomes, such as premature delivery, have a substantially different effect if gestational age is 28 weeks rather than 37 weeks, both of which would be recorded as third trimester. Thus, use of only trimester to record gestational ages is another limitation of pregnant crash investigations in NASS.

The Crash Investigation Research and Engineering Network (CIREN), also administered by the National Highway Traffic Safety Administration, began including a focus on pregnant occupants in 2006. These cases provide detailed injury information for both the mother and fetus, and post-birth follow-up to check on fetal outcome is required. Since the beginning of the CIREN focus to collect pregnant occupant cases, 11 new cases have been added to the CIREN database.

The most comprehensive crash/injury database involving pregnant occupants has been established by the University of Michigan Transportation Research Institute with initial funding from General Motors, Inc. (Klinich et al. 1999). The UMTRI database is the largest collection of crash investigations involving pregnant occupants that include quantitative information on crash severity as well as follow-up on fetal outcome after birth. Cases in this dataset are limited to pregnant occupants beyond 20 weeks gestation, since prior to 20 weeks gestation, the risk of spontaneous abortion is high. As a result, it is difficult to determine the contribution of a vehicle crash to fetal outcome (Pearlman et al. 1990). Details on fifteen cases from the UMTRI database are included in Klinich et al. (2000), while Klinich et al. (2008) summarize key factors for the 42 cases in the original dataset. The latter publication also includes information on some CIREN cases not included in the original study, as well as additional cases collected through the general UMTRI crash investigation program funded by the Alliance of Automobile Manufacturers.

ANALYSIS OF CRASHES INVOLVING PREGNANT OCCUPANTS

Klinich et al. (2008) performed a statistical analysis of 57 crashes compiled from the original UMTRI pregnant crash database, the CIREN program, and the UMTRI crash investigation program. For each case, fetal outcome was coded as acceptable (good outcome or minor complications) or adverse (major complications or fetal loss). Maternal restraint was classified as proper (3-point-belt, with or without airbag) or improper (none or improper use of seat belts). Chi-squared tests were performed to identify factors that have a significant effect on fetal outcome. Maternal injury, maternal restraint, and crash severity all independently had a significant effect on fetal outcome.

Logistic regression analysis was used to predict multivariate risk of adverse fetal outcome. An outside estimate of fetal loss (Weiss et al. 2001) was used to adjust the intercept of the logistic function to account for the bias toward severe crashes in the pregnant crash/injury database. Figure 4 shows the risk of adverse fetal outcome as a function of crash severity and maternal belt restraint. Crash severity is the most important predictor of fetal outcome. The risk of adverse fetal outcome for improperly restrained pregnant occupants is 4.5 times that of properly restrained pregnant occupants. A 50% risk of adverse fetal outcome occurs at a crash severity of 14 mph for improperly restrained pregnant occupants, while it occurs at a crash severity of 29 mph for properly belt-restrained pregnant occupants.

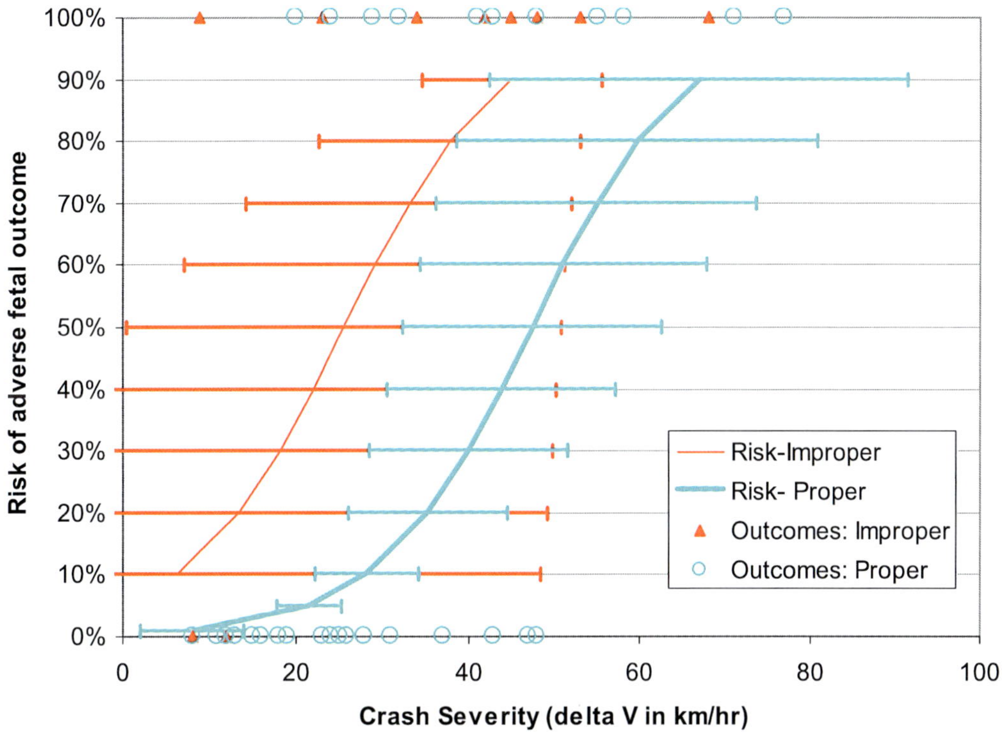

Figure 4: Estimate of adverse fetal outcome as a function of crash severity for improperly restrained and properly restrained pregnant occupants (adapted form Klinich et al. 2008).

Applying these risk curves to the distribution of crash severities experienced by pregnant occupants (derived from NASS) indicates that properly restrained pregnant occupants have a risk of adverse fetal outcome in all crashes of 4.8%, while improperly restrained pregnant occupants have a risk of 30.9%. These risk ratios indicate that proper use of a 3-point belt is 84% effective in reducing adverse fetal outcomes in motor-vehicle crashes. This analysis indicates that approximately half of the fetal losses resulting from motor-vehicle crashes could be prevented if all pregnant women properly used 3-point belts.

MEDICAL CASE STUDIES OF CRASHES INVOLVING PREGNANT OCCUPANTS

In addition to these in-depth crash investigation studies, the medical literature contains many case studies of pregnant women who sustained adverse outcomes resulting from a motor-vehicle crash. These case reports often focus on highly unusual fetal outcomes such as fetal transection, and there are several cases where a seat belt or an airbag is suggested as being the primary source of the adverse fetal outcome (Pepperell et al. 1977, Lane 1989, Van Enk and Van Zwam 1994, Hartl and Ko 1996). Since analysis of in-depth pregnant crash investigations indicates that crash severity has the greatest effect on fetal outcome, case studies without crash severity information must be viewed cautiously. For example, Fusco et al. (2001) report on a near-side impact to a 39-week pregnant driver that was classified as "severe." The authors suggest that the frontal-impact airbag was responsible for the fetal loss, placental abruption, uterine rupture, and fetal skull fracture. However, while it cannot be confirmed without a vehicle inspection, the loading of the driver by the intruding vehicle interior in a severe near-side impact is likely more severe than loading from the frontal-impact airbag.

Klinich et al. (1998) performed a comprehensive study of the medical literature and summarized 93 reported cases of pregnant occupants involved in motor-vehicle crashes with gestational age over 20 weeks by occupant position, crash severity, crash direction, restraint use, gestational age, and fetal outcome. Many of the case reports do not include reliable estimates of crash severity, which is the most important factor related to occupant outcome in crashes. One finding of the literature review is that it is common to have placental abruption as the sole adverse fetal outcome. However, it is much less likely to have uterine rupture/laceration or direct fetal injury in the absence of placental abruption. This finding suggests that the tolerance for placental abruption is lower than for uterine rupture or direct fetal injury, and that countermeasures to prevent placental abruption would also be likely to reduce the occurrence of uterine or direct fetal injury.

Table 1 summarizes more recent cases from the literature since this prior publication by Klinich et al. (1998). The table only includes cases where occupant restraint usage is reported and with gestational age over 20 weeks. None of these cases provide estimates of crash severity, and several do not even describe the impact direction, which is important for understanding mechanisms of injury.

Table 1: Characteristics and fetal outcomes of crashes involving pregnant occupants.

Case	Impact Direction	Impact Severity	Occupant Restraint	Occupant Position	Gest. Age (wk)	Maternal Injuries	Fetal Outcome
Hagmann et al. 2004	Unknown	Unknown	none	Driver	31	Moderate	Contractions, emergency C-S, subdural and intracerebral hemorrhages, early delivery, partial placental abruption
Vaysse et al. 2007	Near-side	Severe	3-point belt	Driver	24	Moderate	Uterine rupture, placental abruption, fetal loss
Karimi 2004	Unknown	Unknown	3-point belt and airbag	Right front	29	Minor	Emergency C-S, clavicle fx, skull fx, multiple brain hemorrhages
Weir 2008	Front	Est. severe	3-point belt	Driver		Minor	Uterine rupture, fetal transection
Bunai 2000	Front	unknown	3-point belt	Driver	24	Minor	Placental abruption, fetal loss
Fusco 2001	Near-side	Est. severe	3-point belt and airbag	Driver	39	Minor	Uterine rupture, placental abruption, fetal skull fx, fetal loss
Aboutanos	Unknown	Unknown	None (ejected)	Driver	22	Severe	
Aboutanos	Unknown	Unknown	none	Passenger	36	Moderate	Uterine rupture, fetal loss
Parida	Unknown	Unknown	3-point belt	Unknown	31	Unknown	Emergency C-section, premature delivery, fetal bowel injury

Several other researchers examined hospital records of pregnant patients involved in motor-vehicle crashes to identify factors contributing to adverse fetal outcome. Reis et al. (2001) performed a case control study of 24 pregnant women involved in motor-vehicle crashes, matching women with and without diagnoses of placental abruption by gestational age. Estimated speed of the vehicle at the time of the crash and maternal injury severity score were higher for the group with placental abruption. Belt restraint usage was not significantly different between the two groups. The study is limited by the use of estimated crash speed as a measure of crash severity.

Metz and Abbot (2006) reviewed hospital records of 30 pregnant women involved in motor-vehicle crashes where airbag deployment occurred. The authors primarily report aggregate data rather than individual results. One-third of the occupants were not using belt restraints. One woman, wearing a three-point belt, sustained placental abruption and fetal loss. A limitation of the study is the use of travel speed prior to the crash as a surrogate for crash severity. The study is also limited by including only pregnant women with airbag deployments, without any control group. The authors conclude that adverse fetal outcome resulting from airbag deployment is rare.

SUMMARY

Conducting detailed investigations of crashes involving pregnant occupants allows assessment of the factors contributing to the outcome of the fetus and the mother. In particular, quantifying crash severity using vehicle crush measurements, confirming use of restraint systems, and identifying occupant contact marks provides insight as to how these specialized occupants are injured during motor-vehicle crashes. While case reports from the medical literature usually present unusual medical outcomes, they seldom contain the information on crash severity, restraint use, and crash type that are essential for understanding how adverse fetal outcomes occur during motor-vehicle crashes. Improved understanding of how adverse fetal outcomes occur provides vehicle and restraint designers with information to develop improved occupant protection systems for pregnant occupants.

REFERENCES

Aboutanos, M. B., S. Z. Aboutanos, D. Dompkowski, T. M. Duane, A. K. Malhotra, and R. R. Ivatury. 2008. "Significance of Motor Vehicle Crashes and Pelvic Injury on Fetal Mortality: A Five-Year Institutional Review." *Journal of Trauma* 65(3):616-520.

Baker, S. P., B. O'Neill, W. Waddon Jr., and W. B. Long. 1974. "The Injury Severity Score: a method for describing patients with multiple injuries and evaluating care systems." *Journal of Trauma* 14:186-7.

Bunai, Y., A. Nagai, I. Nakamura, and I. Ohya. 2000. "Fetal death from abruption placentae associated with incorrect use of a seatbelt." *American Journal of Forensic Medicine Pathology* 21(3):207-209.

Fusco, A., K. Kelly, and J. Winslow. 2001. "Uterine rupture in a motor vehicle crash with airbag deployment." *Journal of Trauma* 15:1192–1194.

Hagmann, C. F., T. Schmitt-Mechelke, J. H. Caduff, and T. M. Berger. 2004. "Fetal intracranial injuries in a preterm infant after maternal motor vehicle accident: a case report." *Pediatric Critical Care Medicine* 5(4):396-398.

Hartl, R. and K. Ko. 1996. "In utero skull fracture: Case report." *Journal of Trauma* 41(3):549-552.

Karimi, P., R. Ramus, J. Urban, and J. M. Perlman. 2004. "Extensive brain injury in a premature infant following a relatively minor maternal motor vehicle accident with airbag deployment." *Journal of Perinatology* 24(7):454-457.

Klinich, K. D., C. A. C. Flannagan, J. R. Rupp, M. D. Sochor, L. W. Schneider, and M. D. Pearlman. 2008. "Fetal outcome in motor-vehicle crashes: effects of crash characteristics and maternal restraint." *American Journal of Obstetrics and Gynecology* 198:450.e1-450.e9.

Klinich, K. D., L. W. Schneider, J. L. Moore, and M. D. Pearlman. 1998. "Injuries to pregnant occupants in automotive crashes." In Proceedings of the Sixteenth International Technical Conference on Experimental Safety Vehicles, Windsor, Ontario, Canada. NHTSA. 2046-2064.

Klinich, K. D., L. W. Schneider, J. L. Moore, and M. D. Pearlman. 1998. "Injuries to pregnant occupants in automotive crashes." In 42nd Annual Proceedings Association for the Advancement of Automotive Medicine, Charlottesville, VA. AAAM. 57-91.

Klinich, K. D., L. W. Schneider, J. L. Moore, and M. D. Pearlman. 2000. "Investigation of crashes involving pregnant occupants." In Proceedings of the Association for the Advancement of Automotive Medicine 44th Annual Conference, Chicago, IL. AAAM. 37-55.

Lane, P. L. 1989. "Traumatic fetal deaths." *Journal of Emergency Medicine* 7(5):433-435.

Metz, T. D. and J. T. Abbott. 2006. "Uterine trauma in pregnancy after motor vehicle crashes with airbag deployment: A 30-case series." *Journal of Trauma* 61(3):658-661.

Parida, S. K., V. M. Kriss, and A. R. Pulito. 1999. "Fetal morbidity and mortality following motor-vehicle accident: two case reports." *Journal of Perinatology* 19(2):144-146.

Pearlman, M., J. E. Tintinalli, and R. P. Lorenz. 1990. "Blunt trauma during pregnancy." *New England Journal of Medicine* 323:1609-1613.

Pepperell, R. J., E. Rubinstein, and I. A. MacIsaac. 1977. "Motor-car accidents during pregnancy." *Medical Journal of Australia* 1(7):203-205.

Reis, P. M., C. M. Sander, and M. D. Pearlman. 2001. "Abruptio placentae after auto accidents. A case-control study."*Journal of Reproductive Medicine* 45(1):6-10.

Van Enk, A. and W. Van Zwam. 1994. "Uterine rupture: a seat belt hazard." *Acta Obstetricia et Gynecologica Scandinavica* 73(5):432-433.

Vaysse, C., F. Mignot, J. P. Benezech, and O. Parant. 2007. "Traumatic uterine rupture: a rare complication of motor vehicle accidents during pregnancy. A case report." *Journal de Gynecologie, Obstetrique et Bioogie de la Reproduction* 36(6):611-614.

Weir, L. F., B. T. Pierce, and J. O. Vazquez. 2008. "Complete fetal transection after a motor vehicle collision." *Obstetrics and Gynecology* 111:530-532.

Weiss, H. B., T. J. Songer, and A. Fabio. 2001. "Fetal deaths related to maternal injury." *Journal of American Medical Association* 286:1863-1868.

The following papers are included in their entirety to augment this chapter:

Klinich, K. D., L. W. Schneider, J. L. Moore, and M. D. Pearlman. 1998. "Injuries to pregnant occupants in automotive crashes." In 42nd Annual Proceedings Association for the Advancement of Automotive Medicine, Charlottesville, VA, October 5-7. 57-91.

Klinich, K. D., L. W. Schneider, J. L. Moore, and M. D. Pearlman. 2000. "Investigation of crashes involving pregnant occupants." In 44th Annual Proceedings Association for the Advancement of Automotive Medicine, Chicago, IL, October 2-4. 37-55.

Klinich, K. D., C. A. C. Flannagan, J. R. Rupp, M. D. Sochor, L. W. Schneider, and M. D. Pearlman. 2008. "Fetal outcome in motor-vehicle crashes: effects of crash characteristics and maternal restraint." *American Journal of Obstetrics and Gynecology* 198:450.e1-450.e9.

Metz, T. D. and J. T. Abbott. 2006. "Uterine trauma in pregnancy after motor vehicle crashes with airbag deployment: A 30-case series." *Journal of Trauma.* 61(3):658-61.

INJURIES TO PREGNANT OCCUPANTS IN AUTOMOTIVE CRASHES

Kathleen DeSantis Klinich, Lawrence W. Schneider,
Jamie L. Moore, Mark D. Pearlman
University of Michigan Transportation Research Institute
Ann Arbor, Michigan, USA

ABSTRACT

Injuries unique to pregnant occupants involved in motor-vehicle crashes include placental abruption, uterine rupture or laceration, and direct fetal injury. The mechanisms and characteristics of these injuries are discussed using examples from a literature review and from recent investigations of crashes involving pregnant occupants. In addition, a review of the relationship between the pregnant driver and automotive restraints and the steering wheel illustrates how injury potential may differ from the non-pregnant occupant.

AUTOMOTIVE TRAUMA IN PREGNANCY

Introduction

Firm statistics on fetal loss resulting from automotive trauma are not available because fetal death certificates do not record recent maternal involvement in crashes as a potential cause of death. In addition, since miscarriage occurs in 10-20% of all pregnancies in the early part of pregnancy, only deaths to fetuses over 20 weeks gestational age are legally defined and recorded. However, based on the frequencies of pregnancies and crash involvement of the general population, it has been estimated that between 1500 and 5000 fetal losses occur each year in the United States as a result of maternal involvement in automotive crashes (Pearlman 1997). Additional uncounted adverse fetal outcomes occur as well, as many children grow up disabled as a result of injuries sustained in utero. Even if a fetus survives, complications arising from early emergency delivery of a premature fetus (such as low birth weight and neonatal respiratory distress syndrome) can lead to long-term negative consequences for the child.

Pregnant Anatomy

Figure 1 illustrates the general anatomy of a pregnant abdomen (Pritchard et al. 1985). The uterus is a muscular organ that grows in capacity from 5 mL to 5 to 10 L over the course of pregnancy. Its wall thickness at seven months is about 2 cm, and at term (40 weeks) it is about 1 cm. The size of the uterus depends more on the size of the fetus than the size of the mother. Since most women generally have babies who weigh 2.7 to 3.6 kg (6 to 8 pounds), uterine size is not expected to vary significantly with the stature or weight of the mother. The uterus is attached at its base to the cervix which leads to the vagina, but is otherwise unattached in the abdominal cavity. As the uterus grows, it pushes the other abdominal organs rearward and upward. The base of the uterus is in close proximity to the lumbar spine.

42nd ANNUAL PROCEEDINGS
ASSOCIATION FOR THE ADVANCEMENT OF AUTOMOTIVE MEDICINE
October 5-7, 1998 · Charlottesville, Virginia

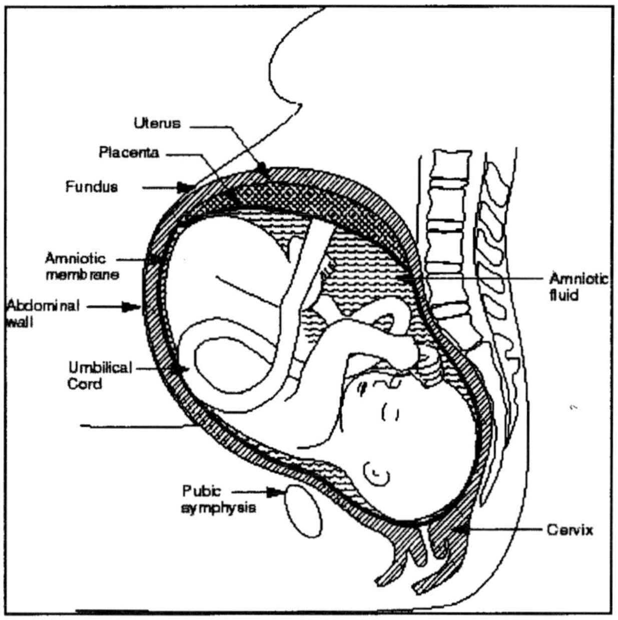

Figure 1. Pregnant anatomy

The placenta is a vascular organ within the uterus that exchanges oxygen, nutrients, and waste between the mother and fetus. Commonly called the "afterbirth," it separates from the uterus and delivers spontaneously after the fetus is born. Eighty percent of placentas are located near the top (fundus) of the uterus by the third trimester. It is roughly circular in shape, and measures 2 to 2.5 cm in thickness in the last trimester of pregnancy. It covers roughly one-fourth of the internal surface area of the uterus throughout pregnancy. The placenta attaches to the uterus by microvilli, which are finger-like projections (diameter of ~ 50 micrometers) that grow into the superficial covering of the uterine wall (decidual layer). The interface between the uterus and placenta is considered to be weaker than either of the two tissues.

The amniotic membrane lines the inner surfaces of the uterus and placenta and contains the fetus and the amniotic fluid. The umbilical cord runs from the placenta through the wall of the membrane to the fetus. As the fetus grows, the relative proportion of the uterine volume filled with amniotic fluid decreases. By the last trimester, the fetus is positioned with its head down in over 95% of pregnancies.

Pregnant Occupants in the Automotive Environment

A research program targeted toward increasing understanding of automotive trauma in pregnancy and developing a second-generation pregnant abdomen for the small female crash dummy is currently underway. As part of this larger program, an anthropometric study of the pregnant automotive occupant is being conducted. Pregnant women in five different stature groups are being tested in laboratory simulations of vehicle interiors. Two seat heights are being used, one representing a mid-size sedan (H30=270 mm) and the other typical of a minivan or light truck (H30=360 mm). Belt anchorage locations are also varied. Subjects are measured in the 3rd, 5th, 7th, and 9th months of pregnancy to study how selected seat position, belt fit, and proximity to the steering wheel change with advancing pregnancy.

Data for fifteen subjects in four different stature groups are currently available. Table 1 shows the group number, mean stature at the initial test session, and number of subjects tested to date. Each subject was tested in four configurations at each visit: two at each seat height with different belt locations.

Table 1.
Subject Information

Group	Mean Stature (mm)	Number of Subjects
2	1580	3
3	1627	5
4	1664	2
5	1708	5

Abdomen-to-wheel clearance is the minimum distance between the bottom of the steering-wheel rim and the subject's midline abdomen contour. Figures 2 and 3 show the abdomen-to-wheel clearance averaged by subject group for each test session. Each point is the average of twice the number of subjects in each group, because each subject was tested twice at each seat height. Different belt conditions are not expected to affect abdomen-to-wheel clearance. Figure 2 illustrates the data for the mid-seat-height test configurations, while Figure 3 shows the data for the high-seat-height test conditions. As expected, all groups show decreasing abdomen-to-wheel clearance with increasing gestational age. The shortest subjects (Group 2) had the least clearance in each test session, while the tallest subjects (Group 5) had the most clearance. Groups 3 and 4 had similar levels of clearance in each test session. The tallest subjects (Group 5) show approximately 100 mm more clearance than the shortest subjects (Group 2) at each test session and at each seat height. Clearances are similar for the two different seat-height configurations.

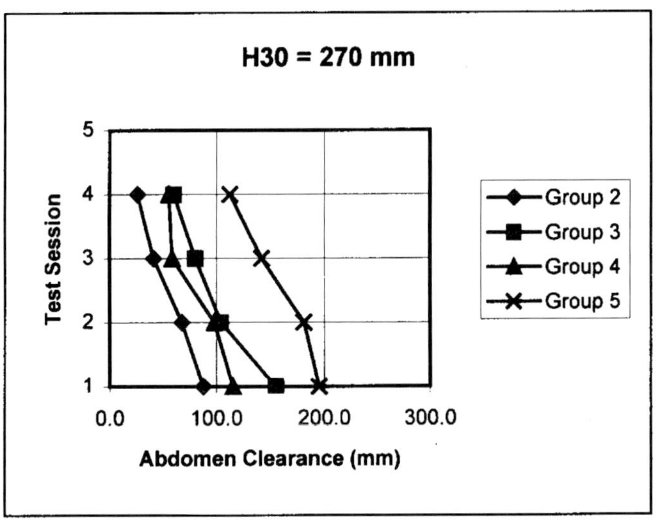

Figure 2. Abdomen-to-wheel-rim clearance averaged by subject group for each test session at mid-seat-height test conditions.

Figure 4 shows a stick-figure representation of the average posture and anthropometry of Group-2 subjects for each test session at mid-seat height conditions. The subjects did not move further away from the steering wheel as pregnancy progressed because of the need to comfortably operate the pedals.

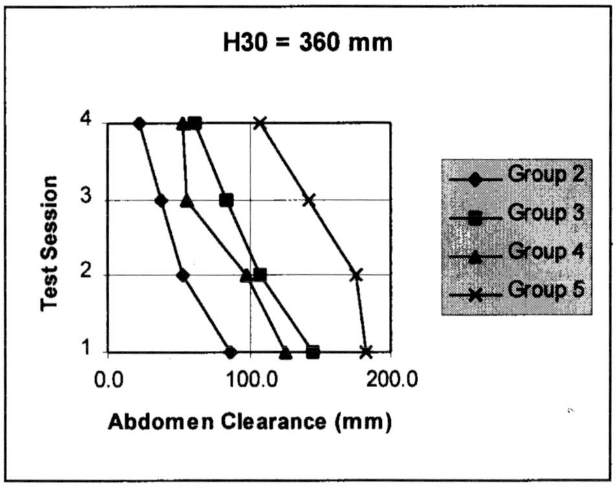

Figure 3. Abdomen-to-wheel-rim clearance averaged by subject group for each test session at high-seat -height test conditions.

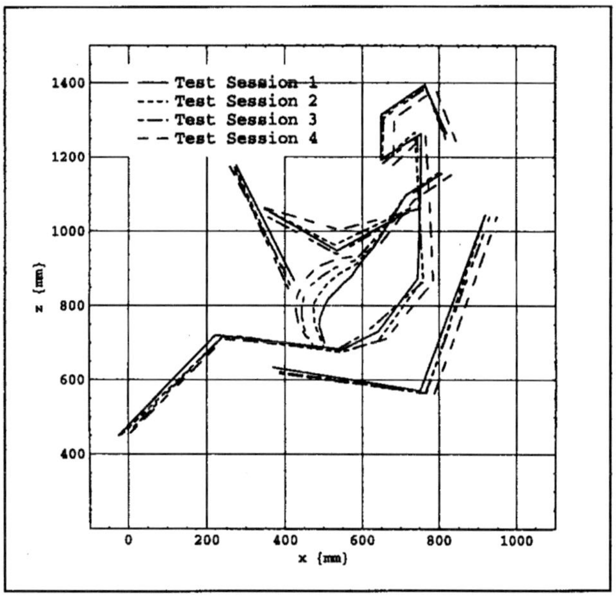

Figure 4. Average posture of Group-2 subjects for each test session at mid-seat-height conditions.

The position of the lap belt relative to the uterus was also measured during testing. Subjects were instructed to position the lap belt below the prominence of their pregnant abdomen, as recommended by the American College of Obstetricians and Gynecologists (ACOG 1991). Ideally, the top of the belt should be at or below the anterior superior iliac spines (ASIS), so the bony structures are loaded during impact rather than the soft tissues. Figure 5 shows a front view of the centerline of a lap belt for one subject. The geometric shape connects the fundus (top of uterus), left ASIS, pubic symphysis, and right ASIS. This subject, who was in her 37th week of pregnancy, positioned the belt so the centerline was very close to the ASIS points.

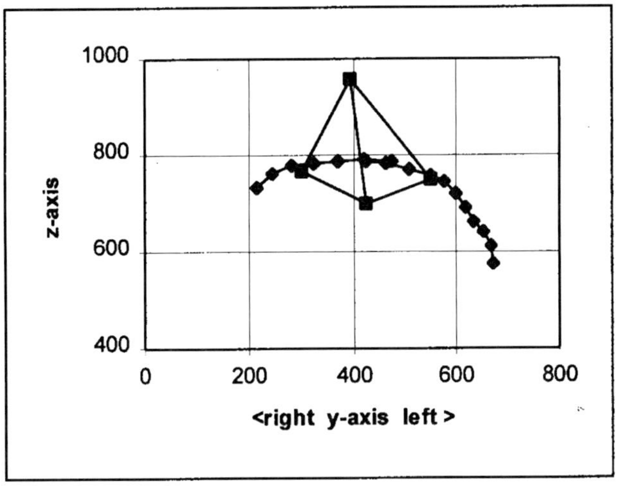

Figure 5. Front view of lap belt centerline (diamonds) relative to fundus, ASIS, and pubic symphysis (squares).

The side view of the lap belt relative to the pregnant abdomen in Figure 6 illustrates the challenges posed by pregnant anatomy to automotive safety engineers. This subject positioned the belt correctly, and it crossed very close to the left and right ASIS as desired. However, in the midline plane, the lap belt crosses over the protruding pregnant abdomen directly over the uterus. In a crash, the uterus would be loaded before the bony structures. For non-pregnant occupants, a correctly positioned lap belt loads the pelvis through the ASIS rather than the soft tissues of the abdomen.

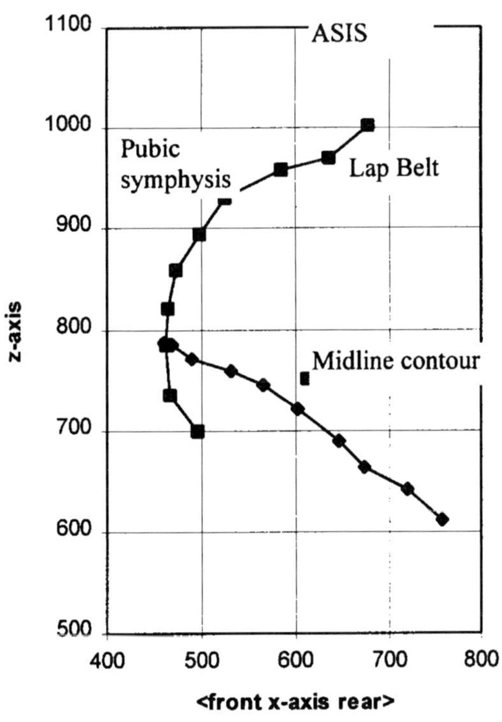

Figure 6. Side view of lap belt centerline (diamonds) showing its intersection with the

INJURIES UNIQUE TO PREGNANT WOMEN IN CAR CRASHES

Overview

Table A.1 in the Appendix provides information on 120 automotive crashes involving pregnant occupants. Most of the cases (93) are taken from the medical and automotive safety literature. The main criteria for inclusion in this review was availability of information on restraint usage by the pregnant occupant and the pregnant occupant being at least 20 weeks pregnant. Since most medical case studies focus on the injuries and treatment and not the crash event, information regarding the impact direction is often unavailable. In addition, the estimate of the crash severity is usually not calculated and is derived from the occupant's or physicians' estimates. The cases from the literature are usually reported because they involve an unusual or serious injury and are not intended to be representative of situations experienced by most pregnant occupants.

The remaining 27 cases in Table A.1 are from UMTRI's ongoing accident investigation study of crashes involving pregnant occupants. Fourteen cases (designated GMP-0XX) are classified as major investigations, while 13 cases (designated GMP-2XX) are considered minor investigations. An investigation is usually considered minor if the vehicle or medical records are not available for inspection or review. However, even in these cases, the vehicle damage information is generally more complete than most of the cases reported in the medical literature. The UMTRI case set is expected to be more representative, as it includes a wider range of outcomes.

The first column in Table A.1 lists a case number for reference in this paper. The second column lists the occupant position in the vehicle, impact direction, and impact severity if the data are available. Some sources from the literature only reported the occupant as being a passenger and did not indicate whether the location was the front or rear seat. The type of crash is classified as front, side, rear, rollover, or multiple impact. Information on whether the crash was a "near" or "far"-side impact is usually not available. Crash severity is based on the following divisions: minor is less than 24 km/h (15 mph) change in velocity, moderate is 24 to 48 km/h (15 to 30 mph), and severe is greater than 48 km/h (30 mph). The values reported for major UMTRI cases are based on input of vehicle crush measurements into crash-reconstruction programs such as SMASH, and are therefore more accurate. The remaining crash-severity estimates were made by the occupants or physicians and probably relate more to the speed of travel before the crash rather than actual change in velocity. In the majority of cases in Table A.1, no estimate of crash severity is made.

The third column in Table A.1 lists the restraint conditions for the pregnant occupant. Some of the sources reported use of a belt restraint, but did not clarify whether a lap belt or 3-point belt was used, or if the belt was positioned properly; these cases are listed only as "restrained." If the pregnant occupant reported wearing the lap belt portion of the belt across the bulge of her pregnant abdomen (crossing near the umbilicus) rather than below it, belt use is documented as improper. The fourth column gives gestational age in weeks, with 40 weeks considered full term. In the text of this paper, gestational ages are grouped into four-week increments and referred to as the 5th month (20-23 weeks), 6th month (24-27 weeks), 7th month (28-31 weeks), 8th month (32-35 weeks), or 9th month (36 weeks or more) of pregnancy.

Maternal injuries are given in the fifth column and are described in two ways. The first term describes the overall severity of injuries that are not related to pregnancy. The categories are none, minor, moderate, and major. These descriptions are estimates of the maximum AIS scores (ignoring uterine or placenta injuries), with minor corresponding to MAIS 1 or 2, moderate, MAIS 3 or 4, and major, MAIS 5 or 6. The remainder of the maternal injury description describes the injuries to the uterus and placenta. Pelvic fractures are specifically noted in the maternal injury column as they are suspected to increase the probability of injury to the pregnant abdomen and fetus. Maternal death is indicated by italic type.

The sixth column describes the fetal outcome, specifically listing any direct injuries or complications. Fetal survival is indicated by italic type.

The final column lists the source of the case data. For cases from the literature, the source lists the authors and year of publication, plus a case number if the source presented more than one case. For UMTRI cases, the case number and year of the crash are given.

Figure 7 documents the distribution of crash types from Table A.1. The majority of crashes are frontal impacts (61%, n=74), followed by side impacts (25%, n=30). The breakdown of crash severity is shown in Figure 8, with over half of the cases not reporting any estimate of crash severity. The distribution of occupant position is shown in Figure 9, with most cases involving drivers (55%, n=66).

Figure 10 shows the distribution of restraint use of the pregnant occupants for the cases in Table A.1. About half (49%, n=59) were unrestrained. Only a few cases (3%, n=4) were confirmed to have improper belt use, with the lap belt positioned across, rather than under, the pregnant abdomen. Since the case dates range back to the 1960's, about 6% (n=7) used a lap belt only, often because it was all that was available. The gestational ages of the pregnant occupants appear in Figure 11, with most cases in the last trimester, particularly the last month.

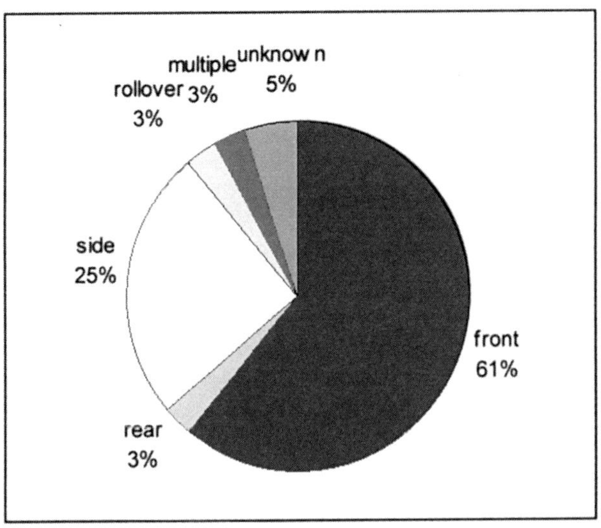

Figure 7. Distribution of impact type for cases in Table A.1.

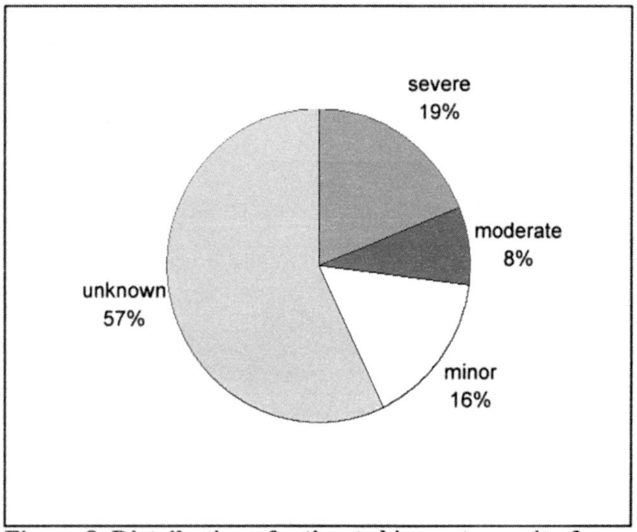

Figure 8. Distribution of estimated impact severity for cases in Table A.1.

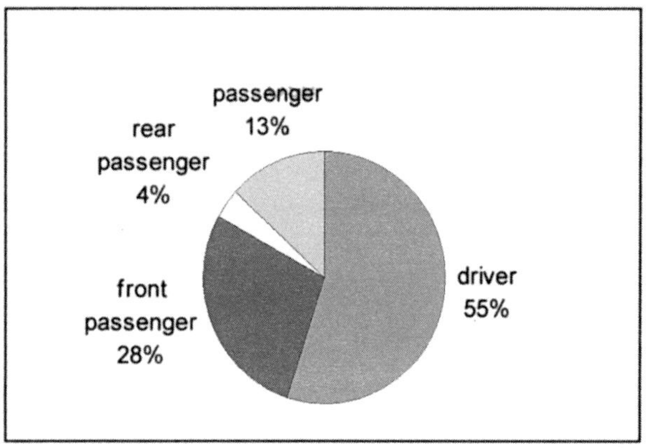

Figure 9. Distribution of seating position of pregnant occupant for cases in Table A.1.

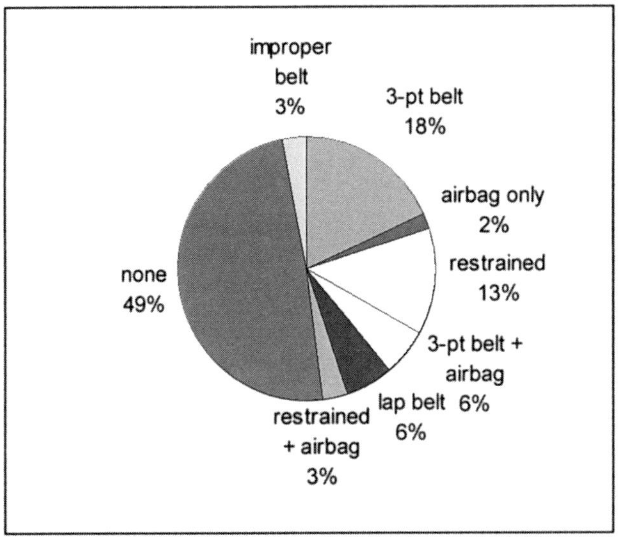

Figure 10. Distribution of restraint use by pregnant occupant for cases in Table A.1.

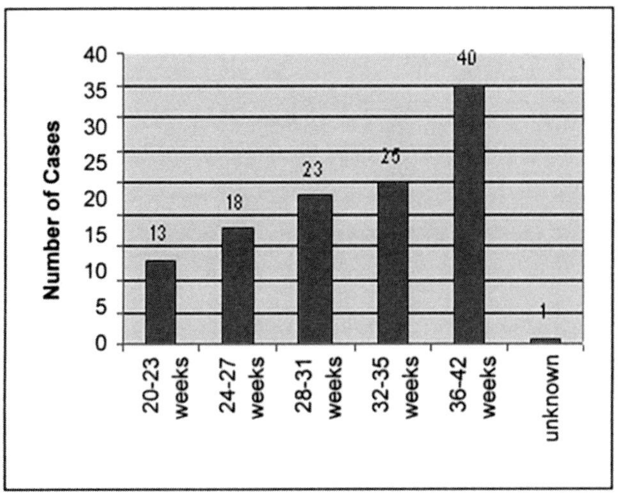

Figure 11. Distribution of gestational age for cases in Table A.1.

The chart in Figure 12 indicates the types of pregnancy-related injuries sustained by the pregnant occupants in Table A.1. The most frequent outcome is placenta injury only; none of the fetuses in these cases survived. There are six instances of placenta injury plus additional complications, with all of these fetuses surviving. Seventeen cases had placenta and direct fetal injuries. Eighteen cases had completely positive outcomes, although two of these mothers sustained pelvic fractures. Five cases involved only complications, and all of these fetuses survived. In eight cases, the fetus was stillborn and no direct injuries to the fetus or reproductive organs were found. However, five of these cases involved maternal death, so there are really just three without additional explanation. Eight cases involve injuries to both the uterus and placenta. In one of these cases, the fetus survived although it experienced complications. Three other of these cases also involved pelvic fracture. The other most common

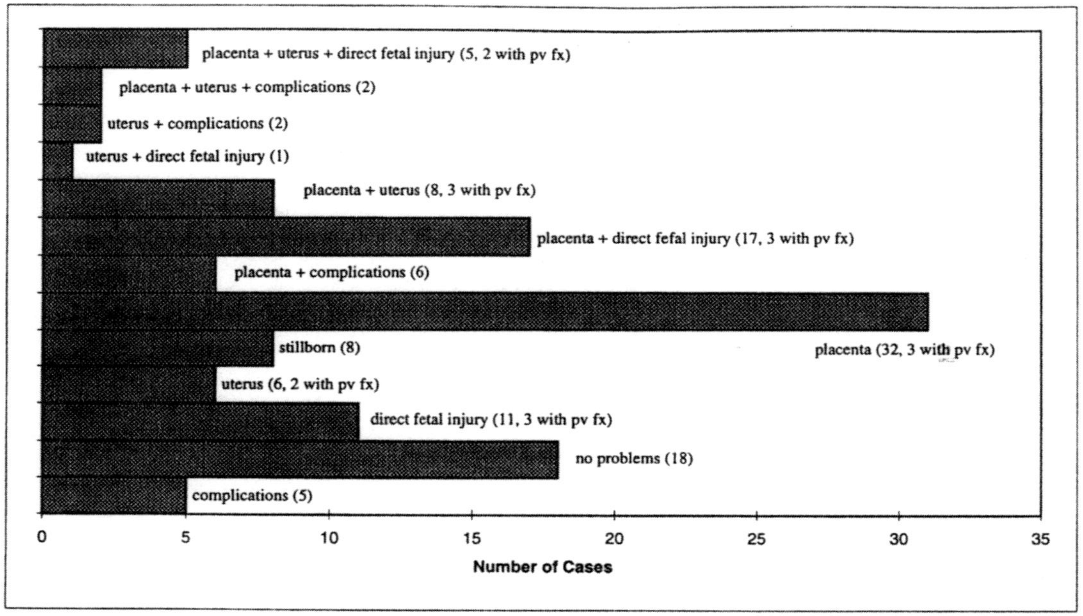

Figure 12. Distribution of injury patterns.

injury patterns are direct fetal injury (11 cases, 3 with pelvic fracture), and uterine injury only (6 cases, two with pelvic fracture). In one of the uterine injury cases, the fetus survived with complications. The remaining bars show the less frequent combinations of placenta, uterine, direct fetal, and pelvic fracture injuries.

Of the 120 cases in Table A.1, nine involve both maternal and fetal death. While the mother survived in the other 111 cases, the fetus did not survive in 74 cases.

The remainder of this paper discusses each type of pregnancy injury and outcome in more detail, and highlights the characteristics of the crashes and injury patterns.

Placental Abruption

Placental abruption is the most common cause of fetal loss in automotive crashes. It occurs in 1 to 5% of minor severity crashes during pregnancy and from 20 to 50% of severe crashes during pregnancy. The injury occurs when the placenta detaches from the uterine wall, disrupting the supply of oxygen and nutrients to the fetus. Partial abruptions can also occur, with the possibility of the pregnancy continuing successfully depending on the degree of placental separation from the uterus. The interface between the placenta and the uterus is considered to be weaker than either the uterus or the placenta, and therefore usually fails before either the uterus or placenta fails.

Pearlman et al. (1990) have noted that the likelihood of placental abruption is independent of the placenta location on the uterus (Pearlman et al. 1990). This may imply that there are several mechanisms of placental abruption, with the particular cause depending on the placental location relative to the loading location and nature of the loading. For example, some placental abruptions may result from high-velocity, low-mass airbag loading, or from compression between the mother's body and the steering wheel, belt, or instrument panel. Abruption may also result from acceleration differentials between the placenta and uterus. Thus, mechanisms of placental abruption may include inertial loading, direct loading by the fetus, and excessive shear or tensile strain.

The first 69 cases in Table A.1 involve some injury to the placenta or the placenta/uterine interface. Of these cases, thirty-nine of the pregnant occupants were drivers, fourteen were front passengers, five were rear passengers, and eleven were passengers of unknown seating position. Most of the impacts were frontal (46), twelve were side, three were rollovers, two each were rear or multiple impacts, and four impact types were not reported. Many of the cases did not estimate impact severity (49), while twelve were severe and four each were estimated to be minor or moderate impacts.

Forty-two cases (1-42) involve complete abruption and fifteen cases (43-57) involve partial abruptions. In eight cases (58-65), lacerations of the placenta occurred, and in the four remaining cases (66-69), the placenta was damaged but not separated or lacerated. Only two cases (16 and 33) resulted in both fetal and maternal death. When the mother survived, the fetus died in 56 cases. Twenty-two of the cases with placental injury also had direct fetal injury, fifteen cases also involve uterine laceration or rupture, and nine of these pregnant occupants sustained pelvic fractures.

In the majority of cases with placental abruption, the pregnant occupant was unrestrained (48 out of 69). One pregnant occupant sustaining a placental abruption was restrained by both a 3-point belt and a deploying airbag, eight were documented as properly using a 3-point belt, and twelve were listed as restrained without further specification. Five women used a lap belt only, and three wore the belt improperly with the lap belt portion positioned across the bulge of their abdomens. While these results show that placental injuries can occur to properly belted pregnant occupants, they also show that avoiding belt use during pregnancy may significantly increase the risk of placental abruption and subsequent fetal loss in a crash.

This compilation of case data also suggests that placental abruption may be more likely to occur later in the term of pregnancy. Five cases of placental injury involve women in their 5th month of pregnancy, 10 in the 6th, 13 in the 7th, 20 in the 8th, and 21 in the 9th month. The pregnant occupants were most likely to have a minor injury in addition to their placental injury (41 cases). Sixteen pregnant occupants sustained moderate injuries, two suffered major injuries, and ten sustained no injuries other that the placental injury. These findings agree with clinical experience in which women with serious pregnancy injuries often appear, at initial exam, to have sustained minor or no other injuries (Pearlman 1990).

Uterine Injury

Uterine rupture and lacerations are rare during pregnancy, occurring in less than 1% of pregnant trauma cases (Pearlman 1990). Injury to the uterus in automotive accidents occurs almost exclusively during pregnancy (compared to any other time) because it is much larger, extends outside the pelvic cavity, and is filled with fluid. While uterine rupture is very rare, the likelihood of fetal death with such an injury is near 100%.

Uterine injury is often reported in the literature as resulting from direct loading from the seat belt. While this may occur, it can also occur to unbelted women who are directly loaded by the steering wheel or instrument panel. Some cases report uterine damage directly beneath the area of seatbelt loading. For example, in case GMP-006 (Case 10), the pregnant driver sustained two lacerations approximately 50 mm apart extending across the lower uterus, corresponding to the width of the seat belt. Other cases (75 and 78) report the site of uterine rupture as being opposite the site of loading, suggesting a "contrecoup" type of injury mechanism.

Twenty-four cases in Table A.1 involve uterine injuries. In eleven of the cases, the pregnant occupant was the driver. Fourteen of the crashes were frontal impacts, seven were side, one was rear, one involved multiple impacts, and one involved unknown crash conditions. Of the seven side impacts, only one pregnant occupant may have been properly restrained (one had the lap belt improperly positioned and five were unrestrained). The impact severity is only estimated for five of the twenty-four uterine injury cases, so no conclusions can be drawn about the frequency of uterine injury and the severity of crashes.

Three different types of uterine injuries are reported: one complete transection (70), thirteen ruptures (1-8, 74-78) and ten lacerations (9-12, 43-44, 58, 71-73). All but three of these experienced fetal loss, and all three of these cases (10, 44, 71) involved fetal complications. Seven of the laceration cases and eight of the rupture cases involved at least a partial placental abruption. Seven of the uterine injury cases also involved direct fetal injury, and six uterine injuries were accompanied by pelvic fractures. (The proportion of cases with uterine injury in Table A.1 is higher than seen clinically, because the literature cases are more likely to report unusual injuries.)

In the twenty-four cases with uterine injury, only four of the pregnant occupants were properly restrained by a three-point belt, and one driver was also restrained by an airbag. Two additional cases reported restraint use, but did not report the type of belt or whether it was positioned properly. Half of the occupants with uterine injury were unbelted. For all the cases in Table A.1, the four cases where improper belt use was confirmed (lap belt portion placed across the dome of the abdomen) all resulted in uterine injury. In the last two cases with uterine injury, the occupants used only a lap belt.

Distribution of gestational ages of pregnant women with uterine injury is fairly uniform, with five occupants each in the 5th, 6th, and 8th months, six in the 7th month, and three in the 9th month. None of the pregnant occupants with uterine injury sustained major injuries, six sustained other moderate injuries, seventeen suffered other minor injuries, and one received no other injuries. This distribution of maternal injuries confirms the trend noted in the literature that pregnant women often sustain a uterine injury that is usually life-threatening to the fetus, while sustaining only minor injuries to themselves.

Direct Fetal Injury

Direct fetal injury (DFI) is also quite rare, occurring in less than 10% of automotive crashes involving pregnant occupants (Pearlman 1990). In the first three months of pregnancy, the uterus is still completely surrounded by the pelvis and is considered an abdominal organ. After the first three months, the uterus protrudes out from the abdomen, but the structure of the pregnant abdomen protects the fetus by encasing it in amniotic fluid which acts as a shock absorber to isolate the fetus. The fetus sometimes sustains injury from direct loading of the abdomen when the protective cavity is compromised by pelvic fractures or uterine rupture. The most frequently injured fetal body region is the head, because it is the largest part of the fetal body and offers the "biggest target." It is thought that skull fracture most often occurs when the fetal head is loaded against the bony structures of the maternal pelvis or spine by a belt, steering wheel, or instrument panel.

Judging from the thirty-four cases involving direct fetal injury in Table A.1, it appears that crashes with minor, moderate, or severe impact severities can all result in direct fetal injuries, although data on the crash severity are limited. Nineteen of the cases involved drivers as the maternal occupant. Nineteen were frontal impacts, ten were side, two were rear, and the remaining three were multiple or unknown types of impacts.

The cases with direct fetal injury in Table A.1 are numbers 1-3, 9, 13-16, 43, 45-53, 65-68, 70, and 79-90. As mentioned previously, the cases taken from the literature tend to focus on more unusual injuries, so the proportion of cases with direct fetal injury in Table A.1 is greater than that seen clinically. Twenty-two cases involve head injuries only, four involve both the head and thorax injuries, six involve thorax/abdomen injury only, and one injury each is documented for the spine and upper extremity. Of the sixteen cases in Table A.1 in which the mother sustained a pelvic fracture, eight of the fetuses sustained direct injuries to the head. Twenty-two of the direct fetal injury cases also involved placenta injuries, suggesting that accident conditions severe enough to result in direct fetal injury often result in placental damage as well. Six occupants sustained uterine injuries with direct fetal injury. Five fetuses with direct injuries survived, with three requiring emergency cesarean section (c/s) deliveries.

Most of the occupants with direct injuries to the fetus were not properly restrained. Nineteen were unbelted, four used a lap belt only, and three had the lap belt improperly positioned across the bulge of the abdomen. Five occupants were wearing the 3-point belt properly, and the remaining three occupants were reported to be using belt restraints, although the type of belt and placement are not reported.

The frequency of direct fetal injury for this sample is slightly skewed toward later gestational ages. Only two DFI cases occurred in the 5th month, six occurred in the 6th month, eight each in the 7th and 8th months, and ten in the 9th month. Intuitively this makes sense, as the fetus is a bigger target as gestational age increases, and the amniotic fluid takes up a relatively smaller proportion of the uterus volume as pregnancy progresses. Direct fetal injuries occur regardless of the injury severity sustained by the mother, with eight of these cases having no maternal injury, sixteen having minor maternal injuries, eight having moderate injuries, and two having major injuries.

Other Negative Outcomes

Some pregnant automotive crash victims suffer negative pregnancy complications whether or not they sustain any placenta, uterus, or direct fetal injuries. One common occurrence is a premature delivery, in which the infant often has a low birth weight and can suffer from neonatal respiratory distress syndrome because its lungs are not fully developed. Both of these factors can lead to health problems and disabilities throughout the child's life.

Even if a fetus is very close to full term, maternal involvement in a crash can lead to an emergency cesarean delivery, which poses higher risks to the mother and neonate. Contractions often begin after an accident, sometimes requiring drug intervention to prevent early delivery. Crashes involving pregnant occupants occasionally result in stillborn births without any visible injury responsible for the loss.

In the sixteen negative outcome cases, nine of the pregnant occupants were drivers. Nine crashes were frontal impacts, two were side, three were rear, and two were unknown. Seven of these crashes were rated as severe, one as moderate, and three as minor in crash severity, with no crash-severity rating available for the remaining five cases.

Table A.1 contains seven cases (90-93, 97, 99,100) in which the crash resulted in a negative pregnancy outcome without any placenta, uterus, or direct fetal injuries. There are also nine cases with placenta or uterine injury (10, 17-19, 44, 54-56, 71) where the fetus survived but experienced some of the additional negative outcomes described above. In three cases, the fetus was stillborn after the crash, although there were no placenta, uterus, or fetal injuries responsible for the loss. In three other cases, the pregnant occupant started having contractions after the crash although no intervention was needed. In one of these cases, the neonate was born months after the crash and diagnosed with brain damage attributed to the crash. In the remaining case, the neonate suffered from hydrocephalus that may have resulted from the crash at 28 weeks gestation. The nine cases with placenta or uterine damage required emergency cesarean delivery of premature infants. Five of these infants suffered from seizures or newborn respiratory distress syndrome.

In these sixteen negative outcome cases, eleven mothers were not using belt restraints. Two used a 3-point belt and airbag, one used a 3-point belt only, and in one case with an unspecified type of belt restraint, the airbag deployed. One pregnant occupant used only a lap belt.

The gestational ages of the cases with negative outcomes is almost evenly distributed over the last three months of pregnancy, with 5 each in the 7th, 8th, and 9th months and one case in the 6th month. Four of the mothers sustained no other injuries, ten had minor injuries only, and two had moderate levels of injury.

Positive Outcomes

Eighteen of the cases in the Table A.1 (103-120) had good fetal outcomes with no complications. The majority of these cases are from the UMTRI study, as the medical literature usually reports only complicated cases. All of the mothers and fetuses in these cases survived.

Twelve of the pregnant occupants were drivers, while the remaining six were front-seat passengers. Ten impacts were frontal, six were side, one was a rear impact, and one impact type was not recorded. Crash severity is available for all of these cases, with thirteen being minor, four moderate, and one severe.

The most obvious difference between the positive- outcome cases and the other cases in Table A.1 is the restraint usage of the maternal occupant: 17 out of 18 of the occupants in the positive outcome cases used a restraint. Seven of these used a 3-point belt, seven used a 3-point belt and the airbag deployed, and two were unbelted but the airbag deployed.

These cases are weighted towards the last month of pregnancy, with ten cases in the 9th month of pregnancy, three in the 5th month, two in the 8th month, and one each in the 6th, 7th, and unknown months. Twelve of the mothers sustained minor injuries, three had none, and three suffered moderate injuries. Two of the mothers had pelvic fractures.

DISCUSSSION

The preceding descriptions of injuries sustained by pregnant women in crashes highlights the conditions under which different injuries occur. To provide an overview, histograms of the distributions of the independent variables impact type, impact severity, seating position, restraint use, and gestational age were developed for all cases in Table A.1, and for subsets of the cases based on injury type or outcome. The different proportions in each distribution were grouped according to the different levels of each independent variable for plotting in Figures 13 through 17. The number of cases in each injury or outcome appears in Table 2. The sum of the numbers of cases in all the different groups does not equal the total number of cases in Table A.1 because cases with more than one injury are included more than once (i.e., a case with both uterus and placenta injuries is counted in both categories).

Table 2.
Number of Cases in Each Outcome

Outcome	Number of Cases
All Cases	120
Placenta	69
Uterus	24
Direct Fetal	34
Negative Outcomes	16
Positive Outcomes	18

The distributions of impact type in Figure 13 show that, in general, the distributions for each outcome are similar to the overall distribution. The negative outcome cases had higher proportions of rear and unknown impacts. The cases with placenta injuries had a slightly lower proportion of side impacts and a slightly higher proportion of front impacts.

Figure 14 shows distributions of estimated impact severity. The negative outcome cases had higher proportions of major and minor impacts and had fewer unknown crash severities. The positive outcome cases had the highest proportions of minor and moderate severity crashes, and had no unknown impact severities.

The distributions of pregnant occupant seating position are found in Figure 15. Most of the outcomes have a seating-position distribution similar to the overall distribution. The uterine injuries have the highest proportion of front passengers and the lowest proportion of drivers. The placenta injury cases have the lowest proportion of front passengers and the highest proportion of rear passengers.

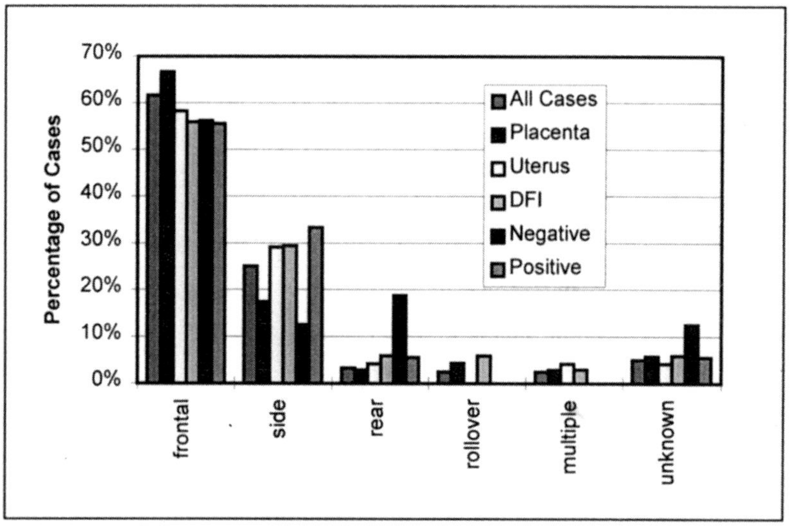

Figure 13. Distributions of impact type for all cases in Table A.1 and by injury or outcome, grouped by impact type.

Figure 16 illustrates the distributions of restraint use by pregnant occupants. Cases with positive outcomes have the highest percentage of 3-point belt use and 3-point belt plus airbag use, and the lowest percentage of no restraint use. Placenta injuries and negative outcomes have relatively high proportions of not using restraints. Uterine injuries have the highest proportion of improper belt use.

The distribution of gestational age for the different outcomes is found in Figure 17. The positive outcome cases had the highest proportion of cases in the 9th month of pregnancy. Since a birth resulting from a crash would not be considered premature in the ninth month, this may contribute to this shift in proportion. Uterus injury cases were the most evenly distributed, and had the highest proportions in the 5th and 6th months of pregnancy. The gestational age, as do direct fetal injuries. Negative outcomes primarily occur in the last trimester of pregnancy.

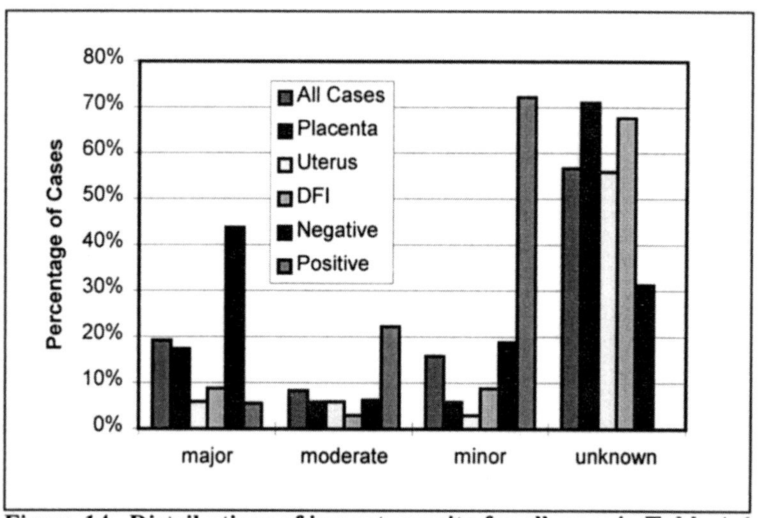

Figure 14. Distributions of impact severity for all cases in Table A.1 and by injury or outcome, grouped by impact severity.

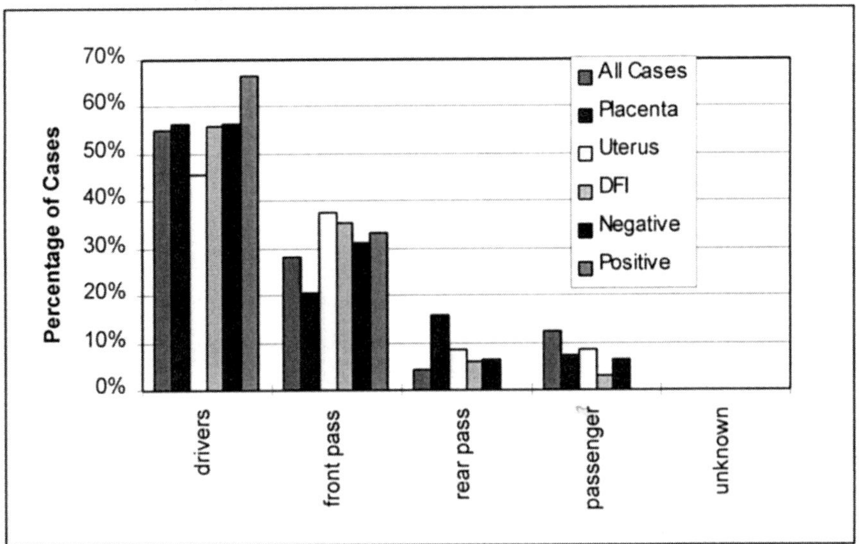

Figure 15. Distributions of pregnant occupant seating position for all cases in Table A.1 and by injury or outcome, grouped by seating position.

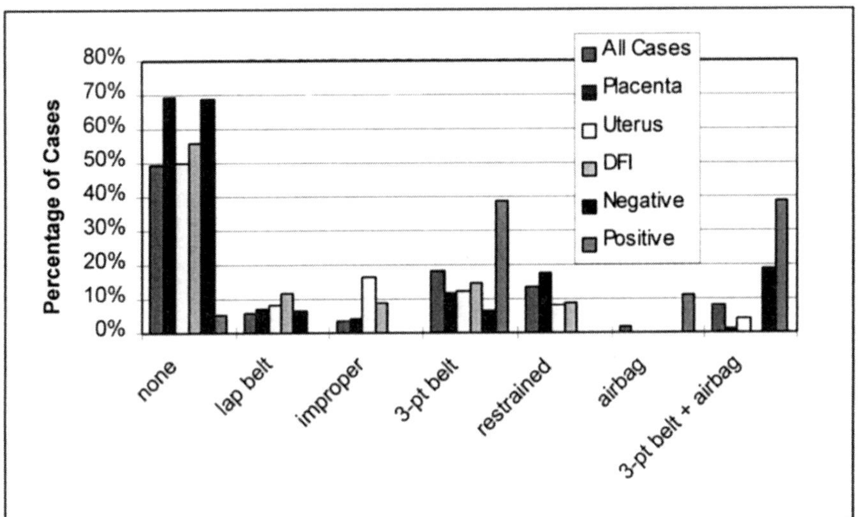

Figure 16. Distributions of pregnant occupant restraint use for all cases in Table A.1 and by injury or outcome, grouped by restraint use.

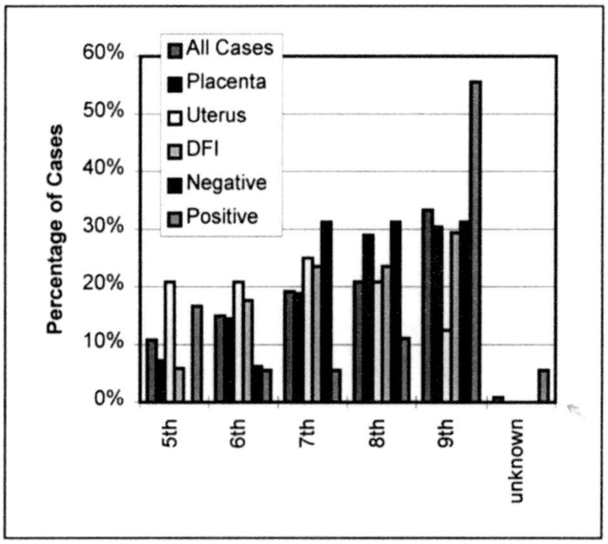

Figure 17. Distributions of gestational age for all cases in Table A.1 and by injury or outcome, grouped by gestational age.

CONCLUSIONS

A review of injuries specific to pregnant motor vehicle occupants illustrates the following points regarding pregnant occupants involved in automotive crashes.

- All nine cases involving maternal death resulted in fetal death, and eight of these women were unbelted. Thus, the most basic way to protect the fetus is to protect the mother through proper use of belt restraints.
- While placental, uterine, and fetal injuries may occur with properly belted occupants, the majority of the cases involving these injuries are to unrestrained occupants.
- The distinguishing characteristic of the cases with completely positive fetal outcomes was proper restraint use by the mother.
- 65% of the cases with direct fetal injury also involved placental injury, as did 63% of the cases with uterine injury. Thus, reducing the likelihood of placental abruption is likely to reduce the probability of uterine and direct fetal injury as well.
- All pregnant women involved in crashes, even minor ones, should immediately seek medical attention. In most cases with fetal loss, the mother sustained only minor or no other injuries other than placental, uterine, or fetal damage.
- The anatomy of pregnant women while seated in automotive posture poses a unique challenge to restraint designers because of difficulty positioning the lap belt and close proximity to the steering wheel and airbag module.

The information presented in this paper will be used to guide the development of the second-generation pregnant crash dummy. A key goal of the next-generation design is to assess the likelihood of placental abruption, the leading cause of fetal death in crashes.

ACKNOWLEDGMENTS

The work covered by this report was financed by GM pursuant to an agreement between GM and the U.S. Department of Transportation. The authors gratefully appreciate their support.

REFERENCES

ACOG Technical Bulletin #161 (1993). Trauma During Pregnancy, *International Journal of Gynecology and Obstetrics*, 40:165-170.

Agran, P. F., Dunkle, D. E., Winn, D. G., Kent, D. (1986). Fetal Death in Motor Vehicle Collisions. *30th Annual AAAM Proceedings*, 285-294.

Astarita, D. C., Feldman, B. (1997). Seat Belt Placement Resulting in Uterine Rupture. *Journal of Trauma*, 42(4):738-740.

Bowdler, N., Faix, R. G., Elkins, T. (1987). Fetal Skull Fracture and Brain Injury After a Maternal Automobile Accident: A Case Report. *Journal of Reproductive Medicine*, 32(5):375-378.

Chetcuti, P., Levene, M. I. (1987). Seat Belts: A Potential Hazard to the Fetus. *Journal of Perinatal Medicine*, 15(2):207-209.

Civil, I. D., Talucci, R. C., Schwab, C. W. (1988). Placental Laceration and Fetal Death as a Result of Blunt Abdominal Trauma. *Journal of Trauma*, 28(5):708-710.

Connor, E., Curran, J. (1976). In Utero Traumatic Intra-Abdominal Deceleration Injury to the Fetus--A Case Report. *American Journal of Obstetrics and Gynecology*, 125(4):567-569.

Cumming, D. C., Wren, F. D. (1978). Fetal Skull Fracture From an Apparently Trivial Motor Vehicle Accident. *American Journal of Obstetrics and Gynecology*, 132(3):342-343.

Dash, N., Lupetin, A. R. (1991). Uterine Rupture Secondary to Trauma: CT Findings. *Journal of Computer Assisted Tomography*, 15(2):329-331.

Dittrich, K. C. (1996). Rupture of the Gravid Uterus Secondary to Motor Vehicle Trauma. *Journal of Emergency Medicine*, 14(2):177-180.

Evard, J. R., Sturner, W. Q., Murray, E. J. (1989). Fetal Skull Fracture from an Automobile Accident. *American Journal of Forensic Medicine and Pathology*, 10(3):232-234.

Fakhoury, G. W., Gibson, J. R. M. (1986). Seat Belt Hazards in Pregnancy: Case Report. *British Journal of Obstetrics and Gynaecology*, 93(4):395-396.

Ford, R. M., Picker, R. H. (1989). Fetal Head Injury Following Motor Vehicle Accident; An Unusual Case of Intrauterine Death. *Australian and New Zealand Journal of Obstetrics and Gynecology*, 29(1):72-73.

Fries, M. H., Hankins, G. D.V. (1989). Motor Vehicle Accident Associated with Minimal Maternal Trauma but Subsequent Fetal Demise. *Annals of Emergency Medicine*, 18(3):301-304.

Galle, P. C., Anderson, S. G. (1979). A Case of Automobile Trauma During Pregnancy. *Obstetrics and Gynecology*, 54(4):467-468.

Griffiths, M., Hillman, G., Userwood, M. (1991). Seat Belt Injury in Pregnancy Resulting in Fetal Death. A Need for Education? Case Reports. *British Journal of Obstetrics and Gynaecology*, 98(3):320-321.

Handel, C. K. (1978). Case Report of Uterine Rupture After an Automobile Accident. *Journal of Reproductive Medicine*, 20(2):90-92.

Harrison, S. D., Nghiem, H. V., Shy, K. (1995). Uterine Rupture with Fetal Death Following Blunt Trauma. *American Journal of Roentgenology*, 165(6):1452.

Hartl, R., Ko, K. (1996). In Utero Skull Fracture: Case Report. *Journal of Trauma*, 41(3):549-552.

Heitzman, E. R., Markarian, B. (1970). Vaginal Bleeding After Trauma in Pregnant Woman. *New York State Journal of Medicine*, 70(18):2338-2340.

Kettel, L. M., Branch, D. W., Scott, J. R. (1988). Occult Placental Abruption After Maternal Trauma. *Obstetrics and Gynecology*, 71(3 pt 2):449-453.

Knuppel, R. A., Salvatore, D. L., Agarwal, R., et al. (1994). Documented Fetal Brain Damage Resulting From a Motor Vehicle Accident. *Journal of Ultrasound in Medicine*, 13(5):402-404.

Landers, D. F., Newland, M., Penney, L. L. (1989). Multiple Uterine Rupture and Crushing Injury of the Fetal Skull After Blunt Maternal Trauma: A Case Report. *Journal of Reproductive Medicine*, 34(12):988-993.

Lavin, J. P., Miodovnik, M. (1981). Delayed Abruption After Maternal Trauma as a Result of an Automobile Accident. *Journal of Reproductive Medicine*, 26(12):621-624.

Levin, R. D., Edson, M. (1994). A Pregnant Pause: Blunt Urologic Trauma in a Young Woman. *Urology*, 44(5):764-767.

Matthews, C. D. (1975). Incorrectly Used Seat Belt Associated with Uterine Rupture Following Vehicular Collision. *American Journal of Obstetrics and Gynecology*, 121(8):1115-1116.

Matthews, G., Hammersley, B. (1997). A Case of Maternal Pelvic Trauma Following a Road Traffic Accident, Associated with Fetal Intracranial Haemorrhage. *Journal of Accident and Emergency Medicine*, 14(2):115-7.

McCormick, D.R. (1968). Seat Belt Injury: Case of Complete Transection of Pregnant Uterus. *Journal of the American Osteopathic Association*, 67(10):1139-1141.

Nichols, M. M., Weedn, V. W. (1986). Placental Laceration and Stillbirth Following Motor Vehicle Accident: with Legal Ramifications. *Texas Medicine*, 82(3):26-28.

Parkinson, E. B. (1964). Perinatal Loss Due to External Trauma To The Uterus. *American Journal of Obstetrics and Gynecology*, 90(1):30-33.

Pearlman, M. D., (1997). Motor Vehicle Crashes, Pregnancy Loss, and Preterm Labor, *International Journal of Gynaecology and Obstetrics*, 57(2):127-132.

Pearlman, M. D., Tintinalli, J. E., Lorenz, R. P. (1990). Blunt Trauma During Pregnancy. *New England Journal of Medicine*, 323(23):1609-1613.

Pearlman, M. D., Tintinalli, J. E., Lorenz, R. P, et al., (1988). Trauma in Pregnancy. *Annals of Emergency Medicine*, 17(8):829-834.

Pepperell, R. J., Rubinstein, E., MacIsaac, I. A. (1977). Motor-Car Accidents During Pregnancy. *Medical Journal of Australia*, 1(7):203-205.

Poulson, Jr., A. M., Gabert, H. A. (1973). Fetal Death Secondary to Nonpenetrating Trauma to the Gravid Uterus. *American Journal of Obstetrics and Gynecology*, 116(4):580-582.

Pritchard, J. A., MacDonald, P. C., Gant, N. F. (1985). *Williams Obstetrics*, Seventeenth Edition. Appleton-Cnetruy-Crofts, Norwalk.

Punnonen, R. (1974). Traumatic Premature Separation of Placenta. *Annales Chirurgiae et Gynaecologiae Fenniae*, 63(6):487-488.

Raney, E. H. (1970). Fetal Death Secondary to Non-Penetrating Trauma to the Gravid Uterus. *American Journal of Obstetrics and Gynecology*, 106(2):313-314.

Ravangard, F., Porter, C. V. (1980). Traumatic Laceration of the Placenta. *West Virginia Medical Journal*, 76(6):125-129.

Rothenberger, D. A., Horringan, T. P., Sturm, J. T. (1981). Neonatal Death Following In Utero Traumatic Splenic Rupture. *Journal of Pediatric Surgery*, 16(5):754-755.

Rowe, T. F., Lafayette, S., Cox, S. (1996). An Unusual Fetal Complication of Traumatic Uterine Rupture. *Journal of Emergency Medicine*, 14(2):173-176.

Rubovits, F. E. (1964). Traumatic Rupture of the Pregnant Uterus From "Seat Belt" Injury. *American Journal of Obstetrics and Gynecology*, 90(6):828-829.

Sherer, D. M., Abramowicz, J. S., Babkowski, R., et al. (1993). Extensive Fetal Intrathoracic Injuries Sustained in a Motor Vehicle Accident. *American Journal of Perinatology*, 10(10):414-416.

Siddal-Allum, J. N., Hughes, J. H., Kaler, S., Reginald, P. W. (1991). Splenic Rupture in Utero Following a Road Traffic Accident: Case Report. *British Journal of Obstetrics and Gynaecology*, 98(3):318-319.

Sims, C. J., Boardman, C. H., Fuller, S. J. (1996). Airbag Deployment Following a Motor Vehicle Accident in Pregnancy. *Obstetrics and Gynecology,* 88(4 pt 2):726.

Smith, A. J., LeMire III, W. A., Hurd, W. W., and Pearlman, M. (1994). Repair of the Traumatically Ruptured Gravid Uterus. *Journal of Reproductive Medicine*, 39(10):825-828.

Stafford, P. A., Biddinger, P. W., Zumwalt, R. E. (1988). Lethal Intrauterine Fetal Trauma. *Obstetrics and Gynecology*, 159(2):485-589.

Stuart, G. C. E., Harding P. G. R., Davies, E. M. (1980). Blunt Abdominal Trauma in Pregnancy. *Canadian Medical Association Journal*, 122(8):901-905.

Svendsen, E., Morild, I. (1988). Fetal Strangulation Following Uterine Rupture. *American Journal of Forensic Medicine and Pathology*, 9(1):54-57.

Van Enk, A., Van Zwam, W. (1994). Uterine Rupture: A Seat Belt Hazard. *Acta Obstetricia et Gynecologica Scandinavcia*, 73(5):432-433.

Weinstein, E. C., Pallais, V. (1968). Rupture of the Pregnant Uterus From Blunt Trauma. *Journal of Trauma*, 8(6):1111-1112.

Whitehouse, D. B. (1972). Hazard to Fetus from Safety Harness. *British Medical Journal*, 1(798):510.

APPENDIX A

TABLE A.1 SUMMARY OF AUTOMOTIVE CRASHES INVOLVING PREGNANT OCCUPANTS

No.	Occupant Position Impact Direction Impact Severity	Restraint	Gest Age (wk)	Maternal Injuries (italics indicate maternal fatality)	Fetal Outcome (italics indicate fetal survival)	Case Source
1	driver side impact	improper 3-pt belt	28	minor, placental abruption, uterine rupture	brain hemorrhages	Handel (1978)
2	driver front impact	none	34	moderate, pelvic fx, placental abruption, uterine rupture	skull fracture	Landers et al. (1989)
3	front passenger front impact	lap belt	22	minor, placental abruption, uterine rupture	decapitation	Rowe et al. (1996)
4	passenger side impact	none	37	minor, placental abruption, uterine rupture	stillborn	Pepperell et al. Case 13 (1977)
5	driver front impact moderate crash	none	34	minor, placental abruption, uterine rupture	stillborn	Agran et al. Case 6 (1986)
6	driver side impact	none	24	minor, pelvic fx, placental abruption, uterine rupture	stillborn	Dash & Leptin (1991)
7	rear passenger front impact severe crash	none	30	moderate, pelvic fx, placental abruption, uterine rupture	stillborn	Dittrich (1996)
8	driver	3-pt belt	22	minor, placental abruption,	stillborn	Harrison et al.

TABLE A.1 SUMMARY OF AUTOMOTIVE CRASHES INVOLVING PREGNANT OCCUPANTS

No.	Occupant Position Impact Direction Impact Severity	Restraint	Gest Age (wk)	Maternal Injuries (italics indicate maternal fatality)	Fetal Outcome (italics indicate fetal survival)	Case Source
	front impact moderate crash			uterine rupture		(1996)
9	rear passenger	improper lap belt	27	minor, placental abruption, uterine laceration	brain hemorrhage, transected aorta & spine, lacerated liver & kidney	Astarita & Feldman (1997)
10	driver front impact minor crash	3-pt belt + airbag	35	minor, placental abruption, uterine laceration	*premature birth, respiratory distress*	UMTRI GMP-006 (1997)
11	front passenger front crash	improper 3-pt belt	25	minor, placental abruption, uterine laceration	stillborn	Matthews (1975)
12	front passenger front crash	none	32	moderate, placental abruption, uterine laceration	stillborn	Smith et al. Case 1 (1994)
13	driver side impact	restrained	26	minor, placental abruption	skull fracture	Pepperell et al. Case 6 (1977)
14	passenger rollover	restrained	28	minor, placental abruption	rib fracture, liver rupture	Pepperell et al. Case 19 (1977)
15	driver front impact	none	35	moderate, placental abruption & laceration	skull fracture	Agran et al. Case 7 (1986)
16	driver front impact	none	34	*major, placental abruption*	skull fracture	Pepperell et al. Case 24 (1977)
17	driver front impact	none	28	minor, placental abruption	*emergency C/S, premature,*	Lavin and Miodovnik (1981)

TABLE A.1 SUMMARY OF AUTOMOTIVE CRASHES INVOLVING PREGNANT OCCUPANTS

No.	Occupant Position Impact Direction Impact Severity	Restraint	Gest Age (wk)	Maternal Injuries (italics indicate maternal fatality)	Fetal Outcome (italics indicate fetal survival)	Case Source
	severe crash				*respiratory distress*	
18	rear passenger rear impact severe crash	none	34	minor, placental abruption	*premature*	UMTRI GMP-207A (1996)
19	front passenger rear impact severe crash	none	32	minor, placental abruption	*premature, possible head injury*	UMTRI GMP-207B (1996)
20	driver front impact severe crash	3-pt belt	34	minor, placental abruption	stillborn	Whitehouse (1974)
21	driver front impact	restrained	37	minor, placental abruption	stillborn	Pepperell et al. Case 4 (1977)
22	driver front impact	restrained	27	minor, placental abruption	stillborn	Pepperell et al. Case 5 (1977)
23	driver side impact	restrained	28	none, placental abruption	stillborn	Pepperell et al. Case 7 (1977)
24	driver side impact	restrained	29	minor, placental abruption	stillborn	Pepperell et al. Case 8 (1977)
25	passenger front impact	none	36	minor, placental abruption	stillborn	Pepperell et al. Case 10 (1977)
26	passenger front impact	none	26	moderate, placental abruption	stillborn	Pepperell et al. Case 11 (1977)

TABLE A.1 SUMMARY OF AUTOMOTIVE CRASHES INVOLVING PREGNANT OCCUPANTS

No.	Occupant Position Impact Direction Impact Severity	Restraint	Gest Age (wk)	Maternal Injuries (italics indicate maternal fatality)	Fetal Outcome (italics indicate fetal survival)	Case Source
27	passenger front impact	restrained	32	none, placental abruption	stillborn	Pepperell et al. Case 15 (1977)
28	passenger side impact	restrained	36	moderate, placental abruption	stillborn	Pepperell et al. Case 16 (1977)
29	passenger side impact	restrained	30	minor, placental abruption	stillborn	Pepperell et al. Case 17 (1977)
30	passenger front impact	restrained	34	minor, placental abruption	stillborn	Pepperell et al. Case 18 (1977)
31	passenger front impact	restrained	34	minor, placental abruption	stillborn	Pepperell et al. Case 20 (1977)
32	passenger front impact	restrained	37	minor, placental abruption	stillborn	Pepperell et al. Case 21 (1977)
33	passenger side impact	none	34	*major, pelvic fx, placental abruption*	stillborn	Pepperell et al. Case 25 (1977)
34	rear passenger rollover severe crash	none	23	moderate, placental abruption	stillborn	Agran et al. Case 2 (1986)
35	driver front impact moderate crash	none	23	minor, placental abruption	stillborn	Agran et al. Case 3 (1986)
36	driver front impact	none	26	minor, placental abruption	stillborn	Agran et al. Case 4 (1986)

TABLE A.1 SUMMARY OF AUTOMOTIVE CRASHES INVOLVING PREGNANT OCCUPANTS

No.	Occupant Position Impact Direction Impact Severity	Restraint	Gest Age (wk)	Maternal Injuries (italics indicate maternal fatality)	Fetal Outcome (italics indicate fetal survival)	Case Source
	severe crash					
37	driver front impact	none	30	minor, placental abruption	stillborn	Agran et al. Case 5 (1986)
38	driver front impact moderate crash	none	38	minor, placental abruption	stillborn	Agran et al. Case 8 (1986)
39	driver front impact	none	39	minor, placental abruption	stillborn	Agran et al. Case 9 (1986)
40	driver multiple impact	3-pt belt	32	moderate, pelvic fx, placental abruption	stillborn	Pearlman et al. (1988)
41	driver front impact severe crash	3-pt belt	40	moderate, placental abruption	stillborn	UMTRI GMP-101 (1997)
42	driver front impact	3-pt belt	24	none, placental abruption	stillborn	UMTRI GMP-208 (1994)
43	front passenger side impact	none	31	minor, pelvic fx, partial placental abruption, uterine laceration	skull fracture, scalp laceration	Poulson & Gabert (1973)
44	driver front impact	none	29	moderate, partial placental abruption, uterine laceration	*emergency C/S, premature*	Kettel et al. Case 2 (1988)
45	front passenger front impact	lap belt	28	minor, partial placental abruption	brain hemorrhages	Raney (1970)

TABLE A.1 SUMMARY OF AUTOMOTIVE CRASHES INVOLVING PREGNANT OCCUPANTS

No.	Occupant Position Impact Direction Impact Severity	Restraint	Gest Age (wk)	Maternal Injuries (italics indicate maternal fatality)	Fetal Outcome (italics indicate fetal survival)	Case Source
46	front passenger side impact	none	37	moderate, pelvic fx, partial placental abruption	*emergency C/S, skull fracture, brain hemorrhages, apnea*	Bowdler et al. (1987)
47	front passenger front impact	none	33	minor, partial placental abruption	brain hemorrhages	Stafford et al. Case 4 (1988)
48	front passenger multiple impact	3-pt belt	34	none, pelvic fx, partial placental abruption	skull fracture, brain contusions and lacerations	Stafford et al. Case 5 (1988)
49	driver front impact	none	27	minor, partial placental abruption	brain hemorrhages	Stafford et al. Case 8 (1988)
50	rear passenger rollover	none	41	minor, partial placental abruption	*emergency C/S, skull fracture*	Hartl and Ko (1996)
51	driver front	none	36	none, partial placental abruption	chest hemorrhages	Stafford et al. Case 2 (1988)
52	driver front impact minor crash	none	39	none, partial placental abruption	lacerated spleen	Rothenberger et al. (1981)
53	driver front impact	none	39	moderate, partial placental abruption	liver, adrenal, and kidney contusions, seizures	Connor & Curran (1976)
54	driver	none	36	none, partial placental	*emergency C/S,*	Punnonan (1974)

TABLE A.1 SUMMARY OF AUTOMOTIVE CRASHES INVOLVING PREGNANT OCCUPANTS

No.	Occupant Position Impact Direction Impact Severity	Restraint	Gest Age (wk)	Maternal Injuries (italics indicate maternal fatality)	Fetal Outcome (italics indicate fetal survival)	Case Source
	front impact minor crash			abruption	*premature*	
55	front passenger front impact severe crash	lap belt	37	none, partial placental abruption	*emergency C/S, seizures, renal failure*	Chetcuti and Levene (1987)
56	front passenger	none	34	none, partial placental abruption	*emergency C/S, premature*	Kettel et al. Case 3 (1988)
57	driver front impact	none	41	minor, partial placental abruption	stillborn	Nichols & Weedn (1986)
58	driver side impact	none	24	moderate, placental laceration, uterine laceration	stillborn	Weinstein & Pallais (1968)
59	driver front impact	none	39	none, placental laceration	stillborn	Pepperell et al. Case 2 (1977)
60	front passenger front impact	none	36	minor, placental laceration	stillborn	Ravangard & Porter (1980)
61	driver front impact severe crash	lap belt	34	moderate, placental laceration	stillborn	Stuart et al. Case 1 (1980)
62	driver front impact severe crash	none	28	minor, placental laceration	stillborn	Stuart et al. Case 2 (1980)
63	driver	none	37	moderate, placental	stillborn	Civil et al. (1988)

TABLE A.1 SUMMARY OF AUTOMOTIVE CRASHES INVOLVING PREGNANT OCCUPANTS

No.	Occupant Position Impact Direction Impact Severity	Restraint	Gest Age (wk)	Maternal Injuries (italics indicate maternal fatality)	Fetal Outcome (italics indicate fetal survival)	Case Source
	front impact severe crash			laceration		
64	driver front impact	none	21	minor, placental laceration	stillborn	Stafford et al. Case 3 (1988)
65	front passenger	3-pt belt	36	moderate, placental laceration	head contusions	Griffiths et al. (1991)
66	driver front impact minor crash	none	37	minor, placental damage	skull fracture, brain hemorrhage	Cumming & Wren (1978)
67	driver front impact	lap belt	33	minor, placental damage	skull fracture, brain hemorrhage	Stafford et al. Case 6 (1988)
68	driver front impact	none	32	minor, pelvic fx, placental damage	brain hemorrhages	Stafford et al. Case 7 (1988)
69	front passenger	3-pt belt	31	minor, placental damage	stillborn	Griffiths et al. (1991)
70	front passenger front impact	improper lap belt	20	none, uterine transection	transection	McCormick (1968)
71	driver rear impact	none	39	minor, uterine laceration	*emergency delivery, seizures, respiratory distress*	Galle & Anderson (1979)
72	driver front impact	none	20	minor, pelvic fx, uterine laceration	stillborn	Heitzman & Markarian (1970)

TABLE A.1 SUMMARY OF AUTOMOTIVE CRASHES INVOLVING PREGNANT OCCUPANTS

No.	Occupant Position Impact Direction Impact Severity	Restraint	Gest Age (wk)	Maternal Injuries (italics indicate maternal fatality)	Fetal Outcome (italics indicate fetal survival)	Case Source
73	front passenger side impact	none	37	minor, pelvic fx, uterine laceration	stillborn	Smith et al. Case 2 (1994)
74	front passenger front impact	3-pt belt	30	minor, uterine rupture	strangled by cord	Svendsen & Morild (1988)
75	front passenger front impact severe crash	lap belt	24	minor, uterine rupture	stilllborn	Rubovits (1964)
76	driver front impact	restrained	22	minor, uterine rupture	stillborn	Pepperell et al. Case 9 (1977)
77	passenger side impact	restrained	30	minor, uterine rupture	stillborn	Pepperell et al. Case 14 (1977)
78	front passenger multiple impact	3-pt belt	32	moderate, uterine rupture	stillborn	Van Enk & Van Zwam (1994)
79	driver front impact	none	36	none	splenic rupture	Siddall-allum et al. (1991)
80	front passenger side impact	lap belt	24	minor	*spinal cord damage/paraplegia*	Weyerts et al. (1992)
81	front passenger side impact moderate crash	none	28	*major*	arm fracture	UMTRI GMP-002 (1996)
82	front passenger front impact severe crash	3-pt belt	30	minor	head, thorax, and abdomen hemorrhages	Fakhoury & Gibson (1986)

TABLE A.1 SUMMARY OF AUTOMOTIVE CRASHES INVOLVING PREGNANT OCCUPANTS

No.	Occupant Position Impact Direction Impact Severity	Restraint	Gest Age (wk)	Maternal Injuries (italics indicate maternal fatality)	Fetal Outcome (italics indicate fetal survival)	Case Source
83	driver front impact severe crash	3-pt belt	27	none	cerebral hemorrhage, hepatic hemorrhage, hemoperitoneum	Fries & Hankins (1989)
84	driver side impact	none	27	pelvic fx	skull fracture, brain hemorrhages, thoracic hemorrhages	Sherer et al. (1993)
85	driver front impact	none	30	none	skull fracture	Pepperell et al. Case 3 (1977)
86	driver side impact severe crash	none	32	*moderate, pelvic fx*	skull fracture	Evrard et al. (1989)
87	driver front impact	restrained	30	moderate	brain hematomas	Ford & Picker (1989)
88	front passenger side impact minor crash	none	37	minor	*brain hemorrhage, apnea, seizures*	UMTRI GMP-001 (1996)
89	driver side impact	3-pt belt	39	moderate, pelvic fx	*emergency C/S, brain contusion, seizures*	Matthews & Hammersly (1997)
90	driver front impact severe crash	none	25	minor	*contractions stopped w/o intervention, brain damage*	Knuppel et al. (1994)
91	front passenger front impact	none	28	minor	*hydrocephalus, possibly from crash*	UMTRI GMP-013 (1997)

TABLE A.1 SUMMARY OF AUTOMOTIVE CRASHES INVOLVING PREGNANT OCCUPANTS

No.	Occupant Position Impact Direction Impact Severity	Restraint	Gest Age (wk)	Maternal Injuries (italics indicate maternal fatality)	Fetal Outcome (italics indicate fetal survival)	Case Source
	severe crash					
92	driver minor crash	restrained + airbag	33	minor	*contractions stopped without intervention*	Sims et al. Case 3 (1996)
93	front passenger front impact moderate crash	3-pt belt + airbag	31	minor	*contractions stopped without intervention*	UMTRI GMP-009 (1997)
94	front passenger front impact	none	38	minor	amniotic membrane rupture	Parkinson Case 3 (1964)
95	front passenger front impact severe crash	none	24	*major*	stillborn	Agran et al. Case 1 (1986)
96	driver side impact severe crash	none	23	*major*	stillborn	UMTRI GMP-205 (1997)
97	driver front impact severe crash	3-pt belt	28	minor	stillborn	UMTRI GMP-206 (1995)
98	front passenger side impact severe crash	3-pt belt	22	*major*	stillborn	UMTRI GMP-212 (1997)
99	driver side impact	none	36	none	stillborn	Pepperell et al. Case 1 (1977)
100	passenger	none	38	moderate	stillborn	Pepperell et al.

TABLE A.1 SUMMARY OF AUTOMOTIVE CRASHES INVOLVING PREGNANT OCCUPANTS

No.	Occupant Position Impact Direction Impact Severity	Restraint	Gest Age (wk)	Maternal Injuries (italics indicate maternal fatality)	Fetal Outcome (italics indicate fetal survival)	Case Source
	side impact					Case 12 (1977)
101	passenger front impact	none	38	*major*	stillborn	Pepperell et al. Case 23 (1977)
102	passenger front impact	none	26	*major*	stillborn	Pepperell et al. Case 26 (1977)
103	front passenger front impact minor crash	none	39	moderate	*no complications*	UMTRI GMP-003 (1996)
104	driver side impact moderate crash	3-pt belt	20	minor, pelvic fx	*no complications*	UMTRI GMP-004 (1997)
105	driver front impact minor crash	airbag	23	minor	*no complications*	UMTRI GMP-005 (1997)
106	driver side impact minor crash	3-pt belt	35	minor	*no complications*	UMTRI GMP-007 (1997)
107	front passenger front impact minor crash	3-pt belt	36	minor	*no complications*	UMTRI GMP-008 (1997)
108	driver	3-pt belt +	22	minor	*no complications*	UMTRI GMP-011

TABLE A.1 SUMMARY OF AUTOMOTIVE CRASHES INVOLVING PREGNANT OCCUPANTS

No.	Occupant Position Impact Direction Impact Severity	Restraint	Gest Age (wk)	Maternal Injuries (italics indicate maternal fatality)	Fetal Outcome (italics indicate fetal survival)	Case Source
	side impact minor crash	airbag				(1997)
109	front passenger side impact minor crash	3-pt belt	31	minor	*no complications*	UMTRI GMP-012 (1997)
110	driver front impact minor crash	3-pt belt + airbag	26	minor	*no complications*	UMTRI GMP-014 (1997)
111	driver front impact minor crash	3-pt belt + airbag	38	minor	*no complications*	UMTRI GMP-201 (1997)
112	driver front impact minor crash	airbag	38	moderate	*no complications*	UMTRI GMP-202 (1996)
113	front passenger front impact moderate crash	3-pt belt	38	minor	*no complications*	UMTRI GMP-203 (1996)
114	front passenger front impact minor crash	3-pt belt + airbag	36	minor	*no complications*	UMTRI GMP-204 (1996)
115	driver side impact minor crash	3-pt belt	40	none	*no complications*	UMTRI GMP-209 (1997)

TABLE A.1 SUMMARY OF AUTOMOTIVE CRASHES INVOLVING PREGNANT OCCUPANTS

No.	Occupant Position Impact Direction Impact Severity	Restraint	Gest Age (wk)	Maternal Injuries (italics indicate maternal fatality)	Fetal Outcome (italics indicate fetal survival)	Case Source
116	driver rear impact minor crash	3-pt belt	37	minor	*no complications*	UMTRI GMP-210 (1997)
117	driver front impact moderate crash	3-pt belt + airbag	39	moderate	*no complications*	UMTRI GMP-211 (1997)
118	front passenger severe crash	restrained	?	none, pelvic fx	*no complications*	Levin & Edson (1994)
119	driver side impact minor crash	restrained + airbag	39	none	*no complications*	Sims et al. Case 1 (1996)
120	driver front impact moderate crash	restrained + airbag	35	minor	*no complications*	Sims et al. Case 2 (1996)

INVESTIGATIONS OF CRASHES INVOLVING PREGNANT OCCUPANTS

Kathleen DeSantis Klinich, Lawrence W. Schneider, Jamie L. Moore
Biosciences Division
University of Michigan Transportation Research Institute
Ann Arbor, Michigan

Mark D. Pearlman
Department of Obstetrics and Gynecology
University of Michigan Health Systems
Ann Arbor, Michigan

ABSTRACT

Case reports of 16 crashes involving pregnant occupants are presented that illustrate the main conclusions of a crash-investigation program that includes 42 crashes investigated to date. Some unusual cases that are exceptions to the overall trends are also described. The study indicates a strong association between adverse fetal outcome and both crash severity and maternal injury. Proper restraint use, with and without airbag deployment, generally leads to acceptable fetal outcomes in lower severity crashes, while it does not affect fetal outcome in high-severity crashes. Compared to properly restrained pregnant occupants, improperly restrained occupants have a higher risk of adverse fetal outcome in lower severity crashes, which comprise the majority of all motor-vehicle collisions.

An understanding of how pregnant women and their fetuses are injured in crashes is necessary to help vehicle restraint designers improve protection for pregnant occupants and reduce adverse fetal outcomes. A summary of the literature on automotive trauma in pregnancy is by presented by Klinich et al. (1998); additional studies not included in this summary are presented by Aitokallio-Tallberg and Halmesmaki (1997), Baethmen, et al. (1996), Crosby and Costiloe (1971), Fildes, et al. (1992), Lane (1989), Schultze, et al. (1998), and Wolf, et al. (1993). These resources describe many case studies and some epidemiological data on pregnant women involved in automotive crashes. However, these studies do not provide reliable estimates of the crash type and severity, or detailed information about the vehicle damage or occupant restraint use, and they are generally biased toward negative fetal outcomes.

A crash-investigation project was undertaken at the University of Michigan to obtain a better understanding of the factors

44th Annual Proceedings
Association for the Advancement of Automotive Medicine
October 2-4, 2000, pp. 37-55, Chicago, Illinois

involved in fetal loss to pregnant occupants in motor-vehicle crashes. In-depth investigations of real-world crashes involving pregnant occupants were conducted and a database with quantitative crash information plus detailed information on maternal and fetal injuries and outcomes has been established. The results help identify safety issues and mechanisms of fetal injury and loss, and provide case data for validating injury criteria for a new pregnant crash dummy with improved anthropometry, biofidelity, and injury assessment capabilities.

The new University of Michigan Transportation Research Institute (UMTRI) database of pregnant occupants involved in crashes is unique because it includes cases with a wide range of fetal outcomes, including many cases with good outcomes that have not generally been reported in the literature. The belt-restraint usage rate of occupants in the database is representative of usage rates in the general occupant population [NHTSA 1998], and the database includes a significant number of pregnant occupants exposed to airbag deployments. A key strength of the database is the inclusion of crash severity estimates based on vehicle damage measurements. In addition, the UMTRI database includes detailed information on the fetal outcome that is missing from the National Automotive Sampling System (NASS) crash data.

This paper provides detailed descriptions of selected crashes that illustrate the main conclusions of the study [Klinich, Schneider, Moore, and Pearlman, 1999]. In addition, some cases describing exceptions to these conclusions are presented.

METHODS

Investigators were notified of crashes involving pregnant occupants from medical staff of trauma centers, law enforcement agencies, or respondents to advertisements placed in medical publications. Eighty-seven notifications of crashes involving pregnant occupants were received over the course of this 30-month project, from which forty-two crashes were investigated. Two-thirds of the notifications were from locations within 50 miles of Ann Arbor, MI. Crashes were excluded from investigation if the case occupant was less than twenty weeks pregnant (n=10) or if the occupants did not agree to participate (n=29). Crashes were also not investigated if they involved multiple impacts or vehicle rollovers (n=6) because of difficulty in determining which interior contact points are associated with particular injuries.

The forty-two investigated crashes, involving forty-three pregnant occupants, were classified into twenty-seven major investigations and fifteen minor investigations. An investigation was considered minor if medical records were not available to confirm maternal and fetal outcome or the vehicle was not available for inspection. Standard UMTRI crash investigation techniques were used for the major investigations. Information about the crash circumstances and conditions was determined, the crash scene and external vehicle and internal vehicle damage were measured and photographed, and detailed occupant and fetal data and injuries were obtained from medical records and subject interviews. Crush measurements were input to the WinSMASH program to estimate impact severity.

For comparison purposes, impact severity was categorized as minor (<24 kph), moderate (24 to 48 kph), or severe (>48 kph), regardless of impact direction. A maternal injury rating was developed based on the injury severity score (ISS) [Baker et al. 1974], excluding injuries to the placenta or uterus, which were considered fetal injuries in this study. A minor maternal injury rating is assigned to ISS scores less than 10, a moderate rating to ISS scores from 10 to 20, and a major rating to ISS scores greater than 20. Fetal outcomes were categorized as good, minor fetal complications, major fetal complications, or fetal loss. Minor complications include early contractions or delivery within 48 hours of the crash at a gestational age of at least 32 weeks. Major fetal complications include placental abruption (separation of placenta from the uterus), uterine laceration, direct fetal injury, or premature delivery within 48 hours of the crash before 32 weeks gestational age. Adverse fetal outcomes are defined as fetal loss or major complications, while acceptable fetal outcomes include good fetal outcomes or minor complications.

RESULTS

OVERALL RESULTS. A complete analysis of the entire database is reported in Klinich et al. (1999) and is briefly summarized here. The distribution of adverse and acceptable fetal outcomes by impact severity and maternal restraint use for 43 pregnant occupants is shown in Figure 1. Figure 2 shows the percent of adverse fetal outcome for each grouping in Figure 1. Crash severity has the strongest effect on fetal outcome, with greater crash severities associated with higher percentages of adverse fetal outcomes. In this study, all of the severe crashes had adverse fetal outcomes. Maternal restraint use also has a significant effect on fetal outcome, with proper restraint use (three-point belt with or without airbag) associated with higher percentages of positive fetal outcomes, especially in lower severity crashes. Although limited by sample size, outcomes to mothers restrained by a three-point belt versus a three-point belt and airbag indicate that airbags may have a positive effect on fetal outcome. As shown in Figure 3, maternal injury level has a significant effect on fetal outcome, with a higher percentage of adverse fetal outcomes occurring with increasing levels of injury to the mother. Impact direction, occupant seating position, gestational age, maternal stature, and maternal weight do not have statistically significant effects on fetal outcome in this database.

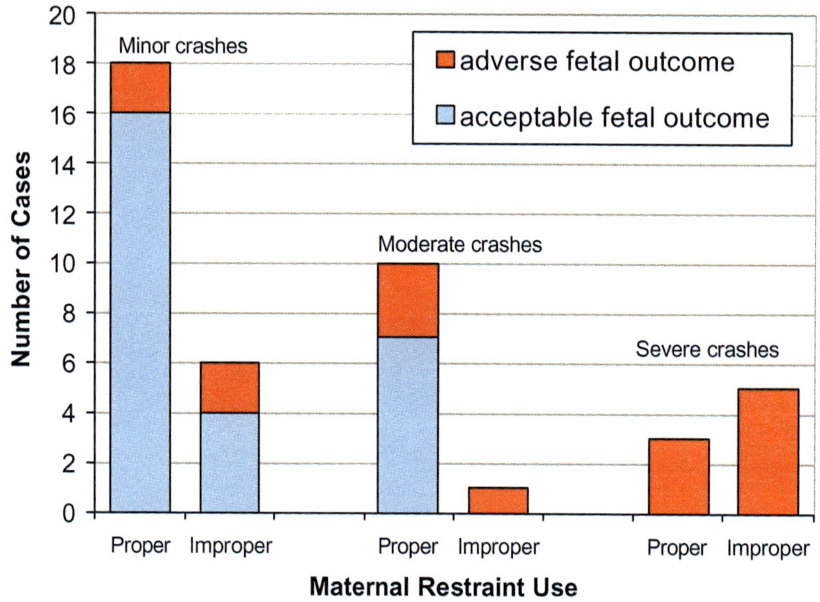

Figure 1. Distribution of fetal outcome by crash impact severity and maternal restraint. Use of a three-point belt, with or without airbag deployment, is considered proper restraint in this study; all other restraint combinations are considered improper.

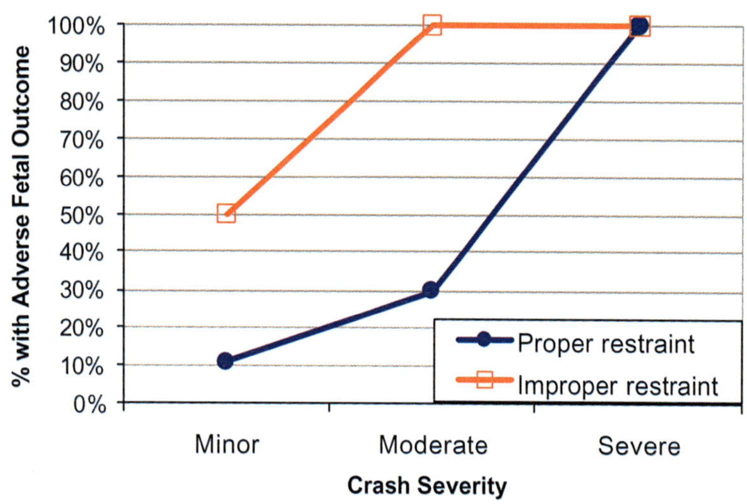

Figure 2. Percentage of cases with adverse fetal outcome at each crash severity level according to maternal restraint.

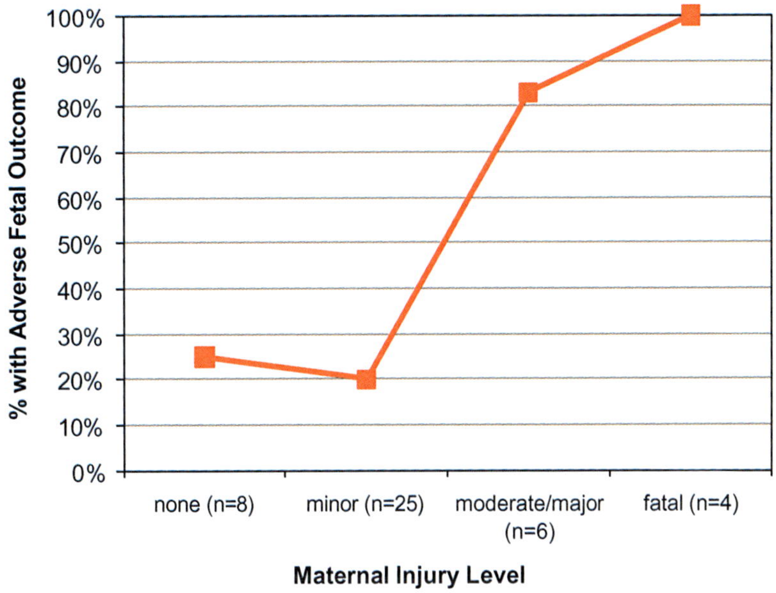

Figure 3. Percentage of cases with adverse fetal outcome at each maternal injury level.

The following case descriptions, ordered by impact direction and crash severity, illustrate these main findings, as well as some exceptions.

SEVERE FRONTAL IMPACT TO BELT-RESTRAINED DRIVER WITH FETAL LOSS (GMP-010). A 1992 Saturn SL2 (Figure 4), driven by a 30-year-old pregnant female, was involved in an extremely severe head-on crash with a 1987 Ford Mustang when the Saturn crossed the centerline and struck the Mustang in an offset mode. The delta V is estimated to have been approximately 71 kph. There was extensive intrusion into the driver's seating area. The 30-year-old driver of the Saturn (175 cm, 90 kg) was 40 weeks pregnant. She was wearing the automatic shoulder belt and the manual lap belt at the time of this severe crash. It took the rescue crew approximately 45 minutes to extricate her from the vehicle. Upon arrival at the hospital, an ultrasound showed no fetal heart tones or activity. The fetus died in utero as a result of placental abruption before arrival at the hospital.

The belt-restrained pregnant driver sustained multiple facial and lower extremity injuries that included fractures. She also had contusions to her chest and abdomen showing evidence of belt use. Maternal ISS was 29. It is suspected that abdomen contact with the steering wheel or lap belt resulted in the placental abruption (AIS 4) and fetal loss. The mother recovered from her injuries.

Figure 4. Damage to 1992 Saturn SL2 in case GMP-010, with roof removed by emergency personnel. CDC: 12-FYEW-5; maximum crush: 125 cm.

MODERATE FRONTAL IMPACT TO BELT- AND AIRBAG-RESTRAINED RIGHT-FRONT PASSENGER WITH GOOD FETAL OUTCOME (GMP-023). At an intersection, a 1991 Chevrolet Caprice made a left turn in front of the case vehicle, a 1996 Nissan Maxima. The pregnant occupant was the right-front passenger in the Maxima, which struck the right side of the Caprice with its front (Figure 5). The equivalent barrier speed is estimated to have been approximately 24 kph. The 24-year-old right-front passenger (163 cm, 64 kg) was 37 weeks pregnant. She was wearing the three-point belt and the airbag deployed, and she reportedly had the seat in the full-rearward seat-track position.

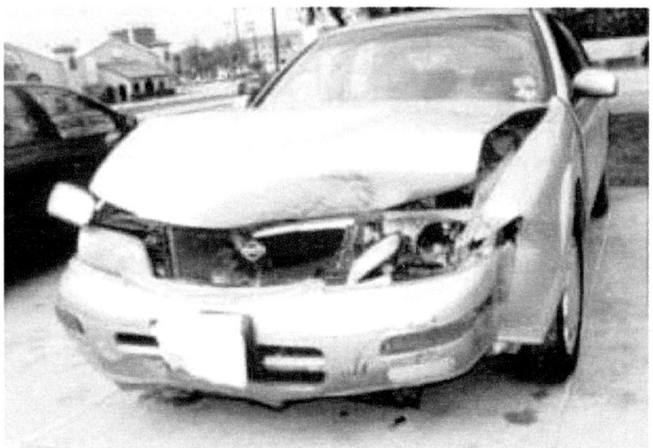

Figure 5. Damage to 1996 Nissan Maxima in case GMP-023. CDC: 12-FDEW-2; maximum crush: 32 cm.

Measurements were taken of the seat position after the impact, and the mother was subsequently photographed and measured in the vehicle at this reported seating position, as shown in Figure 6. The front of her abdomen and her knees were approximately 44 and 28 cm from the instrument panel, respectively. In addition to numerous AIS 1 contusions and abrasions, she sustained an AIS 2 fracture to her sternum from loading by the shoulder portion of the three-point belt.

After the crash, she was transported to the hospital by ambulance. The fetus was monitored and maintained a normal heart rate, so the mother was discharged home about 24 hours after the crash. Approximately 2 weeks after the crash, at a gestational age of 39 weeks, she delivered a healthy baby.

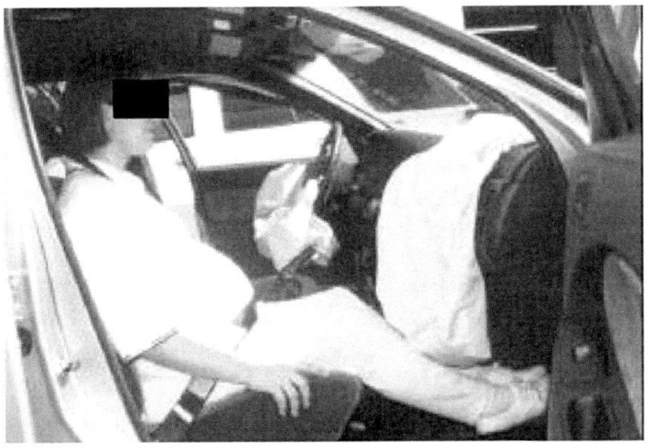

Figure 6. Case occupant in reported approximate position and posture at the time of the crash for case GMP-023.

MODERATE FRONTAL IMPACT TO BELT- AND AIRBAG-RESTRAINED DRIVER WITH GOOD FETAL OUTCOME (GMP-025). Weather conditions were good when the case vehicle, a 1998 Dodge Stratus (Figure 7), struck a stopped 1996 GMC K1500 pickup in the rear. The equivalent barrier speed is estimated to have been approximately 28 kph. The 35-year-old female (163 cm, 66 kg) driver was 39 weeks pregnant and she reportedly had the seat in the midtrack position. She was wearing the three-point belt and the airbag deployed. She sustained AIS 1 contusions to her extremities. After the crash, the fetus maintained a normal heart rate, and the pregnant driver was discharged within 24 hours of the crash. Two weeks after the crash, at a gestational age of 41 weeks, she delivered a healthy baby.

Figure 7. Damage to 1998 Dodge Stratus in case GMP-025. CDC: 12-FDEW-2; maximum crush: 40 cm.

MODERATE FRONTAL IMPACT TO BELT- AND AIRBAG-RESTRAINED RIGHT-FRONT PASSENGER WITH MAJOR FETAL COMPLICATIONS (GMP-026). This crash occurred on a two-lane, rural roadway. A 1994 Mitsubishi Gallant struck the front of the case vehicle shown in Figure 8, a 1998 Chevrolet Cavalier. The delta V is estimated to have been about 41 kph. The 18-year-old female (152 cm, 68 kg), right-front passenger was wearing the three-point belt and the passenger frontal-impact airbag deployed. She was 30 weeks pregnant, and reported that the seat was in the full-rearward track position. When she saw the impending collision, she reportedly braced herself by placing her right hand on the instrument panel and her feet against the toepan. In addition to numerous abrasions and contusions, she suffered an AIS 3 radius fracture and AIS 2 tibia and fibula fractures.

Figure 8. Damage to 1998 Chevrolet Cavalier in case GMP-026. CDC: 12-FDEW-3; maximum crush: 57 cm.

Approximately 9 weeks after the crash, at a gestational age of 39 weeks, the mother delivered a 2.52 kg baby. The infant sustained a right clavicle fracture as a result of difficult birth. The infant was admitted to the hospital after 7 days for evaluation of a small head circumference. A CT scan of the brain revealed ventricular cranial synostosis, right frontal and left parietal encephalomalcia, sagittal and left lambdoidal synostosis, and an enlarged ventricle. These complications are suspected to have resulted from the trauma of the crash based on similarities to a case reported in the literature [Gabis 1997] with more extended documentation of fetal status before and after the crash.

MINOR FRONTAL IMPACT TO UNRESTRAINED RIGHT-FRONT PASSENGER WITH GOOD FETAL OUTCOME (GMP-003). The front of a 1986 Ford Tempo (Figure 9) struck the rear of a 1984 Mercedes-Benz that had stopped for traffic. The equivalent barrier speed for the Tempo is estimated to have been about 12 kph. An unbelted 18-year old female (157 cm, 70 kg), who was 39 weeks pregnant, was seated in the right front of the Tempo, and she sustained AIS 1 head and back injuries. Initial diagnosis of the fetus was good. The infant was delivered 16 days after the crash at 41 weeks gestational age. The infant was discharged in healthy condition after treatment for a cleft palate unrelated to the crash.

Figure 9. Damage to 1986 Ford Tempo in case GMP-003. CDC: 12-FYEW-1; maximum crush: 6 cm.

MINOR FRONTAL IMPACT TO AIRBAG-RESTRAINED DRIVER WITH GOOD FETAL OUTCOME (GMP-005). A 1996 Toyota Corolla, driven by a 23-week pregnant female, struck the rear of a 1989 Ford LTD Crown Victoria that stopped quickly on a wet road. The delta V of the Corolla shown in Figure 10 is estimated to have been approximately 12 kph. The 30-year-old (175 cm, 102 kg) driver was not wearing the three-point belt, but the airbag deployed. Measurements were taken of the seat position after the impact, and the mother was subsequently photographed and measured in the vehicle at this seating position, as shown in Figure 11. With these settings, the front of her abdomen was approximately 14 cm from the bottom of the steering-wheel rim. She sustained AIS 1 injuries to her chest, head, and knee.

Figure 10. Damage to 1996 Toyota Corolla in case GMP-005. CDC: 12-FREW-1; maximum crush: 21 cm.

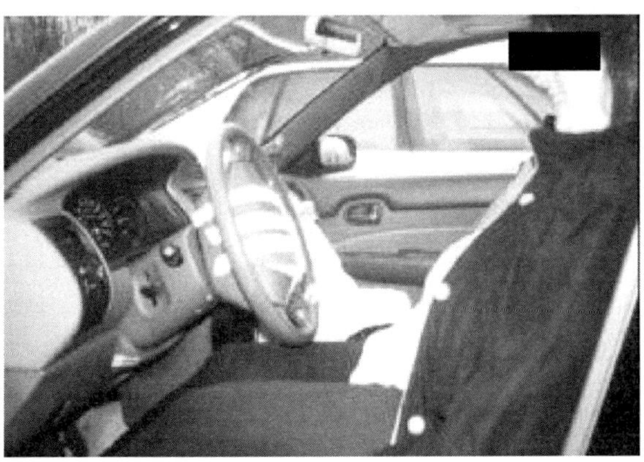

Figure 11. Approximate occupant position at the time of the crash in case GMP-005.

Measurement of fetal heart rate after the crash was normal, and the occupant was discharged within 12 hours of the crash. At 41 weeks gestation, she delivered by cesarean section because of failure to progress. No neonatal injuries resulted from the motor-vehicle crash.

MINOR FRONTAL IMPACT TO BELT- AND AIRBAG-RESTRAINED DRIVER WITH MAJOR FETAL COMPLICATIONS (GMP-006). The front of a 1993 Dodge Caravan was struck by a 1992 Chevrolet Cavalier traveling in the opposite direction when the pregnant driver of the Caravan attempted to make a left turn. At the time of the vehicle inspection, the front bumper, grille and left-front fender were detached from the Caravan and unavailable for inspection (Figure 12). The estimated equivalent barrier speed is less than 24 kph. The 27-year-old female (170 cm, 91 kg) driver was 35 weeks pregnant at the time of the crash. She was wearing the three-point belt and the airbag deployed. In addition to AIS 1 contusions, she sustained two horizontal lacerations approximately 6.4 cm apart on the left posterior side of the uterus (AIS 3), apparently from indirect loading of the lap belt. A partial abruption (33%) was also diagnosed (AIS 4).

Figure 12. Damage to the 1993 Dodge Caravan in case GMP-006, with some parts removed. CDC: 01-FDEW-1; maximum crush: minor.

On hospital admission, the mother was experiencing abdominal pain and contractions approximately 3 minutes apart. Measurement of fetal heart rate indicated a non-reassuring tracing ("fetal distress"); a decision was made to deliver the baby by cesarean section within 3 hours of the crash. The 2.50 kg baby had mild respiratory distress syndrome due to premature birth, and required oxygen therapy during this initial time period. The baby was discharged 10 days after delivery, apparently doing well. No neonatal injuries resulted from the motor-vehicle crash.

MINOR FRONTAL IMPACT TO BELT- AND AIRBAG-RESTRAINED RIGHT-FRONT PASSENGER WITH MINOR FETAL COMPLICATIONS (GMP-009). A 1996 Toyota Camry was turning left when it struck the front of the case vehicle, a 1996 Chevrolet Lumina (Figure 13). The equivalent barrier speed is estimated to have been approximately 16 kph. The 23-year-old female (170 cm, 88 kg), right-front passenger was 31 weeks pregnant. She was wearing the three-point belt and the airbag deployed. She sustained several AIS 1 injuries and was transported to the hospital complaining of contractions. The contractions stopped without intervention and she was discharged within 12 hours of the crash. Nine weeks after the crash, at a gestational age of 40 weeks, the patient delivered a healthy baby.

Figure 13. Damage to 1996 Chevrolet Lumina in case GMP-009. CDC: 12-FDEW-1; maximum crush: 14 cm.

MINOR FRONTAL IMPACT TO BELT-RESTRAINED DRIVER WITH GOOD FETAL OUTCOME (GMP-016). The pregnant driver of a 1994 Suzuki Swift attempted to turn left at an intersection, and her vehicle was struck in the front (Figure 14) by a 1990 Mercury Sable that was traveling in the opposite direction. The equivalent barrier speed is estimated to have been approximately 23 kph. The 26-year-old female (160 cm, 89 kg) driver was 23 weeks pregnant and was wearing the available three-point belt. She reportedly had her seat adjusted to a forward track position because of her short stature. She sustained several AIS 1 injuries. After the crash, the driver was transported to the hospital where the baby was monitored and found to be in good condition. Four months after the crash, at a gestational age of 39 weeks, she delivered a healthy baby.

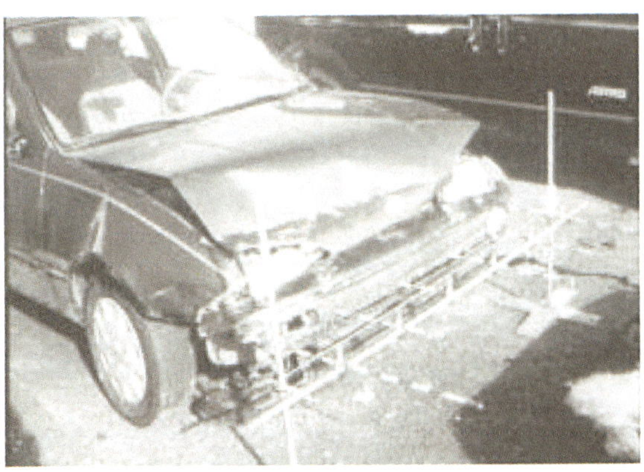

Figure 14. Damage to 1994 Suzuki Swift in case GMP-016. CDC: 01-FDEW-1; maximum crush: 19 cm.

MINOR FRONTAL IMPACT TO SHOULDER-BELT-RESTRAINED RIGHT-FRONT PASSENGER WITH MAJOR FETAL COMPLICATIONS (GMP-022). The driver of a 1990 Chevrolet Beretta attempted to make a left turn in front of a 1990 Ford Tempo at an intersection. The delta V for the Tempo, which was struck in the center-left front as shown in Figure 15, is estimated to have been approximately 23 kph. The 18-year-old female (160 cm, 52 kg), right-front passenger was 28 weeks pregnant. She was wearing the automatic shoulder belt, but not the manual lap belt. The seat appeared to have been in the full-forward position, and she had her left foot on the instrument panel above the glove compartment. An unrestrained passenger in the right-rear seat loaded the seatback of the pregnant occupant.

It was reported that the pregnant occupant was holding her chest and breathing hard after the crash. Her foot and leg were still up on the dashboard, and she was unresponsive. She was pronounced dead on arrival at the hospital. Her most serious injuries were an AIS 5 heart laceration and AIS 4 lung contusions. An obstetrician was in the emergency department when the patient arrived and quickly performed a perimortem cesarean section. There was no placental abruption. The infant developed typical complications resulting from a premature birth, but showed no other problems of trauma. The infant was discharged 10 weeks later.

Figure 15. Damage to 1990 Ford Tempo in case GMP-022. CDC: 11-FYEW-2; maximum crush: 39 cm.

MINOR FRONTAL IMPACT TO BELT- AND AIRBAG-RESTRAINED DRIVER WITH GOOD FETAL OUTCOME (GMP-027). A 1995 Dodge Neon driven by a 38-week pregnant driver struck the rear of a 1991 Ford Escort that had stopped to make a left turn. Vehicle damage shown in Figure 16 was measured and the equivalent barrier speed is estimated to have been 13 kph. The 33-year-old (160 cm, 86 kg) driver was wearing the three-point belt and the frontal-impact airbag deployed. She sustained AIS 1 injuries to her upper extremities and abdomen. Approximately 3 weeks after the crash, the patient delivered a healthy baby by a planned cesarean section.

Figure 16. Damage to 1995 Dodge Neon in case GMP-027. CDC: 12-FDEW-1; maximum crush: 7 cm.

MODERATE NEAR-SIDE IMPACT TO UNRESTRAINED RIGHT-FRONT PASSENGER WITH MATERNAL AND FETAL DEATH (GMP-002). Because of icy road conditions, the driver of a 1989 Ford Tempo (Figure 17) lost control of the vehicle, which rotated into opposing traffic and was struck in the right-side door by a 1993 Ford Ranger. Severe intrusion into the right-passenger compartment resulted, and the delta V for the Tempo is estimated to have been approximately 42 kph. An 18-year-old female (160 cm, 54 kg) was the right-front passenger in the Tempo and was 28 weeks pregnant. She was not wearing the automatic shoulder belt or the manual lap belt. She sustained multiple injuries, including an AIS 4 liver laceration and an AIS 3 right distal femur fracture. She died from respiratory compromise when fluids from the femur fracture passed into the bloodstream and prevented normal lung function (fat embolism), complicated by severe hepatic trauma. The 28-week fetus died as a result of the trauma sustained by the mother. The fetus also sustained a fractured left humerus (AIS 2), probably due to loading by the side-door interior.

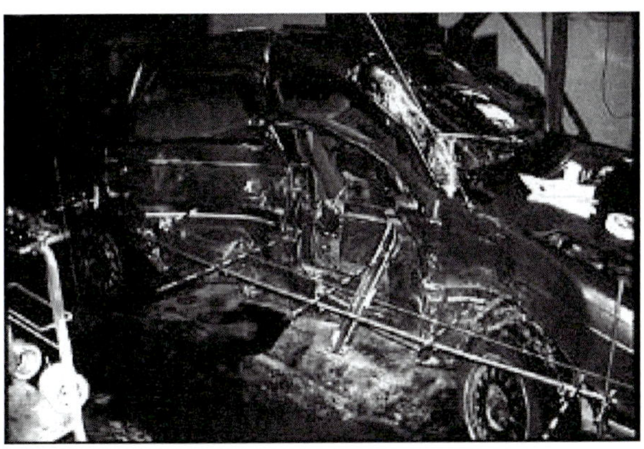

Figure 17. Damage to 1989 Ford Tempo in case GMP-002. CDC: 03-RYAW-4; maximum crush: 68 cm.

MODERATE NEAR-SIDE IMPACT TO BELT-RESTRAINED DRIVER WITH GOOD FETAL OUTCOME (GMP-004). A 1994 Ford Taurus driven by a woman who was 19.5 weeks pregnant was struck in the left-side front door (Figure 18) by a 1996

Ford Crown Victoria. An equivalent barrier speed of approximately 37 kph was estimated for the Taurus. The 28-year-old pregnant driver (173 cm, 59 kg) of the Taurus was wearing the three-point belt. During the left-side impact, she sustained AIS 2 pelvis fractures and an AIS 3 pubic symphysis separation. Initial diagnosis of the fetus was good. The infant was delivered vaginally about 4.5 months after the crash at 38.5 weeks gestational age. The baby was healthy with no problems, and the mother's pelvic injuries healed.

Figure 18. Damage to 1994 Ford Taurus in case GMP-004. CDC: 10-LPEW-3, maximum crush: 50 cm.

MINOR NEAR-SIDE IMPACT TO UNRESTRAINED RIGHT-FRONT PASSENGER WITH MAJOR FETAL COMPLICATIONS (GMP-001). A 1986 Chevrolet Cavalier with a 37-week pregnant right-front passenger was struck in the right-front fender (Figure 19) by the left rear of a 1993 Plymouth Sundance that was merging into the left lane. The estimated equivalent barrier speed of the side impact is 9 kph. The 21-year-old pregnant occupant (155 cm, 85 kg) was not wearing the available three-point belt, and reportedly had her seat in the full-forward position. She sustained AIS 1 contusions and abrasions to her head and knee.

Figure 19. Damage to 1986 Chevrolet Cavalier in case GMP-001. CDC: 03-RFEW-2; maximum crush: 10 cm.

The mother reported cessation of fetal movement after the impact. Spontaneous rupture of membranes occurred 1.5 hours after the crash, and the infant was born vaginally within 18 hours of the crash. In the first 2 days after being born, the infant experienced two episodes of apnea and seizure activity. A CT scan detected a left peridural / tentorial hemorrhage with a subdural hematoma (AIS 4). Given the uncomplicated birth process, it is likely that the infant's injuries were related to the mother's contact with the door armrest during the near-side impact. The infant was discharged after 4 days in stable condition.

MINOR NEAR-SIDE IMPACT TO BELT- AND AIRBAG-RESTRAINED DRIVER WITH GOOD FETAL OUTCOME (GMP-011). A 1995 International tractor-trailer pulled out in front of a 1992 Plymouth Grand Voyager. The 33-year-old female (157 cm, 69 kg) pregnant driver of the Voyager braked and steered to the right, but her vehicle was struck in the driver door and left-front fender by the tractor-trailer (Figure 20). The equivalent barrier speed of the Voyager for this side impact is estimated to have been about 16 kph, but this estimate may be low because of override of the truck on the Voyager. The driver, who was 22 weeks pregnant, reportedly had the seat in a midtrack position. She was wearing the three-point belt and the airbag deployed. She sustained AIS 2 head injuries in addition to multiple AIS 1 integumentary injuries. After the crash, she was transported to the hospital. The fetus was monitored and determined to be in good condition. Four months after the crash, at 40 weeks gestational age, the subject delivered a healthy infant.

Figure 20. Damage to 1992 Plymouth Grand Voyager in case GMP-011. CDC: 10-LYAW-3; maximum crush: 23 cm.

MINOR REAR IMPACT TO BELT-RESTRAINED DRIVER WITH FETAL LOSS (GMP-021). An Oldsmobile Cutlass traveling at low speed struck the rear of a 1990 Honda Civic driven by a 26-week pregnant driver. There was only about 1 cm of crush to the rear bumper of the Civic (Figure 21) and the delta V was calculated to be approximately 14 kph. The 36-year-old female (155 cm, 75 kg) driver of the Civic was wearing the three-point belt. The pregnancy had been uncomplicated to date, with fetal heart tones and movements noted the day before the crash at a prenatal checkup. She did not sustain any injuries, but went to the hospital approximately 3 to 4 hours after the crash. At the emergency department, no fetal heart tones were found, and she was sent to labor and delivery. An ultrasound was conducted and it was determined that the fetus was dead. Autopsy of the 0.94 kg fetus did not indicate any anomaly other than the expected changes of a fetus that died in utero. There was no placental abruption or other specific trauma-related explanation for the fetal loss.

Figure 21. Damage to 1990 Honda Civic in case GMP-021. CDC: 06-BDLW-1; maximum crush: 1 cm.

DISCUSSION

In this study, impact severity and maternal injury are the factors most strongly associated with fetal outcome (Klinich et al.1999). Three cases presented in this paper particularly illustrate these associations. In case GMP-010, the 40-week three-point-belt-restrained pregnant driver sustained fetal loss and maternal injuries (ISS=29) in a severe frontal impact. The unrestrained mother and fetus both died in case GMP-002, a moderate near-side impact. Case GMP-026 involved a pregnant right-front passenger restrained by a three-point belt and airbag in a moderate frontal crash. She sustained moderate injuries, and her infant was diagnosed after birth with complications suspected to have resulted from the crash.

Most of the minor severity impacts presented (GMP-003, GMP-005, GMP-009, GMP-011, GMP-016, GMP-027) led to minimal maternal injury and uncomplicated fetal outcomes. The two remaining moderate frontal impacts presented (GMP-023 and GMP-025) also had good maternal and fetal outcomes to properly restrained occupants. In case GMP-004, the mother sustained several pelvic fractures in a moderate severity near-side impact, but the fetus experienced no problems. It is suspected that the small size of the fetus in this case (gestational age of 19.5 weeks) may have been a significant factor in the lack of fetal injuries despite the compromised pelvic cavity.

Four of the cases presented are exceptions to the general finding that worse fetal outcomes are associated with increasing crash severity and maternal injury, illustrating that occasionally minor maternal injury can be associated with catastrophic fetal outcome. In three minor severity impacts, one frontal (GMP-006), one side (GMP-001), and one rear (GMP-021), the fetus had an adverse outcome although the crash was minor and the mother sustained minor or no injuries. In GMP-022, the fourth case, the mother died in a minor frontal impact, but the fetus was delivered prematurely and survived. The maternal death from internal bleeding from heart and spleen injuries in this case probably resulted from a combination of improperly using only the passive shoulder belt, irregular positioning with feet on the dash, and an unbelted rear passenger loading her seatback. It is also possible that the pregnancy may have contributed to maternal death, as the enlarged pregnant uterus directly contacts other abdominal organs such as the spleen.

Proper restraint of a three-point-belt with or without airbag deployment is associated with better fetal outcomes in low and moderate severity crashes (Klinich et al. 1999). In eleven of the cases presented in this paper, the pregnant occupant was wearing a three-point belt. In seven of these cases, an airbag also deployed. Seven of eleven properly restrained women had acceptable fetal outcomes (GMP-004, GMP-009, GMP-011, GMP-016, GMP-023, GMP-025, and GMP-027). All of these cases were minor or moderate in impact severity. In the four cases with adverse fetal outcomes to properly restrained women, one crash was a severe frontal (GMP-010), one was a moderate frontal (GMP-026), one was a minor frontal (GMP-006), and the fourth was a minor rear impact (GMP-021).

Five of the cases presented in this paper involved pregnant occupants who were improperly restrained and three of these resulted in adverse fetal outcome. As discussed previously, use of only a shoulder belt in case GMP-022 led to maternal death and premature delivery for a minor frontal crash. In the minor near-side impact of case GMP-001, the pregnant right-front passenger was unbelted, and her fetus was delivered early, 18 hours after the crash, with a head injury. Case GMP-002 was a moderate near-side impact to an unrestrained pregnant woman that resulted in maternal and fetal death. The other two cases with improper restraint (GMP-005 with airbag only and GMP-003 with no restraint) led to good fetal outcomes, but were frontal crashes of minor severity.

An interesting finding of this study is the effect of belt-restraint usage in near-side impacts. For the four near-side impacts presented, the two properly belt-restrained pregnant women (Case

GMP-004 and GM-011) had good fetal outcomes, while the unrestrained pregnant occupants of cases GMP-001 and GMP-002 had adverse fetal outcomes. This finding is somewhat unexpected, as belt restraints are generally not as effective at reducing injuries to occupants in near-side impacts [Thomas et al. 1987]. Most of the other near-side impacts in the database that were not described in this paper also have acceptable fetal outcomes for restrained occupants in near-side impacts.

Eight of the cases described in this paper involved airbag deployments. Two frontal crashes, one of minor severity (GMP-006) and the other of moderate severity (GMP-026), resulted in major complications for the fetus, but not in fetal death. The one airbag-only case included (GMP-005) had a good fetal outcome. The five other cases with airbag deployments had acceptable fetal outcomes. There were no cases where the airbag was considered to be the cause of an adverse fetal outcome or serious maternal injury.

Overall, proper restraint use has a positive effect on fetal outcome in low and moderate level crashes. Since crashes of this severity constitute over 95% of all crashes (based on analysis of NASS data on towaway crashes), these results support current recommendations that pregnant women should properly use three-point belts and should not disable the airbags [ACOG 1999]. Although the study found a strong association of poor fetal outcome with severe maternal injury and high crash severity, there are some notable exceptions where adverse fetal outcomes occurred in minor-severity crashes with little or no maternal injury. These exceptions support recommendations from the literature [Pearlman et al.1997, 1990] that all pregnant occupants involved in crashes should seek medical treatment after the crash, regardless of the severity of maternal injury.

ACKNOWLEDGMENTS

The work described in this paper was financed by General Motors (GM) pursuant to an agreement between GM and the U.S. Department of Transportation. The authors would like to acknowledge Joel MacWilliams, Tim Compton, and Dale Halloway, who assisted with the investigations, Carol Flannagan, who provided statistical guidance, and Patricia Reis Van Buren, who helped analyze the medical records.

REFERENCES

American College of Obstetricians and Gynecologists. ACOG educational bulletin: Obstetric aspects of trauma management, Number 251, Sep 1998. Int J Gynaecol Obstet **64**(1):87-94; 1999.

Aitokallio-Tallberg, A., Halmesmaki, E. Motor vehicle accident during the second or third trimester of pregnancy. Acta Obstet et Gynecol Scand **76**:313-7; 1997.

Baethman, M., Kahn, T., Lenard, H.-G., et al. Fetal CNS damage after exposure to maternal trauma during pregnancy. Acta Paediatr **85**(11):1331-8; 1996.

Baker, S.P., O'Neill, B., Waddon, W., Jr, et al. The Injury Severity Score: a method for describing patients with multiple injuries and evaluating care systems. J Trauma **14**:186-7; 1974.

Crosby, W.M., Costilow, J.P. Safety of lap-belt restraint for pregnant victims of automobile collisions. N Engl J Med **284**:632-6; 1971.

Fildes, J., Reed, L., Jones, N., et al. Trauma: the leading cause of maternal death. J Trauma **32**:643-5; 1992.

Gabis, L., Gelman-Kohan, Z., Mogilner, M. Microcephaly due to fetal brain disruption sequence: case report. J Perinat Med 25:213-5; 1997.

Klinich, K. D., Schneider, L. W., Moore, J. L., et al. Investigations of crashes involving pregnant occupants: final report. Ann Arbor, MI; UMTRI, Report No.: UMTRI-99-29. 1999.

Klinich, K. D., Schneider, L. W., Moore, J. L., et al. Injuries to pregnant occupants in automotive crashes. Proc 41st Ann Conf AAAM. Des Plaines, IL; Association for the Advancement of Automotive Medicine, 57-91,1998.

Lane, P. L. Traumatic fetal deaths. J Emerg Med **7**(5):433-5; 1989.

National Highway Traffic Safety Administration. Traffic Safety Facts 1998. Washington, D.C.; NHTSA, Report No.: DOT HS 808 983. 1998.

Pearlman, M. D. Motor vehicle crashes, pregnancy loss, and preterm labor. Intl J Gynaecol and Obstet **57**(2):127-132; 1997.

Pearlman, M. D., Tintinalli, J. E., Lorenz, R. P. Blunt trauma during pregnancy. N Engl J Med**323**(23):1609-13;1990.

Schultze, P.M., Stamm, C.A., Roger, J. Placental abruption and fetal death with airbag deployment in a motor vehicle accident. Obstet Gynecol **92**(4):719;1998.

Thomas, C., Henry, C., Harteman, F., et al. Injury patterns and parameters to assess severity for occupants involved in car-to-car lateral impacts. Proc. 11th ESV 49-61; 1987.

Wolf, M. S., Alexander, B. H., Rivara, F. P., et al. A retrospective cohort study of seatbelt use and pregnancy outcome after a motor vehicle crash. J Trauma **34**(1):116-9;1993.

(*Presenter:* Kathleen Klinich)

Leonard Evans: First, let me commend you on a very nice study most clearly and excellently delivered. It was so easy to follow. Easy though it was, I can't remember all the details, but there was one case where the outcome was bad, the delta V was low, there was a 3 point belt and an airbag and just casually looking at it, do you have some reason for excluding the possibility that it was the airbag that was the cause of the bad outcome?

K. Klinich: We actually think for that instance it was more the belt that induced the injury. There were uterine lacerations in that woman, they weren't directly underneath the belt, but they were on the posterior side of the uterus. We think it was from indirect loading by the belt because there were two lacerations which were about 2 inches apart which kind of indicated that the force was transmitted from the other side so we think for that case it was probably more the belt than the airbag.

Carl Clark: Additional protection in automobiles can be provided by preinflated airbags. It would be my recommendation that the system of having pregnant women ride in a car with preinflated and not deflated airbags would provide much better protection. This should be explored by somebody.

K. Klinich: We do have more airbag cases than have been previously reported with pregnant occupants and we were able to make some slight statistical conclusion that an airbag improves the outcome compared to a 3 point belt alone.

C. Clark: Do you understand what I mean by a preinflated airbag? As she drives out of the driveway, the airbag is preinflated. That ought to be explored.

Stefan Duma: Do you think your results or trends would change if you looked at the fetal morbidity as opposed to just mortality?

K. Klinich: We were counting adverse fetal outcome as pretty much anything with bad longterm consequences or death so our results do group that as an adverse consequence just because we did not have enough cases of fetal loss versus fetal injury.

Fetal outcome in motor-vehicle crashes: effects of crash characteristics and maternal restraint

Kathleen DeSantis Klinich, PhD; Carol A. C. Flannagan, PhD; Jonathan D. Rupp, PhD; Mark Sochor, MD; Lawrence W. Schneider, PhD; Mark D. Pearlman, MD

OBJECTIVE: This project was undertaken to improve understanding of factors associated with adverse fetal outcomes of pregnant occupants involved in motor-vehicle crashes.

STUDY DESIGN: In-depth investigations of crashes involving 57 pregnant occupants were performed. Maternal and fetal injuries, restraint information, measures of external and internal vehicle damage, and details about the crash circumstances were collected. Crash severity was calculated using vehicle crush measurements. Chi-square analysis and logistic regression models were used to determine factors with a significant association with fetal outcome.

RESULTS: Fetal outcome is most strongly associated with crash severity ($P < .001$) and maternal injury ($P = .002$). Proper maternal belt-restraint use (with or without airbag deployment) is associated with acceptable fetal outcome (odds ratio = 4.5, $P = .033$). Approximately half of fetal losses in motor-vehicle crashes could be prevented if all pregnant women properly wore seat belts.

CONCLUSION: Higher crash severity, more severe maternal injury, and lack of proper seat belt use are associated with a higher risk of adverse fetal outcome. These results strongly support recommendations that pregnant women use properly positioned seatbelts.

Key words: airbags, crash investigations, fetal loss, pregnancy, seatbelts, wounds and injuries

Cite this article as: Klinich KD, Flannagan CAC, Rupp JD, et al. Fetal outcome in motor-vehicle crashes: effects of crash characteristics and maternal restraint. Am J Obstet Gynecol 2008;198:450.e1-450.e9.

★ EDITORS' CHOICE ★

The public-health problem of fetal loss and injury from maternal involvement in motor-vehicle crashes is difficult to quantify. Reliable statistics on fetal loss resulting from automotive trauma are not available because maternal involvement in crashes has not been consistently recorded on fetal death certificates. In addition, the pregnancy status of women involved in crashes has only been tracked in national databases since 1995. The annual number of fetal losses resulting from automotive crashes in the United States is estimated to range from 90-369 fetal losses each year.[1] These estimates indicate that the actual number of fetal losses may be greater than the number of deaths to infants under 1 year from motor-vehicle crashes (154 in 2004).[2] Adverse fetal outcomes other than death can also occur as a result of trauma sustained in utero.[3,4] If a fetus survives a crash, complications resulting from the crash including emergency delivery of a premature fetus,[5] such as low birthweight and neonatal respiratory distress syndrome or fetal asphyxia, can lead to long-term physical or neurologic problems for the child.

Abruptio placentae (AP) is the leading cause of fetal loss in motor-vehicle crashes.[6] Although shear and tensile failure have been suggested as mechanisms that cause traumatic AP, the exact mechanism of placental separation caused by trauma is not certain. If reliable information about fetal loss was available, knowledge about the factors and circumstances surrounding fetal loss and injury in motor-vehicle crashes could be used to help vehicle restraint designers improve protection for pregnant occupants, and reduce the frequency and severity of adverse fetal outcomes. Published studies of pregnancy outcome after automobile crashes have generally not included complete and accurate information about crash severity and restraint use.[7-9] Moreover, the case reports in the literature generally emphasize crashes that result in negative fetal outcomes to illustrate particularly unusual or severe injuries, and generally do not reflect the typical spectrum of fetal

From the University of Michigan Transportation Research Institute (Drs Klinich, Flannagan, Rupp, Sochor, and Schneider) and the Department of Biomedical Engineering (Dr Schneider), University of Michigan, and the Departments of Emergency Medicine (Dr Sochor) and Obstetrics and Gynecology and Surgery (Dr Pearlman), University of Michigan Health Systems, Ann Arbor, MI.

This research was presented at the 26th Annual Scientific Meeting of the American Gynecological and Obstetrical Society, Chicago, IL, Sept. 26-29, 2007.

Received April 18, 2007; revised Nov. 25, 2007; accepted Feb. 1, 2008.

Reprints not available from the authors.

The National Highway Transportation Safety Administration and the Alliance of Automobile Manufacturers provided support for the CIREN and UMTRI crash investigation programs, respectively.

0002-9378/$34.00 • © 2008 Mosby, Inc. All rights reserved. • doi: 10.1016/j.ajog.2008.02.009

For Editors' Commentary, see Table of Contents

outcomes for pregnant occupants in crashes.[7,8,10,11] To improve vehicle safety for the pregnant population, cases that illustrate both positive and negative fetal outcomes after crashes provide more useful information than cases that focus only on unusual and severely adverse outcomes. The methodology for this research was documented in a preliminary report, although the current article describes a more detailed analysis of a larger database.[12]

The purpose of this study was to conduct in-depth investigations of motor-vehicle crashes involving pregnant occupants, with a focus on determining how restraint conditions and specific crash characteristics affect fetal outcome. Inclusion of crashes with both positive and adverse fetal outcomes, and with both belt-restrained and unbelted pregnant occupants, will provide a wider variety of cases than previously reported in the literature. In addition, providing quantitative information on the crash and maternal restraint conditions allows for more quantitative analysis of the data with regard to the causes of fetal injury or loss. Previous studies have demonstrated that pregnant women are more likely to properly wear seat belts when instructed to by their caregivers.[13-15] Therefore, the findings may help both medical practitioners and safety engineers reduce the number of fetal losses from motor-vehicle crashes through appropriate advice to pregnant women and improvements in vehicle restraint design.

MATERIALS AND METHODS

Data from the crashes included in this study were collected in 2 phases. For the first part of the study, crashes with pregnant occupants* were identified through notifications from trauma centers and police agencies and responses to advertisements in medical facilities. Eighty-seven notifications were received over 30 months (Aug. 1996-Feb. 1999). Two-thirds of the notifications were from locations in the Southeast Michigan area, with the remaining notifications received from across the United States. Thirty-seven notifications were received from medical contacts, twenty-four from law-enforcement or crash-investigation sources, and the remaining 26 from responses to notices. Before participating in the study, each pregnant occupant (or her legal representative) signed institutional review board-approved consent forms.

Study subjects were included if they were at least 20 weeks'gestation, were involved in a motor-vehicle crash, and agreed to participate. Cases involving a rollover crash were not investigated. No other exclusion criteria were used. Of the 87 crash notifications, 10 were excluded because the fetus was less than 20 weeks' gestation. This exclusion criterion was chosen because the high frequency of spontaneous pregnancy loss before 20 weeks would make association of fetal loss with a motor-vehicle crash this early in pregnancy questionable. Six crashes were not investigated because they involved a rollover of the case vehicle, for which occupant kinematics and injury causations are difficult to determine. Twenty-nine crashes were not investigated because the pregnant occupant declined to participate. These exclusions resulted in a total of 42 crash investigations involving 43 pregnant occupants (1 crash involved 2 pregnant occupants). Each of the 43 occupants is considered a case.

In the second phase of the study, data on 14 crashes involving pregnant occupants were retrieved from 2 other sources. The Crash Injury Research Engineering Network (CIREN) is a database of occupants involved in motor-vehicle crashes that have sustained serious injury, while the University of Michigan Crash Investigation Program collects detailed data on approximately 100 crashes per year in a variety of crash conditions. When reviewing these 2 datasets, any case with a pregnant occupant that met the inclusion criteria and contained a record of fetal outcome was included in this study.

For the 57 cases in the resulting dataset, information about the crash circumstances and scene was obtained, external crush profiles of involved vehicles were measured, internal vehicle intrusions and contact marks were identified, vehicle interior and exterior damage was photographed and detailed information and injuries for the pregnant occupant and fetus were obtained from medical records. Restraint use was determined through subject interviews and examination of physical evidence during the vehicle inspection, such as markings on the belt webbing or D-ring that indicate loading in a crash. Crash severity was estimated by entering vehicle crush measurements into the Winsmash[16] crash-reconstruction program, which calculates the estimated change in velocity (ΔV) or equivalent barrier speed (EBS) of the case vehicle during the impact event. Because crashes with ΔV or EBS greater than 48 km/h are more severe than 95% of all crashes, crashes were classified as severe if the ΔV or EBS was greater than 48 km/h, moderate for ΔV or EBS of 24 to 48 km/h, and minor for ΔV or EBS less than 24 km/h.

Fetal outcome was evaluated within 1 month after birth, and did not include a longer-term evaluation. In some analyses, fetal outcomes were prospectively defined into separate outcome categories, including good, minor complications, major complications, or fetal loss. Minor complications include early contractions or delivery within 48 hours of the crash at a gestational age of at least 32 weeks. Major complications include abruptio placenta (diagnosed by delivering care provider), uterine rupture or laceration, any direct fetal injury, or premature delivery within 48 hours of the crash before 32 weeks' gestational age. In some analyses, fetal outcome was also prospectively defined and grouped into binary categories of either acceptable (good or minor complications) or adverse (major complications or fetal loss). Maternal injury was categorized by using the injury severity score (ISS),[17] excluding injuries to the placenta or uterus. A no-injury rating was assigned to ISS of 0, ISS from

*The rights, welfare, and informed consent of the volunteer subjects who participated in this study were observed under guidelines established by the US Department of Health and Human Services on Protection of Human Subjects and accomplished under medical research design protocol standards approved by the Committee to Review Grants for Clinical Research and Investigation Involving Human Beings, Medical School, The University of Michigan.

1-9 were designated minor, ISS from 10-20 were designated moderate, and a ISS greater than 20 designated as major. A fifth maternal injury category used was maternal death, regardless of ISS.

Chi-square tests were performed to determine whether fetal outcome and the various crash, occupant, and restraint factors are independent. For factors with statistically significant effects on fetal outcome, the pattern of the relationship was determined by inspection of standardized residuals.

For variables with a statistically significant relationship to fetal outcome, risk models were developed by using logistic regression routines in SPSS. For this analysis, restraint was grouped into binary categories of (1) proper (3-point belt or 3-point belt plus airbag) or (2) improper (unrestrained, airbag only, and shoulder belt only, with and without airbag.) None of the 57 cases involved the use of only a lap belt. According to standard practice in crash database analysis, this analysis was limited to frontal (n = 33) and near-side crashes (n = 15) because the number of rear (n = 5) and far-side impacts (n = 3) was small.

Case-control techniques were used to account for the oversampling of severe crashes in the dataset. The National Automotive Sampling System,[18] a stratified sample of crashes occurring in the United States, was used to estimate the number of pregnant women involved in frontal and near-side motor-vehicle crashes each year. This value was used with Weiss' estimate of 369 fetal deaths from motor-vehicle crashes per year[1] to estimate the frequency of fetal loss in motor-vehicle crashes. This estimate was used with case-control techniques[19] to adjust the intercept of the logistic regression models to estimate the risk of adverse fetal outcome as a function of crash severity and maternal restraint use. These data were also used to estimate the number of fetal losses that could be prevented by belt use.

FIGURE 1
Distribution of cases

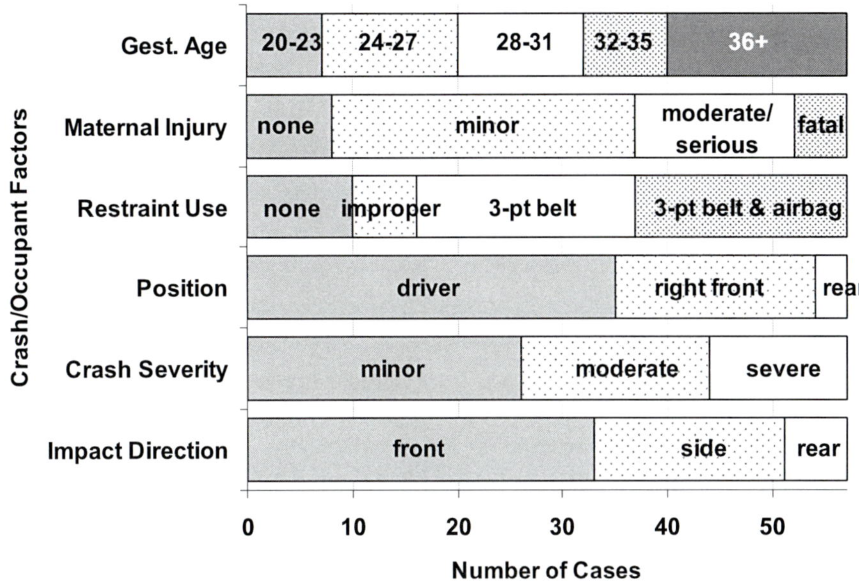

Distribution of cases by key crash and occupant factors.
Klinich. Fetal outcome in motor-vehicle crashes. Am J Obstet Gynecol 2008.

RESULTS
Overview of crashes

Figure 1 shows the distributions of key independent variables in the 57 cases. The Appendix summarizes these variables and the fetal outcome for each case, grouped by impact direction, crash severity, and maternal restraint use because comparisons of fetal outcome according to maternal restraint use are most appropriately made among crashes with similar characteristics.

Twelve cases resulted in fetal loss. Four of these occurred in severe frontal impacts and 1 was in a moderate frontal impact (cases 2, 3, 4, 7, and 8, respectively). AP is thought to be the cause of fetal death in all 5 of these cases. Four of the remaining fetal losses (cases 38, 40, 43, and 44) resulted from maternal death in moderate-to-severe near-side impacts (vehicle struck on the side where the pregnant occupant is seated); 1 of these fetuses also sustained a fractured humerus. Another fetal loss from AP occurred to a seriously injured mother in a severe near-side impact (case 36). One fetal loss from maternal death occurred in a far-side impact (impact to the side opposite to where the occupant is seated) when the unbelted pregnant driver sustained fatal head injuries from contact with the vehicle interior on the passenger side (case 34). The remaining fetal loss (case 55) occurred when the fetus died in utero within 6 hours after a minor rear impact for unknown reasons without any signs of AP or direct fetal injury.

In 11 cases, the fetus survived but experienced major complications. In 1 minor frontal crash (case 22), the pregnant right-front passenger was improperly restrained by only a shoulder belt and exsanguinated from cardiac and splenic injuries. The infant was born by emergency cesarean delivery at 28 weeks and survived, although it suffered from respiratory distress syndrome (long-term outcome is not available). In 3 cases (1, 3, and 46), the fetus sustained in utero head injury as the primary adverse outcome. One of these cases (1) involved a pregnant, unrestrained rear-seat passenger who was lying in a prone position across the seat at the time of the crash. The fetus sustained a trauma-induced hydrocephaly. The other 7 cases with major fetal complications resulted in AP and/or uterine laceration with fetal survival (cases 5, 12, 31, 35, 39, 52, and 53).

All of the remaining 34 cases had acceptable fetal outcomes. One case (10) resulted in delivery of a healthy infant at 40 weeks on the day of the crash, 1 case led to a successful emergency caesarean delivery immediately after the crash (case 50), and 7 cases led to preterm con-

TABLE

Logistic regression models used to evaluate crash severity, maternal injury, and maternal restraint as predictors of fetal outcome

Model	Variable	−2 log (LR)	*P*-value	Odds ratio [95% CI]
1	Crash severity	43.341	<.001	1.097[a] [1.045-1.165]
2	Proper restraint	59.592	.009	4.545 [1.127-18.324]
3	Maternal injury	49.549	.002	3.002 [1.601-7.927]
4	Crash severity	39.491	.006	1.101 [1.028-1.179]
	Proper restraint		.049	9.257[1.006-85.195]
	Maternal injury		.266	1.624 [0.691-3.813]
5	Crash severity	36.767	<.001	1.120 [1.050-1.195]
	Proper restraint		.020	12.046 [1.485-97.658]

[a] Odds ratios for crash severity are for an increase of crash severity of 1 km/h.

Klinich. Fetal outcome in motor-vehicle crashes. Am J Obstet Gynecol 2008.

tractions that did not result in delivery. One case with a good fetal outcome (case 42) involved a moderate near-side impact to a pregnant driver wearing a 3-point belt who sustained multiple pelvic fractures. At the time of the crash, the fetus was 20 weeks' gestation, and sustained no injuries despite the compromised maternal pelvic cavity. She ultimately had an uncomplicated spontaneous vaginal delivery at 38.5 weeks' gestation without evidence of neonatal injury.

Of 41 properly restrained pregnant occupants, 12 (29%) experienced adverse fetal outcomes. Of 6 improperly restrained women, 3 (50%) had adverse fetal outcomes. For unrestrained women, 8 of 10 (80%) had adverse fetal outcomes. The only 2 acceptable fetal outcomes to unrestrained women were in crashes of minor severity. Overall, 11 of the 13 severe crashes resulted in poor fetal outcomes, regardless of restraint type.

Analysis of crashes

For this study, χ^2 tests show that fetal outcome is significantly associated with crash severity (P = .005), such that minor crashes are more likely to result in acceptable fetal outcomes, whereas severe crashes are more likely to produce adverse fetal outcomes. However, there are no significant relationships between fetal outcome and impact direction (P = .782) or fetal outcome and the position of the pregnant occupant in the vehicle (P = .251).

A χ^2 test also shows that fetal outcome is strongly associated with the severity of maternal injury (P = .001), with more serious fetal outcomes associated with more severe maternal injury. In this study, gestational age (grouped by 4-week increments) has a statistically significant effect on fetal outcome (P = .009), with the eighth month of pregnancy having a higher frequency of major complications than expected, and the fifth month having more good fetal outcomes than expected.

For analysis by maternal restraint use, 2 categories of proper and improper restraint were used because of the small number of each type of improper restraint. Proper restraint use is significant (P = .006) with unrestrained occupants less likely to have acceptable fetal outcomes and more likely to experience major complications. Women restrained by 3-point belts or 3-point belts plus airbags are less likely to have major complications than expected statistically. Both proper and improper restraint categories had close to the statistically expected number of fetal losses.

The most significant findings from the χ^2 tests are that fetal outcome is related to crash severity, maternal injury, and maternal restraint. Because these variables are not independent, logistic regression was used to analyze the combined contributions of these predictors to fetal outcome. The table shows 5 different models used to assess fetal outcome. Models 1, 2, and 3 show that crash severity, proper restraint, and maternal injury are all independently good predictors of fetal outcome. However, when all 3 are used as predictors simultaneously in model 4, maternal injury becomes insignificant, because maternal injury is also a function of crash severity (P = .001, odds ratio = 1.066 [1.027-1.105]). Model V predicts probability of adverse fetal outcome as a function of crash severity and proper restraint. (The odds ratio for crash severity indicates the increase in risk per km/h of crash severity; an increase in crash severity of 10 km/h corresponds to an odds ratio of 2.47).

Figure 2 shows the results of model 5 after correcting the intercept in the logist function by using estimates of fetal loss from motor-vehicle crashes (population fetal loss rate 2.5% vs 21% in the current study). At a given crash severity, the risk of adverse fetal outcome for improperly restrained pregnant occupants is higher than that for properly restrained pregnant occupants. For example, at a crash severity of 30 km/h, the risk of adverse fetal outcome is about 12% for properly restrained pregnant occupants and about 70% for improperly restrained pregnant occupants.

These risk curves indicate that an 84% reduction in risk of adverse fetal outcome is obtained by properly wearing a seatbelt. On the basis of this relative risk and an overall belt use rate of 80%, unbelted pregnant occupants sustain an estimated 62% of all fetal losses in motor-vehicle crashes. Considering the annual estimate of 369 fetal losses, increasing the belt use rate of pregnant occupants from 80% to 100% would prevent approximately 192 fetal losses.

COMMENT

The database established in this study is the largest collection of motor-vehicle crashes involving pregnant women that includes detailed quantitative information about both the crash event and fetal outcome. Unlike the majority of case studies presented in the literature, crashes with both positive and negative fetal outcomes were investigated. In addition, the 72% belt usage rate in the sample population of pregnant occu-

FIGURE 2
Risk of adverse fetal outcome by crash severity

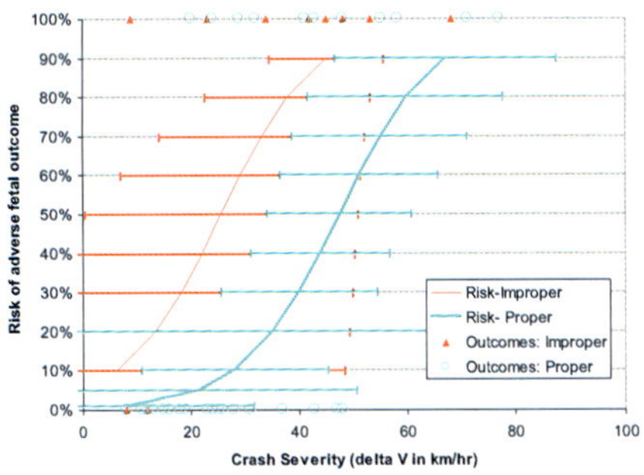

Risk of adverse fetal outcome by crash severity for properly restrained (light) and improperly restrained (dark) pregnant occupants based on logistic regression analysis. Each open circle represents a properly restrained case, whereas each filled triangle represents an improperly restrained case. Data points along the top of the graph (value = 100%) correspond to adverse outcome, whereas data points along the bottom of the graph (value = 0%) represent acceptable (ie, nonadverse) outcomes. Horizontal error bars indicate a 95% confidence interval.

Klinich. Fetal outcome in motor-vehicle crashes. Am J Obstet Gynecol 2008.

pants approaches current belt-restraint usage rates of 82% in the United States.[20] The sample also includes a greater number of pregnant women involved with airbag deployments than previously reported.[21]

Crash severity is the factor most strongly associated with fetal outcome in this study. Some case studies in the literature have suggested that belt or airbag restraints cause adverse fetal outcomes, but have not included information on crash severity.[7,11,22] Claims that restraints cause adverse fetal outcomes cannot be substantiated without reliable information on crash severity. This study also suggests that restraint use protects the fetus by protecting the mother, because maternal injury is predictive of fetal outcome, and proper restraint use reduces maternal injury severity.

The current study has several limitations. Rollovers were excluded because it is difficult to accurately assess belt use during in-depth investigations of rollovers.[23] Although the exclusion of rollover crashes could create a bias in assessment of restraint effects, rollovers account for only 2% of all crashes annually in the United States.[24] The effect of this exclusion is therefore expected to have minimal impact on the study findings. About one-third of the identified pregnant women in crashes declined to participate in the study, creating a potential for selection bias. However, a review of the investigated and excluded crashes, though limited by estimated and missing information in the excluded crashes, suggests that the distributions of key independent variables (listed in Figure 1) are similar for the included and excluded cases.

The limited size of the database, and the nonrandom sampling process are likely the source of some unexpected findings and limit the extent to which conclusions should be applied to the general population. The finding that crash direction is not associated with fetal outcome is unexpected, because the likelihood, severity, and patterns of injury for nonpregnant occupants have been shown to depend on impact direction.[25] This result suggests that the limited number of cases in the study probably does not include a representative selection of side and rear impacts.

In addition, the database is biased toward severe crashes and subsequently has a greater proportion of adverse fetal outcomes than expected in the general population. Less than 1% of all real-world crashes have severities greater than 48 km/h (personal correspondence, Compton, C. University of Michigan Transportation Research Institute, May, 2007), whereas 18.6% of the crashes in this database have severities greater than 48 km/h. However, the method of using case-control techniques to adjust the logistic regression intercept to account for the oversampling of severe crashes can aid in resolving the sampling issue. Breslow[19] has stated that even if a sample is biased toward more severe cases, logistic regression produces unbiased estimates of the odds ratios for predictors. With a sample biased toward more severe cases, the intercept will be biased upward and absolute risk will be overestimated, but relative odds will be unbiased. However, the use of an outside estimate of overall population risk can be used to adjust the logist function intercept and correct the absolute risks.

Within these recognized limitations, it was important to evaluate the effect of airbag deployment on fetal and maternal outcome, because widespread use of this restraint is relatively new. Although limited by the small number of cases, the data from this study suggest that restraint by an airbag plus a 3-point belt leads to fetal outcomes that are as good as restraint only by a 3-point belt. The important point of this finding is that airbags do not appear to worsen fetal outcomes.

The risk of adverse fetal outcomes estimated from the data are higher than, but consistent with, estimated fetal loss rates reported in the literature of 1% to 2% in minor severity crashes[26-28] and 30% to 50%[29-31] in severe crashes. Assuming that the qualitative definitions of minor and severe crashes in the literature are similar to the quantified definitions used in this study, the risk values in the current study include all adverse fetal outcomes, including but not limited to fetal loss from AP, whereas some of the values from the literature are based only on fetal losses resulting from AP. By using the results shown in Figure 2, the risk of adverse fetal outcome for properly restrained pregnant occupants is less than 10% in minor crashes and more than 60% in severe crashes.

A better understanding of the factors contributing to fetal loss in motor-vehicle crashes will help guide medical practitioners in providing appropriate advice to pregnant patients on the proper use of vehicle restraints. Currently, the main reason for recommending that pregnant occupants use three-point belts is that protecting the mother is the first step to protecting the fetus, and that using seat belts significantly reduces injury risk for the general vehicle-occupant population.[32] However, the practice of recommending that pregnant patients be instructed to properly use 3-point belts[33] has been questioned by some practitioners because the medical literature contains isolated cases of fetal injuries allegedly caused by seat belt loading in crashes.[34,35] Also, questions regarding the safety of deploying airbags for pregnant women have been raised. This study shows that proper use of a belt restraint by the pregnant occupant has a significant, positive effect on fetal outcome. Seventy-nine percent of the pregnant women who were properly wearing a 3-point belt, with or without airbag deployment, had acceptable fetal and maternal outcomes in lower severity crashes. An estimated 192 fetal losses could be prevented each year if all pregnant women properly wore seatbelts. The strong association between maternal injury and fetal outcome, particularly for the 4 cases of maternal death to unrestrained or improperly restrained women, supports previous observations that protecting the mother is the first step in protecting the fetus.[36] The results or this study support the current recommendation that pregnant women should properly wear 3-point belts. ■

ACKNOWLEDGMENT

Jamie Moore, who conducted most of the crash investigations.

REFERENCES

1. Weiss HB, Songer TJ, Fabio A. Fetal deaths related to maternal injury. JAMA 2001;286 (15):1863-8.
2. Minino A, Heron M, Smith B. Deaths: preliminary data for 2004. Natl Vital Stat Rep 2006;54(19):1-49.
3. Baethman M, Kahn T, Lenard H-G, Voit T. Fetal CNS damage after exposure to maternal trauma during pregnancy. Acta Paediatr 1996;85(11):1331-8.
4. Weyerts LK, Jones MC, James HE. Paraplegia and congenital contractures as a consequence of intrauterine trauma. Am J Med Gen 1992;43(4):751-2.
5. Kettel LM, Branch DW, Scott JR. Occult placental abruption after maternal trauma. Obstet Gynecol 1988;71(3 pt 2):449-53.
6. Pearlman M, Tintinalli JE, Lorenz RP. Blunt trauma during pregnancy. N Engl J Med 1990;323:1609-13.
7. Pepperell RJ, Rubinstein E, MacIsaac IA. Motor-car accidents during pregnancy. Med J Aust 1977;1(7):203-5.
8. Lane PL. Traumatic fetal deaths. J Emerg Med 1989;7(5):433-5.
9. Wolf ME, Alexander BH, Rivara FP, Hickok DE, Maier RV, Starzyk PM. A retrospective cohort study of seatbelt use and pregnancy outcome after a motor vehicle crash. J Trauma 1993;34(1):116-9.
10. Hartl R, Ko K. In utero skull fracture: case report. J Trauma 1996;41(3):549-52.
11. Van Enk A, Van Zwam W. Uterine rupture: a seat belt hazard. Acta Obstetr Gynecol Scand 1994;73(5):432-3.
12. Pearlman M, Klinich K, Schneider LW, Rupp J, Moss S, Ashton-Miller J. A comprehensive program to improve safety for pregnant women and fetuses in motor vehicle crashes: a preliminary report. Am J Obstet Gynecol 2000;182(6):1554-64.
13. Pearlman M, Phillips ME. Safety belt use during pregnancy. Obstet Gynecol 1996; 88(6):1026-9.
14. McGwin G Jr, Russel SR, Rux RI, Leath CA, Valent F, Rue LW. Knowledge, beliefs, and practices concerning seat belt use during pregnancy. J Trauma 2004;56(3):670-5.
15. McGwin G Jr, Willey P, Ware A, Kohler C, Kirby T, Rue LW 3rd. A focused educational intervention can promote the proper application of seat belts during pregnancy. J Trauma 2004; 56(5):1016-21.
16. WinSmash. Release 2.06x. Washington, DC:NASS; 1999.
17. Baker SP, O'Neill B, Waddon W Jr, Long WB. The injury severity score: a method for describing patients with multiple injuries and evaluating care systems. J Trauma 1974;14:186-7.
18. Weiss HB, Strotmeyer S. Characteristics of pregnant women in motor vehicle crashes. Inj Prev 8(3):207–10.
19. Breslow NE. Statistics in epidemiology: the case-control study. J Am Stat Assoc 1996; 91(433):14-28.
20. Traffic Safety Facts 2005. Washington, DC: National Highway Traffic Safety Administration, NHTSA; 2005. Report No.: DOT HS 810 623.
21. Sims CJ, Boardman CH, Fuller SJ. Airbag deployment following a motor vehicle accident in pregnancy. Obstet Gynecol 1996;88(4 pt 2): 726.
22. Schultze PM, Stamm CA, Roger J. Placental abruption and fetal death with airbag deployment in a motor vehicle accident. Obstet Gynecol 1998;92(4):719.
23. Robertson L. Further reflections on the seatbelt use and effectiveness issue. Inj Prev 2003;(9):284-5.
24. Traffic Safety Facts 1998. Washington, DC: National Highway Traffic Safety Administration, NHTSA; 1988. Report No.: DOT HS 808 983.
25. Thomas C, Henry C, Hartemann F, Patel A, Got C. Injury pattern and parameters to assess severity for occupants involved in car-to-car lateral impacts. Proceedings of the 11th International Conference for the Enhanced Safety of Vehicles; 1988; Washington, DC.
26. Dahmus MA, Sibai BM. Blunt abdominal trauma: are there any predictive factors for abruptio placentae or maternal-fetal distress? Am J Obstet Gynecol 1993;169(4):1054-8.
27. Pearlman MD, Tintinalli JE, Lorenz RP. A prospective controlled study of outcome after trauma during pregnancy. Am J Obstet Gynecol 1990;(162):1502-10.
28. Goodwin TM, Breen MT. Pregnancy outcome and fetomaternal hemorrhage after noncatastrophic trauma. Am J Obstet Gynecol 1990;162(3):665-71.
29. Esposito TJ, Gens DR, Smith LG, Scorpio R, Buchman T. Trauma during pregnancy: a review of 79 cases. Arch Surg 1991;(126): 1073-8.
30. Hoff WS, D'Amelio LF, Tinkoff GK, Lucke JF, Rhodes M, Diamond DL. Maternal predictors of fetal demise in trauma during pregnancy. Surg Gynecol Obstet 1991;172(3):175-80.
31. Rothenberger D, Quattlebaum FW, Perry JF Jr, Zabel J, Fischer RP. Blunt maternal trauma: a review of 103 cases. J Trauma 1978;18 (3):173-9.
32. Viano DC. Restraint effectiveness, availability and use in fatal crashes: implications to injury control. J Trauma 1995;38(4):538-46.
33. American College of Obstetricans and Gynecologists. Obstetric aspects of trauma management. ACOG Educational Bulletin 1998: 251.
34. Astarita DC, Feldman B. Seat belt placement resulting in uterine rupture. J Trauma 1997;42(4):738-40.
35. Chetcuti P, Levene M. Seat belts: a potential hazard to the fetus. J Perinat Med 1987;15(2):207-9.
36. Crosby WM, Costilow JP. Safety of lap-belt restraint for pregnant victims of automobile collisions. N Engl J Med 1971;284:632-6.

APPENDIX

The cases are grouped by impact direction, impact severity, and restraint use. Impact severity is categorized as minor (<24 km/h), moderate (24-48 km/h), or severe (>48 km/h). Gestational age is grouped into 4-week intervals to allow for comparisons based on anatomical differences.

CHARACTERISTICS AND FETAL OUTCOMES OF CRASHES INVOLVING PREGNANT OCCUPANTS

Case	Impact direction	Impact severity	Occupant restraint	Occupant position	Gest. age (wk)	Maternal injuries	Fetal outcome
1	Front	Severe	None	Left rear	28-31	Moderate	Fetal hydrocephalus from intraventricular hemorrhage, probably from crash
2	Front	Severe	Shoulder belt and airbag	Driver	36+	Moderate	Fetal loss, placental abruption
3	Front	Severe	3-pt belt	Driver	28-31	Minor	Fetal loss
4	Front	Severe	3-pt belt	Driver	36+	Moderate	Fetal loss, placental abruption
5	Front	Severe	3-pt belt and airbag	Driver	24-27	Moderate	Placental abruption, emergency C-section, 27-week premature delivery
6	Front	Severe	3-pt belt and airbag	Driver	28-31	Major	No problem
7	Front	Severe	3-pt belt and airbag	Driver	28-31	Moderate	Fetal loss, placental abruption
8	Front	Moderate	3-pt belt	Driver	24-27	None	Fetal loss, placental abruption, early delivery
9	Front	Moderate	3-pt belt	Right front	36+	None	Contractions stopped without intervention
10	Front	Moderate	3-pt belt	Right front	36+	Minor	Delivery within 48 hours
11	Front	Moderate	3-pt belt and airbag	Right front	24-27	Minor	No problem
12	Front	Moderate	3-pt belt and airbag	Driver	24-27	Minor	Premature birth, respiratory distress, placental abruption, apnea, intraventricular hemorrhage
13	Front	Moderate	3-pt belt and airbag	Right front	28-31	Moderate	Fetal head injury, probably from crash
14	Front	Moderate	3-pt belt and airbag	Driver	32-35	Moderate	Premature labor requiring intervention
15	Front	Moderate	3-pt belt and airbag	Right front	36+	Minor	No problems
16	Front	Moderate	3-pt belt and airbag	Driver	36+	Minor	No problems
17	Front	Moderate	3-pt belt and airbag	Driver	36+	Minor	No problems
18	Front	Minor	None	Driver	20-23	Minor	No problems
19	Front	Minor	None	Right front	36+	Minor	No problems
20	Front	Minor	Airbag	Driver	20-23	Minor	No problems
21	Front	Minor	Airbag	Driver	36+	Minor	Contractions stopped at hospital
22	Front	Minor	Shoulder belt	Right front	28-31	Major (fatal)	Premature birth, fetal respiratory distress

(Cont'd)

Case	Impact direction	Impact severity	Occupant restraint	Occupant position	Gest. age (wk)	Maternal injuries	Fetal outcome
23	Front	Minor	3-pt belt	Driver	20-23	Minor	No problems
24	Front	Minor	3-pt belt	Driver	24-27	None	No problems
25	Front	Minor	3-pt belt	Driver	24-27	Minor	No problems
26	Front	Minor	3-pt belt	Rear pass	24-27	None	No problems
27	Front	Minor	3-pt belt and airbag	Driver	24-27	Minor	No problems
28	Front	Minor	3-pt belt and airbag	Driver	24-27	Minor	No problems
29	Front	Minor	3-pt belt and airbag	Driver	28-31	Minor	Contractions stopped without intervention
30	Front	Minor	3-pt belt and airbag	Right front	28-31	Minor	Contractions stopped without intervention
31	Front	Minor	3-pt belt and airbag	Driver	32-35	Minor	Uterine lacerations, placental abruption, fetal respiratory distress, early delivery
32	Front	Minor	3-pt belt and airbag	Driver	36+	Minor	No problems
33	Front	Minor	3-pt belt and airbag	Right front	36+	Minor	Contractions stopped at hospital
34	Side	Severe	None	Driver	20-23	Major (fatal)	Fetal loss (twins)
35	Side	Severe	Airbag	Driver	32-35	Moderate	Uterine laceration, emergency C-section at 33 weeks, premature delivery
36	Side	Severe	3-pt belt	Driver	24-27	Major	Fetal loss, placental abruption
37	Side	Severe	3-pt belt and airbag	Right front	36+	Minor	No problems
38	Side	Moderate	None	Right front	28-31	Major (fatal)	Fetal loss, fetal arm fracture
39	Side	Moderate	None	Right front	32-35	Major	Placental abruption, emergency C-section, premature delivery
40	Side	Moderate	None	Right front	36+	Moderate (fatal)	Fetal loss, placental abruption, emergency C-section
41	Side	Moderate	Airbag	Right front	20-23	Minor	No problems
42	Side	Moderate	3-pt belt	Driver	20-23	Moderate	No problems
43	Side	Moderate	3-pt belt	Right front	24-27	Major (fatal)	Fetal loss
44	Side	Moderate	3-pt belt	Right front	28-31	Major (fatal)	Fetal loss
45	Side	Moderate	3-pt belt and airbag	Driver	28-31	Major	No problems
46	Side	Minor	None	Right front	36+	Minor	Early delivery, fetal head injury
48	Side	Minor	3-pt belt	Driver	32-35	Minor	No problems
49	Side	Minor	3-pt belt	Driver	36+	None	No problems
50	Side	Minor	3-pt belt	Driver	36+	Moderate	Emergency C-section immediately after crash

(Cont'd)

Case	Impact direction	Impact severity	Occupant restraint	Occupant position	Gest. age (wk)	Maternal injuries	Fetal outcome
47	Side	Minor	3-pt belt	Right front	28-31	None	No problems
51	Side	Minor	3-pt belt and airbag	Driver	20-23	Minor	No problems
52	Rear	Severe	None	Right front	32-35	Minor	Placental abruption, early delivery
53	Rear	Severe	None	Rear pass	32-35	Minor	Placental abruption, early delivery, possible fetal head injury
54	Rear	Minor	3-pt belt	Driver	24-27	Minor	No problems
55	Rear	Minor	3-pt belt	Driver	24-27	None	Fetal loss
56	Rear	Minor	3-pt belt	Driver	32-35	None	No problems
57	Rear	Minor	3-pt belt	Driver	36+	Minor	Contractions stopped without intervention

Uterine Trauma in Pregnancy After Motor Vehicle Crashes with Airbag Deployment: A 30-Case Series

Torri D. Metz, MD, and Jean T. Abbott, MD, MH

Background: We describe a cohort of women of at least 20 weeks gestation to determine the frequency of placental abruption or fetal compromise after a motor vehicle crash (MVC) with airbag deployment.

Methods: A retrospective chart review was conducted at two urban emergency departments. All women of at least 20 weeks gestation with MVCs in which their airbag deployed and who were admitted for monitoring or trauma care between 1997 and 2002 were identified by the Trauma Registry for each hospital and included.

Results: Thirty women in the 6-year period were identified. Gestational age (GA) at the time of the MVC ranged from 20 weeks to 37 weeks (median GA 29 weeks). Of the 30 women, 15 (50%) were belted drivers, 5 (17%) were unbelted drivers, 5 (17%) were belted front-seat passengers, and 5 (17%) were unbelted front-seat passengers. Median speed was 35 mph.

One woman experienced placental abruption with subsequent intrauterine fetal demise. Twenty-two (73%, 95% CI: 58–89) experienced uterine contractions. Sixteen (53%, 95% CI: 36–71) reported abdominal pain. Abnormal fetal heart rate was documented in six patients (20%, 95% CI: 6–34). Vaginal bleeding was reported in two patients (7%). Only three women had no abnormal abdominal findings after the MVC.

Conclusions: From this preliminary series, placental abruption does not appear to occur frequently in MVCs with airbag deployment. It is unclear if fetal risk is higher than that without airbags. A significant number of pregnant women do not wear seatbelts.

Key Words: Pregnancy, Airbags, Motor vehicle crashes.

J Trauma. 2006;61:658–661.

Airbags are a relatively new method of passenger restraint. Concerns for injury/mortality caused by airbags have been raised for people of short stature who are positioned too close to the airbag, including the elderly, children less than 12 years of age, and infants in rear-facing car seats. This also raises theoretical concerns for pregnant women, because insufficient distance may exist between the steering wheel or dash and the gravid abdomen to avoid traumatic impact as the airbag expands rapidly.[1] In 1997, the National Highway Traffic Safety Administration (NHTSA) issued their most recent statement which recognized the potential risk of airbags in pregnancy, but recommended that the benefits of airbags outweigh the risks for pregnant women, based on current data. They also recommended that airbags be disconnected if vehicle occupants cannot position themselves with their sternum (or uterine fundus) at least 10 inches back from the center of the airbag cover.[1]

The current literature questions the possibility of an association between airbag deployment and obstetrical complications such as placental abruption, uterine rupture, or direct fetal trauma, but consists primarily of case reports.[1–3] One case report of placental abruption after airbag deployment,[2] and one case of uterine rupture have been reported.[3] The largest published case series to date is a three-patient series in which none of the pregnant women had any obstetrical complications after motor vehicle crashes (MVCs) with airbag deployment.[4] The authors of these studies support further research on the subject to determine the safety and efficacy of airbags for pregnant women.

The purpose of this study was to review a consecutive cohort of pregnant women to determine the frequency of placental abruption or other fetal compromise in women of at least 20 weeks gestation in MVCs with airbag deployment.

PATIENTS AND METHODS

This study was a six-year retrospective chart review from 1997 through 2002 of two teaching hospital emergency departments (EDs) (Level I and Level II). Participants were identified by the Trauma Registry at the individual hospitals. Patients pregnant at less than 20 weeks gestation were not predictably admitted to monitor for fetal health after trauma at the two hospitals, and were excluded from the series. Forty-four patients were identified with the original Trauma Registry query. Upon review of these medical records, nine did not have an airbag deployed, two were passengers in a crash without deployment at their position in the car, and two were at less than 20 weeks gestation. One additional patient was excluded because the reports in the ED record and the obstetrics (OB) consult were discrepant in determining whether or not an airbag was deployed. Ultimately, 30 preg-

Submitted for publication November 11, 2004.

Accepted for publication January 12, 2006.

From the Departments of Obstetrics and Gynecology (T.D.M.) and Surgery (J.T.A.), University of Colorado Health Sciences Center, University Hospital and Denver Health Medical Center, Denver, Colorado.

Abstract was presented at the Society for Academic Emergency Medicine Annual Meeting, Orlando, Florida, May 2004.

Address for reprints: Jean T. Abbott, MD, University of Colorado Health Sciences Center, 4200 East Ninth Avenue, Campus Box 215, Denver, CO, 80262; email: jean.abbott@uchsc.edu.

DOI: 10.1097/01.ta.0000209599.76059.06

nant women of at least 20 weeks gestation who were involved in MVCs with airbag deployment were entered into the study.

A standardized data collection tool was utilized to extract demographic data, mechanism of injury, obstetrical and non-obstetrical injuries/complications, length of stay, birth data, and signs and symptoms of placental abruption, including: uterine contractions, vaginal bleeding, abdominal pain, and abnormal fetal heart rate. The ED housestaff and faculty record, the obstetrical and surgical consults, the paramedic record, and the in-patient obstetrics log were all reviewed. If the investigated signs or symptoms were mentioned in any of these records, they were considered positive for the purposes of the study. All reported speeds in each patient record were averaged to determine the speed of the MVC. Although the total or net impact velocities involved would be a more accurate indication of forces to the uterus, these could not be easily determined from the Trauma Registry documentation. Gestational age was determined by the dating in the obstetrical consult note based on last menstrual period, prior ultrasound, or ultrasound at the hospital after the MVC.

A single author reviewed each chart and abstracted data on the structured worksheet. Ten charts were reviewed by a second investigator to assess interrater variability. Agreement was 90% or better, and the kappa was 0.7 to 1.0, for nine of ten key variables reviewed by both authors; the exception was the variable of frontal impact ($K = 0.33$). The chart abstractors were not blinded to the study purpose.

Statistical analysis of the data (mean, median, frequencies, duration) was performed using Microsoft Excel. Study approval was obtained from the Colorado Multiple Institutional Review Board.

RESULTS

Of the 30 women in the study, 12 (40%) were white, 9 (30%) were Hispanic, 2 (7%) were black, 1 (3%) was Asian, and 6 (20%) were of unknown ethnicity. The gestational age of the fetus ranged from 20 to 37 weeks with a median gestational age of 29 weeks. There was one identified case of placental abruption; fetal death occurred in this case.

In the study population, 15 (50%) of the participants were restrained drivers, 5 (17%) were unrestrained drivers, 5 (17%) were restrained front seat passengers and 5 (17%) were unrestrained front seat passengers. All had an airbag deployed at their position in the vehicle. None of the participants were ejected from the vehicle. The speed of the crash was not recorded in 9 (30%) of the cases. In the remaining cases, the speed of the crash ranged from 5 to 65 mph with a median speed of 35 mph. At least 12 (40%) of the MVCs occurred at a speed greater than 30 mph. Twenty-three (77%) of the cases involved frontal impact. Maternal vital signs upon arrival to the ED were abnormal (systolic blood pressure >180 or <90, or heart rate >100 or <60) in 10 cases (33%). Thirteen (43%) participants received a trauma diagnosis other than "rule-out blunt abdominal trauma." These diagnoses are recorded in Table 1.

Table 1 **Trauma Diagnoses Other Than "Rule-out Blunt Abdominal Trauma"**

Diagnosis	No. of Patients*
Abrasion/contusion/laceration	5
Upper extremity fracture	4
Neck strain	3
Abdominal wall abrasion/contusion	2
Closed head injury	2
None	17

* Several patients had >1 injury.

All of the women in this study were admitted to the hospital for further observation. Most (28) were admitted to the Labor and Delivery floor under the supervision of both the obstetrical and surgical/trauma staff, to optimize continuous fetal monitoring. One of the women went directly to the operating room (for treatment of an open fracture) upon completion of the ED evaluation, and one went to a surgical floor. Length of stay (LOS) from arrival in the ED to discharge from the hospital ranged from 4.8 to 61.0 hours with a median total LOS of 23.9 hours.

The records of the women in the study were reviewed for signs and symptoms of placental abruption, including: vaginal bleeding, abdominal pain, abnormal fetal heart tones, and uterine contractions. These results are listed in Table 2. Twenty-two (73%) women were noted to have uterine contractions during Labor and Delivery monitoring. In one case, these were subjective only. One woman received tocolytics.

Uterine sonography was performed by an obstetrician in all cases. The placenta was anterior in 13 (43%) cases, posterior in 7 (23%), lateral in 7 (23%), and fundal in 8 (27%). There were also two (7%) cases of placenta previa. The placenta could not be visualized in one woman. A retroplacental clot was not seen in any of the women, including the patient with placental abruption. A trauma ultrasound was also performed by either the ED physician or surgical staff in 26 (87%) of the cases. There were no reported cases of free fluid in the abdomen.

The single case of placental abruption was confirmed by pathology report after delivery of the dead fetus and placenta. The fetus was at a gestational age of 28 weeks. This woman was a restrained driver in a vehicle traveling at 50 mph at the time of the crash. There was frontal impact. The mother had an anterior placenta by ultrasound in the emergency department. No retroplacental clot was identified by ultrasound.

Table 2 **Presence of Signs/Symptoms of Placental Abruption in Study Population**

Sign	No. of Patients	95% CI
Uterine contractions	22 (73%)	58–89
Abdominal pain	16 (53%)	36–71
Abnormal fetal heart tones	6 (20%)	6–34
Vaginal bleeding	2 (7%)	0–16
None	3 (10%)	0–21

The patient never complained of abdominal pain or vaginal bleeding. Fetal heart tones of 160 were noted upon arrival to the ED. No maternal injuries were found. The status of the fetus was unclear during the ED evaluation; but, upon admission to Labor and Delivery 4 hours later, there was no fetal cardiac activity. The dead fetus was delivered vaginally the following day. This case has been reported previously.[2]

Information about the subsequent delivery of the fetus after discharge was only available for 9 (30%) of the women. The other women delivered at other area hospitals and outcome was not available. In the 9 (30%) cases where delivery information was obtained, there were 7 (78%) vaginal deliveries, 1 (11%) nonemergent cesarean section, and 1 (11%) emergent cesarean section (c-section). The two c-sections were performed for indications unrelated to the prior MVCs. One patient with trauma at 33 weeks gestation had a c-section 3 weeks later for a compound presentation precluding vaginal delivery, and the other, seen at 32 weeks for trauma, had operative delivery at 39 weeks for placenta previa. With the exception of the one case of intrauterine fetal demise at 28 weeks, all other known deliveries were at full-term and without complications.

DISCUSSION

Trauma accounts for 3.7 fetal deaths (after 20 weeks) per 100,000 live births, with 75% to 80% of injuries caused by MVCs. Abruptio placenta (AP) is the mechanism of fetal demise in almost half of these cases.[5,6] There has been, in the past, some controversy about the timing of fetal demise after AP and the length of time needed to exclude AP after injury. Two large recent studies have recently determined that AP is usually manifest early after a blow to the uterus, and cardiotocographic monitoring for 4 to 6 hours with no evidence of uterine irritability is sufficient to virtually exclude abruptio placenta after injury.[7,8] Based on these studies, a protocol of monitoring pregnant patients of at least 20 weeks gestation for at least 4 to 6 hours has been adopted in trauma centers throughout the country. Although one recent case-control study of 12 women diagnosed with abruptio placenta after MVCs indicated a higher incidence of preterm delivery, stillbirth, and lower mean birth weight,[9] these later adverse outcomes could not be measured in this study.

Seat belt use in pregnancy decreases maternal mortality and, therefore, protects the fetus, particularly if worn properly (low over the anterior pelvis with the shoulder strap between the breasts and above the uterine fundus).[1] However, two studies of pregnant women in the past 10 years indicate there is a substantial group of women, ranging from 25 to 33%, who do not wear restraints properly.[10,11] Likewise, in the current study, only two thirds of pregnant victims were wearing seat belts at the time of airbag deployment. Teaching about the safety and importance of seat belt use is still uncommon, and physician counseling during pregnancy was reported by only 37% of pregnant women in a recent survey.[11] In Colorado, a weak secondary-enforcement seat belt law (ticketing only if stopped for another traffic violation) has been in effect from 1998 to the present, and the rate of seat belt use is thought to be below national averages.

The installation of airbags in newer cars is designed to provide passive restraint of occupants in the event of a MVC. However, the safety and efficacy of airbag deployment in pregnant women has not been clearly demonstrated. Concerns about potential obstetrical complications, including placental abruption, uterine rupture, and direct fetal injury, have been raised.[1–3] This is higher in the second half of pregnancy, when the uterus is a leading point of impact and the baby is potentially viable. This study attempts to increase our understanding of the likelihood of placental abruption or other fetal compromise after a MVC with airbag deployment in pregnant women of at least 20 weeks gestation.

Placental abruption is a potential complication of airbag impact with the gravid abdomen because of the proximity of the gravid uterus to the rapidly and forcefully deploying airbag. Forces on the gravid uterus have been studied using pregnancy crash test dummies and computer modeling.[12,13] Pearlman et al. performed a study with crash test dummies which indicated a possible increased risk of placental abruption or direct fetal injury with airbag deployment when the dummy was out-of-position or unbelted.[12] The authors advocated further studies with modeling of forces on the uteroplacental interface.[12] A computer modeling study by Moorcroft et al. included an examination of the uterus, placenta, and amniotic fluid during simulated crashes at 30 weeks gestation at varying speeds.[13] The computer model did not consider airbag deployment in an unbelted passenger, but did conclude that the safest restraint for pregnant passengers appeared to be the use of both a three-point restraint and airbag.[13] This combination resulted in the lowest amount of uterine strain, and should therefore correlate with a lower incidence of placental abruption.[13] Their tests assumed a superiorly-located placenta and a distance of 38 mm between the steering wheel and uterus.[13]

In this case series of 30 women, only one woman had a placental abruption and subsequent intrauterine fetal demise. This would suggest a low frequency of placental abruption with airbag deployment in MVCs. Abruptio placenta is also well-described as a complication of steering wheel impact without restraints; however, further research is needed to determine the safety and efficacy of airbag deployment with pregnant women relative to risk without airbags. It may be that computer modeling is the most efficient and only way to answer this important safety question.

Although this is the largest case series to date, the study has several limitations. Because this is a small chart review-based study, we were unable to analyze confounding factors: including oncoming vehicle forces, gestational age, placental location, and type of seat belt. It is also retrospective, and there were no control cases to compare fetal risk with and without airbags. The abstractors were not blinded to the study purpose.

Our study was in a single city and relied on the Trauma Registry to identify patients to enroll in the study. By potentially missing eligible study participants that were not seriously injured, or did not present to the ED, we recognize a selection bias for more seriously injured patients, with increasing risk of adverse outcome. Although we have a small number of patients in our hospital system who go directly to the Labor Deck for evaluation or fetal concern (and this may occur after minor trauma), these are less severely injured and primarily are seen after non-MVC injuries. Our prehospital system does not allow release without EMS transport and ED evaluation of known pregnant patients after MVCs to which they respond. This bias should, if anything, have led to an increased risk of significant uterine trauma and enriched our sample.

The additional limitation is the poor long-term follow-up of our patient population. Large prospective studies have established the current standard of care for pregnant trauma victims by demonstrating that almost all adverse fetal outcome is detected by 4 hours of continuous monitoring after trauma.[7,8,14] In our cohort, the median length of stay was 23.9 hours, so fetal distress in virtually all cases should have been discovered before discharge. Although one recent study has suggested that long-term outcome may be affected by placental injury remote from delivery,[9] our methodology could only assess acute effects of airbag deployment on fetal well-being. Most importantly, because the study was retrospective, the relative risk of adverse fetal or maternal outcome (higher or lower) to women could not be determined. A larger, prospective study with controls would help to determine whether airbag deployment is a risk factor for obstetrical complications including placental abruption.

Although abruptio placenta is a theoretical concern for women and their fetuses in the second half of pregnancy involved in MVCs, it appears to be an infrequent occurrence in a case series of significant MVC cases. A considerable number of pregnant women still do not wear seatbelts. Further studies are needed to clarify the risk of airbags to the gravid uterus relative to the risk of three-point restraints or impact to the steering wheel or dashboard themselves.

REFERENCES

1. National Highway Traffic Safety Administration. National Conference on Medical Indications for Airbag Disconnection Conducted by The Ronald Reagan Institute of Emergency Medicine, Department of Emergency Medicine and The National Crash Analysis Center, The George Washington University Medical Center, Washington, DC, July 16–18, 1997. Available at: www.nhtsa.dot.gov. Accessed February 1, 2004.
2. Schultze PM, Stamm CA, Roger J. Placental abruption and fetal death with airbag deployment in a motor vehicle accident. *Obstet Gynecol.* 1998;92(4 Pt 2):719.
3. Fusco A, Kelly K, Winslow J. Uterine rupture in a motor vehicle crash with airbag deployment. *Trauma.* 2001;51:1192–1194.
4. Sims CJ, Boardman CH, Fuller SJ. Airbag deployment following a motor vehicle accident in pregnancy. *Obstet Gynecol.* 1996;88:726.
5. Pearlman MD. Motor vehicle crashes, pregnancy loss and preterm labor. *International J Gynec Obstet.* 1997;57:127–132.
6. Weiss HB, Songer TJ, Fabio A. Fetal deaths related to maternal injury. *JAMA.* 2001;286:1863–8.
7. Pearlman MD, Tintinalli JE, Lorenz RP. A prospective controlled study of outcome after trauma during pregnancy. *Am J Obstet Gynecol.* 1990;162:1502–1510.
8. Goodwin TM, Breen MT. Pregnancy outcome and fetomaternal hemorrhage after noncatastrophic trauma. *Am J Obstet Gynecol.* 1990;162:665–671.
9. Reis PM, Sander CM, Pearlman MD. Abruptio placentae after auto accidents. A case-control study. *J Reprod Med.* 2000;45:6–10.
10. Pearlman MD, Phillips ME. Safety belt use during pregnancy. *Obstet Gynecol.* 1996;88:1026–1029.
11. McGwin G, Russell SR, Rux FL, et al. Knowledge, beliefs, and practices concerning seat belt use during pregnancy. *J Trauma.* 2004;56:670–675.
12. Pearlman MD, Viano D. Automobile crash simulation with the first pregnant crash test dummy. *Am J Obstet Gynecol.* 1996;175: 977–981.
13. Moorcroft DM, Stitzel JD, Duma GG, et al. Computational model of the pregnant occupant: Predicting the risk of injury in automobile crashes. *Am J Obstet Gynecol.* 2003;189:540–544.
14. Pearlman MD. Motor vehicle crashes, pregnancy loss and preterm labor. *Int J Gynecol Obstet.* 1997;57:127–132.

CHAPTER 4

PREGNANT CRASH TEST DUMMIES AND PREGNANT OCCUPANT ANTHROPOMETRY

CHAPTER 4: PREGNANT CRASH TEST DUMMIES AND PREGNANT OCCUPANT ANTHROPOMETRY

Jonathan D. Rupp, Kathleen D. Klinich, and Lawrence W. Schneider
University of Michigan, University of Michigan Transportation Research Institute

BACKGROUND

Crash test dummies are a subset of a larger group of human surrogates called anthropomorphic test devices (ATDs) that are used to assess the ability of vehicle safety systems to mitigate the risk of injury to an occupant in a staged motor-vehicle crash. To successfully fill this role, they must do two things—interact with restraint systems and vehicle components in a humanlike manner and be able to assess the risk of injury to a human. The former of these requires that crash test dummies have humanlike size, shape, and inertial characteristics and that dummies move and deform in a humanlike manner when loaded in a motor-vehicle crash. The latter requires that crash test dummies have instrumentation that measures a mechanical response to applied loading that can be related to the probability of injury to a body region or body component (e.g., chest compression of the chest can be related to the probability of serious thoracic injury). In addition to these requirements, dummies must be durable and produce repeatable and reproducible responses.

Crash test dummies are generally designed for use in a specific crash type and to represent specific sizes of vehicle occupants. The first crash dummies were designed to represent midsize (nominally 50th percentile in stature and body mass based on survey data for the U.S. adult population) and large (95th percentile) males. Later, a small female crash test dummy that was ~5th percentile in stature and mass was developed. The rationale for the sizes of these crash test dummies is that, when combined, there is a crash test dummy representing each major component of the adult vehicle occupant population for which differences in size, shape, and mechanical response have the potential to result in significant increases in injury potential in a motor-vehicle crash.

In the early 1990s, the set of available adult crash-test dummies was expanded by the development of the first pregnant crash test dummy. This ATD was developed in response to data suggesting that fetal loss and adverse fetal outcomes resulting from motor-vehicle crashes were substantially more common than was expected and that existing ATDs were unable to evaluate the potential for seatbelts and airbags to protect the pregnant occupant and the fetus in frontal motor-vehicle crashes (Pearlman and Viano 1996, Viano et al. 1996). The first pregnant dummy was developed by modifying the small female Hybrid III crash test dummy (CFR 47 part 572, Subpart O) to add a new pelvis, a pregnant abdominal insert sized to match the abdominal contour at 30 weeks gestation reported by Culver and Viano (1990), and a chest jacket that contained ballast weights. Inside the abdominal insert were urethane rubber representations of the amniotic fluid and a vinyl fetus. Instrumentation in the abdomen included a transducer to measure force transmitted through the abdomen, as well as accelerometers that were attached to the fetal insert.

In response to deficiencies in this first pregnant dummy, including an unrealistic abdominal contour, overly stiff force-deflection response, and lack of criteria relating force transmitted through the abdomen and fetal acceleration to the likelihood of adverse fetal outcomes, a second-generation pregnant dummy was developed (Rupp et al. 2001a, b). This dummy is called the Maternal Anthropomorphic Measurement Apparatus Version 2B, or MAMA-2B. Like the first-generation pregnant dummy, the MAMA-2B was designed to represent a pregnant woman at 7 months gestation, and was implemented in the Hybrid III small-female crash test dummy. The rationale for this choice was that the Hybrid III small female was the only widely accepted female crash test dummy, and since small women sit closer to the steering wheel, they are therefore more likely to load the steering-wheel rim with their abdomens in a frontal crash (Klinich et al. 1999). The MAMA-2B employs a water-filled silicone rubber abdomen with force-deflection response characteristics that fall within corridors developed from tests of nonpregnant cadavers (Hardy et al. 2001a, b) that have been scaled to account for differences in response between pregnant and nonpregnant occupants.

The MAMA-2B abdomen is instrumented with internal pressure transducers attached to the anterior and posterior surfaces. Relationships between peak internal pressures measured by these transducers and the risk of injury to the pregnant abdomen were developed by reconstructing generic crashes in which the risk of adverse fetal outcome was known based on previously established relationships between crash severity (delta V) restraint use (belted/unbelted) and the risk of adverse fetal outcome (Klinich et al. 1999, Klinich et al. 2008).

Several papers included in this chapter describe the specific anthropometric, response, and injury criteria data used in the development of the MAMA-2B in further detail. Since the size, shape, and impact response of nonpregnant women of child-bearing age have been described elsewhere (e.g., Schneider et al. 1983, Mertz et al. 1989, Mertz 2002), this chapter only focuses on anthropometric factors and impact-response characteristics that differentiate pregnant women from their nonpregnant counterparts, and how this information has been used to develop pregnant crash test dummies for use in frontal impacts.

The following sections provide brief overviews of the papers contained in this chapter and the associated literature, as well as recommendations for the future of pregnant dummy development.

ANTHROPOMETRY

The first study to report on the size and shape of the pregnant abdomen in a seated posture was by Culver and Viano (1990), who scaled abdominal contour measurements from 5th percentile Japanese pregnant women reported by Yamana et al. (1984) to their larger 5th percentile American counterparts. Abdominal contours for American females who are 50th and 95th percentile in stature and weight were also developed using length-based scaling factors that were based on the ratio of the seated heights of 5th, 50th, and 95th percentile females. This study also reports data on fetal size and weight as a function of gestational age that was originally published by Kapit and Elson (1977). Results of this study were used to define the abdomen size and shape of the first-generation pregnant crash test dummy.

Klinich et al. (1999) measured the seated anthropometry of twenty-two pregnant women, at 3, 5, 7, and 9 months gestation in a laboratory vehicle buck in subject-selected postures and fore-aft seat position. In testing at each gestational age, shoulder belt and lap belt anchorage locations were varied to quantify the effects of changing lap and shoulder belt angles on belt fit and positioning with respect to the pregnant abdomen. Results of this study were used to define the size and shape of the pregnant abdomen in a seated posture. Importantly, results of this study demonstrate that maternal stature (and seated height) are unrelated to the size and shape of the pregnant abdomen and suggest that previous efforts to use scaling based on ratios of seated height to determine pregnant abdomen shape were flawed. Results of this study also indicate that small women sit closer to the steering wheel rim than their taller counterparts, and that pregnant women do not adjust their seat position rearward during the course of their pregnancy to compensate for their increasing abdomen size. As a result, the abdomen of a small pregnant driver in the third trimester will be very close to the lower steering-wheel rim and thus be likely to interact with the steering-wheel rim in a frontal crash. Results of the Klinich et al. study are supported by subsequent studies of seated anthropometry during pregnancy (Acar et al. 2004, Motozawa et al. 2008).

Jensen et al. (1995) measured the changes in the mass and mass distribution of sixteen body segments from 15 pregnant women over the courses of their pregnancies by measuring the changes in external shapes using photogrammetry. These shapes were applied to a simple mathematical model of the body that described each body segment as a series of 2-cm-thick elliptical cylinders, each with a density determined from values reported in literature. The net changes in mass and mass distribution of each body segment were reported as a series of regression equations.

ABDOMEN IMPACT RESPONSE

One of the core difficulties in developing pregnant ATDs (and pregnant finite element models) is that there are no data on the dynamic force-deflection response of the pregnant abdomen in response to the types of loading that are likely to occur in a motor-vehicle crash. As a result, the force-deflection characteristics of the pregnant abdomen have been estimated by applying equal-stress equal-velocity scaling (Eppinger et al. 1989) to response corridors derived from tests in which nonpregnant cadaver abdomens were loaded with a 1-in.-diameter rigid bar, a lap belt, and a device that simulated close-proximity abdominal loading by a deploying airbag (Hardy et al. 2001a, b). Although such scaling may seem questionable, evidence that the quasi-static force-deflection responses of pregnant and nonpregnant humans are similar at low compression levels supports using scaled corridors for nonpregnant occupants as response specifications for pregnant occupants (Rupp et al. 2001 a).

ABDOMINAL INJURY MECHANISMS, INJURY CRITERIA, AND DUMMY INSTRUMENTATION

Investigations of crashes involving pregnant occupants (Klinich et al. 1998, Klinich et al. 2008) and analyses of various trauma registries (Pearlman 1997) demonstrate that the most common type of severe injury to the pregnant abdomen is placental abruption, which is failure of the uteroplacental interface (UPI), resulting in the separation of the placenta from the uterine wall. The high frequency of placental abruption relative to other injuries to the pregnant abdomen, coupled with the observation that placental abruption is commonly observed in conjunction with other injuries to the pregnant abdomen and fetus suggests that placental abruption requires a less severe mechanical insult (Pearlman et al. 2000, Rupp et al. 2001a). Regardless of which of these is true, these observations suggest that a pregnant dummy designed to assess the risk of placental abruption will also reasonably assess the combined risk of placental abruption, uterine laceration, and direct fetal injury in motor-vehicle crashes.

Two potential mechanisms of placental abruption of have been hypothesized. The first of these involves failure of the villous uteroplacental interface from strains produced by direct loading of the pregnant abdomen in the area where the placenta is attached. The second mechanism involves the UPI failing in tension produced by negative pressures generated by decelerating the skeleton, the inertia of the amniotic fluid, and the viscous coupling between the posterior uterus and the spine (Rupp et al. 2001b). The applicability of each of these mechanisms depends on the location in the uterus where the placenta is attached. A placenta that is attached to the anterior uterus is more likely to fail from direct loading by the seatbelt or steering wheel, since all direct loading is applied to the anterior surface of the abdomen in a frontal crash. In contrast, a posterior placenta is not likely to be directly loaded in a frontal crash and is therefore more likely to fail from tension in the UPI generated by inertially induced negative pressures. Since ~70 to 80% of placentas are attached to the fundal (top) region of the uterus (Freid 1978), direct loading by objects that can contact the superior uterus, such as the steering-wheel rim, likely plays a role in many placental abruptions in frontal crashes.

Results of simulations of frontal crashes with a computational model of a 30-week pregnant occupant suggest that anterior-to-posterior compression of the pregnant abdomen at the level where the placenta is attached is a good injury criterion, since it is highly correlated to strain at the uteroplacental interface from direct loading (Duma et al. 2004, Duma et al. 2005). However, the MAMA-2B does not contain instrumentation to measure abdominal deformation because technology to accurately measure the dynamic compression of a fluid-filled abdomen was not available at the time the MAMA-2B was developed (Rupp et al. 2001a).

In place of instrumentation to measure abdominal deflection, the MAMA-2B employs transducers to measure intra-abdominal pressure, which is related to abdominal compression. An injury criterion for the MAMA-2B was developed by performing regressions of peak intraabdominal pressures measured in sled test reconstructions of generic frontal crashes for which the risk of adverse fetal outcome was estimated based on relationships between belt use, crash severity (delta V), and the risk of adverse fetal outcome reported by Klinich et al. (1999).

LIMITATIONS OF THE MAMA-2B AND THE FUTURE OF PREGNANT CRASH TEST DUMMIES

The MAMA-2B has several limitations pertaining to the peak abdominal pressure-based injury criterion that should be addressed by future work. Specifically, abdominal pressure measurements cannot distinguish between

- large amounts of abdominal compression applied over a small area and a smaller amount of compression applied over a larger area of the abdomen (Rupp et al. 2001a, b)
- pressures caused by contact loading of the abdomen and pressures caused by the restraint of the pelvis and the resulting motion of the fluid in the abdomen (Rupp et al. 2001b, Motozawa et al. 2009)
- the locations on the abdomen where a compression is applied (Rupp et al. 2001a, b).

As discussed by Rupp et al. (2001a, b), the first of these limitations is not likely to be meaningful because objects with small surface areas are unlikely to load the abdomen in frontal crashes. However, the latter two limitations are more problematic because the risk of adverse fetal outcome likely depends on the location of the abdomen that is loaded and may depend on whether abdominal loading is primarily inertial.

One way in which these limitations could be addressed is by adapting the Kelvin-probe abdominal deflection measurement technology introduced by Rouhana et al. (2001) for use with the MAMA-2B and developing an injury criterion that relates abdominal defection to the risk of adverse fetal outcome. Such an injury criterion could be based on relationships between abdominal deflection and risk of adverse fetal outcome determined from simulations where a validated pregnant occupant FE model is loaded using conditions that are representative of those that occur in frontal crashes. This FE model should incorporate recently collected data on uterus and placenta material properties and failure characteristics under dynamic loading (Manoogian et al. 2007a, Manoogian et al. 2007b, Hu et al. 2009).

Prior to applying a deflection-based injury criterion derived from FE analysis to the MAMA-2B, the force-deflection response of the MAMA-2B should be revisited by comparing it to the force-deflection response predicted by the same validated FE model of the pregnant abdomen. The performance of the injury criterion should also be verified by reconstructing frontal crashes with the MAMA-2B in which the risk of adverse fetal outcome is known based on the injury risk curves developed by Klinich et al. (2008).

CONCLUSION

In conclusion, pregnant occupants and other vulnerable populations should be considered when designing and testing restraint systems and other vehicle interior components. Despite its limitations, the MAMA-2B is the only available crash dummy that provides for objective evaluations of new restraint systems designed to protect pregnant occupants and reduce the risk of adverse fetal outcomes. Examples of how the MAMA-2B can be used to evaluate new restraint systems can be found in Rouhana et al. (2003) and Motozawa (2007, 2009, 2010a, 2010b).

REFERENCES

Acar, B. S., A. M. Weekes. 2004. "Vehicle Entry and Egress in Pregnancy: A Systems Approach." Paper No. ESDA2004-58149. In Proceedings of ESDA04, 7th Biennial Conference on Engineering Systems Design and Analysis. ASME, Manchester, England.

CFR 47 part 572. Subpart O. Hybrid III 5th Percentile Female, Alpha Version.

Culver, C. C. and D. C. Viano. 1990. "Anthropometry of seated women during pregnancy: Defining a fetal region for crash protection research." *Human Factors* 32(6): 625-636.

Duma, S. M., D. Moorcroft, J. D. Stitzel, and G. G. Duma. 2004. "Evaluating pregnant occupant restraints: The effect of local uterine compression on the risk of fetal injury." In 48th Annual Proceedings, Association for the Advancement of Automotive Medicine. AAAM, Barrington, IL. 103-114.

Duma, S. M., D. Moorcroft, J. D. Stitzel, and G. G. Duma. 2005. "A computational model of the pregnant occupant: Effects of restraint usage and occupant position on fetal injury risk." Paper No. 05-0367. In Proceedings of the 19th International Technical Conference on the Enhanced Safety of Vehicles. National Highway Traffic Safety Administration, Washington, DC.

Eppinger, R. H. 1989. "Prediction of thoracic injury using measurable parameters." In Proceedings of the 6th International Technical Conference on Experimental Safety Vehicles. National Highway Traffic Safety Administration, Washington, DC. 770-780.

Freid, A. M. 1978. "Distribution of the bulk of the normal placenta: Review and classification of 800 cases by ultrasonography." *American Journal of Obstetrics & Gynecology* 132(6): 675-80.

Hardy, W. N. and L. W. Schneider. 2001a. "Development and Refinement of Abdominal-Response Corridors, Project D2a - Development of a Reusable, Rate-Sensitive Abdomen." Final Report No. UMTRI-2001-10. University of Michigan Transportation Research Institute, Ann Arbor, MI.

Hardy, W. N. and L. W. Schneider. 2001b. "Development and refinement of abdominal-response corridors." *Stapp Car Crash Journal* 45: 1-32.

Hu, J. W., K. D. Klinich, C. S. Miller, G. Nazmi, M. D. Pearlman, L. W. Schneider, and J. D. Rupp. 2009. "A new method for quantifying dynamic mechanical properties of human placenta tissue using uniaxial testing, specimen-specific finite-element modeling, and optimization techniques." *Journal of Biomechanics* 42: 2528-2534.

Jensen, R. K., S. Doucet, and T. Treitz. 1996. "Changes in segment mass and mass distribution during pregnancy." *Journal of Biomechanics* 29(2): 251-256.

Kapit, W. and L. Elson. 1977. *The anatomy coloring book.* New York: Harper & Row.

Klinich, K. D., L. W. Schneider, J. L. Moore, and M. D. Pearlman. 1998. "Injuries to pregnant occupants in automotive crashes." In 16th ESV Conference Proceedings. National Highway Traffic Safety Administration, Washington, DC.

Klinich, K. D., L. W. Schneider, J. D. Rupp, B. Eby, and M. D. Pearlman. 1999. "Challenges in frontal crash protection of pregnant drivers based on anthropometric considerations." SAE Paper No. 1999-01-0711. In *Occupant Protection*, SAE-SP-1432. SAE International, Warrendale, PA. 105-127.

Klinich, K. D., C. A. Flannagan, J. D. Rupp, M. P. Sochor, L. W. Schneider, and M. D. Pearlman. 2008. "Fetal outcome in motor-vehicle crashes: Effects of crash characteristics and maternal restraint." *American Journal of Obstetrics and Gynecology* 198: 450.e1-450.e9.

Manoogian, S. J., J. C. Bisplinghoff, C. McNally, A. R. Kemper, A. C. Santago, and S. M. Duma. 2008. "Dynamic tensile properties of human placenta." *Journal of Biomechanics* 41(16): 3436-3440.

Manoogian, S. J., C. McNally, J. Stitzel, and S. M. Duma. 2008. "Dynamic Biaxial Tissue Properties of Pregnant Porcine Uterine Tissue." *Stapp Car Crash Journal* 52: 167-185.

Mertz, H. J., A. L. Irwin, R. L. Stalnaker, and M. S. Beebe. 1989. "Size, weight, and biomechanical impact response requirements for adult size small female and large male dummies." SAE Paper No. 890756. SAE International, Warrendale, PA.

Mertz, H. J. 2002. "Anthropomorphic test devices." In *Accidental Injury–Biomechanics and Prevention.* Springer-Verlag, New York.

Motozawa, Y. 2007. "Sled Experiments Using a Pregnant Dummy—Analysis of the Abdominal Pressure During Rear Collision." Paper No. 20075227. Japan Society of Automotive Engineers, Tokyo, Japan.

Motozawa, Y., M. Hitosugi., and S. Tokudome. 2008. "Analysis of Seating Position and Anthropometric Parameters of Pregnant Japanese Drivers." *Traffic Injury Prevention* 9(1): 77-82.

Motozawa, Y., M. Hitosugi, T. Abe, and S. Tokudome. 2009. "Analysis of Overall Kinematics and Abdominal Response of Pregnant Drivers during Frontal Vehicle Collisions." SAE Paper No. 2009-01-0380. SAE International, Warrendale, PA.

Motozawa, Y., M. Hitosugi, T. Abe, and S. Tokudome. 2010a. "Analysis of the kinematics of pregnant drivers during low-speed frontal vehicle collisions." *International Journal of Crashworthiness* 15(3): 235-239.

Motozawa, Y., M. Hitosugi, T. Abe, and S. Tokudome. 2010b. "Effects of seat belts worn by pregnant drivers during low-impact collisions." *American Journal of Obstetrics and Gynecology* 203: 62, 1-8.

Pearlman, M. D. and D. Viano. 1996. "Automobile crash simulation with the first pregnant crash test dummy." *American Journal of Obstetrics and Gynecology* 175: 977-981.

Pearlman, M. D. 1997. "Motor vehicle crashes, pregnancy loss and preterm labor." *International Journal of Gynecology & Obstetrics* 57: 127-32.

Pearlman, M. D., K. D. Klinich, L. W. Schneider, J. D. Rupp, and S. Moss. 2000. "A comprehensive program to improve automotive safety for pregnant women and their fetuses in motor vehicle crashes: A preliminary report." *American Journal of Obstetrics and Gynecology* 182(6): 1554-1564.

Rouhana, S. W., A. M. Elhagediab, and T. Walbridge. 2001. "Development of a reusable rate-sensitive abdominal insert for the Hybrid III family of dummies." In Proceedings of the 45th Stapp Car Crash Conference. SAE International, Warrendale, PA.

Rouhana, S. W. 2002. "Abdominal injury biomechanics." In *Accidental Injury–Biomechanics and Prevention*. Springer-Verlag, New York.

Rouhana, S. W., P. Bedewi, S. V. Kankanala, P. Prasad, J. J. Zwolinski, A. Medufsky, J. D. Rupp, T. J. Jeffreys, and L. W. Schneider. 2003. "Biomechanics of 4-point seat belt systems in frontal impacts." *Stapp Car Crash Journal* 47: 367-400.

Rupp, J. D., K. D. Klinich, S. Moss, J. Zhou, M. D. Pearlman, and L. W. Schneider. 2001a. "Development and testing of a prototype pregnant abdomen for the small-female Hybrid III ATD." *Stapp Car Crash Journal* 45: 61-77.

Rupp, J. D., L. W. Schneider, K. D. Klinich, S. Moss, J. Zhou, and M. D. Pearlman. 2001b. "Design, Development, and Testing of a New Pregnant Abdomen for the Small Female Hybrid III Crash Test Dummy." Final Report No. UMTRI 2001-07-11. University of Michigan Transportation Research Institute, Ann Arbor, MI.

Schneider, L. W., D. H. Robbins, M. A. Pflüg, and R. G. Snyder. 1983. "Development of anthropometrically based design specifications for an advanced adult anthropomorphic dummy family." Report No. DOT-HS-806-715, Vol. 1. U.S. Department of Transportation, National Highway Traffic Safety Administration, Washington, DC.

Viano, D., J. Smrcka, E. Jedrzejczak, B. Deng, P. Kempf, and M. D. Pearlman. 1996. "Belt and airbag testing with a pregnant Hybrid III female dummy." Paper No. 96-S1-O-03. Proceedings of the 15th Experimental Safety Vehicles Conference. National Highway Traffic Safety Administration, Washington, DC.

Yamana, Y., K. Okabe, C. Nanako, B. Zenitani, and T. Saita. 1984. "Body form of pregnant women on monthly transitions." *Japanese Journal of Ergonomics*, 2, 171-178.

The following papers are included in their entirety to augment this chapter:

Motozawa, Y., T. Abe, M. Hitosugi, and S. Tokudome. 2009. "Analysis of Overall Kinematics and Abdominal Response of Pregnant Drivers during Frontal Vehicle Collisions." SAE Paper No. 2009-01-0380. SAE International, Warrendale, PA. .

Rouhana, S. W., A. M. Elhagediab, A. Walbridge, W. N. Hardy, and L. W. Schneider. 2001. "Development of a Reusable, Rate-Sensitive Abdomen for the Hybrid III Family of Dummies." *Stapp Car Crash Journal* 45:33-60.

Rupp, J. D., K. D. Klinich, S. Moss, J. Zhou, M. D. Pearlman, and L. W. Schneider. 2001. "Development and Testing of a Prototype Pregnant Abdomen for the Small-Female Hybrid III ATD." *Stapp Car Crash Journal* 45:61-77.

2009-01-0380

Analysis of Overall Kinematics and Abdominal Response of Pregnant Drivers during Frontal Vehicle Collisions

Yasuki Motozawa
Honda R&D Co., Ltd. / Department of Legal Medicine, Dokkyo Medical University School of Medicine

Takeshi Abe
Honda R&D Co., Ltd.

Masahito Hitosugi and Shogo Tokudome
Department of Legal Medicine, Dokkyo Medical University School of Medicine

ABSTRACT

The authors conducted sled tests using the MAMA-2B pregnant dummy to analyze the mechanisms of injuries to pregnant drivers during frontal collisions. The seating position of the dummy was determined from the means of anthropometric values obtained from pregnant Japanese women. Deceleration waveforms with a delta-V of 13, 26 and 40 km/h were applied to the sled and tests were run with and without a seatbelt. Without a seatbelt, at a delta-V of 13 km/h, abdominal pressure peaked at the point where the dummy contacted the steering wheel. However, tests without a seatbelt and with a higher delta-V, the dummy slipped down the seat, consequently its torso slid below the airbag without a significant peak of abdominal pressure. This finding indicated that fetal outcome should be predicted based on more than just internal pressure of pregnant drivers, taking the overall status of the mother's health into account.

INTRODUCTION

Pregnancy or fetal safety is likely to become an important issue in efforts to reduce traffic accident fatalities. Connolly et al. [1] reported that 6 - 7% of pregnant women undergo some type of traumatic injury during pregnancy, and approximately two-thirds of such injuries occur during traffic accidents. Klinich et al. [2] also reported that approximately 130,000 women in late-term pregnancy are involved in traffic accidents in the United States annually, and the annual estimate of abortions or stillbirths ranges from 300 to 3,800. On the other hand, since nationwide statistics regarding pregnant women involved in traffic accidents are not available in Japan, the actual number of injured pregnant women is unknown. The authors previously analyzed the relationship between the severity of injuries sustained by pregnant women and fetal outcomes, using data from Japanese automobile insurance claims and associated accidents, as well as from injuries to pregnant women relative to claims for their fetuses [3, 4]. Based on these findings, most claims (46%) were submitted by women who were driving automobiles. However, the results of those studies indicated difficulties with predicting negative fetal outcomes based on injury severity to the mother after a slight (low-energy frontal or rear) impact, which is the most frequent type of vehicle accident. Injury severity scores did not significantly differ between pregnant occupants with subsequent spontaneous abortion and those whose pregnancies were uneventful, although the scores were significantly higher among pregnant occupants whose fetuses died, which included still birth and newborn birth within one month of birth. Therefore, the injury mechanisms to pregnant women and fetuses need to be analyzed under the various circumstances experienced by pregnant women during collisions not only in high- but also in low-energy impact. Japan and some other countries exempt pregnant women from mandatory seat belt use [5, 6]. Thus, pregnant Japanese occupants of vehicles might not be wearing seatbelts at the time of involvement in a vehicle accident. Therefore, the possibility of secondary contact between occupants and vehicle interior components should be examined. However, due to the absence of a microanalysis database, detailed information about traffic accidents involving pregnant women in Japan is limited. Thus, the injury mechanisms to pregnant women and fetuses, especially during low-energy impact, have not yet been completely explained. A previous study analyzed the outcomes of a series of automotive rear impact sled tests using an anthropomorphic dummy of a pregnant woman to understand the injury mechanisms [7, 8]. The results of that study showed that the unbelted dummy experienced secondary contact to the steering wheel due to rebound, resulting in increased pressure on the abdomen, even during relatively minor rear-end collisions with slight impact. These findings suggested that pregnant drivers would contact the steering wheel (or other interior components) during low-speed frontal as well as rear collisions. Duma et al. [9] assessed pregnant passengers using finite element model simulation to estimate the amount of stress exerted on the abdomen. They analyzed the injury mechanisms under various delta-V experienced by pregnant women during collisions. However, the seating postures applied in these studies were based on measurements using the interior buck at the University of Michigan Transportation Research Institute (UMTRI) [14], which differs from current Japanese vehicles, in terms of inclination of the cushion surface which is assumed to affect the overall kinematics of the occupant. Moreover, their assessments included the simplified interior buck model, which did not represent the detailed contour and features of interior components such as the instrument panel. Therefore, secondary contact with each component was not sufficiently examined. No other studies have examined the responses of pregnant drivers during low-speed impact, especially when not wearing a seatbelt. To understand the injury mechanisms to pregnant drivers and associated fetal outcomes, the authors conducted a series of sled tests to examine the effect of vehicular frontal impact upon an anthropomorphic dummy of a pregnant woman. The seated posture of the dummy was based on seat adjustments made by pregnant Japanese volunteers in an actual passenger vehicle. Based on the results, the mechanisms of injury to pregnant drivers are discussed.

METHODS

DUMMY - The dummy used herein is the most current version of the Maternal Anthropometric Measurement Apparatus, version 2B (MAMA-2B), developed by First Technology Safety Systems (FTSS) and UMTRI [10, 11, 12]. This dummy is based on the Hybrid-3 AF5th percentile dummy, with the pelvis, sternum and ribcage modified to accommodate a silicone rubber bladder representing the uterus at 30 weeks of gestation (Figure 1).

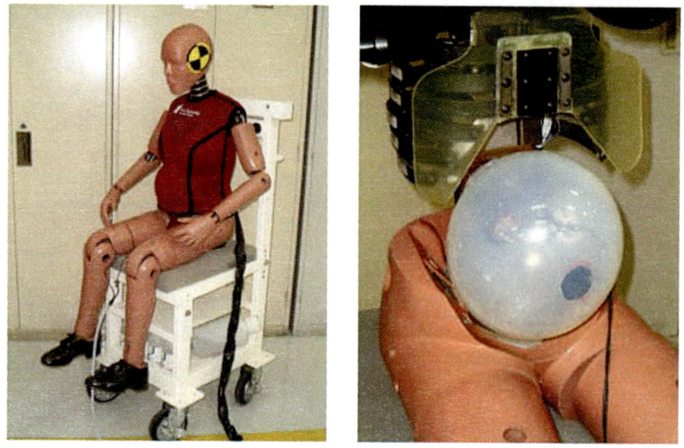

Figure 1. MAMA-2B anthropometric pregnant dummy and its abdominal bladder.

The bladder is roughly spherical, approximately 200 mm in diameter, and contains 3,000 cm^3 of pure water representing the amniotic fluid when in use. Two pressure sensors are installed on the anterior and posterior surfaces inside the bladder. To measure chest deflection, a pair of the 2-D Thorax for Multi-Point and High rate (THMPHR) devices was installed in the ribcage parallel to the No. 2 Rib of the dummy. Each comprised the InfraRed Telescoping Rod for Assessment of Chest Compression (IR-TRACC) device and rotary potentiometer. The IR-TRACC measures absolute length using infrared light and a photo sensor. Other implementation of accelerometers and load cells are identical to the conventional Hybrid-3 AF5th percentile dummy.

TEST SETUP - The authors previously investigated the effect of secondary contact to a pregnant driver within a vehicle using the smallest available adult female dummy (American Female (AF) 5th percentile dummy) [7, 8]. The authors measured the seated posture of volunteers at approximately 30 weeks of gestation who were regular drivers (all results are shown as means ± SD unless otherwise stated) [13]. The seating position of the dummy was determined from the means of anthropometric values obtained from seven pregnant women aged 28.7 ± 2.4 years with similar dimensions to the dummy: height, 152.6 ± 3.0 cm; weight, 54.7 ± 2.8 kg; gestation, 30.8 ± 1.0 weeks; abdominal circumference, 86.4 ± 4.2 cm. The mean seat slide was placed at 65 ± 30 mm from the full-forward position and reclined at 7.4 ± 3.0° from the full-forward position (torso angle, 13°). The horizontal distance from the lower rim of the steering wheel to the abdomen was 107 ± 4.2 mm. Considering the minimum increment of the seat adjuster devices (10mm for the seat slide and 2° for the seat reclining), the authors determined the posture of the dummy at a seat slide position of 70 mm from the full-forward position (50 mm forward from the neutral position) and reclined at 8° (torso angle, 21°), which was the closest position to the mean value of measurement. The seating position of the dummy was the same in the present and previous studies. Figure 2 shows the test setup using the Instron servo sled apparatus. The sled buck comprised the same body of the passenger vehicle that was used for postural measurements. The seat, seatbelt, instrument panel, steering wheel and steering column installed in the sled buck were identical to those in the vehicle and placed in the same relative position.

To measure deceleration applied to the sled buck, two single axis accelerometers were installed on the floor of sled buck and on the stage of sled respectively. Four high-speed video cameras were installed on the ground. Two of them were for close shot and the others for long shot respectively. Two were on the right-side and the others were on the left-side of the sled. The secondary contacts to the interior components and the timing of the maximum excursion of the dummy were examined mainly by the close shot from the right side.

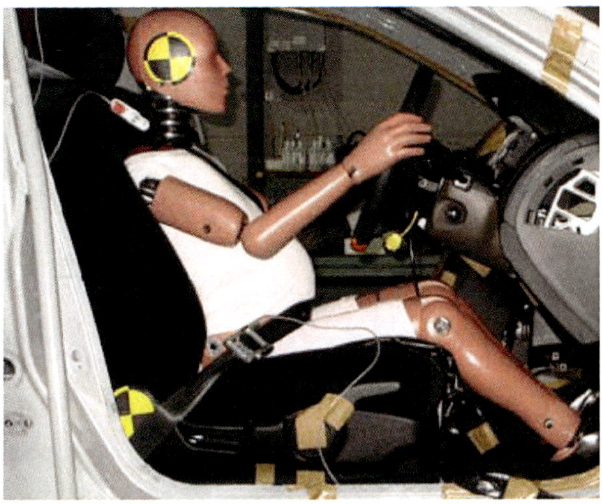

Figure 2. The MAMA-2B dummy sitting in test setup.

TEST CONDITIONS – Six tests were performed. To represent real-world frontal impact to a passenger vehicle driven by a pregnant woman, deceleration waveforms of the identical passenger vehicle that was used for postural measurements were applied to the sled. The waveforms actually measured during flat barrier tests with a delta-V of 13, 26 and 40 km/h were applied. Figure 3 shows the sled pulses applied to the sled buck.

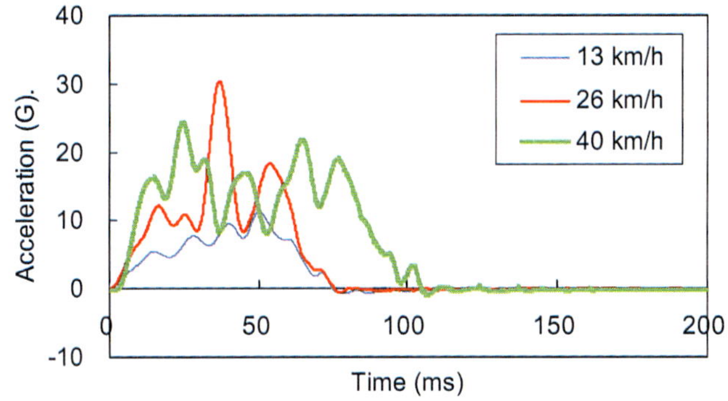

Figure 3. Time history of the sled pulses applied to the sled buck

Tests were run with (Tests 1, 2 and 3) and without (Tests 4, 5 and 6) the seatbelt, and the airbag and seatbelt pre-tensioner (PT) were activated in tests with a delta-V of 26 (Tests 2 and 5) and 40 km/h (Tests 3 and 6). Since the authors mainly focused on secondary contacts by the occupants, the overall kinematics of the dummy such as trajectory during impacts were examined. However, the quantitative dummy responses such as the time history of acceleration and pressure on the abdominal bladder (uterus) during impact were also measured and analyzed. Table 1 shows a matrix summarizing the test conditions described above.

Table 1. Test matrix

	Seatbelt use	Airbag and PT status	Delta V (km/h)
Test 1	Belted	-	13
Test 2	Belted	Activated	26
Test 3	Belted	Activated	40
Test 4	Unbelted	-	13
Test 5	Unbelted	Activated	26
Test 6	Unbelted	Activated	40

RESULTS

Figures 4 and 5 show the kinematics of the dummy at 100 ms from the initiation of impact (onset of sled pulse) taken by high-speed video camera form the right side of the sled. Figures 6 and 7 show the resultant acceleration of the head and chest of the dummy respectively. Figures 8, 9 and 10 show the horizontal acceleration of the pelvis, the time history of the femoral load and the chest deflection of the dummy respectively. The maximal value of the chest deflection could not be measured in Test 6 because it exceeded the pre-determined range. Figure 11 shows the anterior and posterior pressure on the abdominal bladder of the dummy (hereafter referred to as abdominal pressure). Table 2 shows the comparison of the values of HIC15, peak acceleration of the chest (residual) and pelvis (X: horizontal axis), the maximal femoral load, the maximal chest deflection, the maximal chest VC and the maximal abdominal pressure.

In tests with seatbelts (Tests 1, 2 and 3), the dummy moved forward by 100 mm from its initial position at 60 - 70 ms from the initiation of impact (onset of sled pulse), and in Test 1, the abdomen slightly contacted the rim of the steering wheel. The abdomen did not directly contact the steering wheel but contacted the airbag in Tests 2 and 3, subsequently causing the dummy to move downwards to the bottom.

The abdomen of the dummy in Test 4 (without a seatbelt) contacted the steering wheel at 70 ms, while the chest did not. The face of the dummy contacted the upper rim of the steering wheel at 130 ms. Tests 5 and 6 (without seatbelts, with airbags) showed that the dummy contacted the airbag, but its inferior sternum contacted the steering wheel; the pelvis of the dummy subsequently slipped down the seat and finally the entire dummy torso slid under the instrument panel.

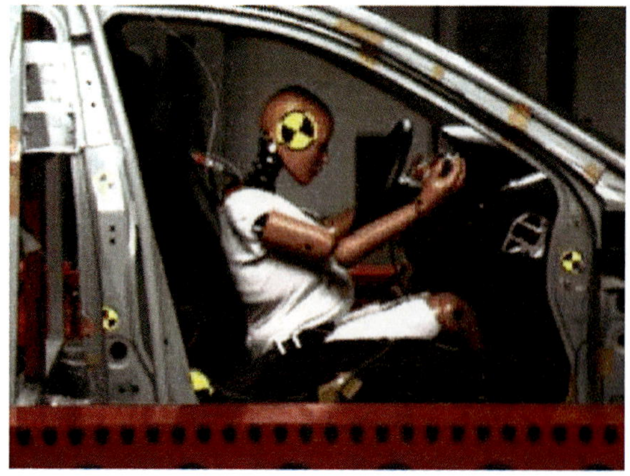

Test 1: belted, with a delta-V of 13 km/h

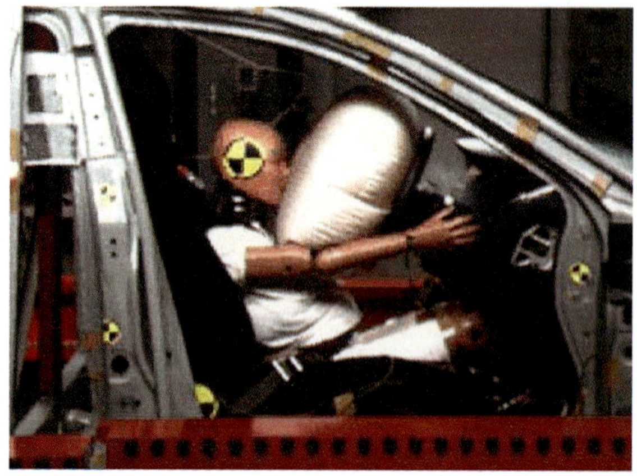

Test 2: belted, with a delta-V of 26 km/h

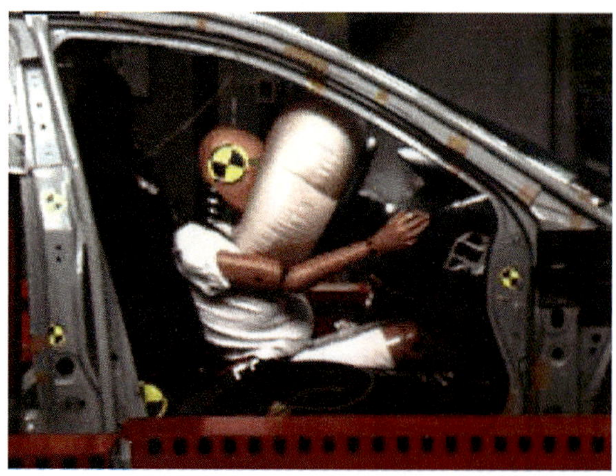

Test 3: belted, with a delta-V of 40 km/h

Figure 4. Kinematics of the belted dummy observed in Test 1, 2 and 3 at 100 ms from the initiation of impact

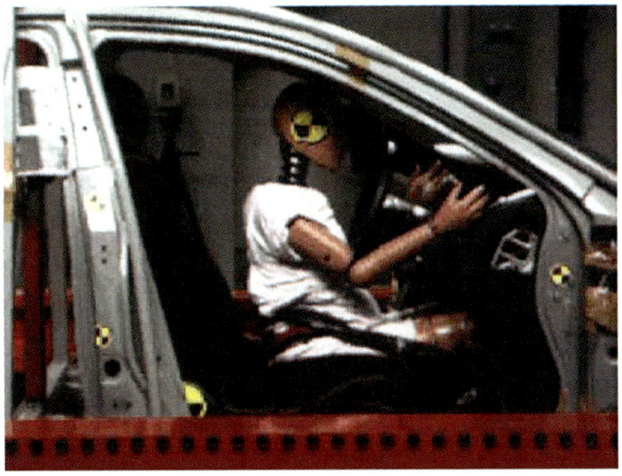

Test 4: unbelted, with a delta-V of 13 km/h

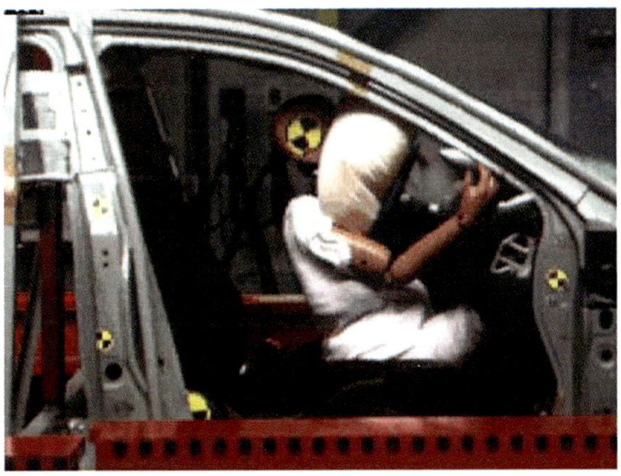

Test 5: unbelted, with a delta-V of 26 km/h

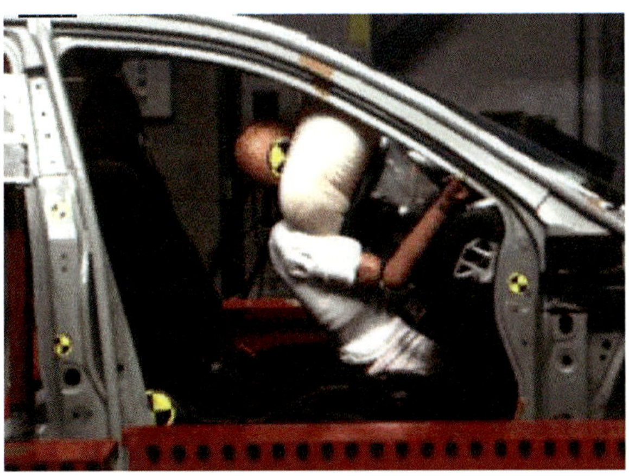

Test 6: unbelted, with a delta-V of 40 km/h

Figure 5. Kinematics of the unbelted dummy observed in Test 4, 5 and 6 at 100 ms from the initiation of impact

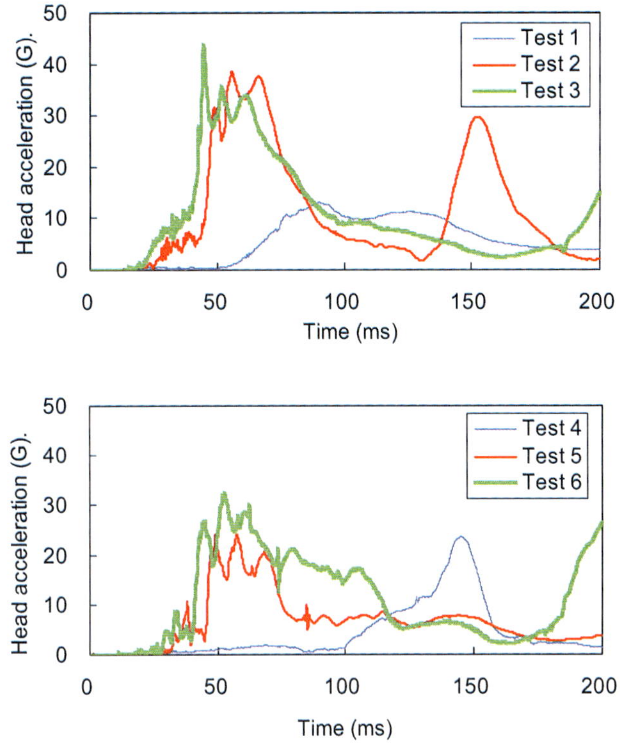

Figure 6. Time history of resultant head acceleration.

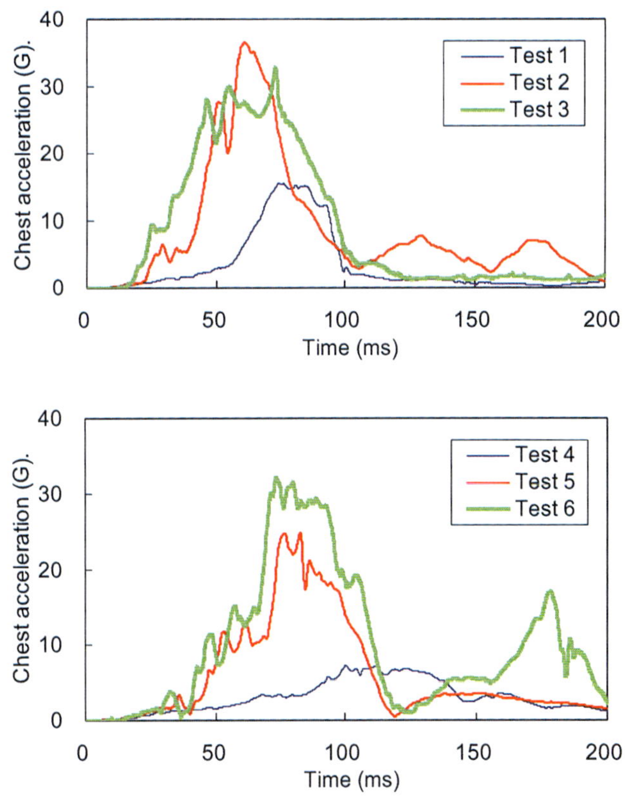

Figure 7. Time history of resultant chest acceleration.

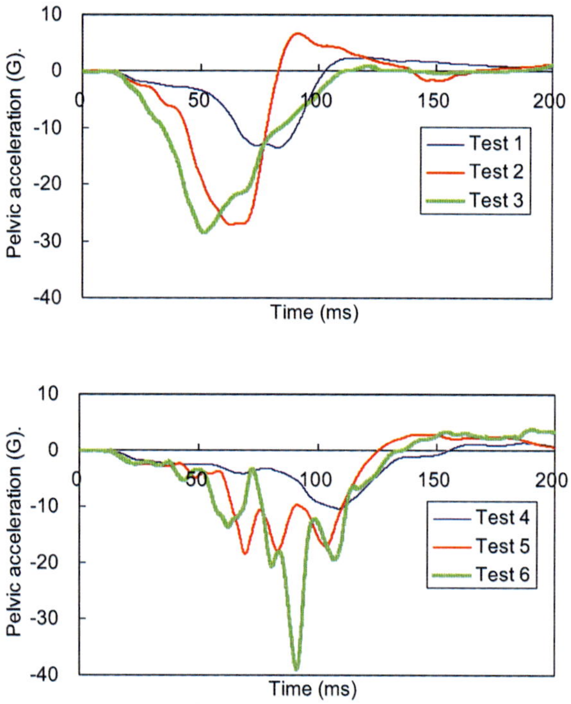

Figure 8. Time history of horizontal pelvic acceleration.

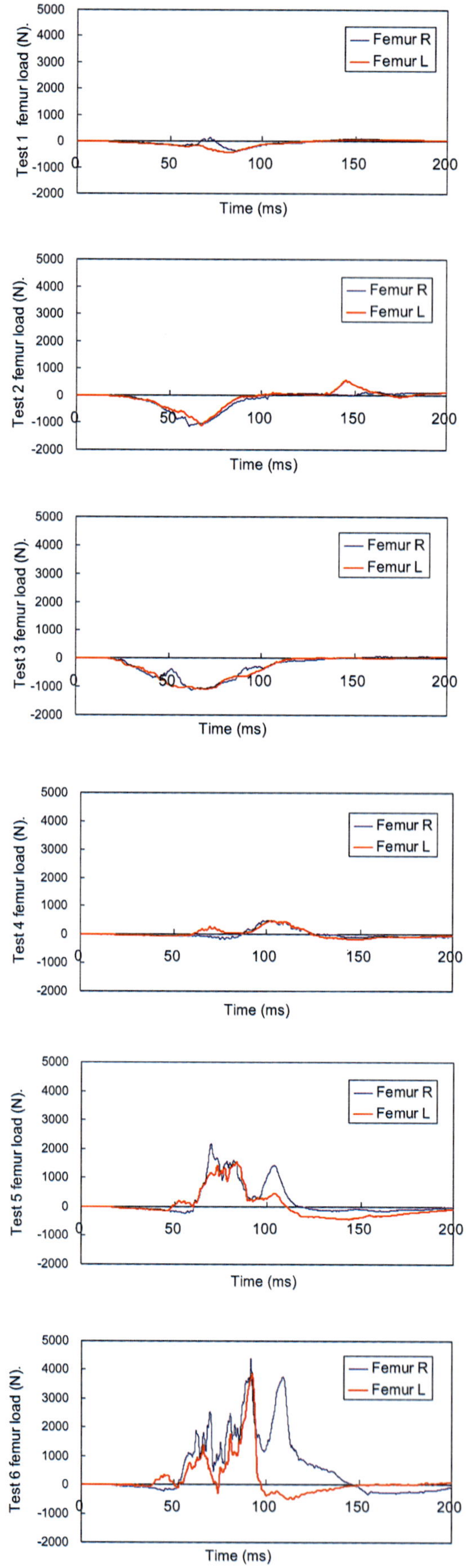

Figure 9. Time history of femoral load.

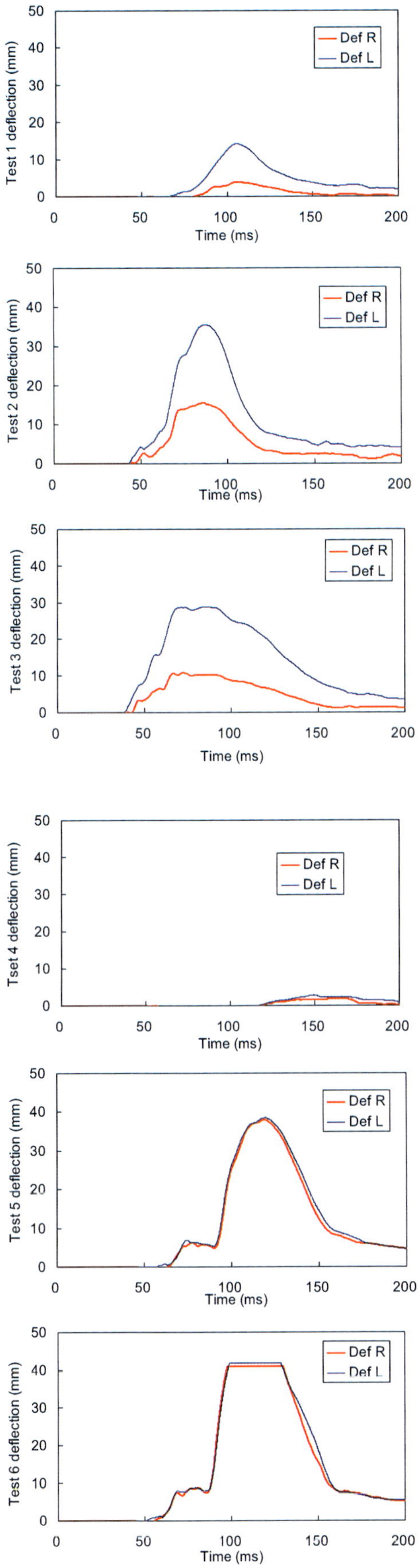

Figure 10. Time history of chest deflection. Maximal value in Test 6 was not measurable as it exceeded pre-determined range.

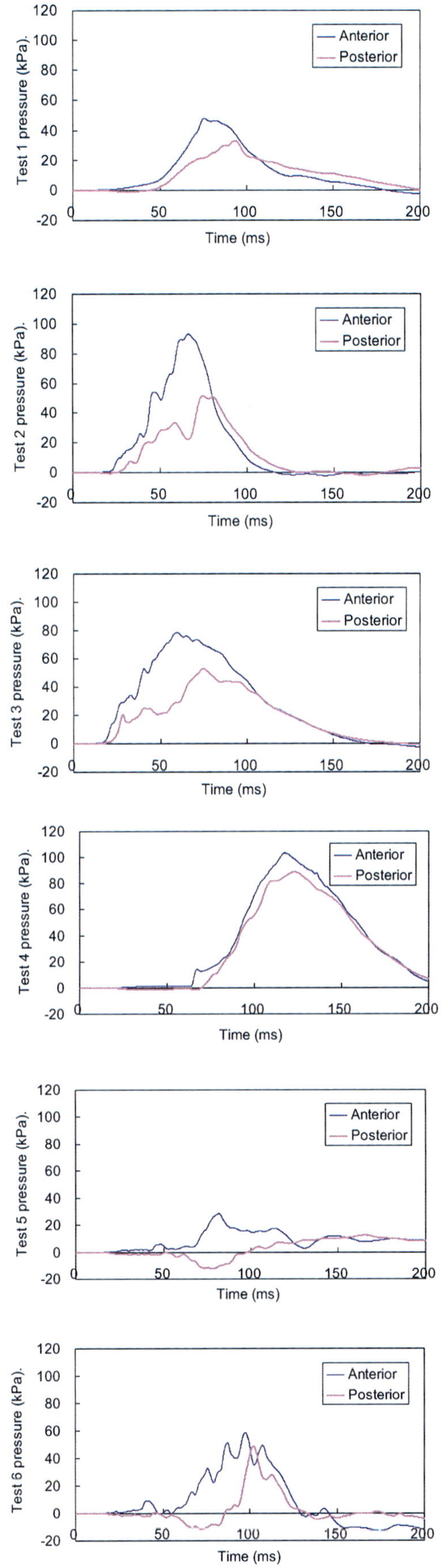

Figure 11. Time history of anterior and posterior pressure on abdominal bladder.

Table 2. Comparison of the values of HIC15, the peak acceleration of the chest (residual) and the pelvis (horizontal axis), the maximal femoral load, the maximal chest deflection, the maximal chest VC and the maximal abdominal pressure

	HIC 15	Chest res (3ms G)	Pelvis X (3ms G)	Femur load (R or L, kN)	Chest def (mm)	Chest VC	P Abd an (kPa)	P Abd po (kPa)
Test 1	7	15.3	13.3	0.15	9.0	0.05	47.6	33.2
Test 2	113	35.8	27.0	0.58	25.5	0.47	92.9	51.6
Test 3	86	28.7	28.1	0.85	19.8	0.33	78.3	52.6
Test 4	27	6.9	10.2	0.52	2.3	0.01	103.3	98.9
Test 5	15	23.8	16.7	2.18	38.2	0.77	28.3	12.5
Test 6	60	29.9	35.5	4.39	No data (>40.9)	No data (>1.45)	58.6	48.7

The authors applied the mean value of the right-side and the left-side measurement from IR-TRACC as overall chest deflection.

Femur load indicates larger value between the right-side and the left-side measurement.

Maximal value of the chest deflection in Test 6 was not measurable as it exceeded pre-determined range.

DISCUSSION

According to the comparison based on the results shown in Table 2, the author tried to examine the injury mechanism of the pregnant driver and to discuss the appropriate predictor of the negative fetal outcome.

The HIC15 values were much lower than the limit for AF5 dummy (700) through a series of tests. Therefore the HIC does not contribute to the prediction of the negative fetal outcome in these tests. The maximal values of the chest acceleration were smaller than the limit for AF5 (60G), although relatively larger value exceeded 35G was observed in Test 3.

The values of femoral load, in tests with seatbelts, were small or negative, which did not correspond with other indices. However, in tests without seatbelts, relatively large values of the femoral load were observed, corresponding with the chest and pelvic acceleration. In Test 6, the right femoral load indicated the value exceeded 4.3kN but was lower than the limit for AF5 (6.8kN).

Regarding the thoracic response, the maximal chest VC values in tests with seatbelts corresponded with the maximal chest acceleration, maximal chest deflection and maximal abdominal pressure. However, in test without seatbelts, they did not correspond significantly with other indices except the maximal chest deflection. In Tests 5 and 6, larger VC values were observed. Although the maximal value of the chest deflection in Test 6 was not measurable, it obviously exceeded 40.9, and VC value exceeded 1.45. However, the values of the abdominal pressure were smaller than that of Test 4.

Then the authors took a look at the abdominal pressure in detail. In the tests with seatbelts, the anterior abdominal pressure peaked at about 60 ~ 70 ms from the initiation of impact. At this point, the anterior abdominal pressure reached 1.4 - 2-fold that of the posterior abdominal pressure. However, no severe contact of the abdomen to the steering rim was observed. The responses of the anterior abdominal pressure closely corresponded with the time history of pelvic acceleration. The similar response was observed during the initial phase of rear impact sled tests which the authors previously conducted [8]. The analogy between the present results and the previous study suggest that inertial loading due to restraint force applied to the dummy was the dominant contributor to the abdominal pressure rather than direct compression in the tests with seatbelts. However, since the pressure peak corresponded with the timing of the contact with the airbag or steering wheel, the peak pressure was considered to be affected by the secondary contact.

In Test 4, the abdominal protrusion of the dummy contacted the steering wheel at 70 ms, while the chest did not. The anterior and posterior abdominal pressure indicated similar peak values and time histories. The similar response was observed during the secondary phase (after rebound) of rear impact sled tests previously conducted by the authors [8]. From these observations, it was obvious that the compression to the abdomen directly from the steering rim was the major contributor of the maximal abdominal pressure. This might also explain the smaller value of the chest deceleration, because the kinetic energy of the dummy was thought to be absorbed mainly by its abdomen.

Rupp et al. [11] conducted a series of sled tests using the MAMA - 2B dummy during its development, and examined the correlation between abdominal pressure and delta-V during vehicle collisions. They consequently derived a correlation between peak anterior abdominal pressure and adverse fetal outcome from actual case analyses conducted by Klinich et al. [10]. To the best of the present authors' knowledge, this is the only risk curve proposed with regard to adverse fetal outcomes currently applicable in dummy tests (Figure 12).

This curve predicts less than 20% of the probability of the negative fetal outcome in Test 1, and more than 60% in Test 2 and 4. However, the curve was derived from the abdominal compression tests to the dummy using a rigid bar representing steering rim not from the inertial loading without secondary contact. Therefore in Test 3, this curve will be applicable. Whereas in tests with seatbelt, the correspondence between the peak pressure value and the negative fetal outcome should be examined separately, because the yielding mechanism of the abdominal pressure was inertial loading, which was different from that of unbelted tests.

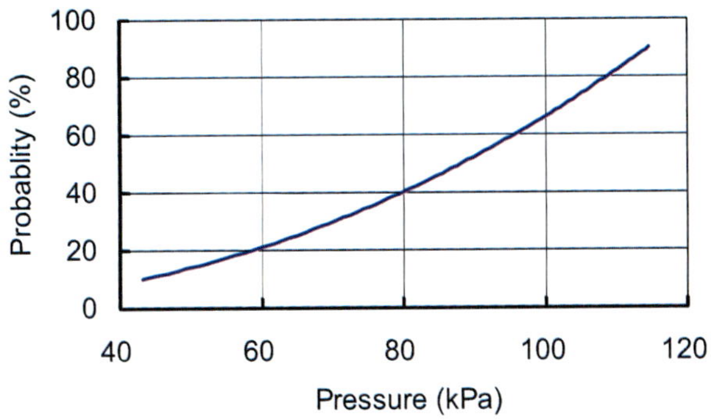

Figure 12. Correlation between peak anterior abdominal pressure and adverse fetal outcome derived by Rupp et al.

In Tests 5 and 6, the abdomen of the dummy did not make significant secondary contact and abdominal pressure did not significantly peak, indicated smaller values than those of belted tests while the maximal chest deflection and maximal VC indicated large values. In these tests, the abdominal pressure does not seem to be an appropriate predictor of the negative fetal outcome.

These findings indicated that the peak value of the abdominal pressure, or any other indices with regard to the uterus itself, may not be a sole predictor of adverse fetal outcome. In other words, the abdominal pressure may have a high positive predictive value, but may not have high sensitivity to all cases of adverse fetal outcome. If a certain value would be lower than the tolerance determined by certain risk curves, maternal health and wellbeing could be affected by sustained injuries and consequently affect fetuses. Therefore, fetal outcome should be predicted taking the overall status of the mother's health into account.

The kinematics of the dummy was unstable in tests without seatbelts. In Test 4, both the anterior and posterior abdominal pressure curves indicated a steep onset at about 70 ms from impact initiation due to secondary contact to the steering rim, whereas the dummy indicated different kinematics in Tests 5 and 6. The dummy torso slid below the deployed airbag or instrument panel with unstable trajectories and secondary contacts. Two hypotheses might explain these phenomena. One is a difference in the center of the gravity (COG) of the occupants. Due to the abdominal bladder, the COG of the pregnant dummy is lower than that of the non-pregnant dummy. Therefore, the loading from the airbag possibly yields backward rotation of the upper torso observed in Figure 6. The other is a difference in the seating position. The seating posture in the present study was based on measurements from Japanese volunteers at a seat slide position of 70 mm from the full-forward position, whereas that of the AF5 dummy determined in the FMVSS test protocol is based on the full-forward position. Therefore, the knee joint of the dummy in the present study was less bent than that in the FMVSS test protocol. Thus, the knee of the dummy was not constrained by the lower surface of the instrument panel possibly due to a shallower bend. A comparison with tests using the non-pregnant AF5 dummy is required to determine the dominant factor.

The average longitudinal distance between the steering wheel and the abdomen of pregnant drivers is approximately 100 mm less than that of non-pregnant women [4, 13]. The present study represented low or moderate speed frontal impact to a passenger vehicle. Secondary contact with the steering wheel or airbag occurred in tests with and without seatbelts. However, the dummy tested without a seatbelt and with a higher delta-V indicated unstable kinematics which resulted in the pelvis of the dummy slipping down the seat without a significant peak of abdominal pressure. Consequently, fetal outcome and the mother's health could be affected by more severe secondary contacts. These findings indicate that a seatbelt can reduce the overall severity of injury to pregnant drivers during low-energy as well as high-energy impact. Pregnant women in Japan are exempt from mandatory seat belt use for health reasons. These regulations are not intended to relieve pregnant women of the requirement to wear their seat belts under all circumstances. For example, a low rate of seatbelt use among pregnant women at all terms of pregnancy is a nationwide concern in Japan [4, 15]. One conceivable reason for this is insufficient understanding of the effectiveness of seatbelts as well as of the law. Since the severity of injuries sustained by occupants might be considered minimal in low or moderate speed collisions, pregnant as well as non-pregnant drivers might not correctly understand the value of wearing a seatbelt. However, the findings of the present study suggest that seatbelts protect pregnant women during various types of accidents, and thus pregnant Japanese drivers and passengers should be encouraged to use them.

CONCLUSION

The authors conducted frontal sled tests using a pregnant dummy (MAMA-2B) that represents the anthropometry of a pregnant woman at the gestational age of 30 weeks. Deceleration waveforms with a delta-V of 13, 26 and 40 km/h were applied to the sled. Tests were run with and without a seatbelt. The overall kinematics of the dummy such as the HIC, the acceleration of the chest and pelvis, the femoral load, the chest deflection, the chest VC and the abdominal pressure were analyzed as well as the trajectory of the dummy. In tests with seatbelts, abdominal pressure peaked where the dummy reached maximal forward excursion without severe secondary contacts. In the test without a seatbelt and with a delta-V of 13 km/h, abdominal pressure peaked at the point when the dummy contacted the steering wheel. However, in tests with a delta-V of 26 and 40 km/h, the dummy slipped down the seat, consequently the dummy torso slid below the deployed airbag or instrument panel, despite significant secondary contact to interior parts, abdominal pressure did not significantly peak. This finding indicated that fetal outcome should be predicted based on more than just internal pressure of pregnant drivers, taking the overall status of the mother's health into account. The present findings suggest that seatbelts protect pregnant drivers during various types of accidents, and thus their use should be encouraged.

REFERENCES

1. Connolly AM, et al. (1997). Trauma and Pregnancy. Am J Perinatol, 14

2. Klinich D, et al. (1999). Investigations of Crashes Involving Pregnant Occupants, Final Report. No. UMTRI 99-29

3. Hitosugi M, et al. (2006). Traffic Injures of the Pregnant Women and Fetal or Neonatal Outcomes. Forensic Science International, 159

4. Hitosugi M, et al. (2007). The Benefits of Seatbelt Use in Pregnant Drivers. Forensic Science International, 169

5. Weiss HB, et al. (2006). International Survey of Seat Belt Use Exemptions. Injury Prev., 12

6. McGwin G, et al. (2004). A Focused Educational Intervention can Promote the Proper Application of Seat Belts during Pregnancy. J. Trauma, 56

7. Motozawa Y, et al., An Experimental Analysis of the Abdominal Pressure of the Pregnant Occupants during Automotive Collision using AF5 Pregnant Dummy, presented at 20th ESV Conf., 07-0094, 2007

8. Motozawa Y, et al. (2008). Analysis of Pregnant Dummy Responses during Automotive Rear Collisions, Review of Automotive Engineering, 29(2)

9. Duma SM, et al., A computational Model of the Pregnant Occupant: Effects of Restraint Usage and Occupant Position of Fatal Injury Risk, presented at 19th ESV Conf., 05-0367, 2005

10. Klinich D, et al., Challenges in Frontal Crash Protection of Pregnant Drivers Based on Anthropometric Considerations, presented at SAE World Congress, 1999-01-0711, 1999

11. Rupp JD, et al., (2001) Development and Testing of a Prototype Pregnant Abdomen for the Small-Female Hybrid III ATD. Stapp Car Crash Journal, 45

12. Zhao Y, et al., Enhancement and Evaluation of a Small Female Hybrid III Pregnant Dummy, presented at International IRCOBI Conference, 2006

13. Motozawa Y, et al., (2008). Analysis of Seating Position and Anthropometric Parameters of Pregnant Japanese Drivers. Traffic Injury Prevention, 9(1)

14. Klinich D, et al., (1999). Seated Anthropometry during Pregnancy, Final Report. No. UMTRI 99-16

15. Ichikawa M, et al., (2003). Car Seatbelt Use during Pregnancy in Japan Determinants and Policy Implications. Injury Prev., 9

CONTACT

Yasuki Motozawa; Automotive R & D Center, Honda R&D Co., Ltd., 4630 Shimotakanezawa, Haga-machi, Haga-gun, Tochigi, Japan

yasuki_motozawa@n.t.rd.honda.co.jp

2001-22-0002

Stapp Car Crash Journal, Vol. 45 (November 2001), pp.

Development of a Reusable, Rate-Sensitive Abdomen for the Hybrid III Family of Dummies

Stephen W. Rouhana
Ford Motor Company

Ali M. Elhagediab
First Technology Safety Systems

Anthony Walbridge
Aerotek Scientific

Warren N. Hardy
Lawrence W. Schneider
University of Michigan Transportation Research Institute

ABSTRACT –The objective of this work was to develop a reusable, rate-sensitive dummy abdomen with abdominal injury assessment capability. The primary goal for the abdomen developed was to have good biofidelity in a variety of loading situations that might be encountered in an automotive collision. This paper presents a review of previous designs for crash dummy abdomens, a description of the development of the new abdomen, results of testing with the new abdomen and instrumentation, and suggestions for future work. The biomechanical response targets for the new abdomen were determined from tests of the mid abdomen done in a companion biomechanical study. The response of the abdominal insert is an aggregate response of the dummy's entire abdominal area and does not address differences in upper versus lower abdominal response, solid versus hollow organs, or organ position or mobility. While the abdomen developed has demonstrated good biofidelity in rigid bar, seat belt and airbag loading situations, some work remains to be done before it can be used in crash testing.

KEYWORDS – Abdomen, abdominal injury, abdominal response, ATD, crash dummy, Hybrid III

INTRODUCTION

Blunt abdominal trauma constitutes as much as 20% of all AIS $\geq$ 5 injuries from motor vehicle-related collisions. This trauma can have many forms and can be associated with different contact points within the vehicle. Yet the ability to assess abdominal injury risk has lagged behind other body regions in anthropomorphic test devices (ATDs) for vehicle development crash testing.

One of the key characteristics of ATDs is the degree of their mechanical equivalence to human beings, called biofidelity. Biofidelity is often assessed using the force-time, force-deflection, moment-angle or some other appropriate relationship for a given dummy component. The data from the dummy is compared to a published corridor or envelope that encompasses force-deflection or other data that has been obtained through laboratory experiments with human volunteers or human surrogates (typically cadavers or animals) under the same test conditions. If all of the data for a given dummy component fit into the corridor for the human, that component is said to have excellent biofidelity.

The importance of biofidelity cannot be understated since it gives the automotive engineer more confidence that injury assessment measures made with that dummy component are applicable to living humans. For example, if the injury criterion for a body region is the amount of compression and the corresponding region in the dummy is made from a soft foam, then the dummy may register a very large amount of compression under many automotive test conditions. In this case, the automotive engineer could be led to believe that the design does not meet established assessment values when in fact the dummy response was not realistic. Similarly, if the dummy component is overly stiff, the engineer may be led to the conclusion that the design is far better than established assessment values when, again, the dummy response was not realistic.

Likewise, it is important that the dummy be capable of measuring physical quantities that are relevant to injury risk as identified through established injury criteria. If there are no transducers in a particular region of an ATD and no measurements are made, then, loading inadvertently applied to that region can go unnoticed.

Objective

The objective of this research was to develop a new abdomen for the Hybrid III family of crash dummies. While there have been a number of attempts in the past to develop a dummy abdomen, the goals of this project were aimed at developing an abdomen that is reusable, has similar force-deflection characteristics and rate sensitivity to the human abdomen under a variety of loading conditions, and enables the measurement of biomechanically relevant parameters for assessment of injury risk. The loading conditions considered were those deemed likely to be encountered under some circumstances in automotive collisions.

This was a multifaceted research effort encompassing biomechanics, instrumentation and mechanical design. The program included analysis of field accident data to set priorities for instrumentation efforts; experiments to expand the knowledge of abdominal biomechanical response for steering wheel, seat belt and airbag loading (see Hardy et al. 2001); parallel development efforts for materials from which to make the abdominal insert; and multiple instrumentation efforts aimed at developing or proving out new methods to measure deformation.

Comparison of Human and Dummy Anatomy

Human Anatomy - The abdomen is the largest "cavity" in the body, yet it can hardly be considered a cavity because the organs within normally fill the entire space (Gray 1977). The abdomen is bounded above by the diaphragm, which forms a dome over the viscera, extending high into the thorax, up to the level of the junction of the fourth costal cartilage with the sternum (Zuidema 1977). It is bounded below by the bony pelvis and the muscles attached to it. The upper abdomen is bounded anteriorly and laterally mainly by the lower rib cage, and posteriorly by the ribs and vertebral column (which is not usually considered as part of the abdomen). The mid-abdomen is bounded anteriorly and laterally by musculature, and posteriorly by musculature and the vertebral column. The lower abdomen is bounded posteriorly by the sacrum and coccyx, laterally by the ilia, and anteriorly by musculature.

The abdominal viscera include two main types of organs, "solid" and "hollow" organs, which behave quite differently under various types of mechanical loading. The solid organs include the liver, spleen, pancreas, kidneys, adrenal glands, and ovaries; the hollow organs include the stomach, small and large intestines, urinary bladder, and the uterus.

Note that the liver occupies most of the space enclosed by the lower rib cage on right side of the body, while the spleen and stomach occupy the space within the lower rib cage on the left side. This has led some researchers to classify the upper abdomen as the "Hard Thorax" (Eppinger et al. 1984), but it is probably more appropriately called the "Soft Thorax" because of the increased compliance relative to the upper rib cage, or the "Hard Abdomen" because of the more protected environment relative to the lower abdomen. Either way, there is a clear anatomical difference between the upper and lower abdomen.

Hybrid III Dummy "Anatomy"- The Hybrid III dummy abdomen is made of urethane foam covered with vinyl skin. Unlike the thorax it does not have an established biofidelity corridor. It is somewhat spherical to elliptical in shape and has a large notch at the top to accommodate the chest deflection instrumentation. The standard Hybrid III has no instrumentation to measure abdominal injury, although one can purchase iliac wing load cells for help in determining belt-to-pelvis interactions and timing of occupant submarining, if it occurs (Uriot et al. 1994). Since the Hybrid III abdomen does not have biofidelity in its mechanical response (Mooney and Collins 1986; Begeman et al. 1990), even if it had instrumentation, it would be unable to accurately assess any deflection-based injury criteria (Rouhana 1989).

"Real World" Data

In an effort to understand the distribution of abdominal injuries in the "real world" a literature review of field accident data and clinical data was performed (Rouhana 2001). While the details are beyond the scope of this paper, a summary of those findings is appropriate.

Reports as to which organs are injured most frequently vary as a function of the source. From field accident databases, the organs injured, in decreasing order of frequency, were liver, spleen and digestive system (Nahum 1970; Danner and

Langweider 1979; Ricci 1980; Hobbs 1980; Galer 1985; Rouhana and Foster 1985; Mackay 1993; Klinich and Burton 1993; Khaewpong 1995). From clinical databases of blunt abdominal trauma, the organs injured, in decreasing order of frequency, were the spleen, liver, and kidneys (Griswold and Collier 1961; Baxter and Williams 1961; Perry 1965; Cox 1984; Bergqvist 1985; Bond 1991; Tso 1993, Augenstein 1995,1996; Gupta 1996).

Further, Elhagediab and Rouhana (1998) used the NASS database for the years 1988 through 1994 to analyze abdominal injuries in non-rollover frontal impacts to non-ejected drivers and right front passengers. They found that, in frontal impact, abdominal injuries constituted 8% of all injuries of AIS ≥ 3, 16.5% of all injuries of AIS ≥ 4, and 20.5% of all injuries of AIS ≥ 5. Most injuries were to the solid abdominal organs (65%). In general, the organs injured most frequently were the liver (38%), spleen (23%) and digestive system (17%). For lap-shoulder belted occupants the organs injured most frequently were digestive (48%), arteries (7%) and kidneys (7%). The objects within the vehicle, which were most often associated with the abdominal injuries, were the steering wheel (68%), belt (17%) and airbag/other interior objects (14%). However, only 10% of the vehicles in the study contained airbags.

In summary, these studies have shown that blunt abdominal trauma can constitute as much as 20% of all AIS ≥ 5 injuries from motor vehicle-related collisions. There were generally more abdominal injuries to the solid organs (liver, spleen, etc.) compared to hollow organs (intestines, stomach, etc.). When age was taken into account, children and the elderly were over-represented in terms of incidence of serious abdominal injuries. While safety belt use was associated with an increase in minor abdominal injury compared to the unbelted occupant, there was a clear decline in serious injury to the abdomen when safety belts were used.

These studies highlight the need for a crash dummy abdomen with the ability to evaluate interactions with steering wheels and with deploying airbags. Likewise, the ability to unambiguously detect occupant submarining and to assess the risk of injury, if submarining were to occur would be helpful to a vehicle program in the development of occupant restraint systems.

Past Work on Abdominal Injury Criteria

Rouhana (1993, 2001) has extensively reviewed the injury criteria proposed for abdominal injury assessment. Abdominal compression as a function of time and the loaded area on the anterior abdominal surface were chosen as the primary focus of instrumentation activities. The compression measurement would allow calculation of four proposed abdominal injury criteria, namely, maximum abdominal compression (Miller 1989; Rouhana et al. 1989, 1990), rate of abdominal compression (Mertz and Weber 1982), the Abdominal Injury Criterion (AIC = $V_{max}*C_{max}$; Rouhana et al. 1984, 1985) and Viscous Criterion ($V*C = \{V(t)*C(t)\}_{max}$; Lau and Viano 1986).

The abdominal compression measurement has been shown to be appropriate for belt loading during submarining which occurs at a velocity of approximately 3 m/s (Miller 1989; Rouhana 1989, 1990). The other criteria are applicable to higher velocity loading such as that from a deploying airbag or an unbelted occupant interacting with a steering wheel rim.

Past Work on Abdomens for Crash Dummies

Many researchers have expressed the need to develop more human-like abdominal regions for dummies (Begeman et al. 1990; Chamouard et al. 1996; Digges et al. 1996). In fact, over the years, there have been numerous attempts to develop surrogate abdomens (Leung et al. 1979; Maltha and Stalnaker 1981; Daniel et al. 1982; Melvin and Weber 1986; Mooney and Collins 1986; Czernakowski and Klanner 1987; Rouhana et al. 1989, 1990; Beebe 1990; Schneider et al. 1992; Biard et al. 1993, 1997; Haffner et al. 1994; Ishiyama et al. 1994; Uriot et al. 1994; Rangarajan et al. 1996). However, it has been shown that for a surrogate abdomen to provide biofidelity under different loading conditions it must have a rate-sensitive force-deflection characteristic (Cavanaugh et al. 1986; Ishiyama et al. 1994; Nusholtz et al. 1988). Few of the previous efforts were able to achieve that end.

Leung et al. (1979) used silicone foam to match the static load-deflection properties of the human abdomen in the Hybrid II dummy. They also used transducers mounted on top of the iliac crests to measure the force associated with belt loading. Uriot et al. (1994) refined the force-sensing technique by modifying the pelvis to make the iliac spines behave like cantilevered beams to measure the load.

Maltha and Stalnaker (1981) used a weighted foam wrapped around a metal drum surrounding the spine to get biomechanical response in lateral impact. They used force-sensitive switches mounted to the rigid metal drum to indicate that a critical load and deflection had been reached simu ltaneously. This system did not give deflection-time or load-time. It only gave an output if the critical load and deflection values were met simultaneously. This served as the basis for the Euro-SID abdomen (Walfisch et al. 1980; Jansen and Vermissen 1988).

Daniel et al. (1982) used Styrofoam™ to measure penetration in a side impact dummy. Whereas this device was used only as an indicator of abdominal loading, Rouhana et al. (1989, 1990) used Styrofoam™ that had been designed to mimic human abdominal response at 3 m/s to assess submarining injury risk in frontal impact belt loading in the Hybrid III family of dummies. This device, called the Frangible Abdomen, measured maximum deformation which could be directly related to injury risk using tables provided by Rouhana et al. (1990). Schneider et al. (1992) used a modified version of the Frangible Abdomen in the TAD-50M dummy.

Melvin and Weber (1986) used fluid-filled tubes wrapped around a child dummy spine to estimate abdominal injury risk from the measured pressure. Czernakowski and Klanner (1987) extended this concept to two-dimensions, but did not design a human-like human force-deflection response.

Mooney and Collins (1986) used a foam-filled air bladder in which they measured pressure to estimate the abdominal deformation. They designed the biomechanical response based on the corridor used by Maltha and Stalnaker (1981). While they did observe rate-sensitivity in the response of the foam/air bladder, they found that the response was dependent on the shape of the impactor and the location of the impact. Schneider et al. (1992b) performed an extensive investigation into fluid-filled bladders and other concepts to achieve rate sensitivity for the thorax of the TAD-50M. They abandoned those efforts, however, in favor of a modified rib concept.

Biard et al. (1993, 1997) used a pressure transducer in a fluid-filled tube to estimate the abdominal deflection-time history. The tube was positioned between a polyurethane foam insert and the spine of the dummy. They also used a rigid plate load cell in place of the tube. Both transducers gave similar results, but the results were found to be configuration dependent and did not provide the ability to assess the occurrence of submarining with certainty. In a similar manner, van Ratingen (1997) et al. used a load plate behind a deformable element for the Q3 crash dummy abdomen.

Ishiyama et al. (1994) used elastic foam rubber (Neoprene 15) to match human force-deflection response given by Cavanaugh at 6 m/s. They used a thin stainless steel band instrumented with 25 strain gauges to get curvature from which penetration is calculated. The authors noted that their sensor did not function well if some permanent deformation occurred to the steel band. They also indicated that they did not have the information with which to design the force-deflection characteristics of the abdomen over a wide range of velocities. And, they stated the need to include measurements made with steering wheels or seat belts in their design.

Rangarajan et al. (1996, 1998) developed a two-segment abdomen for the THOR dummy. Each abdomen segment was made of two layers of foam inside of a stiff Kevlar™ bag. In the lower abdomen, deflection was measured using two Double-Gimbaled String Potentiometers (DGSPs) originally developed by Schneider et al (1992) for the TAD-50M (predecessor to the THOR). The upper abdomen instrumentation consists of a high-tension string potentiometer and an accelerometer.

The main issue that has remained unresolved even after the aforementioned efforts is that none of the abdomens developed have demonstrated the rate-sensitivity of the human abdomen. That is, no abdomen has been developed which has force-deflection characteristics that are applicable across the range of loading conditions experienced in automotive collisions. Thus, the need for further development is apparent.

METHODS

Design Goals

Given the need for rate-sensitivity to achieve the biofidelity goals discerned in previous biomechanical studies, and the promising results of previous efforts with fluid-filled elements, a fluid-filled bladder concept was selected for the abdominal insert. The criteria listed below were established to provide a metric for judging the success of this effort.

Abdominal Insert Design Criteria

Material Design Criteria for the Shell

- The material used in the construction of the bladder or shell of the abdominal insert should be commercially available with a well-defined process for reproducibility

- The material should be durable enough to allow a minimum of 30 abdominal impacts without failure

- The mechanical properties of the shell should be reproducible from batch to batch and within a batch for a given formulation

- The shell should be easy to seal once filled with fluid and the seal should be durable

Material Design Criteria for the Fluid

- The fluid must be containable within the shell
- The fluid must be stable and not interact chemically with the shell material
- In the event that the shell ruptures, the fluid must not cause damage to the sled, instrumentation or other dummy parts

Abdomen/Dummy Interface Design Criteria

- There should be minimal modifications to the existing Hybrid III family of dummies to accommodate the new insert
- Any mass or inertial changes should be documented so that reballasting of the dummy can be achieved
- There should be no effect on dummy posture
- Performance of the dummy, as measured by dummy kinematics should be unaffected when there is no submarining. If submarining occurs, the kinematics of the dummy with the rate-sensitive abdomen may be different since the belt will interact with a new mechanical structure
- Physical modifications to the dummy should not require major machining
- The insert should remain in position throughout any whole dummy testing such as sled or full-scale tests

Force-Deflection Response Design Criteria

- The mechanical response of the combined shell and fluid inside the dummy must meet the force-deflection corridors of the human abdomen for rigid bar, lap belt and out-of-position airbag interactions
- The mechanical properties of the combined shell and fluid should not be altered by the loading that is exerted on them in crash tests
- The mechanical properties of the combined shell and fluid should not change over time due to aging

Instrumentation Design Criteria

At the beginning of the project a search was made of commercially available instrumentation that might be useful for the abdomen. The instrumentation design goals are given in Table I.

TABLE I. Instrumentation Design Goals

Parameter	Value
Acceleration	Up to 50 g
Accuracy	± 1 mm
Displacement Range	0 to 150 mm (0 to 6 in.)
Response Time	50 msec or better
Velocity Range	0 to 25 m/s

In addition to the goals in Table I, the sensor was expected to be able to operate within the dummy and in the abdominal cavity, but could have either onboard or offboard processing of data. It could have an umbilical cable to offboard instrumentation if necessary. The sensor was also expected to be capable of measuring 3-dimensional motion of multiple points on the abdominal surface and have an output that is a linear function of displacement. Finally, the system was expected to be easy to calibrate and use. These design goals eliminated many of the available technologies during the initial review.

While measurement of pressure was considered (Melvin and Weber 1986), it was not pursued because ribcage impingement on the dummy abdomen could increase pressure without antero-posterior deformation. This could give the impression that an injurious event had taken place from purely superior-to-inferior loading by the ribcage. Such loading might not be distinguishable from pressure that was exerted by lap belt submarining, for example.

Many different techniques and systems were evaluated including capacitive, electromagnetic (Radar, especially, Micropower Impulse Radar or MIR), fiber optic, fluid pressure, ultrasonic, laser, magnetic and fluid resistivity (using mercury gauges and other approaches). Information on the most promising of these is included in Appendix I. Significant issues arose with most of these technologies due to the fluid-filled bladder design approach. Given the above, a deflection measurement sensor became the primary focus of these instrumentation efforts.

Abdominal Insert Development

Hardware – Two approaches were selected as semi-final designs. While both were filled with silicone gel, one used urethane as the shell and the other used silicone rubber. The final abdominal insert design is a shell made of non-medical grade liquid silicone rubber (Applied Silicone, Inc.). Information relative to the urethane design can be found in Appendix I.

To generate the 3-dimensional shape of the inserts, First Technology Safety Systems (FTSS - Plymouth, MI) was contracted to develop mathematical data based on detailed measurements of the abdominal cavity of the Hybrid III dummy from the lower margin of the bottom rib to the inside surface of the pelvis. This mathematical data (Figure 1) was then used to create a prototype of the insert using a Laminate Object Manufacturing process (LOM). In the LOM process, the math data are used to generate two-dimensional slices of an object that are laser cut from a piece of special paper. These slices are cut sequentially from the bottom of the object to the top, with each successive piece glued to the one below it. In this way a three-dimensional object with very complex shapes can be created (Figure 1). This prototype was then used to develop a mandrel.

Liquid silicone rubber (LSR) bladders were made, using the fluoropolymer-coated cast aluminum mandrel by Specialty Manufacturing (Saginaw, MI). To fabricate the abdominal inserts, the mandrel was repeatedly dipped into a vat of liquid silicone rubber. This was done until the desired wall thickness was achieved. After curing, the bladder was peeled off the mandrel and filled with either silicone gel (Dow Corning) or ultra-high viscosity silicone fluid (1.5 Mcps). The bladder was sealed using a patch of silicone sheet bonded with silicone adhesive (Figure 2). The resulting bond was as strong as the bladder material and never failed in our tests.

The silicone shell can be specified using standard durometer hardness, elongation and/or tear strength, among other parameters (Table II). This design had a nominal volume of 6000 cc and a 6.5 mm (.25") shell wall thickness. All but LSR # 3 were filled with gel that gave an approximate penetrometer depth of 5.0 mm.

Instrumentation - Since it became clear that there were no "off-the-shelf" technologies available, ways to instrument the abdomen were examined through several small contracts. The various efforts aimed at developing deflection sensors are discussed in Appendix I.

The final instrumentation design selected was a method to determine the abdominal displacement by measuring the electrical resistance across the abdomen. This method was developed under contract by Hughes Research Laboratories (HRL). In this design six anode electrodes were bonded to the front interior surface of the abdominal insert and a single common cathode was bonded to the rear interior surface (Figure 3). The electrical resistance between the electrodes was measured as a function of time.

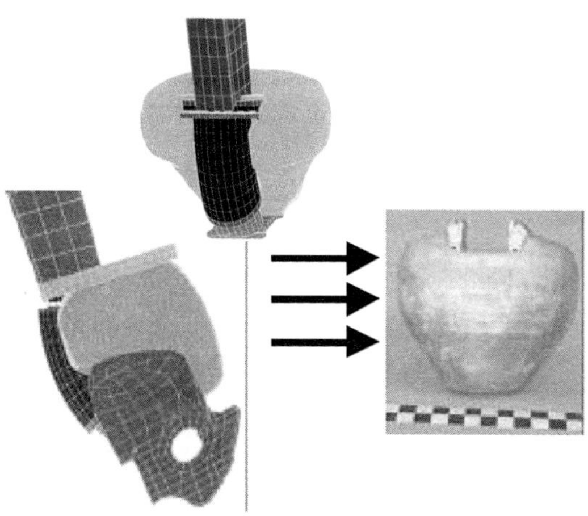

FIGURE 1. Mathematical data representing the abdominal cavity (left top and left bottom) was used to generate a 3-dimensional prototype (right) using a Laminate Object Manufacturing process (LOM).

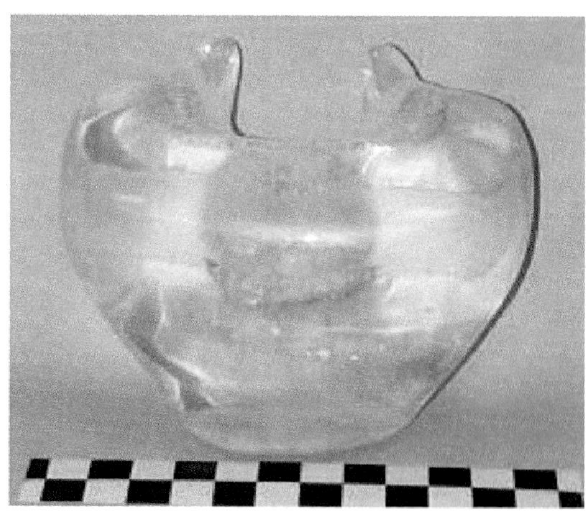

FIGURE 2. Front view of final abdominal insert design – a liquid silicone rubber shell, filled with silicone gel.

TABLE II. Liquid Silicone Rubber Shell Materials

	Shell Material						Silicone Gel Filling
Abdominal Insert ID	Material	Durometer (Shore A)	Tensile Strength (psi)	Elongation (%)	Tear Strength (lb/in)	Average Shell Thickness (in)	Gel Penetration Depth (mm)
LSR #1	Liquid Silicone Rubber 30	30	1205	720	155	0.14	5.2
LSR #2	Liquid Silicone Rubber 30	30	1205	720	155	0.221	5.1
LSR #3	Liquid Silicone Rubber 30	30	1205	720	155	0.225	17.8
LSR #4	Liquid Silicone Rubber 40	42	1084	530	154	0.244	4.9
LSR #5	Liquid Silicone Rubber 40	42	1084	530	154	0.093	4.7

The resistance of the abdominal insert should be proportional to the distance between the electrodes, but was typically very small in the abdominal inserts because conducting fluids were used to fill them. The resistance across the inserts was, therefore, very difficult to measure because the contact resistance of the measuring probes (the electrodes) was on the same order of magnitude as the resistance of the fluid.

To improve the accuracy of the resistance measurement, HRL developed a Kelvin probe that eliminated the contact resistance of the measuring probe. This method used 15 mm diameter, disk-shaped electrodes which were bonded to the silicone shell using the same medical grade silicone adhesive that was used to seal the shell after filling. For this method to work, there must be a conductive fluid inside the bladder. Preliminary tests were performed in a bladder filled with ECG Electrode gel. In the prototype system, up to six anode electrodes could be bonded to the inside of the anterior surface of the bladder, allowing for six independent deflection measurement points.

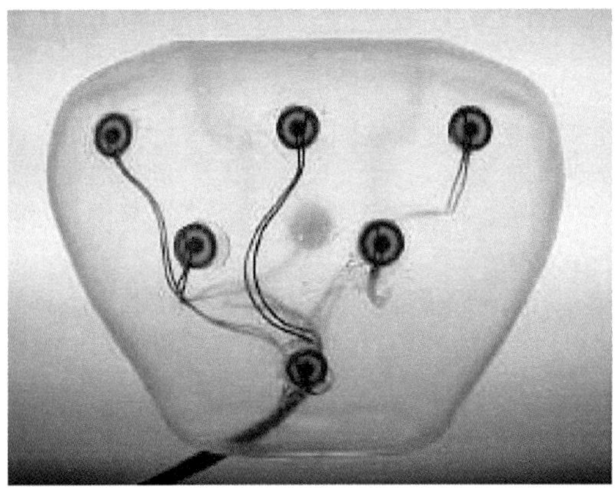

FIGURE 3A. Front view of liquid silicone rubber abdominal insert with anodes bonded to anterior wall on inside of bladder.

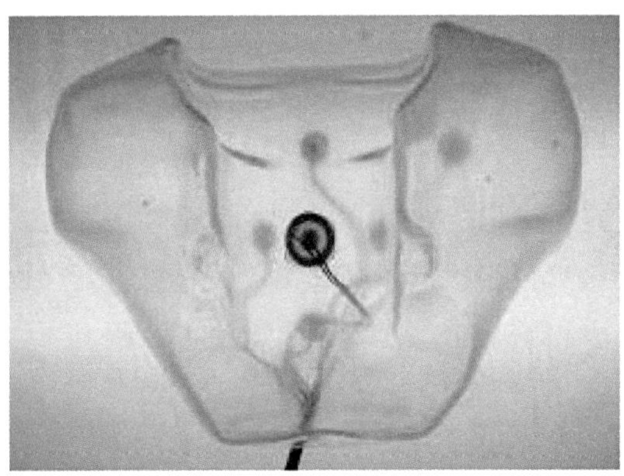

FIGURE 3B. Rear view of liquid silicone rubber abdominal insert with common cathode bonded to posterior wall on inside of bladder.

Test Methods

The five abdominal inserts described in this paper were tested on the bench and *in situ* in Hybrid III dummies. The force-deflection corridors used to design the inserts were obtained from a parallel study performed under contract at the Biosciences Division of the University of Michigan Transportation Research Institute (UMTRI) and reported in a companion paper to this (Hardy et al. 2001). Most of the in-dummy tests reported here were performed at UMTRI under identical test conditions as the cadaver study utilizing the powered pendulum, pneumatic lap-belt loading device and/or airbag test conditions. The test methodology of the UMTRI cadaver tests is reported in detail in Hardy et al. (2001) and therefore, will not be described in as much detail as the other tests done.

Some in-dummy tests were performed at the GM R&D Center (GMR) using a gravity pendulum, an MTS Systems™ servohydraulic high rate test machine (hereafter, MTS), or a HYGE™ crash sled (hereafter, sled). GMR test data was acquired using a MicroVAX data acquisition board as a front end to a VAX 6220 mainframe computer. Data was acquired at 10 kHz, passed through a 2 kHz anti-alias filter and stored on the VAX for analysis.

Dummy Modifications - In-dummy tests required some modifications to the Hybrid III mid-sized male to accommodate the abdominal insert. These modifications included: removal of Cerobase™ alloy from the lumbar spine bracket (Hybrid III Drawing Number 78051-53; see Figure 4), removal of the chest deflection potentiometer and front portion of the adaptor mount bracket that holds the chest pot bearing (Hybrid III Drawing Number 78051-78) and padding of the plates that join the lumbar spine to the thoracic spine box (Figures 5A and 5B). No other modifications were made to the dummy.

Gel Material Property Tests -A measurement procedure using a penetrometer was used to obtain repeatable mechanical properties of the gel used in the final abdominal insert design. The penetrometer probe weighed 21.5 grams and consisted of a 0.125" diameter plunger with a radiused tip. Gel samples to be tested were cured in a 4-fluid ounce hard plastic cup, which was placed in the penetrometer test fixture. The fixture contained the probe, which was held in a locking block. The probe was positioned so that it was just touching the sample to be evaluated. The lock was then released and the probe was allowed to "free-fall" into the test sample for 15 seconds. The distance traveled was measured and used for repeatability control by comparison with other samples.

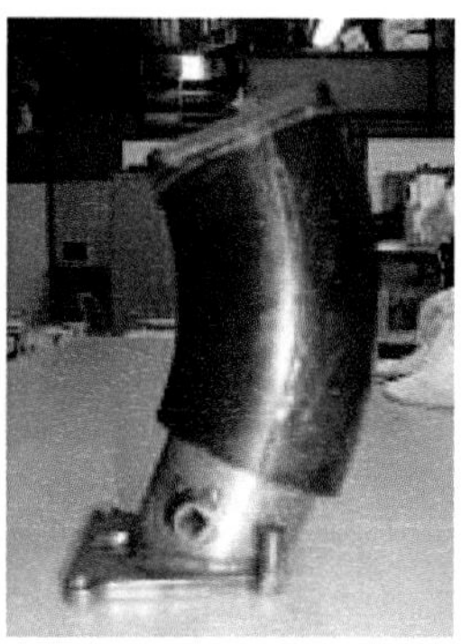

FIGURE 4A. Side view of lumbar spine adapter bracket without the Cerobase™ alloy.

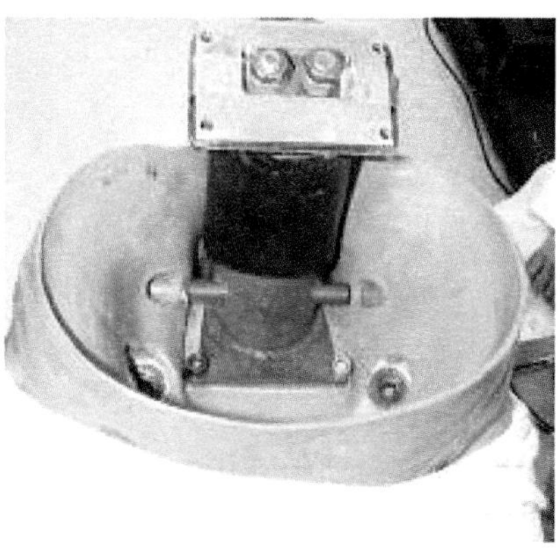

FIGURE 4B. Front view of lumbar spine adapter bracket *in situ* in the Hybrid III dummy pelvis.

FIGURE 5A. Steel plates at junction of thoracic and lumbar spines after removing bearing plate for chest deflection potentiometer.

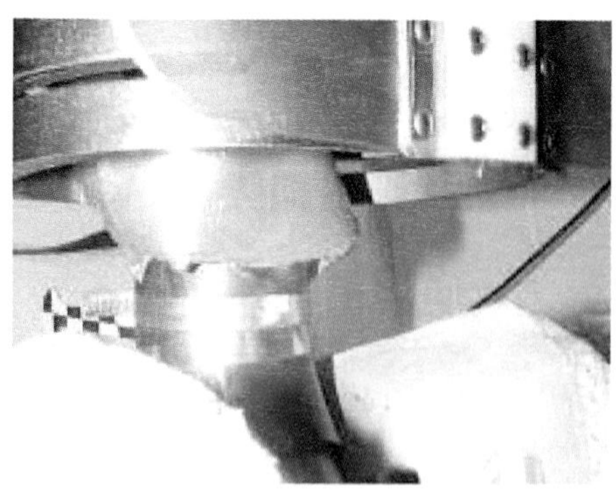

FIGURE 5B. Padding applied to thoraco-lumbar junction of Hybrid III dummy.

Gravity Pendulum Tests – Some in-dummy rigid bar tests were performed using a 24 kg gravity pendulum or the UMTRI 48 kg combination air cannon/gravity pendulum. The gravity pendulum was capable of reaching speeds of 6 m/s and the UMTRI pendulum was capable of reaching speeds of 9 m/s. The rigid bar was a 25 mm (1 inch) diameter steel bar, approximately 457 mm (18 inches) wide. Gussets from the edges of the bar to the central shaft helped to provide structural rigidity. The tests were performed with the dummy seated upright on Teflon™ sheets on a height adjustable platform (Figure 6). The setup was done by raising or lowering the platform to predetermine the impact location of the rigid bar on the dummy.

Pendulum velocity was set by the pendulum height and verified by an optical velocity trap. Pendulum force was obtained using the acceleration signal and the pendulum mass. Penetration was tracked using high-speed video analysis. The pendulum and dummy had targets affixed to accommodate the video motion analysis (Figure 6).

A-Tests and P-Tests – To examine the boundary condition presented by the chest jacket and anterior pelvic skin, tests were performed in the Hybrid III with and without the chest jacket on the dummy. In some tests, the chest jacket from NHTSA's advanced dummy (THOR) was used in an attempt to unitize the thorax, abdomen and pelvis.

In some tests the rigid bar was set up to impact just above the anterior pelvic skin without making contact with the skin (A-Test). In other tests the rigid bar was set up so that the bar impacted the anterior pelvic skin with ½ of the bar above the pelvic skin and ½ of the bar in contact with the pelvic skin (P-Test). Note: only ½ the bar diameter contacted the pelvic skin because if the whole bar made contact, the bar would strike the iliac spines. Some seat belt tests were also carried out at UMTRI using A-Test or P-Test set up.

MTS Tests - Bench tests and some in-dummy tests of the inserts were done on the MTS using a belt and yoke device described in Rouhana et al. (1989, 1990). Some of the tests used a 25 mm diameter rigid bar. The primary focus of these tests was a determination of the force-deflection properties of the insert. In the MTS tests, the dummy was placed on its back under the belt loading yoke or rigid bar (Figures 7A and 7B) and positioned such that the abdominal area to be loaded was directly beneath the belt or bar, with the belt or bar oriented in a transverse plane of the dummy at a level just above the pelvic flesh or with prescribed amount of overlap with the pelvic flesh (see section on A-Tests and P-Tests above).

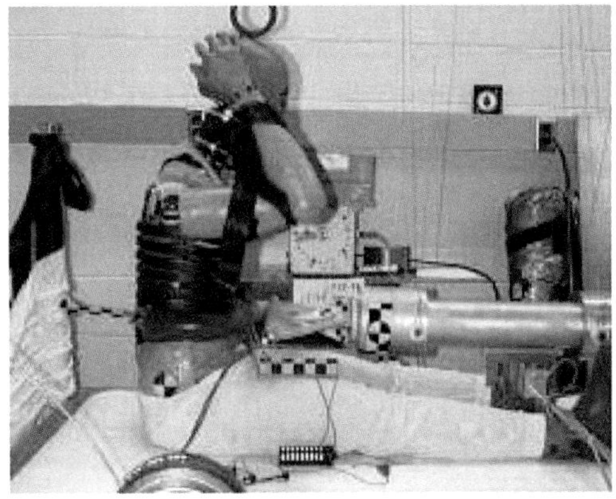

FIGURE 6. Hybrid III dummy with urethane abdomen in gravity pendulum test pre-impact position.

FIGURE 7A. MTS belt test setup. Note, belt webbing positioned horizontally under yoke (arrow) and above 'v-shaped' support used in original swine tests.

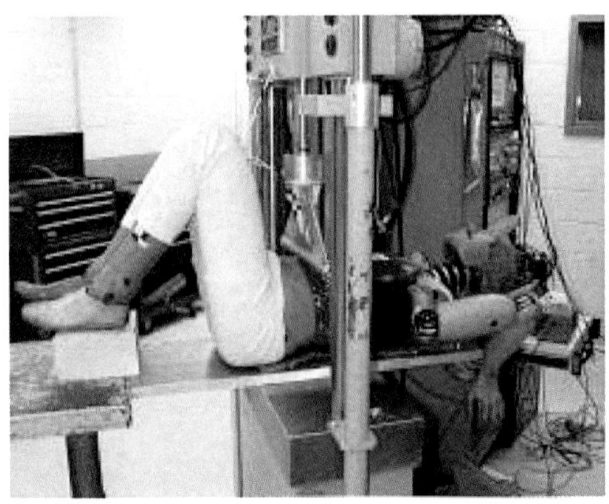

FIGURE 7B. MTS rigid bar test setup with dummy in supine position on table.

Prior to impact the piston was lowered to the point where all belt slack had been removed or the bar was in contact with the abdomen, but negligible pre-load was applied. The MTS piston was then run through a programmed displacement stroke and velocity. Total axial force developed and piston stroke were recorded. The belt loading profile applied was nominally identical to that used in the development of the Frangible Abdomen (Rouhana et al. 1989, 1990) and consisted of a haversine pulse with a peak velocity of 3 m/s. The rigid bar tests performed on the MTS were run at nominal peak velocities of 3 m/s and 6 m/s.

Pneumatic Belt Pull Tests – Some tests were performed using the UMTRI pneumatic lap-belt loading device (Hardy et al. 2001) that pulled a length of standard seatbelt webbing into the dummy abdomen from behind (Figure 8). The webbing was in contact with the anterior aspect of the abdomen, and fastened behind the dummy to a T-shaped bar that was connected to a pneumatic cylinder. The top of the T was positioned about 25 cm behind the dummy, oriented right-to-left. The seatbelt webbing was routed straight back from the sides of the dummy. The webbing was attached to the T via clamps whose position was adjusted right-to left. The T was used to pull the webbing into the abdomen. The device was designed to provide a peak loading rate of approximately 3 m/s with an approximately haversine-shaped penetration speed-time history.

The dummy was seated upright with the legs outstretched, supported by a Teflon skid, with the hands positioned above the head. Three string potentiometers were used to measure the displacement of the webbing and a laser tracked the displacement of the spine relative to the T. A linear potentiometer, an inductive speed transducer, and an accelerometer measured the motion of the ram. Two seat-belt load cells and an inertia-compensated load cell in the T mechanism measured the forces developed in the webbing.

Airbag Tests – Some vehicles contain "mid-mount" passenger side airbags that can interact with the occupant's abdominal area. It is also conceivable that some tilt steering wheels could be positioned such that the airbag is in contact with the occupant's abdominal region before deployment.

Hardy et al. (2001) performed tests of airbags into the abdomens of out-of-position human cadavers to develop a force-deflection corridor for airbag loading of the abdomen. They also developed a surrogate airbag to help assess the biofidelity of a dummy abdomen to airbag loading. It was felt that production airbags should not be used because of the variability in airbags from different vehicles and manufacturers. A surrogate airbag should allow for a repeatable test procedure that could be performed in any dummy laboratory.

The design specifications for the surrogate airbag included a very low mass impactor with dimensions approximately the same as a rolled up passenger airbag and with the ability to impact at speeds representative of real airbags (around 12 m/s). This speed was chosen based on the penetration speeds measured in the biomechanics tests in the companion paper (Hardy et al. 2001). As shown in that paper, the surrogate airbag designed was able to achieve those goals. Therefore, tests conducted with the dummy to examine the response of the abdomen to direct airbag loading used the surrogate airbag.

The surrogate airbag was designed to penetrate the cadaver and Hybrid III dummy abdomen without contacting the ribs or pelvis (Figure 9). It uses a pneumatic firing mechanism and a lightweight aluminum impactor constructed of welded thin-wall

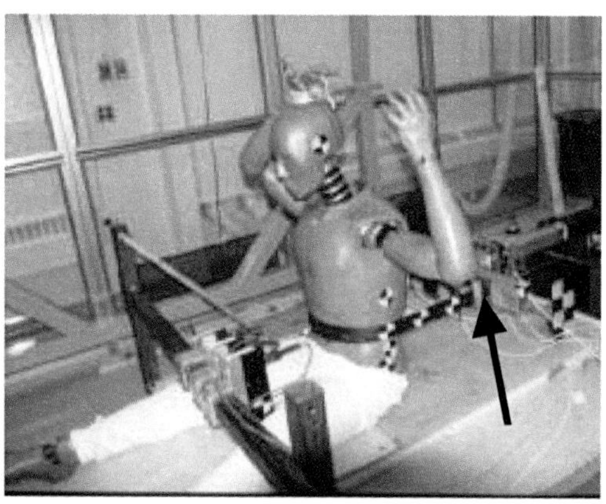

FIGURE 8. UMTRI lap belt test setup with T-bar behind dummy (arrow) and pneumatic impactor to right.

FIGURE 9. UMTRI surrogate airbag setup with a liquid silicone rubber abdominal insert in a Hybrid III mid sized male dummy.

tubing. The face of the impactor is the sidewall of a 7.6-cm diameter tube that is 20 cm long. The ends of the cylinder are fitted with rounded caps to help eliminate edge loading of the abdomen as penetration increases. A solenoid valve actuates a pneumatic cylinder that drives the device. At the end of the surrogate airbag's travel, the pressure driving the device is rapidly exhausted through vents.

Because the subjects moved very little during the early phases of deployment in preliminary tests with real airbags, a fixed-back condition was used to help compensate for the relatively small amount of energy possessed by the surrogate airbag. For a subject with a fixed back, only the motion of the surrogate airbag device needed to be measured to be able to determine abdominal compression. This measurement was accomplished using a high-speed, high-resolution laser displacement device.

The surrogate airbag tests employed the same fixture as that used to position the cadaver in the fixed-back rigid-bar tests described by Hardy et al. (2001). The dummy was seated upright with its legs outstretched and abdomen exposed. A single strap was wrapped around the spine and tightened against the seating fixture using a ratcheting mechanism. The arms were placed at the sides of the dummy, out of the way of the surrogate airbag device. The surrogate airbag contact with the abdomen was preset to avoid contact with the ribs and pelvis.

HYGE Sled Tests - Sled tests were performed to assess the kinematics of the dummy and to determine the need for tethering the insert in the dummy. Only 5 tests of a 48-test matrix were performed since the GM/DOT program officially ended before the test matrix could be completed.

These tests were performed with a mid-sized male Hybrid III dummy in the driver position of a generic body buck on a HYGE™ sled. The dummy was positioned in a modified production split bench seat that had been rigidized to allow multiple runs on a single seat without structural deformation. For all tests, the seat pan angle was set to 4.5°, the seat back angle was set to 7°, femur angle was set to 11°, tibia angle was set to 14° and the pelvis angle was set to 22°. The retractors were pre-locked to reduce test-to-test variability.

The dummy was clothed in cotton thermal underwear, but the shirt was trimmed from the axilla down to improve visualization of the interaction of the belt with the abdomen. A lap-shoulder belt was fastened around the dummy in all tests. In some tests, the lap belt was positioned low on the pelvis and 25 mm of slack was preset in the shoulder belt. In some tests, a pre-submarined position was sought, in which the lap belt was positioned low on the pelvis, but it had 75 mm of slack in it; or the lap belt was clipped or taped to chest jacket above pelvis. Sled parameters were set to provide a generic square wave crash pulse with a 16 g peak and 30 mph delta-v.

An overall view of the test was recorded at 1000 fr/s from the right side using an onboard Kodak high-speed ruggedized video camera. A close up view of the abdominal area was recorded from the left side using a second onboard Kodak high-speed ruggedized video camera. Another overall view was recorded at 500 fr/s off board using a NAC high-speed video camera.

The intention of these tests was to assess dummy kinematics and retention of the abdominal insert in non-submarining and in submarining conditions. Therefore, minimal dummy and sled instrumentation was utilized. The instrumentation included a uniaxial sled accelerometer, head, chest and pelvis triaxial accelerometers, and outboard lap and shoulder belt load cells. Pre-test position of the lap belt and pre-set amounts of slack in the lap and shoulder belts were varied as in Table III to achieve the desired kinematic results.

Electrical Resistance Test Method - Six anode electrodes were bonded in the shape of an inverted triangle with three electrodes on each side (one at each vertex and one at each mid-point) to the inside of the anterior bladder surface to serve as multi-point deflection sensors (Figures 3A and 3B). Each anode was operated at a different frequency, which was band-pass filtered to prevent cross-talk between the sensors. A single electrode was bonded to the inside of the posterior bladder surface directly in front of the

TABLE III. Sled Test Conditions

Test Number	Test Condition	Dummy Submarined ?
4397	Lap belt tight on pelvis; 25 mm slack in shoulder belt	No
4398	75 mm slack in lap belt, shoulder belt tight	No
4399	Top edge of lap belt taped to top edge of anterior pelvic skin; shoulder belt tight	No
4400	Bottom edge of lap belt clipped to bottom edge of chest jacket; shoulder belt tight	Yes
4401	Bottom edge of lap belt clipped to bottom edge of chest jacket; shoulder belt tight	Yes

lumbar spine to serve as the common cathode for all the anodes. The signal was notch filtered and rectified to provide a DC output that was representative of the antero-posterior abdominal depth at any point in time.

The first whole dummy tests of this technology were performed using a pendulum and the MTS. The system was tested for cross-talk and noise generation with all sensors active, and was tested with the dummy in the supine and seated positions at speeds ranging from quasi-static to 6 m/s.

RESULTS

Abdominal Insert Tests

Rigid Bar Tests - Figure 10A shows the biofidelity test for LSR #5 (40 durometer silicone shell with 5 mm silicone gel) in a 48 kg pendulum test with a 25 mm diameter rigid bar at 6.6 and 9.7 m/s compared with the Cavanaugh 6.1 and 10.4 m/s corridors. Some rate sensitivity is demonstrated in his test. Figure 10B shows the same data after being scaled to account for the difference in velocity between the actual test speed and the speed of the Cavanaugh corridors. The scaling consisted of a linear extrapolation of the average stiffness of the curves. That is, the average slopes of the loading portions of the measured force-deflection curves were obtained at 6.6 m/s and 9.7 m/s (excluding the region where the impactor bottomed out). These slopes were then used to extrapolate what the average slopes of the scaled curves at 6.1 m/s and 10.4 m/s would be. The original curves were then adjusted by the scale factors.

The lower speed response is within the low-speed corridor until a deflection of approximately 110 mm after which it is slightly above the upper boundary of the corridor until approximately 125 mm at which point the pendulum bottoms out on the lumbar spine of the dummy. The higher speed response is within the high-speed corridor until about 65 mm after which it is just slightly below the lower boundary of the corridor until around 125 mm of penetration at which point the pendulum again bottoms out on the lumbar spine of the dummy.

Figures 11A and 11B show the human abdominal response from the UMTRI rigid bar cadaver tests compared with the insert response tested in the dummy using the same pendulum. Figure 11A shows the response of the insert (dashed line) at 6.6 m/s compared with three cadaver tests (solid lines) at 6.3 m/s. Figure 11B shows the response of the insert (dashed line) compared with three cadaver tests (solid lines) at 9.7 m/s. In general, the low speed data shows very good agreement until the bar approaches the spine of the dummy. The high-speed data shows very good agreement in the early phase of loading, but shows lower forces for the insert after about 50 mm of penetration, while still in generally acceptable agreement with the cadaver data.

It was noted that under otherwise identical test conditions, the presence of the chest jacket had an effect on the force-deflection response of the insert probably due to the load distribution caused by the chest jacket. Figures 12A and 12B show the force-deflection response of the abdominal insert in a dummy without the chest jacket and with the chest jacket, respectively, with all other test conditions held constant. The effect of the chest jacket appears to increase the apparent stiffness (slope of curves) and the rate sensitivity (separation of the curves).

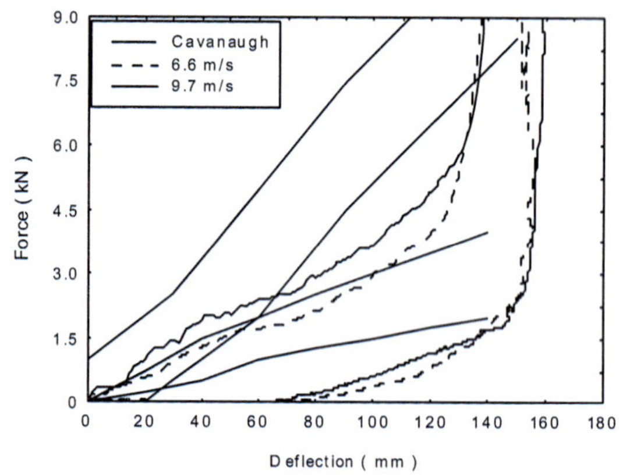

FIGURE 10A. Force-deflection from 48 kg pendulum rigid bar tests of LSR#5 insert at 6.6 and 9.7 m/s compared with Cavanaugh corridors (chest jacket on dummy).

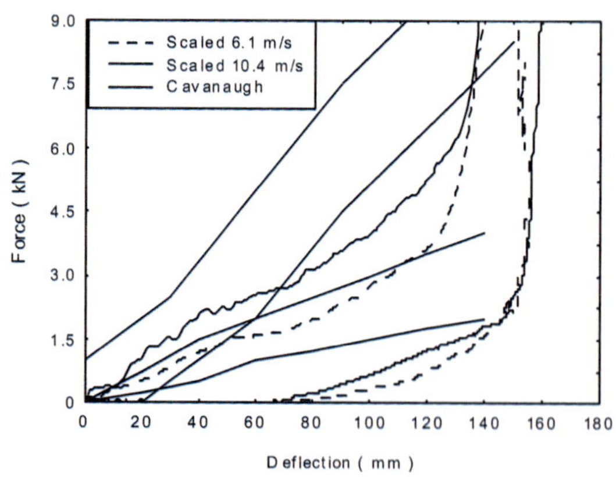

FIGURE 10B. Force-deflection from curves from Figure 10A, scaled to 6.1 and 10.4 m/s, respectively, to better compare with Cavanaugh corridors.

Seat Belt Tests - It should be noted that Hardy et al (2001) found that the force-deflection response corridor of the upright seated cadavers tested at UMTRI was different from that defined using supine swine (Rouhana et al. 1989). It was assumed that this was due to differences in test conditions. Therefore, to examine that hypothesis, the abdominal inserts were tested under both conditions.

MTS Tests- Figure 13 shows the force-deflection response of abdominal inserts tested in the dummy at 3 m/s on the MTS compared with the corridor established by Rouhana et al (1989). Four tests are shown which include two LS-30 urethane shells and two silicone shells (all filled with silicone gel). The tests were performed using the same belt yoke apparatus used in the swine tests to define the original corridor. Excellent agreement is noted in this test mode.

UMTRI Tests – Figure 14 shows the results of a P-Test and an A-Test of a liquid silicone rubber abdominal insert filled with gel compared with the UMTRI belt corridor. These tests were done using the UMTRI test setup and fixture with the insert in the dummy and the chest jacket on. Again there is excellent agreement between the abdominal insert and the cadaver force-deflection responses even though tested in a completely different test protocol.

Airbag Tests – Hardy et al. (2001) developed a force-deflection corridor for airbag interaction with an out-of-position cadaver abdomen. The force-deflection response from 3 tests of the LSR #5 abdominal insert under similar conditions is shown in Figure 15. The insert response again shows excellent agreement with the airbag corridor.

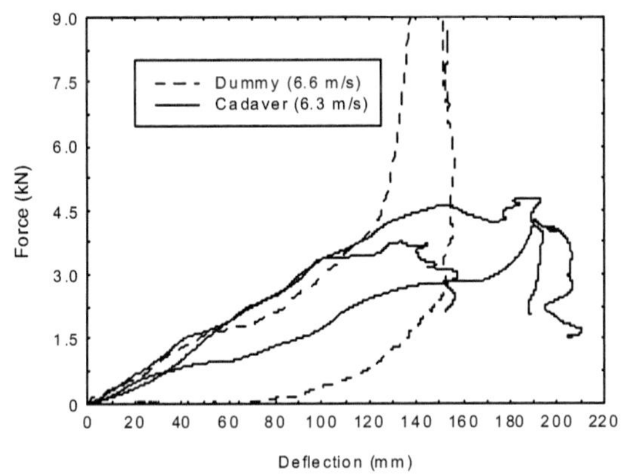

FIGURE 11A. Force-deflection from 48 kg pendulum rigid bar tests of LSR #5 insert at 6.5 m/s with chest jacket on dummy compared with three cadaver tests at the same speeds.

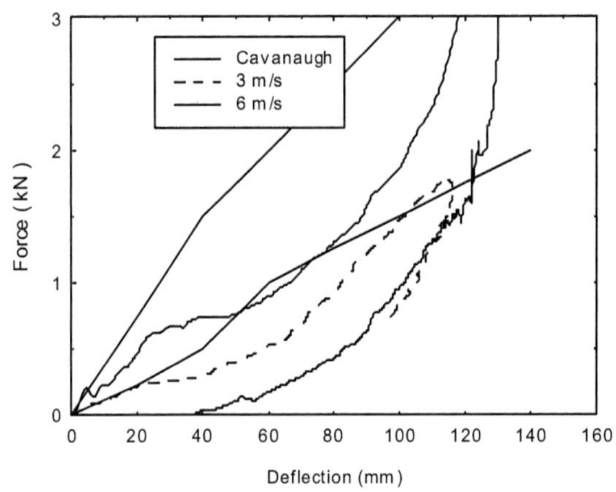

FIGURE 12A. Force-deflection of LSR #5 **without** chest jacket, in rigid bar 24 kg pendulum tests at 3 and 6 m/s, compared to the Cavanaugh low speed corridor.

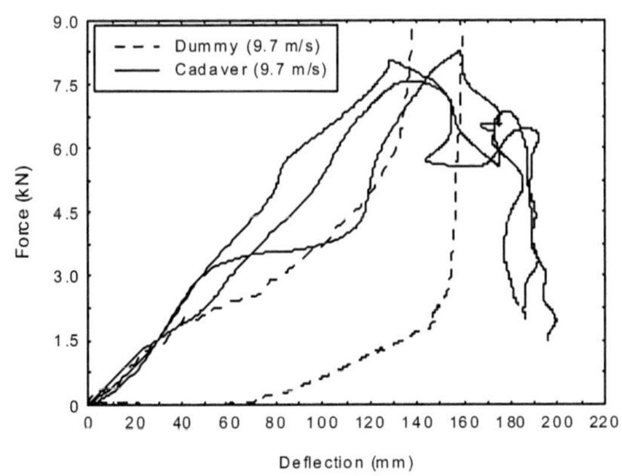

FIGURE 11B. Force-deflection from 48 kg pendulum rigid bar tests of LSR #5 insert at 9.7 m/s with chest jacket on dummy compared with three cadaver tests at the same speeds.

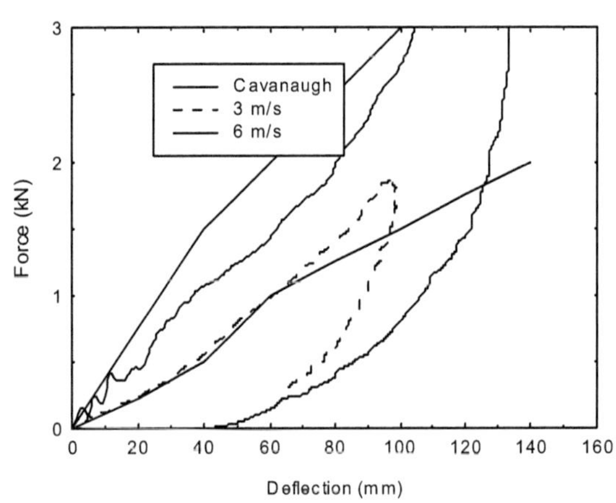

FIGURE 12B. Force-deflection of LSR #5 **with** chest jacket, in rigid bar 24 kg pendulum tests at 3 and 6 m/s, compared to the Cavanaugh low speed corridor.

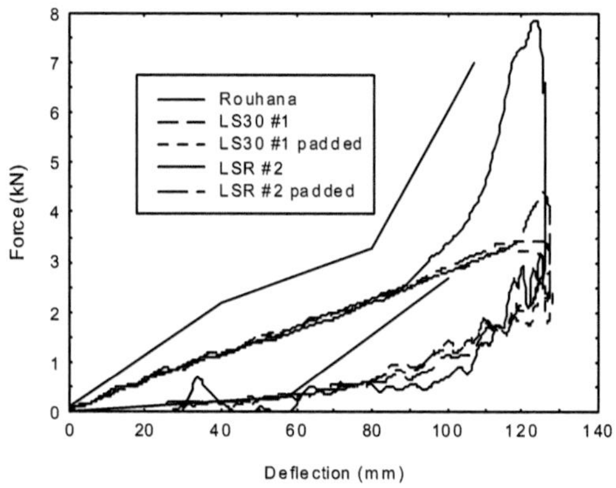

FIGURE 13. Force-deflection response of urethane and LSR #2 inserts in MTS belt tests compared with corridor from Rouhana et al. (1989).

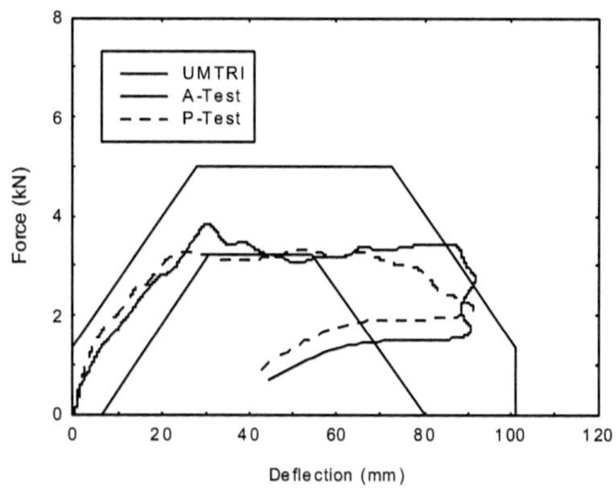

FIGURE 14. Force-deflection response of LSR #5 insert in UMTRI belt test compared with UMTRI belt corridor (A and P-tests are defined in text).

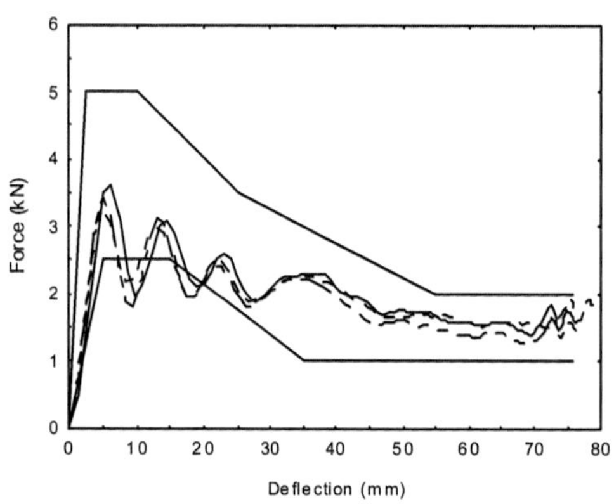

FIGURE 15. Force-deflection response of LSR #5 insert in surrogate airbag tests compared with corridor from Hardy et al. (2001).

Instrumentation Tests

The results of tests with the instrumentation methods investigated are outlined in Appendix Table A.II. None of the methods examined met all of the requirements for crash testing. Each method had advantages and disadvantages that became apparent during these tests. The method chosen for the final design is the HRL method.

Electrical Resistance Measurements - Negligible cross-talk and electrical noise was observed. However, a number of other issues did surface during these tests. First, in the MTS tests, with the dummy in the supine position, a non-linear relationship was observed between the MTS displacement and the electrode voltage output (Figure 16A). In the pendulum tests, with the dummy in the seated position, a linear relationship was observed for most of the displacement range until near the maximum penetration at which point the curve became highly non-linear (Figure 16B). These results are probably

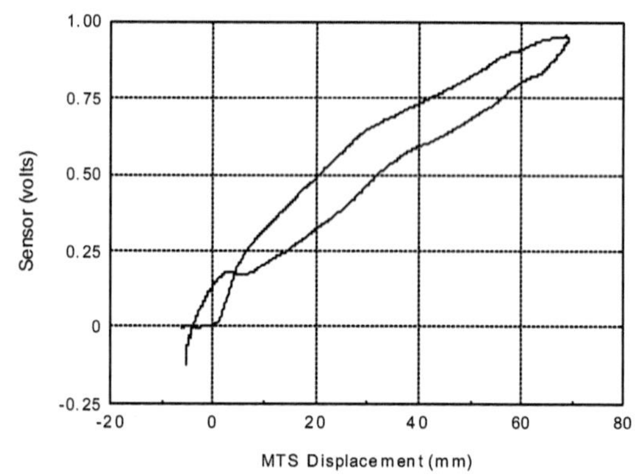

FIGURE 16A. Voltage output from HRL sensor in MTS test of LSR #6 insert at 3.0 m/s.

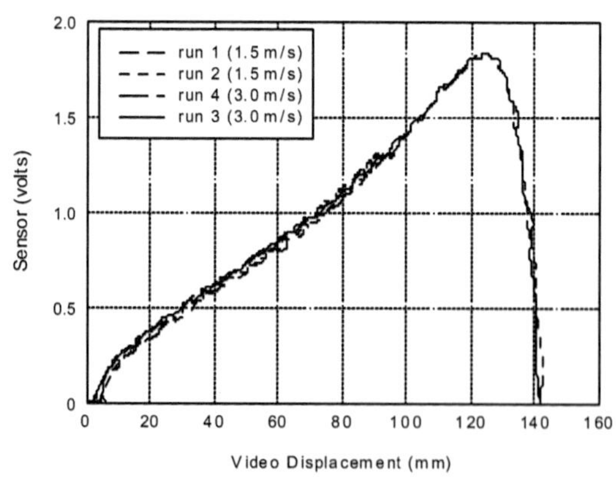

FIGURE 16B. Voltage output from HRL sensor in pendulum tests of LSR #6 insert. Note non-linearity of initial contact phase and as pendulum nears spine.

due to the boundary conditions of the bladder in the different positions.

As shown, the HRL method does not provide a linear output as a function of abdominal compression. It was chosen as the final design because it was considered to have the best overall performance and might be able to be tuned for crash testing with the least additional work as will be discussed later.

DISCUSSION

As mentioned previously, the final design for the abdominal insert is a liquid silicone rubber shell with 5.0 mm penetrometer reading silicone gel and the HRL Kelvin probe for deflection measurements. An ultra-high viscosity silicone fluid could be used instead of the silicone gel with equivalent force-deflection response and performance. An evaluation of how well this system met the original design criteria follows.

Design Criteria

Material Design Criteria

Reproducibility – Figures 17A and 17B show the reproducible nature of the abdominal insert design. Figure 17A is the force-deflection response of the LSR#5 insert, which had an average shell thickness of 2.5 mm (0.1"), a shell hardness of 40 durometer and 5.0 mm penetrometer gel. Figure 17B is the force-deflection response of the LSR#2 insert which had an average shell thickness of 5.1 mm (.2"), a shell hardness of 30 durometer and 5.0 mm penetrometer gel. The 30-durometer material allowed somewhat more penetration, but the force-deflection response during loading was very similar to the 40-durometer material.

Likewise, Figure 13 shown previously (under the belt testing heading) shows that the design appears to be very reproducible across similar materials. Two of the curves were obtained from an LS30 urethane shell prototype and two were obtained from a liquid silicone rubber shell prototype (all gel filled).

Repeatability – Figure 13 also shows one example of the repeatability of the abdominal inserts. In the loading phase, the curves virtually overlay one another through the first 90 to 120 mm of deflection. Similarly, Figure 15 shown previously (airbag force-deflection corridor) shows 3 tests of the abdominal insert relative to the cadaver airbag force-deflection corridor. These curves show excellent agreement through the entire loading event.

Durability – The inserts were very durable under the test conditions to which they were exposed. One LS-30 Urethane prototype was subjected to 24 pendulum and MTS tests at speeds up to 10 m/s without failure. One silicone insert was subjected to more than 65 tests without failure. One silicone insert failed in a pendulum test with a rigid bar when the padding around the steel plates that join the lumbar and thoracic spines was inadvertently left out of the dummy. Therefore, the lumbar spine padding appears to be a necessity for these inserts. In general, the molded urethane inserts appeared to be more vulnerable to tears resulting from sharp edge contacts than the liquid silicone rubber inserts. This was an important factor in the decision to use liquid silicone rubber in the final design.

Sealing of the inserts – The liquid silicone rubber inserts were sealed with sheet silicone and medical grade silicone adhesive (Figure 18). Failure was never observed in any of the liquid silicone rubber insert seals. This was another important factor in the decision to use silicone rubber in the final design.

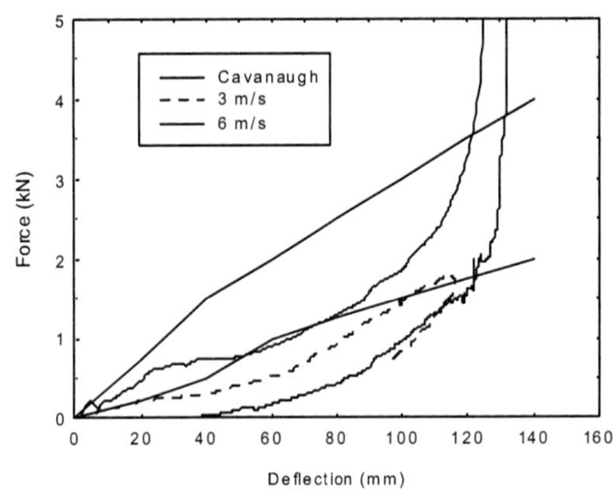

FIGURE 17A. Force-deflection from rigid bar pendulum tests of LSR #2 compared with low speed Cavanaugh corridor.

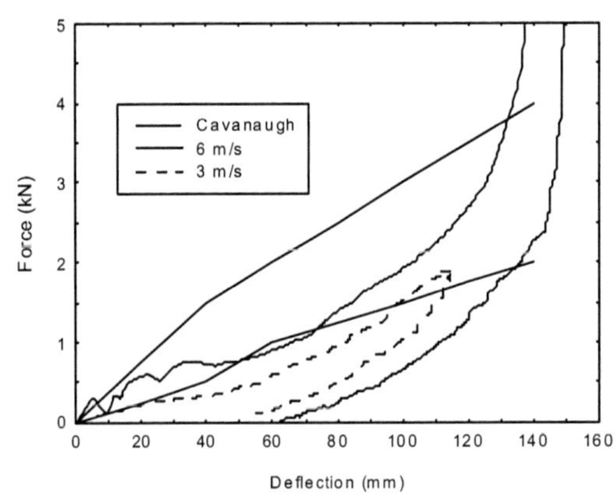

FIGURE 17B. Force-deflection from rigid bar pendulum tests of LSR #5 compared with low speed Cavanaugh corridor.

Effect of failure on other dummy systems – In the case of the insert failure mentioned above, caused by contact of the insert with the unpadded metal edges of the thoraco-lumbar junction plates, there were no deleterious effects on other dummy systems because the gel did not leave the insert.

Chemical interactions – Early studies examined the chemical stability of urethane and silicone inserts with various potential filler materials including water, motor oils, vegetable oils, glycerol solutions and silicone gels. No chemical reactions were observed over a period of approximately six months. Inserts were weighed, measured and tested on the MTS machine. Water was able to evaporate through the urethane inserts, but the other materials did not cross the boundary. Measurements indicated approximately 1 mm change in dimension of the circular bladders as measured from the table surface to the top of the bladder. MTS tests indicated that the mechanical properties as measured by the slope of a force-deflection measurement of the urethane inserts did not change by more than 10% over the 6-month period.

There did not appear to be any measurable changes in the liquid silicone rubber inserts over the time of this project when filled with silicone gel.

Abdomen Dummy Interface Design Criteria

Dummy Mass – As mentioned previously, the Hybrid III dummy required a small number of changes to accommodate the new abdominal insert. The chest deflection potentiometer and its bracket hardware (.85 kg) and the Cerobase™ from the pelvic adapter block (5.9 kg) had to be removed. The result of these changes was a total reduction in the weight of the dummy by 6.8 kg (14.9 pounds). The weight of the silicone gel-filled liquid silicone rubber insert was approximately 5.7 kg (12.6 pounds). Therefore, the amount of reballasting that is necessary is approximately 0.9 kg (2.3 pounds). This reballasting may be done by adding lead shot in the pelvic basin or a block of metal in the thoracic spine. Moments of inertia of the abdominal and pelvic regions were not measured in this effort, but could guide the placement of these masses.

It is likely that the mass distribution in the abdominal and pelvic areas is improved over the baseline Hybrid III. This is because the abdominal and pelvic mass is more distributed with the new abdomen than in the dummy. The human abdominal mass is also more distributed than the Hybrid III.

Chest Deflection Measurement – Although the chest deflection potentiometer and bracket were removed, the measurement of chest deflection can still be achieved through the use of the IR-TRACC device (InfraRed - Telescoping Rod for the Assessment of Chest Deflection; Rouhana et al., 1998). This device allows measurement of chest deflection at speeds up to 30 m/s and is available as a retrofit kit from FTSS.

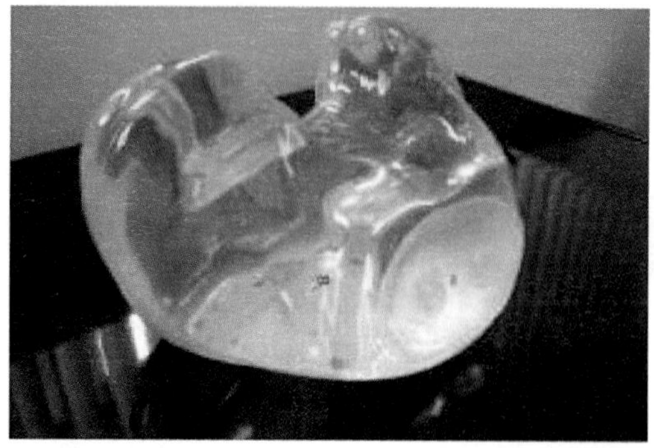

FIGURE 18. LSR #8 abdominal insert after being sealed with silicone sheet and silicone adhesive.

Posture – The posture of the dummy did not appear to be changed with the addition of the abdominal insert. However, the lower rib of the ribcage typically rested right on top of the insert with the dummy in the seated position. There was a clear effect of the ribcage on rate sensitivity of the dummy as will be discussed later.

Sled Test Kinematics – Since the sled test matrix was never completed, it was not possible to evaluate the comparative kinematics of the dummy with and without the abdominal insert. However, in the five sled tests performed to examine dummy performance in submarining and non-submarining situations, the abdominal insert always remained in-position in the dummy.

Force-deflection Response criteria

Load Altering Response – There was no indication that loading by pendulum, seat belt or airbag altered the force-deflection response of the abdomen in any of the tests performed. As shown in previous figures, the repeatability and reproducibility of the inserts was excellent. Tests were conducted with LSR prototype number 1 from August of 1998 through November of 1999. The insert showed no changes in properties over that 15-month period.

Instrumentation

Resistance measured in a wire is a function of both the length of the wire and its cross-sectional area. Likewise, with the four-point Kelvin probe measurements, the measured resistance was expected to be a function of both the linear distance between the anodes and the common cathode and the cross-sectional area of the bladder ("the wire"). In the MTS tests, with the dummy on its back, the abdomen was constrained posteriorly by the spine and MTS table with little constraint anteriorly. In the pendulum tests, with the dummy seated upright, the abdomen was only constrained posteriorly by the lumbar spine. So, the abdominal insert was able to conform around the lumbar spine and move to the posterior part of the pelvis more than in the MTS tests. Paradoxically, this led to a greater initial abdominal depth in the MTS tests with the dummy on its back than in the pendulum tests with the dummy seated.

Similarly, because of the rigid rear surface in the MTS tests, the abdomen had no direction to "flow" during compression except to the top and sides. This constraint almost ensured that there would be a change in cross-sectional area. Thus, the abdomen tended to push out to the sides and to push up into the thoracic cavity. In contrast, in the pendulum tests with the dummy in the seated position, the abdomen could "flow" rearward around the lumbar spine and therefore, the cross-sectional area was not forced to change.

It is anticipated that, in further development, two of the seven channels could be repositioned to the sides of the bladder and used to estimate the lateral dimensional changes. This might in turn be used to calculate cross-sectional area changes to correct the resistance measurements.

Another issue noted in the pendulum tests was that the sensor output changed sharply at the time of pendulum contact with the anterior abdominal surface and again as the pendulum approached contact with the spine. It is believed that when the initial contact with the anterior surface of the abdominal insert occurred, the electrodes within changed their orientation slightly as they aligned with the impacting surface. This was not expected to alter the resistance but apparently did. A fix for this might be to use smaller or spherical electrodes in a future design. The sharp change in output as the pendulum neared the spine was believed to be associated with near-field effects of the electrodes, which could be reduced by choosing sensor locations that minimized the probability of contact between anode and cathode.

Finally, it was noted that a signal discontinuity occurred if the anterior surface of the abdomen separated from the gel as might occur if the surface moves outward relative to the spine.

Force-Deflection Response

The force-deflection response of the abdominal insert is an aggregate response of the dummy's entire abdominal area and does not address differences in upper versus lower abdominal response, solid versus hollow organs, or organ position or mobility.

It was stated above that the abdominal inserts meet both the Rouhana belt corridor and the UMTRI belt corridor, yet Hardy et al. (2001) note that the two corridors are different. The reason the corridors are different is because they were obtained under different test conditions. The tests that formed the basis of Rouhana's corridor (1989; Miller 1989) used a flat 50 mm wide seat belt that was pushed into abdomen by a yoke attached to an MTS machine (see Figure 7A). This resulted in loading that was mainly from the front with nearly no abdominal loading from the side. The abdomen, because of the relatively high mobility of its organs, is able to squeeze sideways as a load is applied from the front. Therefore, there is a lower force for given deflection in this test setup.

In contrast, the UMTRI tests had the belt wrapped around abdomen (Figure 8) so, the abdomen could move sideways. This boundary condition, therefore, caused higher forces for a given penetration. This highlights the necessity to do biofidelity verification tests for dummies under the same test conditions as the original data was obtained. The fact that the liquid silicone rubber abdominal inserts met both corridors helps to demonstrate the degree of biofidelity obtained.

It was also stated that the abdominal insert's force-deflection response for airbag loading using the surrogate airbag (Figure 9) was in excellent agreement with the data obtained with cadavers. To assess the similarity of the test conditions, the penetration-time and speed-time histories of the real and surrogate airbags into the cadaver abdomen and the surrogate airbag into the dummy abdomen were compared in Figures 19 and 20, respectively. As can be seen in Figure 19, the penetration-time history of the real airbag into the cadaver abdomen, the surrogate airbag into the cadaver abdomen and the surrogate airbag into the dummy abdomen are very similar for the loading phase up to the peak deflections at approximately 10 ms. After the peak deflection is reached, the curves for the surrogate airbag into the dummy abdomen and the real airbag into the cadaver abdomen remain very similar and show some rebound. The penetration of the surrogate airbag into the cadaver remains constant.

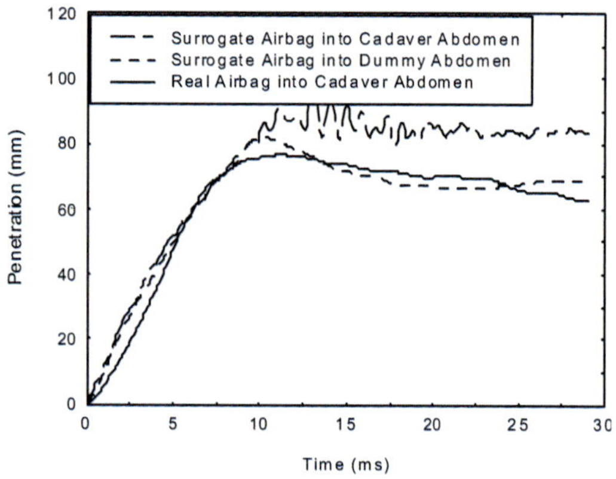

FIGURE 19. Comparison of penetration time from real airbag into human cadaver (solid line), surrogate airbag into cadaver (big dashed line) and surrogate airbag into LSR #5 dummy abdomen (small dashed line)

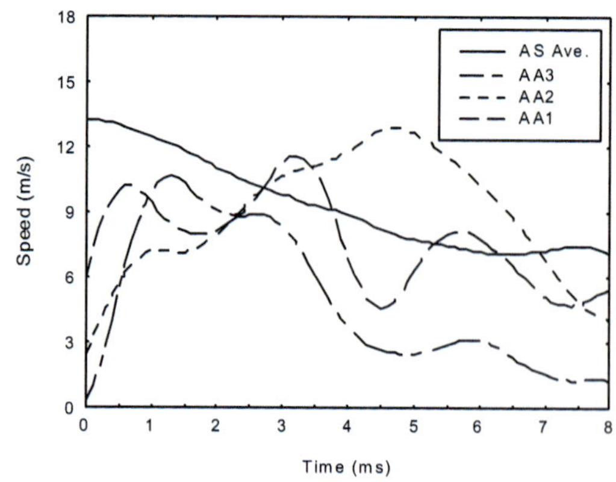

FIGURE 20. Penetration speed versus time for three real airbags (AAx) into human cadaver abdomens compared with that of the surrogate airbag into an LSR #5 abdominal insert *in situ* (AS Average).

Figure 20 shows that, in general, the speed-time history of real airbag into the cadaver abdomen is also very similar to the speed-time history of the surrogate airbag into the dummy abdomen. A difference is observed in the initial loading phase (approximately the first 1 ms) due to the fact that the airbag started out in contact with the abdomen with no relative velocity, but the surrogate airbag had to be accelerated up to speed before impact and thus had its peak velocity at the time of impact. Nevertheless, the differences are confined to the initial region (< 1 ms) and after that show excellent agreement. Thus, since the test conditions appear to be nominally identical, the conclusion of biofidelity for airbag loading appears to be justified.

Abdominal Rate-Sensitivity

While the goals of a rate-sensitive abdomen have been mostly achieved, the abdomen developed did not exhibit all of the rate-sensitivity observed in the human cadaver tests. The degree of rate sensitivity was measured by the increase in slope of the force-deflection curves as the abdomen was tested at progressively higher velocities. In particular, while the force-deflection response of the abdomen in the low speed rigid bar tests was well within the low speed corridor, the response did not increase enough in the higher speed tests to stay within the high-speed rigid bar corridor. The fact that the abdomen did not stay within the high-speed corridor may not be important since the presence of airbags in vehicles makes it unlikely that an occupant would impact a steering system at 9 m/s. Thus, the high speed rigid bar test may no longer be applicable to the automotive environment. Nevertheless, the parameters affecting the rate-sensitivity of the abdominal inserts were explored further.

In general, it was noted that the rate-sensitivity of the abdominal inserts was affected by the boundary conditions of the tests performed. For example, there was a marked increase in apparent rate-sensitivity when the dummy was tested after removing the four lower ribs and allowing the abdomen to deform to a larger extent upward into the chest cavity. Similarly, when the abdominal insert was tested in the dummy without the chest jacket, the apparent rate-sensitivity was reduced compared to tests with the chest jacket on.

In addition, as noted previously, the dummy abdomen did not fit into the higher speed rigid bar force-deflection corridor as well as it fit into the lower speed corridor. This was apparent in tests with the second prototype liquid silicone rubber abdomen (LSR#2) when tested with the 48 kg UMTRI pendulum with a rigid bar impactor. As Figure 21 shows, at 10.6 m/s there was an initial increase in

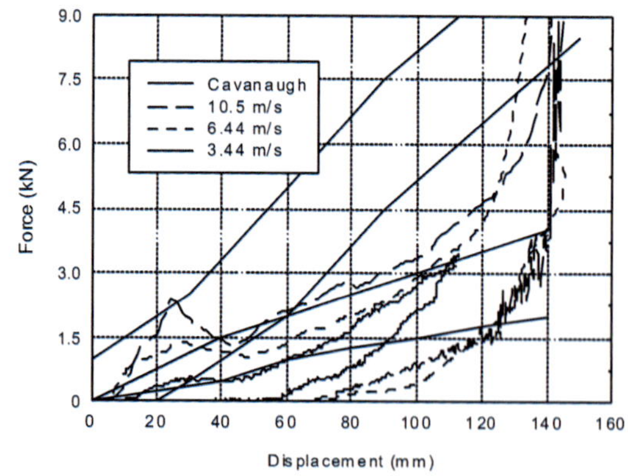

FIGURE 21. Force-deflection response of LSR #2 in 48 kg rigid bar pendulum tests showing drop in force after 25 mm of penetration in 10 m/s test.

load followed by a marked decrease as the deflection exceeded 20 mm. This was only marginally apparent at 6.4 m/s and still less apparent at 3.5 m/s.

It is not clear whether this diminished rate-sensitivity in the dummy was due to differences in the actual rate sensitivity of the abdominal system, differences in effective mass during impact, or structural differences.

One hypothesis considered was that the rate-sensitivity of the human lumbar spine might affect the rate-sensitivity of the measured human abdominal response since the impact to the human cadaver subjects included the abdomen and lumbar spine as a system. There are many obvious structural differences between the human and dummy lumbar spines.

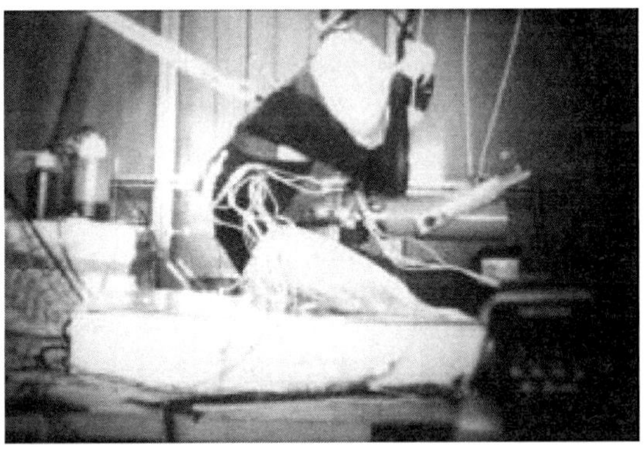

FIGURE 22A. Cadaver kinematic response in 48 kg rigid bar pendulum test showing "C-shape" of cadaver during impact.

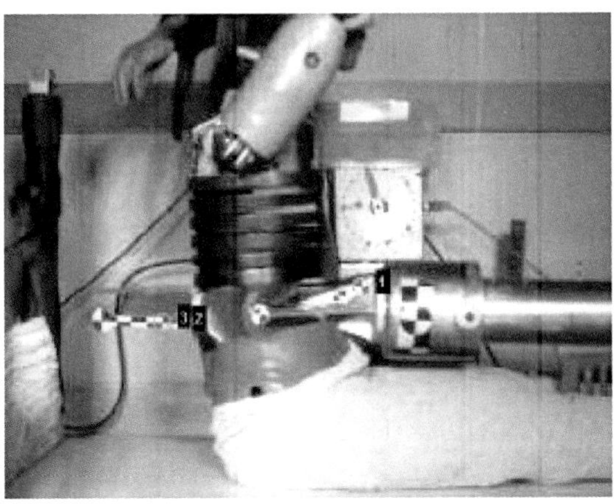

FIGURE 22B. Dummy kinematic response in 24 kg rigid bar pendulum test showing lack of "C-shaped" response.

Another hypothesis was generated by the observation that, during the rigid bar testing, the kinematics after the initial contact were markedly different between the cadavers and the dummy. As seen in Figure 22A, the cadaver started to bend over in a "c-shape" during the impact, while in contrast, Figure 22B shows that the dummy remained upright in its seated posture through nearly the entire impact. Since the time scale of this bending over in the cadaver was in milliseconds, it was presumed to be more than the gravitational force that caused this bending.

It is hypothesized that the presence of the rectus abdominus muscles and the other soft tissues that unite or integrate the thoracic, abdominal and pelvic regions of the human are responsible for the bending over of the cadaver. Specifically, it could be that the presence of the rectus abdominus allowed the impactor to recruit more and more of the subject's mass into the impact as the penetration proceeded into the cadaver due to the tension built up in the muscle. Because the rectus abdominus insertions are into the ribcage and pelvis, this would have the effect of coupling those masses into the abdominal impact.

To test these observations and hypotheses, the force-deflection response of the free-back cadaver tests were compared with the response from the fixed-back cadaver test. The fixed-back test was done to remove the contribution of the lumbar spine so that a "purely abdominal" response could be observed. The test setup for the fixed-back tests included clamping the spine of the cadaver to a rigid support and mechanically "pulling the punch" to limit the absolute penetration into the cadaver to 200 mm (Hardy et al. 2001). The clamped spine prevented the c-shaped response of the free-back impacts and likely reduced the mass recruitment into the impact. Instead, the soft tissues probably yielded more although this was not tested at the time. As shown in Figure 23 from Hardy et al. (2001; Figure 10B), the cadaver fixed-back force-deflection response falls below the free-back response corridor after approximately 60 mm of deflection.

To examine the hypothesis further, the dummy was modified to simulate the presence of rectus abdominus muscles with an LS-30 urethane abdominal insert *in situ*. This simulation was achieved by wrapping two pieces of standard 50 mm seat belt webbing around the torso and pelvis in the sagittal plane. The pieces were positioned to the left and right of the midsagittal plane and were wrapped over the shoulders, behind the back, under the buttocks, between the legs and up to the shoulders again where the two ends were ties together.

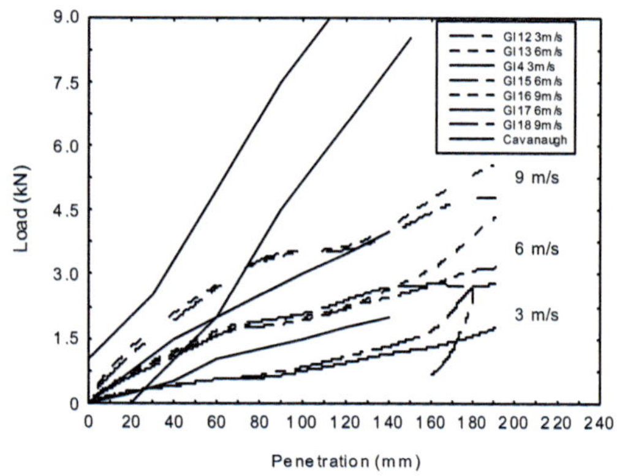

FIGURE 23. Fixed back cadaver response from Hardy et al. (2001) compared with Cavanaugh free-back response corridors showing drop in force after approximately 50 ms in 9 m/s tests.

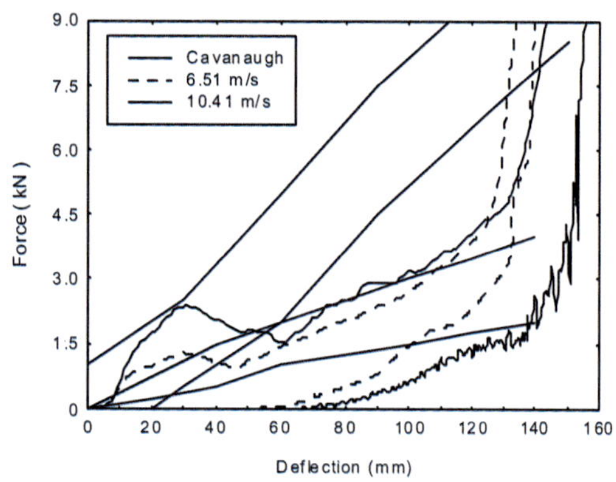

FIGURE 24A. Force-deflection response of LSR #1 abdominal insert in 48 kg rigid bar pendulum tests.

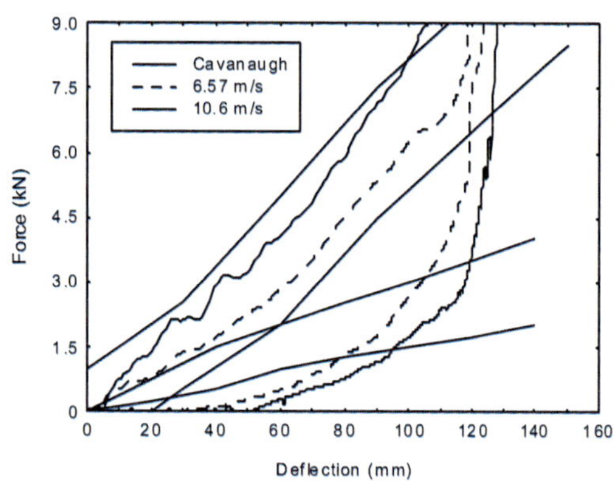

FIGURE 24B. Force-deflection response of LSR #1 abdominal insert in 48 kg rigid bar pendulum tests with simulated *rectus abdominus* muscles.

Figure 24 shows the results of two tests at 6.5 m/s and two tests at 10.5 m/s. At each speed, one test was run with the simulated rectus abdominus and one test was run without it. There is no visible difference in force-deflection up to 40 mm, at which point it is assumed that the thorax becomes drawn into the impact requiring more force to continue penetration as seen in the figure. The rate sensitivity of the abdomen as measured by the increase in slope of the force-deflection curves appears to have increased dramatically. In fact, with the simulated muscles, the response in the 6 m/s test was too stiff. The kinematics of the dummy with the simulated rectus abdominus began to resemble those of the cadaver.

Without more extensive testing, the reason for less rate-sensitivity in the rigid bar dummy abdomen tests remains a hypothesis. It appears to be mostly influenced by the mass recruitment, but is probably also affected by the lumbar spine involvement.

It is interesting to note that a similar phenomenon has been observed in thoracic impacts. In Figure 16 of Foster et al. (1977) the force-deflection curves for the Hybrid III dummy in a 6.7 m/s pendulum impact show an initial peak at approximately 8 mm of deflection. This is followed by a reduction in force as penetration continues until approximately 25 mm of deflection. After that, the force increases again to its maximum value around the time of maximum deflection. If the Hybrid III had an integrated thorax, abdomen and pelvis, it is possible that this drop in the force-deflection curve would be eliminated by the mass recruitment discussed above, bringing the response closer to that of the Kroell corridors which were obtained from human cadaver tests.

Injury Assessment Strategy (Criteria)

Given the fact that none of the instrumentation methods could meet the needs of automotive crash testing without further development, other strategies for injury assessment were explored. The assessment of submarining injury risk may be the most amenable to use with the abdomen in the absence of deflection instrumentation.

Leung et al. (1982) performed tests with belted human cadavers to study injury from submarining. Their data have been analyzed further to evaluate the relationship between belt force and abdominal injury. Figure 25 shows the abdominal AIS as a function of peak normalized lap belt force after submarining occurred. This figure was created using data from Figure 19 in Leung's paper. There is a clear relationship between abdominal AIS and lap belt force after submarining.

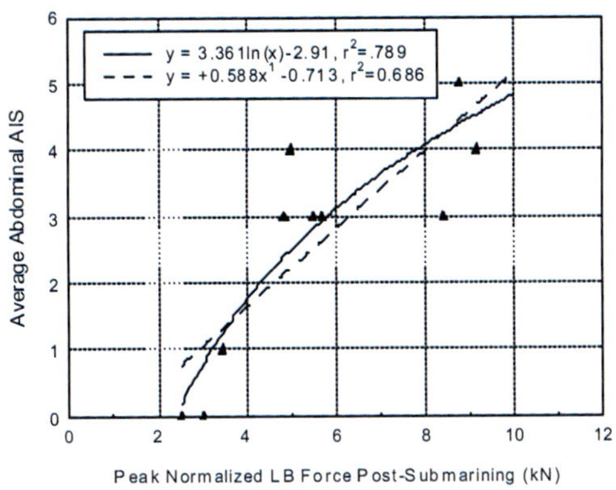

FIGURE 25. Average abdominal AIS in cadaver subjects versus peak normalized lap belt force after submarining has occurred (from reanalysis of Leung et al. 1982).

The effects of impact force on abdominal injury have been studied previously and shown to be predictive of some types of abdominal injury (Rouhana et al. 1986). Logistic regression analysis of injury as a function of belt force by Miller (1989) showed maximum force to be as well correlated to probability of AIS >= 3 injury as V*C and other criteria (R=.668, X2=12.8, p=.0003). The ED25 and ED50 were 2.93 kN and 3.96 kN, respectively. These are in good agreement with the Leung study (Figure 25), which clearly shows a transition to serious injury around a force of 4 kN.

Since the abdominal insert developed in this work has biofidelity in belt loading, it can be used in conjunction with the iliac spine load cells now available for most Hybrid III dummy pelves to provide injury assessment for submarining. The iliac spine transducers can be used to determine whether submarining has occurred, and if so, the belt force-abdominal AIS relationship can be used to evaluate the risk of abdominal injury even in the absence of a deflection transducer.

Future Work

While the force-deflection response of the abdomen developed did not stay fully within the high-speed rigid bar corridor, the abdomen has demonstrated good biofidelity in the low speed rigid bar, seat belt and airbag loading situations. This should be adequate for automotive crash testing needs.

Still, some work remains to be done before this abdomen can be used in crash testing. This work includes improving the instrumentation in order to have injury assessment capability in more situations than just submarining. While the four-point Kelvin probe system appears to have the best near term potential, the magnetic position sensor may have better potential in the long term as an easier system to manufacture since it can be bonded to the outside surface of the abdominal insert.

Similarly, in order to understand the influence of the abdomen on other dummy measurements and kinematics, a full set of sled tests under various loading conditions should be done. These tests could compare a standard Hybrid III and THOR dummy's kinematics and measured responses to those of a Hybrid III with the new rate-sensitive abdominal insert. Particular effects to examine include the effects of ribcage impingement on measurements made inside abdomen and the effects the presence of the abdomen might have on the kinematics of the dummy and chest deflection measurements.

CONCLUSION

A program to develop a rate-sensitive, reusable abdomen for crash test dummies has been partially successful. While the amount of rate-sensitivity achieved was not large, when tested as a system with abdomen *in situ* and the dummy chest jacket on, the rate sensitivity was adequate and appropriate for the automotive environment. Moreover, the biofidelity of the abdominal insert in rigid-bar, seat belt and airbag loading conditions is good. Most of the design criteria have been assessed and accomplished.

Although the electrical resistance method of deflection measurement shows promise, injury assessment in steering wheel impacts and airbag loading cannot be assessed at this time. However, when coupled with a measurement of lap belt force and iliac wing load cells, the abdomen should enable the assessment of the occurrence of submarining and abdominal injury risk assessment in submarining situations. Before it can be used, however, a series of sled tests is necessary to further examine abdomen/dummy interactions, especially the effects of the abdomen on measurements of chest deflection and the kinematics of the dummy

ACKNOWLEDGMENTS

This work was supported in part by General Motors pursuant to an agreement between GM and the Department of Transportation.

Many people contributed to this work by helping out in the laboratories in which the testing took place or by engaging in thoughtful discussion over the course of the project. While the course of time and pressures of today may have caused some to be overlooked, all who contributed are sincerely thanked and the authors express regret over anyone who should be included in this list but is not.

The authors' gratitude goes out to Ken Desaele, Brian Frantz, Janis Georgen, Kit Green, John Hiben, Annette Irwin, Ed Jedrzejczak, Doug Kononen, Ian Lau, Joe McCleary, John Melvin, Deb Nowak-Vanderhoef, Jim Pywell and Bob Wooley at GM; Brian Eby, Tom Jeffreys, Kathy Klinich, Matt Reed, Leda Ricci, Jon Rupp and Stewart Simonett at UMTRI.

In addition, the work of John Pasiecznik and David Rensch of Hughes Research Labs, Carl Nelson and Brian Jacobs of Johns Hopkins University - Applied Physics Labs, Caroline VanIngen-Dunn and Phil Schaefer of Simula Safety Systems, Larry Lee and Frank Davies of Ktech Corp, Measurand Corp, and Sonometrics, Inc. was truly appreciated.

Finally, the authors express gratitude to Rolf Eppinger and Mark Haffner at the NHTSA for many stimulating and helpful discussions over the course of this project.

REFERENCES

Augenstein, J. S., Digges, K. H., Lombardo, L.V., Perdeck, E. B., Stratton, J.E., Malliaris, A. C., Quigley, C. V., Craythorne, A. K., Young, P. E. (1995) Occult abdominal injuries to airbag-protected crash victims: A challenge to trauma systems. J. Trauma 38(4): 502-508.

Augenstein, J. S., Perdeck, E. B., Digges, K. H., Lombardo, L.V., Malliaris, A. C. (1996) Injuries sustained by airbag protected drivers. In Occupant Protection Technologies for Frontal Impact, SP-1144, SAE Technical Paper No. 960660. Society of Automotive Engineers, Warrendale, PA.

Baxter, C. F., Williams, R. D. (1961) Blunt abdominal trauma. J. Trauma 1: 241-247.

Beebe, M. S. (1990) What is BioSID? SAE Technical Paper No. 900377. Society of Automotive Engineers, Warrendale, PA.

Begeman, P.C., Kopacz, J.M., King, A.I. (1990) Steering assembly impacts using cadavers and dummies. Proc. 34th Stapp Car Crash Conference, pp. 123-144, SAE Technical Paper No. 902316. Society of Automotive Engineers, Warrendale, PA.

Bergqvist, D., Hedelin, H., Lindblad, B., Matzsch, T. (1985) Abdominal injuries in children: an analysis of 348 cases. British J. Accident Surgery 16(4): 217-220.

Biard, R., Alonzo, F., Russo, J.-L., Bloch, J. (1993) Child abdominal protection: presentation of a new transducer for child dummies. Proc. 1st Child Occupant Protection Symposium, SP-986, pp. 261-273, SAE Technical Paper No. 933106. Society of Automotive Engineers, Warrendale, PA.

Biard, R., Alonzo, F., Muge, A.-C., Dejeammes, M., Chamouard, F., Tarriere, C. (1997) Torso improvements in child dummies used for certification tests in europe. Proc. 2nd Child Occupant Protection Symposium, P-316, pp. 231-242, SAE Technical Paper No. 973315. Society of Automotive Engineers, Warrendale, PA.

Bond, S. J., Gotschall, C. S., Eichelberger, M. R. (1991) Predictors of abdominal injury in children with pelvic fracture. J. Trauma 31(8): 1169-1173.

Cavanaugh, J. M., Nyquist, G. W., Goldberg, S. J., King, A. I. (1986) Lower abdominal tolerance and response. Proc. 30th Stapp Car Crash Conference, pp. 41-63, SAE Technical Paper No. 861878. Society of Automotive Engineers, Warrendale, PA.

Chamouard, F., Tarriere, C., Baudrit, P. (1996) Protection of children on board vehicles – influence of pelvis design and thigh and abdomen stiffness on the submarining risk for dummies installed on a booster. Proc. 15th International ESV Conference, pp. 1063-1075, SAE Technical Paper No. 976088. Society of Automotive Engineers, Warrendale, PA.

Cox, E. (1984) Blunt abdominal trauma - a 5-year analysis of 870 patients requiring celiotomy. Annals of Surgery 199(4): 467-474.

Czernakowski, W., Klanner, W. (1987) Development of a two-dimensional sensor determining abdominal loading on tno-dummies. Proc. 11th International ESV Conference, pp. 323-332, SAE Technical Paper No. 876044. Society of Automotive Engineers, Warrendale, PA.

Daniel, R. P., Koga, M. S., Prasad, P., Yost, C. D. (1982) A force measuring mechanical test device for estimating and comparing the energy absorbing characteristics of vehicle interior side panels. Proc. 9th International ESV Conference, pp. 465-469, SAE Technical Paper No. 826058. Society of Automotive Engineers, Warrendale, PA.

Danner, M., Langwieder, K. (1979a) Collision characteristics and injuries to pedestrians in real accidents. Proc. 7th International ESV Conference, pp.587-610. Society of Automotive Engineers, Warrendale, PA.

Danner, M., Langwieder, K., Wachter, W. (1979b) Injuries to pedestrians in real accidents and their relation to collision and car characteristics. Proc. 23rd Stapp Car Crash Conference, pp. 161-198. Society of Automotive Engineers, Warrendale, PA.

Digges, K., Haffner, M., Lombardo, L., Stucki, L., Malliaris, A., Augenstein, J., Perdeck, E. (1996) Challenges in injury measurement technology for testing of driver airbag systems. Proc. 15th International ESV Conference, pp. 1578-1583, SAE Technical Paper No. 976153. Society of Automotive Engineers, Warrendale, PA.

Elhagediab, A. M., Rouhana, S. W. (1998) Patterns of abdominal injury in frontal automotive crashes. Proc. 16th International ESV Conference, pp.327-337. Society of Automotive Engineers, Warrendale, PA.

Eppinger, R. H., Marcus, J. H., Morgan, R. M. (1984) Development of dummy and injury index for nhtsa's thoracic side impact protection Research Program. SAE Technical Paper No. 840885. Society of Automotive Engineers, Warrendale, PA.

Foster, J.K., Kortge, J.O., Wolanin, M.J. (1977) Hybrid III – A biomechanically-based crash test dummy. Proc. 21st Stapp Car Crash Conference, pp. 975-1014, SAE Technical Paper No. 770938. Society of Automotive Engineers, Warrendale, PA

Galer, M., Clark, S., Mackay, G. M., Ashton, S. J. (1985) The causes of injury in car accidents - an overview of a major study currently underway in britain. Proc. 10th International ESV Conference, pp. 513-525. Society of Automotive Engineers, Warrendale, PA

Gray, H. (1977) Anatomy: Descriptive and Surgical. Bounty Books, New York.

Griswold, R. A., Collier, H. S. (1961) Blunt abdominal trauma: Collective review. Surg. Gynec. Obstet. 112: 309.

Gupta, S., Talwar, S., Sharma, R. K., Gupta, P., Goyal, A., Prasad, P. (1996) Blunt trauma to the abdomen: A study of 63 cases. Indian J. Medical Sciences 50(8): 272-276.

Haffner, M., Kleinberger, M., Eppinger, R., Hennessey, B. C., Pritz, H., Beebe, M., Hagedorn, A. V., Klinich, K., Ore, L., Tanner, C. B., Kuppa, S. (1994) Progress in the development of new frontal dummy components for the NHTSA advanced frontal protection program. Proc. 14th International ESV Conference, pp. 1405-1419. Society of Automotive Engineers, Warrendale, PA.

Hardy, W. N., Schneider, L. W., Rouhana, S. W. (2001) Abdominal impact response to rigid-bar, seatbelt, and airbag loading. Stapp Car Crash Journal 45.

Hobbs, C. A. (1980) Car occupant injury patterns and mechanisms. Proc. 8th International ESV Conference, pp. 755-768. Society of Automotive Engineers, Warrendale, PA.

Ishiyama, S., Tsukada, K., Nishigaki, H., Ikeda, Y., Sakuma, S., Matsuoka, F., Kanno, Y., Hayashi, S. (1994) Development of an abdominal deformation measuring system for hybrid iii dummy. Proc. 34th Stapp Car Crash Conference, pp. 265-279, SAE Technical Paper No. 942223. Society of Automotive Engineers, Warrendale, PA.

Janssen, E. G., Vermissen, A. C. M. (1988) Biofidelity of the european side impact dummy - EUROSID. Proc. 32nd Stapp Car Crash Conference, pp. 101-124. Society of Automotive Engineers, Warrendale, PA.

Khaewpong, N., Nguyen, T. T., Bents, F. D., Eichelberger, M. R., Gotschall, C. S., Morrissey, R. (1995) Injury severity in restrained children in motor vehicle crashes. SAE Technical Paper No. 952711. Society of Automotive Engineers, Warrendale, PA.

Klinich, K. D., Burton, R. W. (1993) Injury patterns of older children in automotive accidents. Proc. 1st Child Occupant Protection Symposium, SP-986, pp. 17-24, SAE Technical Paper No. 933082. Society of Automotive Engineers, Warrendale, PA.

Lau, I. V., Viano, D. C. (1986) The Viscous Criterion - Bases and applications of an injury severity index for soft tissues. Proc. 30th Stapp Car Crash Conference, pp. 123-142, SAE Technical Paper No. 861882. Society of Automotive Engineers, Warrendale, PA.

Leung, Y. C., Tarrier, C., Fayon, A., Mairesse, P., Delmas, A., Banzet, P. (1979) A comparison between part 572 dummy and human subject in the problem of submarining. Proc. 23rd Stapp Car Crash Conference, pp. 675-720, SAE Technical Paper No. 791026. Society of Automotive Engineers, Warrendale, PA.

Leung, Y. C., Tarriere, C., Lestrelin, D. (1982) Submarining injuries of 3 pt. belted occupants in frontal collisions - description, mechanisms and protection. Proc. 26th Stapp Car Crash Conference, pp. 173-205, SAE Technical Paper No. 821158. Society of Automotive Engineers, Warrendale, PA.

Mackay, G. M., Hill, J., Parkin, S., Munns, J. A. R. (1993) Restrained occupants on the nonstruck side in lateral collisions. Accid. Anal. & Prev. 25(2): pp. 147-152.

Maltha, J., Stalnaker, R. L. (1981) Development of a dummy abdomen capable of injury detection in side impacts. Proc. 25th Stapp Car Crash Conference, pp.651-682, SAE Technical Paper No. 811019. Society of Automotive Engineers, Warrendale, PA.

Melvin, J. W., Weber, K. (1986) Abdominal intrusion sensor for evaluating child restraint systems. In Passenger Comfort, Convenience and Safety: Test Tools and Procedures, P-174, pp. 249-256, SAE Technical Paper No. 860370. Society of Automotive Engineers, Warrendale, PA.

Mertz, H. J., Weber, D. A. (1982) Interpretations of impact responses of a 3-year-old child dummy relative to child injury potential. Proc 9th International ESV Conference, pp. 368-376, SAE Technical Paper No. 826048. Society of Automotive Engineers, Warrendale, PA.

Miller, M. A. (1989) The biomechanical response of the lower abdomen to belt restraint loading. J. Trauma 29(11): 1571-1584.

Mooney, M. T., Collins, J. A. (1986) Abdominal penetration measurement insert for the hybrid iii dummy. In Passenger Comfort, Convenience and Safety: Test Tools and Procedures, P-174, pp. 285-289, SAE Technical Paper No. 860653. Society of Automotive Engineers, Warrendale, PA.

Nahum, A. M., Siegel, A. W., Brooks, S. (1970) The reduction of collision injuries: past, present and future. Proc 14th Stapp Car Crash Conference, SAE Technical Paper No. 700895. Society of Automotive Engineers, Warrendale, PA.

Nusholtz, G. S., Kaiker, P. S., Lehman, R. J. (1988) Steering system abdominal impact trauma. UMTRI Report No. 88-19. MVMA, Detroit. MI.

Perry, J. F. (1965) A five-year survey of 152 acute abdominal injuries. J. Trauma 5: 53-61.

Rangarajan, N., Shams, T., White, R. P., Zhao, Y-M., Beach, D., Haffner, M., Eppinger, R. H., Digges, K. H. (1996) Design and evaluation of an instrumented abdomen for the nhtsa advanced dummy. Proc. 15th International ESV Conference, pp. 1622-1631, SAE Technical Paper No. 976157. Society of Automotive Engineers, Warrendale, PA.

Rangarajan, N., White, R. P., Shams, T., Beach, D., Fullerton, J., Haffner, M. P., Eppinger, R. H., Rhule, D., Pritz, H., Dalmotas, D., Fournier, E.: (1998) Design and performance of the thor advanced frontal crash test dummy thorax and abdomen assemblies. Proc. 16th International ESV Conference, pp. 1999-2010, Technical Paper No. 98-S9-O-12. Society of Automotive Engineers, Warrendale, PA.

Ricci, L. (1980) NCSS statistics: passenger cars. UM-HSRI Report 80-36. UMTRI, Ann Arbor, MI.

Rouhana, S.W.: (1993-1st Ed.) (2001-2nd Ed.) Biomechanics of abdominal trauma, In Accidental Injury: Biomechanics & Prevention, A.M. Nahum and J.W. Melvin, pp. 391-428. Springer-Verlag, New York.

Rouhana, Stephen W., Elhagediab, Ali M., Chapp, Jeffrey J. (1998) A high-speed sensor for measuring chest deflection in crash test dummies. Proc. 16th International ESV Conference, SAE Technical Paper No. 98-S9-O-15. Society of Automotive Engineers, Warrendale, PA.

Rouhana, S. W., Lau, I. V., Ridella, S. A. (1984) Influence of velocity and forced compression on the severity of abdominal injury in blunt, nonpenetrating lateral impact. GMR Research Publication No. 4763. General Motors Research Labs, Warren, MI.

Rouhana, S. W., Foster, M. E. (1985a) Lateral impact - an analysis of the statistics in the NCSS. Proc. 29th Stapp Car Crash Conference, pp. 79-98, SAE Technical Paper No. 851727. Society of Automotive Engineers, Warrendale, PA.

Rouhana, S. W., Lau, I. V., Ridella, S. A. (1985b) Influence of velocity and forced compression on the severity of abdominal injury in blunt, nonpenetrating lateral impact. J. Trauma 25(6): 490-500.

Rouhana, S. W., Ridella, S. A., Viano, D. C. (1986) The effect of limiting impact force on abdominal injury: a preliminary study. Proc. 30th Stapp Car Crash Conference, pp. 65-79, SAE Technical Paper No. 861879. Society of Automotive Engineers, Warrendale, PA.

Rouhana, S. W. (1987) Abdominal injury prediction in lateral impact - an analysis of the biofidelity of the euro-sid abdomen. Proc. 31st Stapp Car Crash

Conference, pp. 95-104, SAE Technical Paper No. 872203. Society of Automotive Engineers, Warrendale, PA.

Rouhana, S. W., Viano, D. C., Jedrzejczak, E. A., McCleary, J. D. (1989) Assessing submarining and abdominal injury risk in the hybrid iii family of dummies. Proc. 33rd Stapp Car Crash Conference, pp. 257-279, SAE Technical Paper No. 892440. Society of Automotive Engineers, Warrendale, PA.

Rouhana, S. W., Jedrzejczak, E. A., McCleary, J. D. (1990) Assessing submarining and abdominal injury risk in the Hybrid III family of dummies: Part II - Development of the small female Frangible Abdomen. Proc. 34th Stapp Car Crash Conference, pp. 145-173, SAE Technical Paper No. 902317. Society of Automotive Engineers, Warrendale, PA.

Schneider, L. W., Haffner, M. P., Eppinger, R. H., Salloum, M. J., Beebe, M. S., Rouhana, S. W., King, A. I., Hardy, W. N., Neathery, R. F. (1992) Development of and advanced atd thorax system for improved injury assessment in frontal crash environments. Proc. 36th Stapp Car Crash Conference, pp. 129-155, SAE Technical Paper No. 922520. Society of Automotive Engineers, Warrendale, PA.

Schneider, L. W., Ricci, L.L., King, A. I., Neathery, R. F., Beebe, M. S. (1992b) Design and development of and advanced ATD thorax system for frontal crash environments, Volume 2: Exploration of alternative design approaches. Final Report DOT HS 808 139. UMTRI, Ann Arbor, MI.

Tso, E. L., Beaver, B. L., Haller, J. A. (1993) Abdominal injuries in restrained pediatric passengers. J. Pediatric Surgery 28(7): 915-919.

Uriot, J., Page, M., Tarriere, C., Bendjellal, F., Loiseleux, F., Saloum, M., Fournier, P. (1994) Measurement of submarining on Hybrid III 50th percentile and 5th percentile dummies. Proc. 14th International ESV Conference, pp. 289-298, SAE Technical Paper No. 946040. Society of Automotive Engineers, Warrendale, PA.

Van Ratingen, M. R., Twisk, D., Schrooten, M., Beusenberg, M.C., Barnes, A., Platten, G. (1997) Biomechanically-based design and performance targets for a 3-year old child crash dummy for frontal and side impact. Proc. 2nd Child Occupant Protection Symposium, P-316, pp. 243-260, SAE Technical Paper No. 973316. Society of Automotive Engineers, Warrendale, PA.

Walfisch, G., Fayon, A., Tarriere, C., Rosey, J. P., Guillon, F., Got, C., Patel, A., Stalnaker, R.L. (1980) Designing of a dummy's abdomen for detecting injuries in side impact collisions. Proc. 5th International IRCOBI Conference, pp. 149-164.

Zuidema, G.D., ed. (1977) The Johns Hopkins Atlas of Human Functional Anatomy. The Johns Hopkins University Press, Baltimore, MD.

APPENDIX I. – ALTERNATIVE DESIGNS

Hardware

Early work on the abdominal inserts involved semi-extensible bladders to investigate the rate-sensitivity of a fluid-filled bladder. These were primarily water-filled polyurethane football bladders. The bladders were made of two single sheets of polyurethane that were electronically (dielectrically) welded together. The bladder had a one-way plastic valve as a fill port. These showed some promise of rate-sensitivity; so more work was done with circular bladders since the simpler geometry was expected to be easier to model analytically. These circular bladders were filled with various fluids to assess chemical interactions. The fluids tested include: water, motor oils, vegetable oils, glycerol/water solutions, electrode and silicone gels.

When confidence was gained with the two-dimensional bladders, a three-dimensional bladder was constructed. The three-dimensional bladders proved to be impossible to seal using the electronic welding technique so a different material was sought.

A search was made of commercially available foam and solid elastomeric systems, but material tests on these did not show much promise. This fact, and the early positive results with the semi-extensible bladders led to an investigation of extensible bladder systems.

After considering many commercially available materials for the shell of the abdominal insert, two materials were selected for parallel development. The first material was a urethane (BJB Enterprises, Inc.) similar to what is already used in crash dummies. The second material was a non-medical grade liquid silicone rubber (Applied Silicone, Inc.), with mechanical properties that are similar to what is used in the medical implant industry. Various formulations were tried before one of each was chosen for the semi-final design (see Table II in the main text and Table A.I in this Appendix).

TABLE A.I. Urethane Shell Materials

Product	Mixed Viscosity (cps)	Shrinkage (in/in)	Tensile Strength (psi)	Elongation (%)	Tear Strength (psi)	Hardness (Shore A)	Working Life (min)	Demold Time (hrs)	Product Highlights
IE-35A (Innovative Polymers)	975	0.001	400	470	50	30-40	10-15	24	- Tough - Abrasion resist. - Allows tinting
Skinflex III (BJB Enterprises) A/B only			510	930	36	25	15		
Skinflex III (BJB Enterprises) A/B + 70% part C			180	820	17	5	15		- Plasticizer can leach out - 5 day cure
LS-15 A/B (BJB Enterprises)	275	0.008	325	>750	22	15+/-2	90	Over night	- Full cure 3-5 days ambient - Resists moisture absorption
LS-25 A/B (BJB Enterprises)	990	0.009	340	1100	40	25	30	16-24	Same as LS-15
LS-30 A/B (BJB Enterprises)	500	.009	400	800	65	30	25	16-24	Same as LS-15

Urethane and liquid silicone rubber materials utilize two different manufacturing methods. The urethane bladders were fabricated using a lost wax molding process by FTSS. The lost wax method is similar to that used for other dummy components. In this process a male and female mold of the part were used. The male mold was made of wax. Liquid urethane material was poured into a gap between the male and female molds which was held constant by small stand-offs. Once cured, the bladder was heated up to the melting point of the wax, which then ran out of the bladder via the filling port.

The standoff holes in the LS-30 Urethane inserts were sealed with cyanoacrylic glue. The fill port was made out of a hard plastic that was clamped by several screws onto the bladder and was also sealed with cyanoacrylic glue. These seals worked well in the gel-filled bladders, but were prone to leakage with the water-filled bladders.

Instrumentation

Since it became clear that there were no "off-the-shelf" technologies available, methods to instrument the abdomen to measure deflection were examined through several small contracts. These efforts were done in parallel to maximize the likelihood that a sensor would be available at the end of the project.

Fiberoptic Sensor - One effort was started to attempt development of a deflection sensor that would measure the amount of light that passed through a fiberoptic cable. The cable would be attached to the dummy abdomen and would bend as a result of abdominal deflection. In principal, the amount of light would change as a function of how much the cable was bent. While this effort looked like it might have promise, it was suspended when it became apparent that it would require more time to develop than was available to the project.

Magnetic Measurements - A program was also funded at Johns Hopkins University Applied Physics Laboratory (JHU-APL) on a 3D magnetic position system. This work continued the development of a system that JHU-APL had previously prototyped under contract to the NHTSA. The magnetic method involved using three orthogonal electromagnetic coils to create a magnetic field within the pelvis (Figure A.1). A fourth coil of wires was used to act as a magnetic field sensor. As this coil moved through the stationary magnetic field in the pelvis, its three-dimensional position could be tracked using an algorithm developed by JHU-APL.

This method showed very good accuracy with +/- 4 mm maximum error, but it was very difficult to calibrate, had a very small calibrated range and required significant post-processing. This method has potential to be the most accurate of all those tried, but requires further development to be useful in a typical crash lab.

Ultrasonic Method - Work was also pursued at GMR using an immersible ultrasonic sensor developed by Sonometrics, Inc. In contrast with typical ultrasonic transducers, this transducer generated a spherical ultrasonic wave front (Figure A.2).

The spherical ultrasonic transducer gave very good results in quasi-static tests. However, the dynamic performance exhibited large amounts of electrical noise, either due to the impact or whipping of the transducer wires. The output was not usable for crash testing applications. Again, further development would be necessary to utilize this technology.

Force-Sensing Array –The measurement of loaded area on the front of the abdomen could enable the calculation of a fifth proposed injury criterion, $F_{max}*C_{max}$ (Rouhana 1987, 1993, 2001). It might also enable determination of the impacting object in a crash or sled test. This could be accomplished by measuring the force over an array of sensors on the front surface of the abdominal insert. For example, one might question whether a dummy's abdomen had any steering wheel rim loading during a crash test of an airbag development program. Since an airbag would likely obscure the view of any abdomen-to-steering wheel rim interaction that might occur, the force-sensing array might provide the force distribution as a contour plot that could show an image of increased load concentration.

Development of a force-sensing array that operates on a deformable membrane was pursued at Ktech Corporation. This effort used Polyvinylidene Fluoride (PVDF), a thin piezoelectric film, to measure the force on a small area. A novel technique was developed which gave an output proportional only to forces normal to the deformable membrane and allowed cancellation of in-plane or shear stresses. The program ended before this concept could be tested in the dummy abdomen.

The results of these efforts are summarized in Table A.II.

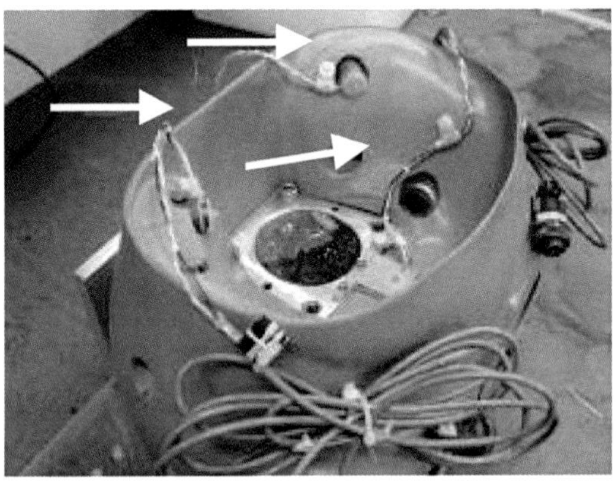

FIGURE A1. Magnetic field transmitters (arrows) inside Hybrid III dummy pelvis.

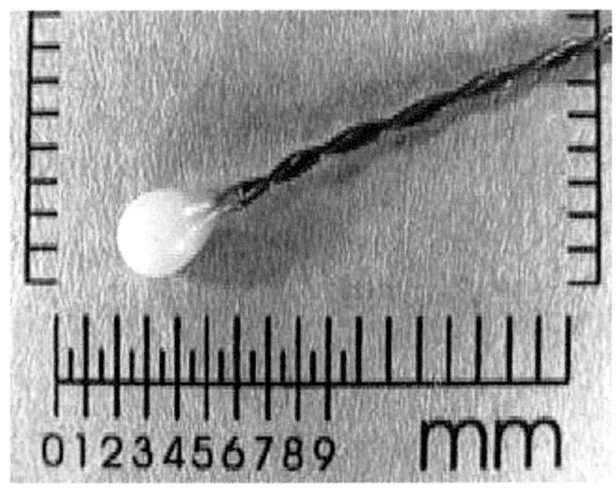

FIGURE A2. Spherical ultrasonic transducer from Sonometrics, Inc.

TABLE A.II. Instrumentation Results

Sensor	Test Results	Challenges	Countermeasures
Four Point Kelvin (Hughes Research Lab)	- Non-linear in MTS tests - Non-linear at start of impact and end of stroke in pendulum - Low susceptibility to noise - Multipoint measurement	- Near field effect - Sensor rotation - Dehydration ? (electrode gel is water-based) - Insert shape effects	- Rearrangement of sensors - Smaller diameter sensor - Other conductive fluids and greases.
Ultrasound (Sonometrics, Inc.)	- Linear & repeatable in quasi-static bench and dummy tests - Works in any fluid - Small transducer size - Easy calibration	- Noise from impact to wiring or acceleration in dynamic tests - Sensor external to bladder - Poor frequency response (1700 Hz)	- Mount sensor internally
3D Magnetic (John Hopkins University - Applied Physics Lab)	- Early results promising: +/- 4 mm maximum error - Low noise - 3D data - Can be linearized - Independent of dynamic insert shape changes	- Tedious calibration - Small calibrated range (30 mm square box) - Extensive post-processing - Affected by foreign metal objects nearby - 9 channels/sensor	TBD
Force Array (Ktech Corp.)	- Satisfactory in controlled bench tests	- Not tested in dummy	TBD

2001-22-0003

Stapp Car Crash Journal, Vol. 45 (November 2001), pp.

Development and Testing of a Prototype Pregnant Abdomen for the Small-Female Hybrid III ATD

Jonathan D. Rupp, Kathleen DeSantis Klinich
University of Michigan Transportation Research Institute

Steve Moss, Jennifer Zhou
First Technology Safety Systems

Mark D. Pearlman
University of Michigan Medical School

Lawrence W. Schneider
University of Michigan Transportation Research Institute

ABSTRACT – A new prototype pregnant abdomen for the Hybrid III small-female ATD is being developed and has been evaluated in a series of component and whole-dummy tests. The new abdomen uses a fluid-filled silicone-rubber bladder to represent the human uterus at 30-weeks gestation, and incorporates anthropometry based on measurements of pregnant women in an automotive driving posture. The response of the new pregnant abdomen to rigid-bar, belt, and close-proximity airbag loading closely matches the human cadaver response, which is thought to be representative to the response of the pregnant abdomen. In the current prototype, known as MAMA-2B (Maternal Anthropomorphic Measurement Apparatus, version 2B), the risk of adverse fetal outcome is determined by measuring the peak anterior pressure within the fluid-filled bladder. Peak internal bladder pressures measured in a series of sled-test simulations of frontal crashes of different severities and occupant-restraint conditions have been correlated to the likelihood of adverse fetal outcome based on risk curves developed from in-depth investigations of real-world crashes involving pregnant occupants. Compared to the original pregnant abdomen, the new prototype has improved geometry and improved impact response to a range of potential in-vehicle loading conditions, However, additional instrumentation development and more rigorous testing are needed before the MAMA-2B can be confidently used to assess restraint system performance with regard to reducing the likelihood of adverse fetal outcome in motor-vehicle crashes.

KEYWORDS – Trauma in Pregnancy, Anthropomorphic Test Device, Crash Test Dummy, Pregnant Anthropometry, Placental Abruption, Injury Criteria.

INTRODUCTION

Based on crash statistics and birth rates, it is estimated that about 130,000 women in the second half of pregnancy are involved in motor-vehicle crashes annually in the United States. Approximately 100,000 are uninjured, 30,000 sustain treatable injuries, and 160 are killed. For the pregnant women who survive, it is estimated that 300 to 3800 will experience fetal loss (Klinich et al. 1999a). While the estimates for crash-induced fetal loss span a wide range, even the lower estimate of fetal deaths is greater than the number of infants killed (n=185) in motor-vehicle crashes in the U.S. each year (National Vital Statistics Reports 1999).

In the past two decades, automotive safety engineers have made significant improvements in restraint systems that are designed to protect the general automotive occupant population. However, relatively little has been done to address the special circumstances associated with restraint of pregnant occupants. One reason for this is that safety engineers have not had the tools to assess the unique injuries sustained by pregnant vehicle occupants and the unique interaction of the pregnant abdomen with restraint systems. In response to this need, Viano et al. (1996) and Pearlman and Viano (1996) developed the first pregnant crash test dummy and used it to evaluate the effect of restraint-system configuration and restraint misuse on the relative injury potential to the pregnant occupant. However, the results of these studies are limited by the unrealistic abdominal contour of the original pregnant dummy (see Figure 4) and its non-biofidelic loading response, which is too stiff (see Figure 14), which make the dynamic interaction of the dummy with restraint systems and vehicle interior components questionable.

OVERVIEW OF PREGNANT ANATOMY

Figure 1 illustrates the general anatomy of a pregnant abdomen near full term (Pritchard et al. 1985). The fetus, placenta, and amniotic fluid are contained within the uterus, which is a muscular organ that grows in volume from about 5 mL to 5 to 10 L over the course of pregnancy. Ligaments attach the cervical portion of the uterus to the sacrum and pelvic wall, but the uterus is otherwise unattached within the abdominal cavity. As the uterus increases in size, it displaces other abdominal organs and is in contact with both the anterior abdominal wall and the lumbar and sacral spines. The size of the uterus depends largely on the size of the fetus, and, since the weight of most newborns carried to full term is generally between 2.7 and 3.6 kg (6 and 9 lb), uterus size does not significantly vary with the stature or weight of the mother (Klinich et al. 1999b and 1999c).

The placenta is roughly circular in shape, about 25 cm across, and about 2 to 2.5 cm thick in the last trimester of pregnancy. It covers about one fourth of the internal surface area of the uterus, and is attached by means of microvilli that form the uterine-placental interface, or UPI. In most pregnancies, the placenta is located near the top or fundal portion of the uterus, and is covered internally by the amniotic membrane that lines the entire inner wall of the uterus. Partial or complete separation of the placenta from the uterus, known as placental abruption, can result in loss of blood supply to the fetus through the connecting umbilical cord, and is a common cause of traumatic fetal loss.

THE ORIGINAL PREGNANT DUMMY

Viano et al. (1996) and Pearlman and Viano (1996) developed an abdominal insert for the Hybrid III small-female dummy to simulate pregnancy at a gestational age of approximately 30 weeks, and assess the potential of fetal injury. This adaptation of the small-female ATD was designed to evaluate the

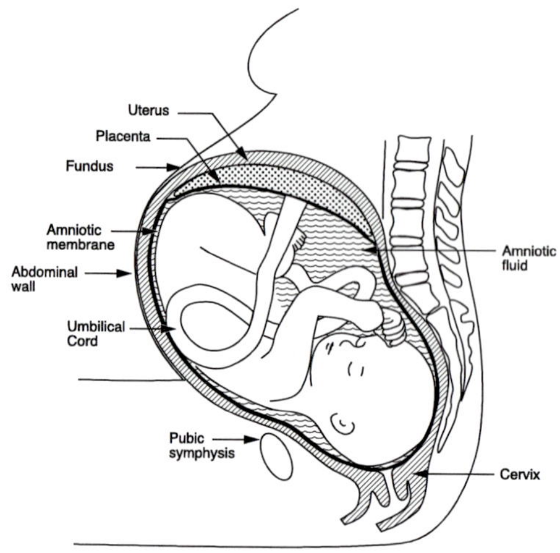

FIGURE 1. Anatomy of the pregnant abdomen.

performance of different restraint systems by measuring forces applied directly to the pregnant abdomen. Figure 2 shows the original pregnant abdomen in a modified Hybrid III pelvis and the separate pregnant abdomen components, which include a surrogate urethane fetus that fits inside a solid-urethane ellipsoidal casing contained within a larger urethane uterus. These nested components are covered anteriorly by the external dummy torso/abdomen jacket. A modified version of the Hybrid III small-female chest skin was contoured so that it blends with the superior surface of the abdominal insert.

As shown in Figure 3, a crescent-shaped block of material was removed from the pelvic cavity of the standard small-female Hybrid III to accommodate the lower portion of the pregnant abdomen. Ballast was added to pockets sewn to the interior surface of the posterior chest jacket to bring total mass of the original pregnant dummy to 58 kg, or approximately 9 kg more than the standard small-female Hybrid III. The fetus is equipped with triaxial accelerometers in the head and thorax. A force transducer is mounted to the lumbar spine behind the pregnant uterus to measure forces transmitted through the abdomen.

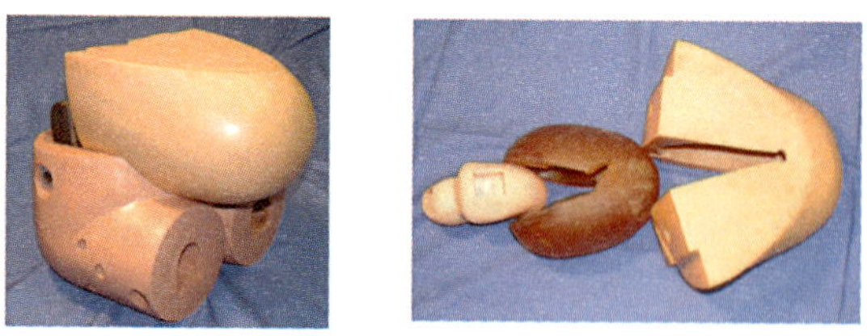

FIGURE 2. Original pregnant abdomen installed in pregnant dummy pelvis (left) and separated into fetal component, urethane casing, and urethane uterus (right).

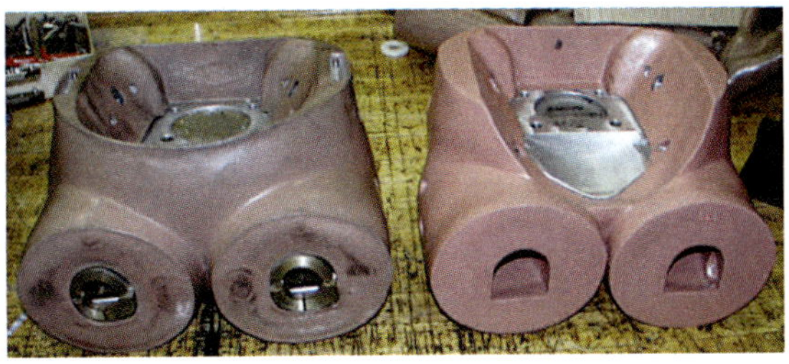

FIGURE 3. Standard Hybrid III small-female pelvis (left) and original pregnant dummy pelvis (right).

While the original pregnant dummy can measure acceleration of the fetal component and forces transmitted through the abdomen, the relationships between these measures and the likelihood of fetal injury and loss are unknown. Furthermore, the unrealistically high stiffness of the urethane abdomen components, the exaggerated anterior contour of the abdomen, and the tendency of the ballast weights to decouple from the dummy torso during impact will result in unrealistic interactions of the pregnant abdomen with restraint systems and vehicle components. This will affect overall dummy kinematics, and result in questionable injury assessment by the abdomen and other body regions.

OBJECTIVES

Based on the limitations of the original pregnant crash test dummy, three main objectives were identified for the design of the new pregnant dummy, which is referred to as the Maternal Anthropomorphic Measurement Apparatus, version 2B (MAMA-2B). These objectives were to develop and test a new pregnant abdomen with:

1. more realistic size and shape,
2. more humanlike force-deflection response, and
3. instrumentation for assessing the risk of fetal loss.

The new pregnant abdomen was targeted for use with the Hybrid III small-female dummy because this is the only frontal crash dummy that represents an adult female. Since smaller women tend to sit closer to the steering wheel, the small pregnant female maybe more likely to experience adverse fetal outcome than larger women because of an increased likelihood of abdomen loading by the steering wheel (Klinich et al. 1999b). Therefore, the small female may be the most appropriate dummy size for pregnant occupant testing.

As with the original pregnant dummy, the new pregnant abdomen was designed for use in the small-female Hybrid III ATD with minimal modifications to standard dummy components and with no significant changes in any of the non-abdominal response characteristics. The new abdomen was designed to comply with the same durability, repeatability, reproducibility, and usability standards as is expected for other components of Hybrid III type ATDs.

ANTHROPOMETRY

As noted previously, the geometry of the pregnant abdomen is important for achieving representative belt fit and for appropriate interactions of the ATD with restraint systems and steering wheels during crash testing. Culver and Viano (1990) provided the first estimate of pregnant abdomen size and shape by scaling contour measurements from 5th percentile Japanese pregnant women to their larger 5th percentile American counterparts. This scaling assumed that abdominal dimensions are proportional to maternal stature and weight, and was used to determine the abdominal geometry of the original pregnant dummy. However, data from a recent study of the seated anthropometry of twenty-two pregnant women conducted by Klinich et al. (1999b, 1999c) indicate that the size and shape of the pregnant abdomen are relatively independent of maternal stature. Abdominal contours at thirty weeks gestation from this study were averaged to develop the midsagittal contour for the MAMA-2B. A gestational age of thirty weeks was selected because it corresponds to the largest abdominal component that can fit into the small-female Hybrid III without a complete redesign of the ribs and sternum. Figure 4 shows the midsagittal contour of the MAMA-2B relative to that of the original pregnant dummy, which protrudes much too far forward.

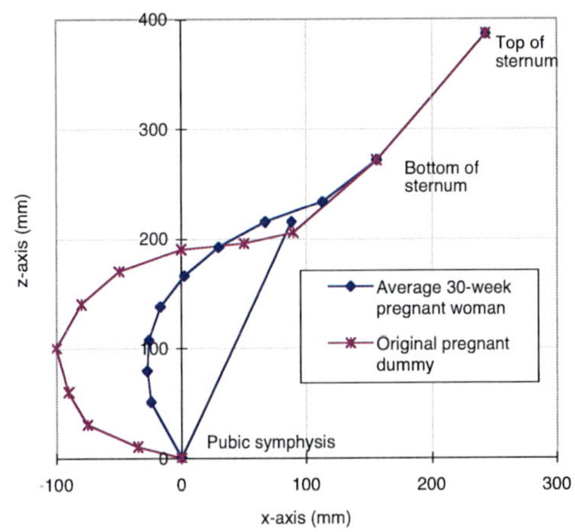

FIGURE 4. Comparison of the midsagittal abdomen contour of the MAMA-2B to that of the original pregnant dummy.

Dummy Mass

The target mass for the MAMA-2B is based on the mass of the standard small adult female dummy with instrumentation (49 kg or 108 lb) plus the recommended mass gain at 30 weeks gestation, which is approximately 9 kg (Institute of Medicine 1990). The target mass was therefore selected to be approximately 58 kg (49 kg plus 9 kg). This is close to the weight selected by Viano et al. (1996) for the original pregnant dummy. Since only about 40% of the weight gain is from volume increases of the uterus and its contents (fetus, placenta, and amniotic fluid), approximately 3.6 kg (8 lb) of the additional mass should be placed in the pregnant abdomen. The remaining 5.5 kg (12 lb) of weight gain is distributed among blood (other than the volume contained in the abdomen), fluid, breasts, and maternal fat reserves, and should therefore be placed in body regions other than the abdomen.

The desired location for the seated center of gravity of the MAMA-2B was based on a study of the changes in segment masses of pregnant women throughout pregnancy (Jensen et al. 1996). These researchers estimated an increase in mass for each body segment that is linearly proportional to gestational age. These relations were applied to the segment masses of the Hybrid III small-female dummy to determine an appropriate seated center of gravity location and for a 9 kg (20 lb) increase in dummy mass. Using this approach, the seated whole-body center of gravity of the MAMA-2B is 227 mm above and 169 mm forward of the H-point. This location is 8 mm forward and 10 mm down from the standard small female Hybrid III seated center of gravity location.

To obtain the desired center of gravity location while minimizing differences between the Hybrid III small female and the MAMA-2B, 5.3 kg (11.7 lb) of ballast were added to the pelvis and 2.5 kg (5.6 lb) were distributed along the posterior spine box. The remainder of the additional mass is included in the abdomen component and its attachment hardware. The added ballast brings the total mass of the pregnant dummy to 58 kg (128 lb) and compensates for mass removed in the redesign of dummy components described below. Rigidly incorporating ballast in the pelvis and the posterior spine box couples the ballast securely to the dummy and consequently eliminates the spikes in chest acceleration that resulted from ballast in the posterior chest jacket decoupling from and subsequently impacting the torso of the first pregnant dummy.

RIBCAGE REDESIGN

Because the anterior portions of ribs 4 through 6 on the small-female Hybrid III cover part of what would be the upper abdomen on a human of similar size, properly positioning an anthropometrically correct abdomen in a small-female dummy required removal of half of the sternum plate and varying amounts of the 4th through 6th ribs to create an inverted "V"-shaped opening. Figure 5 shows the modified ribcage. To stabilize the ends of ribs 4 through 6, and to stiffen the ribcage response in frontal impact, spring-steel stiffeners were added to the anterior ends of the ribs on each side of the ribcage. The performance of the redesigned rib set with the spring-steel stiffeners was determined through both computer modeling and pendulum testing to be within the performance specification for the small female Hybrid III dummy. The measured response of the redesigned ribcage in a standard 6.7 m/s thorax calibration test (Dummy Testing Equipment Subcommittee 1998) is shown in Figure 6, along with the thoracic response for an unmodified small-female thorax that meets the specification.

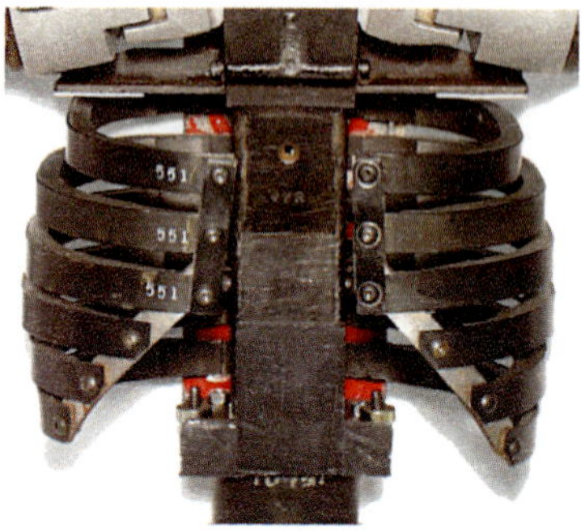

FIGURE 5. Modified Hybrid III ribcage with spring-steel stiffeners connecting the ends of ribs 4 through 6 on each side (redesigned sternum plate not shown).

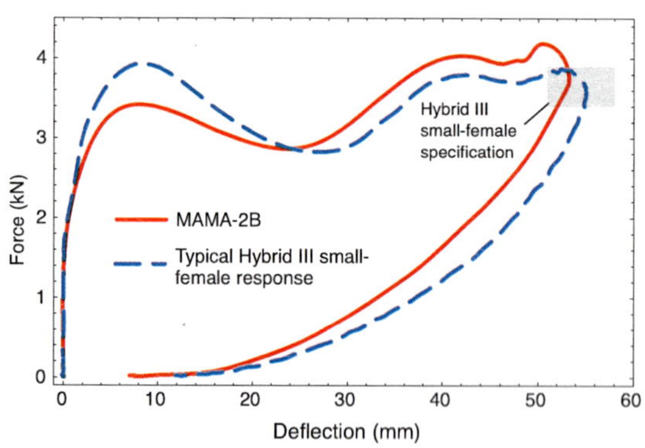

FIGURE 6. Force-deflection thorax responses of the MAMA-2B and standard small-female Hybrid III to 6.7-m/s calibration tests.

CHEST JACKET DESIGN

Figure 7 shows the assembled prototype MAMA-2B with the redesigned chest jacket that replaces the vinyl torso skin of the standard Hybrid III dummy. The jacket is a vest fabricated from neoprene wetsuit material with a zipper in the back. A crotch strap secures the jacket in place and positions it snugly against the anterior contour of the abdomen. The jacket contains interior pockets for silicone breast prostheses, which improve mass distribution and shoulder-belt routing. The prosthetic breasts have proved to be durable in sled tests. As shown in Figure 6, the effect of the breast prostheses on thoracic response is small because the 15.2-cm diameter impactor in the standard chest calibration makes minimal contact with the breast prostheses.

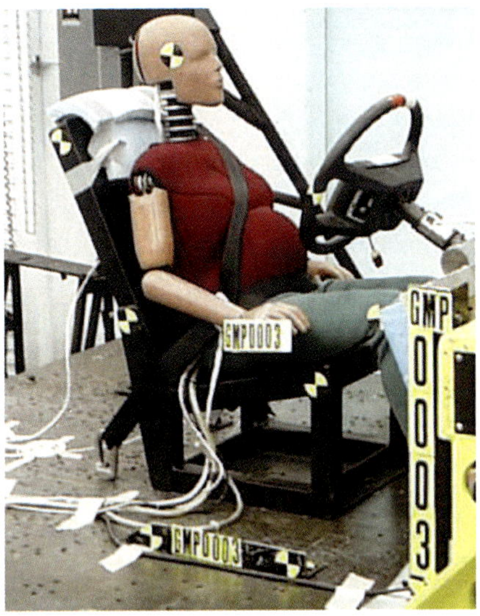

FIGURE 7. Prototype MAMA-2B with neoprene jacket.

DESIGN AND DEVELOPMENT OF THE ABDOMINAL COMPONENT

Abdominal Component Design

In the past, crash-dummy torso components have been designed using elastic urethane foams, damped metal ribs, air-filled bladders, and frangible floatation foams. Melvin et al. (1985) proposed the use of fluid-filled bladders for the design of both thoracic and abdomen components of an advanced multi-directional crash dummy. Schneider et al. (1994) further explored the concept of fluid-filled bladders in the development of a thorax for the NHTSA's advanced frontal-impact dummy. More recently, Rouhana et al. (2001) expanded on the fluid-filled bladder concept in the development of a reusable, rate-sensitive abdomen for the Hybrid III midsize-male dummy. This abdomen consists of a silicone rubber bladder that has a dynamic force-deflection response within the cadaver response corridors developed by Cavanaugh et al. (1986) and Hardy and Schneider (2001).

Based on previous research performed by Schneider et al. (1994), Viano et al. (1996), and Rouhana et al. (2001), two design options were considered for the uterine component of the pregnant abdomen. One approach was to use a solid insert similar to that of the original pregnant abdomen. The other approach was to use a fluid-filled bladder insert. The fluid-filled bladder concept was ultimately selected for three reasons. First, it has the potential to provide greater rate sensitivity to impact loading with minimal temperature dependency. Second, the success of Rouhana et al. (2001) in creating a durable fluid-filled abdomen component with a repeatable response provided confidence that this approach is feasible. Finally, a fluid-filled abdomen component allows for internal pressure measurement, which was hypothesized to relate to the likelihood of fetal loss in both direct and inertial loading of the abdomen. Water was used as the fluid in the MAMA-2B abdomen because it is similar to amniotic fluid and is therefore most likely to produce realistic pressure distributions under dynamic loading.

The primary abdomen component for the pregnant dummy is an ellipsoidal 50-durometer-silicone rubber bladder that is sized to represent a 30-week pregnant uterus when positioned in the modified pregnant dummy pelvis. The nominal wall thickness of the bladder is 6 mm and the major and minor axes of the interior of the bladder have lengths of 196 and 172 mm, respectively. The bladder was fabricated using a mandrel-dipping process similar to that used by Rouhana et al. (2001). Like Rouhana's non-pregnant abdomen, the rate sensitivity in the bladder response is obtained from the viscoelastic characteristics of the bladder material and the viscosity of the fluid within the bladder. The silicone rubber bladder material has several advantages over elastomeric materials commonly used in dummy components (e.g., vinyl and urethane). These include high durability, the ability to bond components to the bladder, and the option to cut into and reseal the bladder, using silicone-based glues, with minimal loss in performance.

To fasten the bladder to the ATD, the bladder was glued to two urethane attachment "cradles" as illustrated in Figure 8. The lower cradle is bolted to the pelvis and represents the anatomical connection between the base of the uterus and the cervix. The upper cradle is attached to the inferior spine box and represents the viscous resistance to forward motion that is hypothesized to occur when restraint forces applied to the pelvis decelerate the spine. That is, any forward displacement of the uterus relative to the spine requires the viscous tissues that surround the uterus to flow in behind it. At the loading rates that can be expected in frontal crashes, this viscous resistance to tissue movement will act to constrain the posterior aspects of the uterus from forward movement.

Figure 9 illustrates the position of the bladder within the MAMA-2B. Installation of the prototype abdomen components required removal of the Hybrid III lumbar load cell and replacement of the standard Hybrid III pelvis with the modified pelvis used in the original pregnant dummy shown in Figure 3.

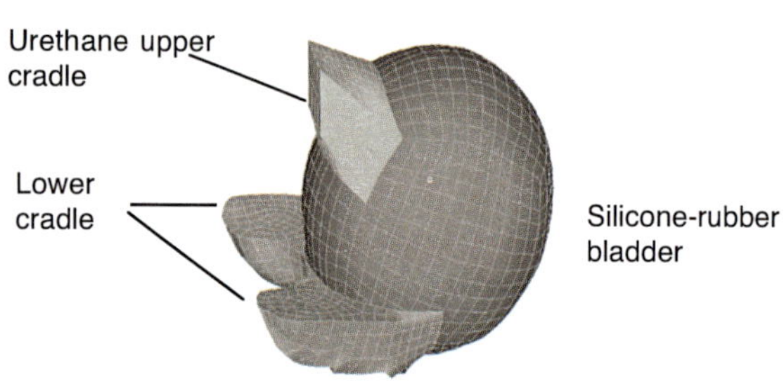

FIGURE 8. Computer model of the prototype bladder and urethane attachment cradles.

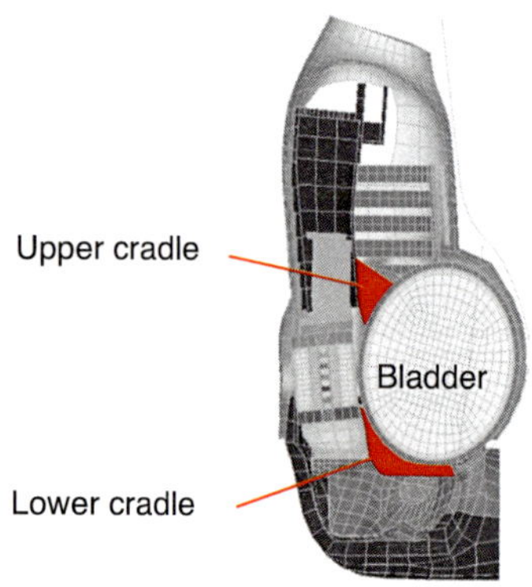

FIGURE 9. FEM of prototype uterus installed in dummy, showing upper and lower mounting "cradles."

During the development of the prototype abdomen, consideration was given to including physical representations of the fetus and placenta in the new abdomen. However, while these components would be visually appealing, they were not considered necessary to achieving the desired response and injury-assessment goals, and could be expected to reduce the desired biofidelity and repeatability. The new pregnant abdomen was therefore constructed without these anatomical features.

Estimating the Force-Deflection Response of the Pregnant Abdomen

The mechanical response of a dummy component is typically based on force-deflection data collected during dynamic testing of unembalmed cadavers. However, such data are not available for pregnant women. Consequently, the dynamic response of the pregnant abdomen was based on dynamic response corridors developed for the non-pregnant cadaver abdomen.

To estimate how the dynamic response of the pregnant abdomen might relate to that of the non-pregnant abdomen, quasi-static testing of pregnant and non-pregnant abdomens was performed using the test apparatus shown in Figure 10.[1] This apparatus consists of a rigid seat and a stationary loading surface, which for these tests was either a 2.54-cm diameter rigid bar or a 7.6-cm diameter rigid disc. For a typical test, subjects were positioned in the seat and the height of the loading surface was adjusted to either ASIS or umbilicus level. Quasi-static loading was applied by the subject at a rate of 7 mm/s using a normally-open switch that activated a motor that moved the seat toward the loading surface. To prevent injury to the pregnant abdomen, applied force was limited to 45 N and penetration into the abdomen was limited to 5 cm. These force and deflection limits were judged by an obstetrician to be well below the threshold for injury to the pregnant abdomen. Also, if the subject released the switch, the motor automatically reversed, thereby unloading the abdomen.

Because the loading response of the abdomen was found to be fairly linear over the tested range of abdominal penetrations, abdomen stiffness was defined as the slope of a line connecting the point on the force-deflection curve where force first surpassed 5 N to the peak force. The results of the quasi-static testing shown in Figure 11 indicate that the quasi-static stiffness of the pregnant abdomen is not significantly different from that of the non-pregnant abdomen male and female abdomen. The impact-response corridors for the non-pregnant cadaver

---

[1] The rights, welfare, and informed consent of the volunteer subjects who participated in this study were observed under guidelines established by the U.S. Department of Heath and Human Services on Protection of Human Subjects and accomplished under medical research design protocol standards approved by the Committee to Review Grants for Clinical Research and Investigation Involving Human Beings, Medical School, The University of Michigan.

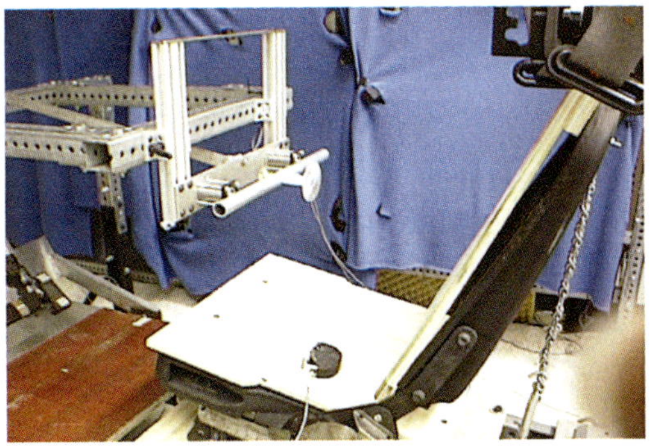

FIGURE 10. Test apparatus used to conduct quasi-static loading of pregnant and non-pregnant abdomens, showing rigid movable seat and stationary loading bar.

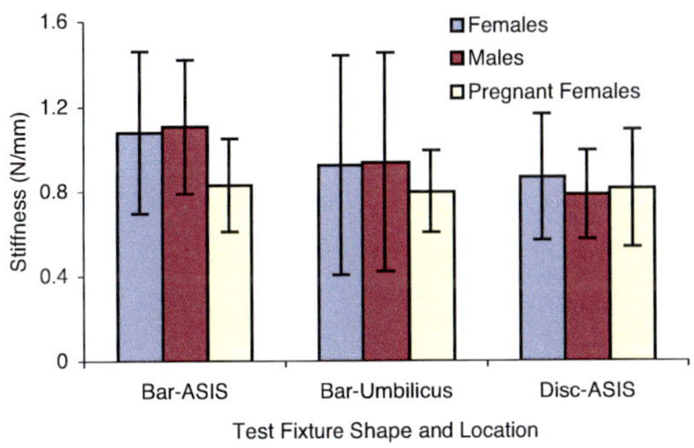

FIGURE 11. Quasi-static abdominal response of pregnant and non-pregnant volunteers for rigid-bar loading at the umbilicus and ASIS level and disc loading at the ASIS level.

abdomen were therefore assumed to be valid for the pregnant abdomen.

Testing of the MAMA-2B Abdomen

Three types of cadaveric abdominal response corridors were used to develop and validate the MAMA-2B. These include corridors developed from rigid-bar impacts, belt loading, and simulated close-proximity airbag loading.

Rigid-Bar Impact Response The rigid-bar impact response corridor developed by Cavanaugh et al. (1986) and later confirmed by Hardy and Schneider (2001) was used to define the 6-m/s loading response of the MAMA-2B abdomen. This corridor is based on the ±1 SD force-deflection response of five unembalmed cadaver abdomens impacted at approximately 6 m/s. Cavanaugh et al. also performed similar rigid-bar impacts to develop a response corridor at a velocity of approximately 10 m/s, and found that the cadaveric human abdomen has a rate-sensitive response. Because of the prevalence of driver airbags, 10-m/s loading by a steering-wheel-shaped object is unlikely to occur in the majority of frontal crashes. Consequently, the MAMA-2B abdomen response was designed to meet the 6-m/s Cavanaugh corridor, but not the 10-m/s corridor. To evaluate the rate sensitivity of the MAMA-2B abdomen, the response to 4.3-m/s rigid-bar impacts was also compared to an average 3-m/s rigid-bar response developed by Hardy and Schneider (2001). The Cavanaugh corridor and the average 3-m/s response were equal-stress, equal-velocity scaled (Eppinger 1978) from the reference mass used to develop the corridors (76 kg for the Cavanaugh corridors, and 78 kg for the typical 3-m/s response), to the pregnant dummy mass of 58 kg. This was done to allow unscaled MAMA-2B abdominal response data to be compared directly to the response corridors

Figure 12 shows the rigid-bar impact test fixture, which is the same 48-kg ballistic pendulum used by Hardy and Schneider to verify the 6-m/s Cavanaugh corridor and to develop the average 3-m/s abdomen response. During a typical test, the dummy was seated on a Teflon skid-plate. The height of the plate was adjusted so that the impactor was located at the level of the anterior-most point of the abdomen. A pneumatic accelerator drove the ballistic impactor and the attached rigid-bar impactor to the desired pre-impact speed. The fore-aft position of the dummy was adjusted so that the ballistic pendulum was traveling at the bottom of its arc when the rigid bar first contacted the dummy abdomen.

Force applied to the abdomen was measured by a load cell positioned between the ballistic mass and the impactor. A single accelerometer measured x-axis acceleration at the pendulum center of gravity during impact. Additional accelerometers were placed on the forward face of the load cell to measure x- and y-axis accelerations. Data from one of the x-axis accelerometers were used to mass correct the force measured by the load cell. Acceleration data, along with the mass of the impactor, were also used to verify forces measured by the load cell. For most tests, impactor motion was arrested by a series of deformable bars connected to the pendulum center of gravity by steel cables. To prevent damage to the instrumentation contained within the bladder and to prevent overloading of the impactor load cell from contact with the lumbar spine, the length of the cables was set so that impactor motion began to be arrested at 75 to 80 mm of abdomen penetration.

Force data were sampled at 10 kHz and filtered at SAE channel class 1000. Abdominal deflection was calculated by digitizing targets on the lumbar spine and pendulum using high-speed video or film data

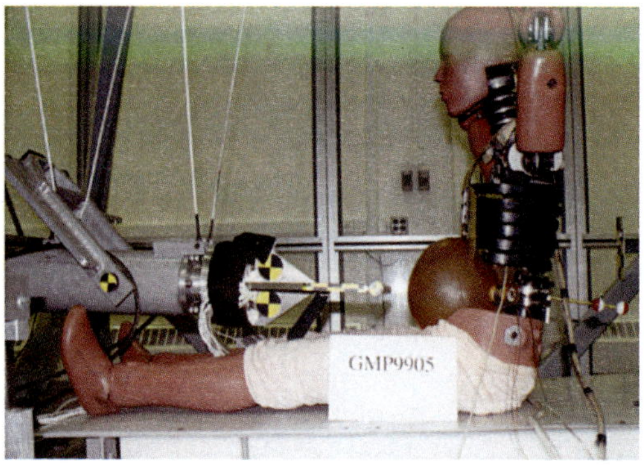

FIGURE 12. Rigid-bar impact test setup.

collected at 1000 frames/s. Digitized deflection data were filtered at SAE channel class 180.

Figure 13 shows the rigid-bar response of the MAMA-2B abdomen to 6.7- and 4.3-m/s impacts with and without the chest jacket, relative to the 6-m/s Cavanaugh corridor and the average 3-m/s response. The MAMA-2B response is within the 6-m/s Cavanaugh corridor and the 4.3-m/s response is close to the average 3-m/s response for deflections less than approximately 50 mm. These test results also indicate that the MAMA-2B abdomen demonstrates some rate sensitivity, but not as much as was observed in rigid-bar cadaver impacts conducted by Hardy and Schneider (2001). The chest jacket has a minimal effect on force-deflection response.

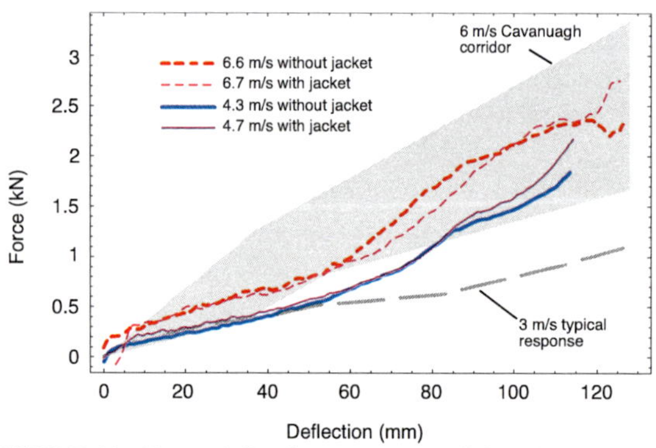

FIGURE 13. Force-deflection responses of the MAMA-2B abdomen to rigid-bar impacts at 4.3 and 6.7 m/s relative to scaled 6-m/s Cavanaugh corridor and typical 3-m/s response.

For comparison, Figure 14 shows the response of the original pregnant dummy to that of the MAMA-2B and the 6-m/s Cavanaugh corridor. The dynamic force-deflection response of the original pregnant abdomen is more than twice as stiff as the upper bound of the 6-m/s Cavanaugh corridor. Because the rigid-bar test damaged the urethane components of the original pregnant abdomen, additional

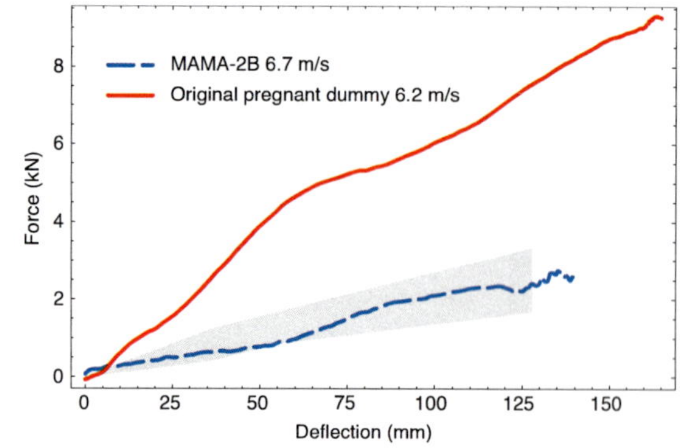

FIGURE 14. Rigid-bar impact response of the original pregnant abdomen compared to the scaled 6-m/s abdominal response corridor (shaded) and the MAMA-2B 6.7-m/s response.

comparison testing with belts and simulated close-proximity airbag loading was not possible.

Belt-Loading Response The response of the porcine abdomen to belt loading has been characterized in a corridor developed by Miller (1989). Hardy and Schneider (2001) compared this corridor to cadaveric abdominal belt-loading response and found that the corridor proposed by Miller is four times stiffer than the human cadaver response in the early loading phase. This difference was attributed to differences in the contact area between the belt and the abdomen. In the tests conducted by Miller, a taut, fairly flat, belt was pressed into the abdomen, while in the tests by Hardy and Schneider, the belt was wrapped around the sides of the abdomen prior to dynamic loading.

Because the belt-loading methodology used by Hardy and Schneider was thought to be more representative of real-world loading than that used by Miller, and because Hardy and Schneider used human cadavers rather than anesthetized-porcine subjects, the belt-loading response corridor generated by Hardy and Schneider was used to develop and validate the MAMA-2B abdomen. This corridor was developed from force-deflection data collected during six simulated belt-loading tests on the abdomens of three cadavers, and was equal-stress, equal-velocity scaled from the reference mass used in its development (78 kg) to the pregnant dummy mass (58 kg) so that unscaled MAMA-2B response data could be compared to the scaled corridor.

The belt-loading apparatus is shown in Figure 15. This is the same device used by Hardy and Schneider to develop the belt-loading corridor. This device was used to apply a horizontal load to the abdomen of the seated dummy through a length of belt webbing connected at both ends to a yoke-fixture that was pneumatically driven rearward relative to the dummy. Force applied to the dummy abdomen was measured by two belt load cells located near the belt attachments to the yoke fixture. These force measurements were verified using a single-axis load cell attached between the yoke and the pneumatic accelerator. Force data from the yoke load cell were mass corrected using data from an accelerometer mounted on the forward face of the load cell. Prior to each test, the belt was in contact with, but was not loading, the abdomen. Belt displacement at the center of the abdomen was measured using a string potentiometer mounted to a stationary fixture, with the string connected to the center of the belt webbing. The height of the potentiometer was adjusted so that the motion of the string was horizontal. The motion of the dummy was measured using a laser-based displacement transducer mounted in the yoke, with the laser beam focused on the lumbar spine. The displacement of the yoke was measured using an LVDT. Abdomen penetration was calculated by determining the motion of the belt at the center of the abdomen relative to the position of the lumbar spine, which was measured relative to the yoke. All measurements were independent of the stretch of the belt webbing and, consequently, belt stretch was not included in any analyses.

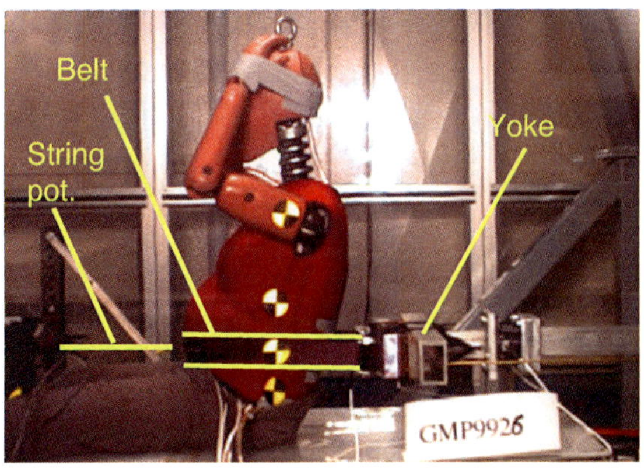

FIGURE 15. Setup for dynamic belt-loading tests. Lines added to illustrate string pot and belt outline.

All data from belt-loading tests were sampled at 10 kHz and filtered using SAE channel class 1000. The loading velocity pulse was approximately sinusoidal with a peak loading velocity between 3 and 4 m/s. This was comparable to the pulse used by Hardy and Schneider to develop the 3-m/s belt-loading corridors.

Figure 16 compares the belt-loading response of the MAMA-2B abdomen to the 3-m/s belt-loading corridor. The MAMA-2B response is within the 3-m/s belt corridor except for a window between 20 and 40 mm of penetration where the response drops below the corridor. The effect of the chest jacket on the belt-loading response is small.

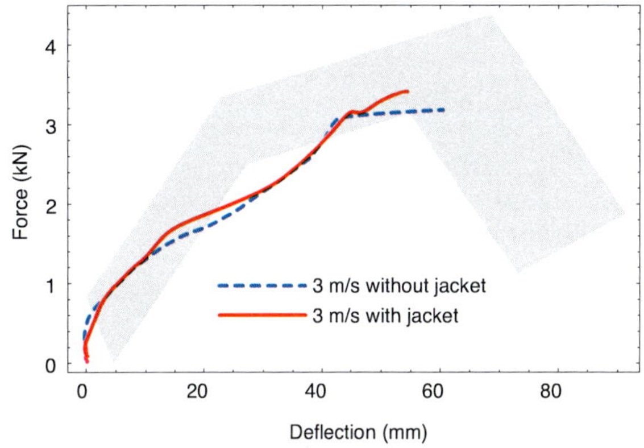

FIGURE 16. Force-deflection response of MAMA-2B abdomen with and without the chest jacket, to belt loading compared to scaled belt-loading corridor (shaded) developed by Hardy and Schneider (2001).

Simulated Close-Proximity Airbag Loading Response
Hardy and Schneider (2001) also characterized the abdominal response to loading by a deploying passenger airbag in close proximity to the abdomen using a surrogate airbag device. The performance of this device has been shown to closely approximate the loading produced in cadaver abdomens impacted by typical passenger airbags during the early stages of deployment. Based on data collected from six surrogate-airbag impacts into three cadaver abdomens, Hardy and Schneider developed an out-of-position airbag-loading corridor. To allow unscaled MAMA-2B response data to be compared directly to the corridor, the corridor was equal-stress, equal-velocity scaled from the reference mass used in its development (78 kg) to the pregnant dummy mass (58 kg).

The response of the new pregnant abdomen to simulated out-of-position airbag loading was measured using the surrogate-airbag device developed by Hardy and Schneider. The device consists of a low-mass 7.6-cm diameter, 20-cm long cylindrical impact surface that is connected to a piston by a hollow connecting rod. The piston travels in a larger diameter tube that is pressurized to drive the piston, connecting rod, and cylindrical impactor to a velocity of approximately 13 m/s just prior to contact with the abdomen. Force applied to the impactor during contact with the abdomen is calculated using the area of the piston and data from an acceleration-insensitive pressure transducer mounted so that it measures pressure applied to the surface of the piston inside the tube. Acceleration of the impactor is measured by an accelerometer bolted to the interior of the cylinder and is used to mass correct the force applied to the cylindrical impactor, to obtain force applied to the abdomen. Penetration into the abdomen is measured using a laser-based displacement transducer.

Figure 17 shows a typical test setup for close-proximity surrogate airbag testing. For these tests, the dummy's lumbar spine was anchored to a rigid seatback. All data were collected at a rate of 20 kHz and filtered at SAE channel class 1000.

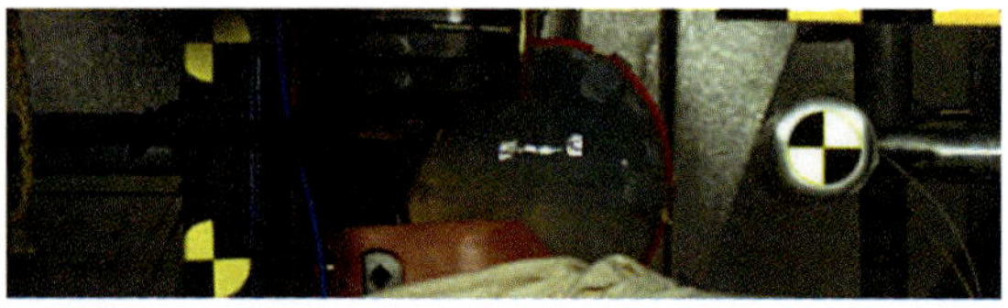

FIGURE 17. Test setup for surrogate airbag tests showing a side view of the low-mass cylindrical impactor (right with contrast target) and the new pregnant abdomen without the chest jacket (center).

Figure 18 shows the force-deflection response from tests with and without the chest jacket relative to the scaled surrogate airbag loading response corridor. The bladder response without the chest jacket is slightly less stiff during initial loading than the surrogate airbag corridor. However, after approximately 5 mm of penetration, the MAMA-2B response falls within the corridor. The reduced initial stiffness with the chest jacket is due to compression of the relatively soft neoprene jacket material.

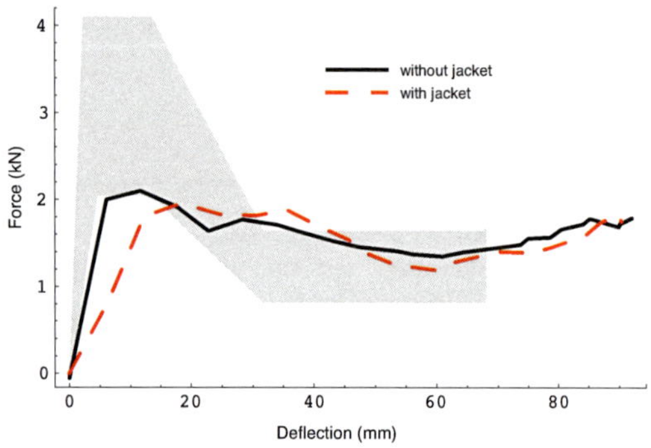

FIGURE 18. Responses of MAMA-2B abdomen to out-of-position surrogate airbag loading response with and without chest jacket compared to scaled out-of-position airbag loading corridor (shaded) developed by Hardy and Schneider (2001).

INSTRUMENTATION

All of the standard small-female Hybrid III instrumentation can be used with the MAMA-2B with the exception of the chest deflection instrumentation and the lumbar load cell. The standard chest-deflection measurement system cannot be used because of the shortened sternum that precludes installation of the slider assembly. However, a linear deflection measurement system such as the IR-TRACC (Rouhana et al. 1998) could be implemented in future versions of the MAMA-2B.

The MAMA-2B abdomen uses submersible miniature transducers (Precision Measurements, Ann Arbor, MI) to measure pressure histories at the anterior and posterior surfaces of the inside of the silicone rubber bladder. Peak intra-abdominal pressures are related to both the peak force applied to the abdomen and the peak accelerations of the abdomen, and have been suggested as potential predictors of injury based on sled testing of pregnant baboons (Crosby et al. 1968, King et al. 1972) and computer models (Rupp et al. 2001).

INJURY CRITERION

Fetal Injury Risk from Crash Investigations

The MAMA-2B injury criterion was developed by relating abdominal pressure measurements to the likelihood of adverse fetal outcome (fetal loss or other major complications to pregnancy) under test conditions that have been associated with both positive and negative fetal outcomes in real-world crashes. These test conditions were developed from relationships between the risk of adverse fetal outcome and crash severity (delta V or EBS) developed by Klinich et al. (1999a). In this in-depth investigation of crashes involving 43 pregnant occupants, variables including crash severity, maternal injury, maternal restraint usage, occupant seating location, maternal stature, gestational age, and impact direction were analyzed to determine their association with fetal outcome. Crash severity was found to have the strongest effect on fetal outcome, with more severe crashes associated with negative fetal outcomes ($p<0.0001$). Proper maternal restraint use (three-point belt or three-point belt plus airbag) was found to result in a significant improvement in fetal outcome ($p=0.0136$). When comparing outcomes for mothers restrained by a three-point belt versus a three-point belt and airbag, the data suggest that airbags have a positive effect on fetal outcome. Impact direction was not found to be associated with fetal outcome ($p=0.649$).

Because the combination of crash severity and proper restraint was found to best predict fetal outcome, a risk analysis was performed by Klinich et al. to quantify the relationships between the probability of adverse fetal outcome and both delta V and restraint usage. Logistic regressions were performed to establish risk curves for adverse fetal outcome for properly restrained occupants (three-point belts or three-point belts plus airbags, n=31) and improperly restrained occupants (unrestrained, airbag only, shoulder belt only, with and without airbag, n=12). The regressions are shown in Figure 19 and were found to be reasonably predictive of fetal outcome ($p<0.0001$, Nagelkerke $R^2=0.547$, Goodman-Kruskal $\gamma=0.628$). At crash severities below 50-kph delta V, unrestrained pregnant occupants have a much greater risk of adverse fetal outcome than do properly restrained occupants. For example, at 20-kph delta V, the estimated risk of adverse fetal outcome for properly restrained pregnant occupants is less than 20%, while it is near 55% for improperly restrained pregnant occupants.

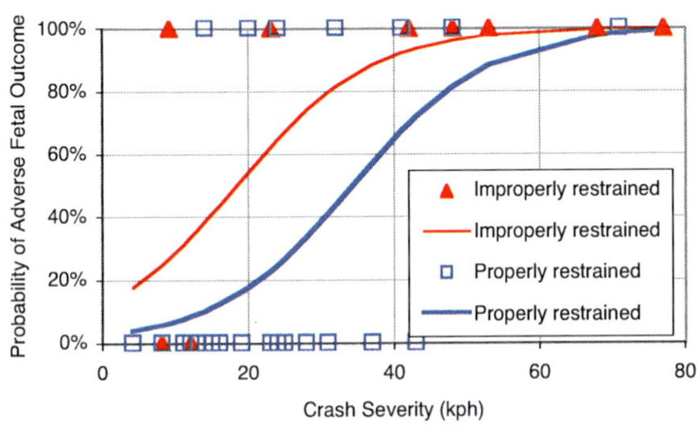

FIGURE 19. Logistic regressions relating adverse fetal outcome to delta V for properly restrained (i.e., three-point belt plus airbag and three-point belt) and improperly restrained (unbelted, airbag only, or shoulder belt only). Adapted from Klinich et al. (1999a).

Injury Criterion Development

Using the relationships between crash severity and restraint use to the likelihood of adverse fetal outcome, a matrix of sled tests was developed to test the MAMA-2B over a range of restraint conditions and crash severities associated with probabilities of adverse fetal outcome ranging from 10-90%. This test matrix is shown in Table 1. Pressure measurements from the MAMA-2B abdomen were correlated to the probability of adverse fetal outcome associated with each of these tests and used to develop an injury criterion that predicts fetal injury regardless of injury mechanism.

The test matrix includes five tests in a driver configuration, restrained by a three-point belt, that were run at 13, 25, 35, 45, and 55 kph delta V to approximate risks of about 10, 25, 50, 75, and 90% on the properly restrained curve. Three unrestrained driver tests were conducted at nominal delta V of 13, 20, and 35 kph to approximate risks of 36, 54, and 86% on the improperly restrained curve. Two tests were conducted at 35 kph in a three-point-belt-restrained driver configuration as a check for repeatability.

TABLE 1.
Test Matrix for Developing MAMA-2B Injury Criterion

Test No.	Δ V (kph)	Position	Restraint	Risk %
1	13	driver	none	36
2	13	driver	three-point belt	9
3	25	driver	three-point belt	26
4	25	driver	three-point belt+ airbag	26
5	25*	driver	three-point belt+ airbag	26
6	20	driver	none	54
7	13	pass	three-point belt	9
8	35	pass	three-point belt	51
9	35	driver	three-point belt	51
10	35	driver	three-point belt	51
11	35	driver	none	86
12	45	driver	three-point belt	76
13	45	driver	three-point belt+ airbag	76
14	55	driver	three-point belt	90
15	55	pass.	three-point belt	90

*Repeat test to account for unexpected steering wheel rotation.

Two pairs of tests were conducted at nominal impact velocities of 25 kph (Tests 3 and 5) and 45 kph (Tests 12 and 13) with the MAMA-2B in a driver configuration and restrained by a three-point belt or by a three-point belt plus airbag. These sets of tests were compared to examine the response of the new abdomen to airbag loading. Steering-wheel loading was also explored using three paired tests conducted in passenger (no steering wheel) and driver positions at delta Vs of 13 kph (Tests 2 and 7), 35 kph (Tests 8 and 9), and 55 kph (Tests 14 and 15).

Figure 20 shows a typical setup for these sled tests. Lap and shoulder belt angles, fore/aft seat position, seatback angle, steering-wheel fore-aft position, and steering-wheel angle were based on average conditions chosen by small pregnant women positioned in an automotive-seated posture (Klinich et al. 1999b). For all tests, the seatpan angle was 14° and the seatback angle was 65°. The seat height (i.e., H-point to heel surface) was 320 mm. For all driver tests, the seat was positioned so that knee-to-knee-bolster clearance was 70 mm. The steering-wheel angle was set to 30° from vertical and the steering-wheel fore-aft position and height were adjusted so the bottom of the steering-wheel rim was 30 mm from the most forward point on the MAMA-2B abdomen. For all passenger tests, the steering wheel and knee bolster were removed to simulate a passenger at a rearward seating position. Nominal values for side-view lap and shoulder belt angles were 45° and 25° from horizontal, respectively. In tests where airbags were used, the airbag was set to deploy 15 ms after the time at which the sled deceleration first rose above 0.5 g.

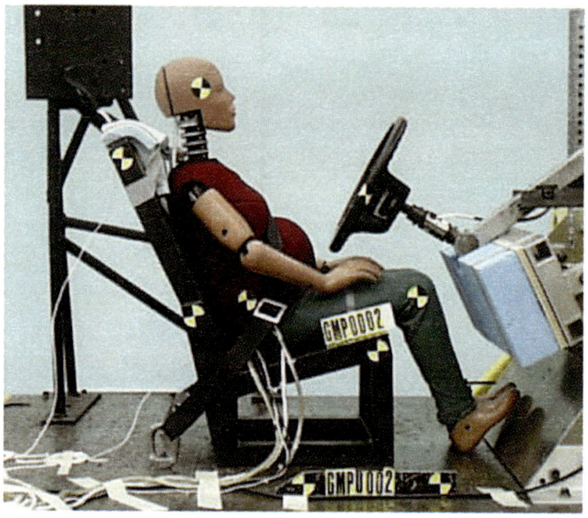

FIGURE 20. Typical setup for injury-criterion development test showing MAMA-2B in a seated driver configuration.

Both anterior and posterior bladder pressure measurements were considered in the search for an abdominal response parameter that correlates with the probability of adverse fetal outcome. However, the anterior pressure measurement was the only measure that showed reasonable correlation with the risk of adverse fetal outcome as defined by the logistic regressions in Figure 19.

Figure 21 shows the relationship between the log of peak anterior internal bladder pressure and the estimated risk of adverse fetal outcome across all test conditions. Impacts associated with a greater likelihood of adverse fetal outcome resulted in higher positive anterior pressures. Using a power-law relationship, 20, 50 and 80% risks of adverse fetal outcome were found to correspond to anterior pressures of 59, 88, and 109 kPa, respectively (R^2 = 0.66, p=0.003). The form of the power-law relationship between the peak anterior pressure and the risk of adverse fetal outcome is shown in Equation 1.

$$\theta = 0.0021\ P^{2.2512} \qquad [1]$$

where,

θ is risk of adverse fetal outcome, and
P is peak anterior pressure (in kPa)

Figure 22 shows that, for each restraint type, peak anterior pressure increased with increasing delta V. Consistent with trends expected from Figure 19, restraint by a three-point belt resulted in lower peak anterior pressures than for unrestrained conditions. Also, adding an airbag further decreased peak anterior pressure. The lowest positive peak pressures at a given delta V were observed when the MAMA-2B was in a passenger position.

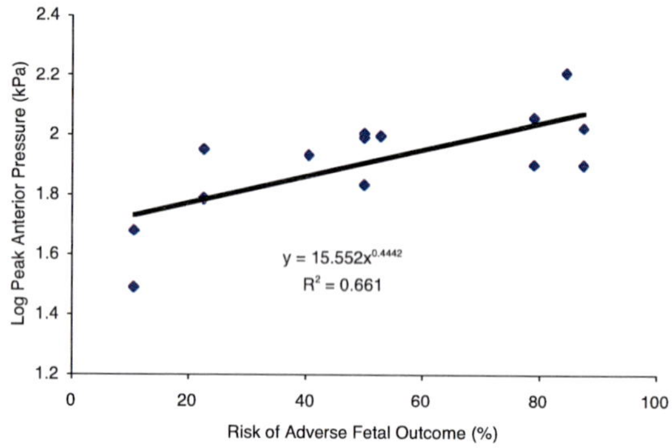

FIGURE 21. Correlation between peak anterior positive pressure and risk of adverse fetal outcome.

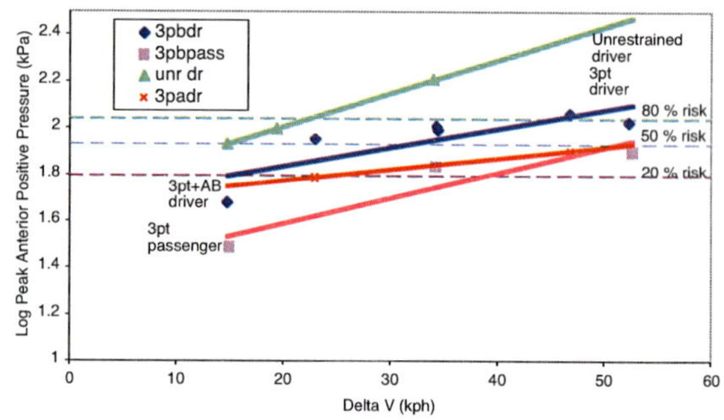

FIGURE 22. Relationship between peak anterior positive pressure and delta V for three-point-belted drivers (3pbdr), three-point-belted passengers (3pbpass), three-point-belted and airbag-restrained drivers (3padr), and unrestrained drivers (unr dr).

Repeatability of Pressure Response

Two identical three-point-belted driver tests at 35-kph (Tests 9 and 10) were used to test the repeatability of the MAMA-2B pressure-based injury criterion. Figure 23 shows the results of these comparisons for anterior pressure. The anterior pressure responses in these two tests differ by less than 4% at all points, and peak pressures differ by less than 1%.

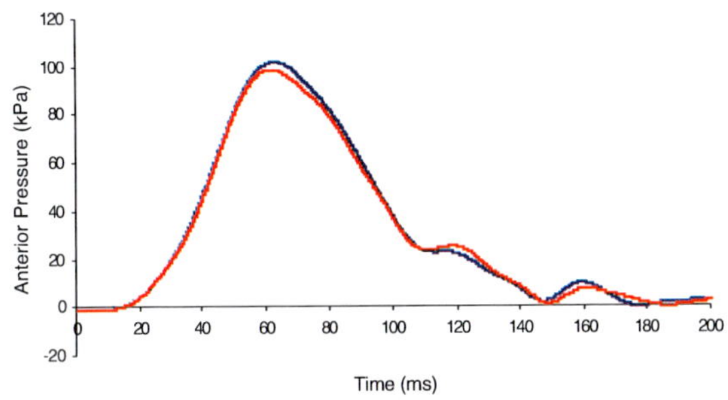

FIGURE 23. Anterior pressure transducer signals from repeated tests at 35 kph with the MAMA-2B in a three-point belted driver configuration.

Repeatability of anterior pressure measurements was also checked using pressure-time histories from similar pairs of rigid-bar impacts, belt-loading tests, and surrogate airbag tests described previously. In general, the differences between similar tests were small, and repeatability was within 10%.

DISCUSSION

Compared to the original pregnant abdomen, the new prototype has improved geometry and improved impact response for a range of potential driver and passenger loading conditions. Although it is still in the prototype stage, results of whole-dummy testing under various restraint and crash-severity conditions indicate that the new design is relatively durable, repeatable, and easy to use. However, additional instrumentation development and more rigorous testing are needed before this version of the pregnant abdomen can be used to confidently assess vehicle and restraint system performance with regard to reducing the likelihood of adverse fetal outcome in motor-vehicle crashes.

Abdominal Response

In the absence of data on the response of the pregnant abdomen, it was assumed that the dynamic response of the pregnant abdomen is similar to that of the non-pregnant abdomen, and that existing dynamic response corridors developed from tests with human cadavers apply. Support for this assumption was obtained from results of quasi-static force-deflection measurements of pregnant and non-pregnant abdomens. While it is possible that the dynamic response of the pregnant abdomen may be different from that of the non-pregnant abdomen, empirical data describing these differences are not likely to ever be available. Computer modeling could possibly be used to estimate potential differences and should be explored, but was beyond the scope of the current project.

The new pregnant abdomen was therefore designed and tested to comply with dynamic abdominal response corridors for rigid-bar, belt, and close-proximity airbag loading of unembalmed cadavers, and is believed to represent a significant improvement in dynamic response to the original pregnant abdomen. The response of the MAMA-2B abdomen falls within force-deflection corridors established for each type of test, and the dynamic stiffness to rigid-bar loading is substantially reduced from that of the original pregnant abdomen. The abdomen also demonstrates rate sensitivity in its dynamic response, although additional rate sensitivity is needed to better match that of the human abdomen.

In developing the MAMA-2B, response corridors from cadaver tests were equal-stress, equal-velocity scaled to the MAMA-2B mass of 58-kg from the reference masses used to develop the corridors. In doing this, it was assumed that the response of the pregnant abdomen is the same as that of a cadaver of identical mass, and that pregnant and non-pregnant abdomens are geometrically similar. Consequently, the scaled cadaver response corridors may not be entirely appropriate for the MAMA-2B, which likely has a greater abdominal depth than a non-pregnant cadaver of similar mass. Although the effects of geometric differences are not expected to be large, especially for the first 75 mm of penetration when injury is expected to occur, future efforts should examine other methods for scaling the corridors that consider the geometric differences.

Use of Anterior Internal Bladder Pressure as a Criterion for Adverse Fetal Outcome

Assessment of the risk of fetal trauma in the current prototype pregnant abdomen is based solely on measurement of peak anterior internal bladder pressure. This measure has the advantage of being responsive to both direct abdominal loading by belts, steering-wheel rims, and airbags, and to inertially induced loading. An injury criterion that relates peak anterior pressure in the MAMA-2B bladder to the likelihood of adverse fetal outcome, without regard for specific injuries or injury mechanisms, was developed from peak internal pressure data collected in a series of frontal sled tests that spanned a wide range of crash severities and maternal restraint conditions.

Peak anterior bladder pressures in these tests ranged from 31 kPa for a three-point belted passenger tested at a delta V of 13 kph (Test 7) to 162 kPa for an unrestrained driver tested at a delta V of 35 kph (Test 11). Although the restraint and test conditions are different, these results are comparable to the peak intrauterine pressures measured in impact testing of pregnant baboons by Crosby et al. (1968), which averaged 66 kPa for lap and three-point belt restrained passengers tested with delta V varying from 39 to 82 kph. Pressures from restraint conditions in the current test series closest to those used by Crosby et al. (i.e., 3-point belted passengers over the same range of delta V) were 69 and 80 kPa for Tests 8 and 15, respectively.

Peak anterior bladder pressures in the sled tests used for injury criterion development show reasonable correlation with the estimated risk of adverse fetal outcome for belt-restrained and unbelted pregnant occupants in real-world crashes. However, the accuracy of this relationship and the efficacy of fetal injury prediction using peak anterior internal pressure may be affected by limitations of the dataset used to generate the logistic regressions that describe the likelihood of adverse fetal outcome. These include errors in estimating crash severities using standard crash reconstruction techniques and computer programs, and the relatively low number of crash investigations used to develop the regressions (n=31 for properly restrained, n=12 for improperly restrained). The latter limits the ability to distinguish between effects of potentially important real-world factors on fetal injury risk other than delta V and restraint usage.

For example, the conditions for three sled tests in Table 1 represent approximately 50% risk of adverse fetal outcome based on the logistic regressions for properly restrained occupants in Figure 19. These three tests include two three-point-belt-restrained driver tests (Tests 9 and 10) and one three-point-belt-restrained passenger test (Test No. 8), all conducted at a delta V of 35-kph. However, because the abdomen of the small-female dummy is positioned 30 mm from the bottom rim of the steering wheel in the driver tests, it is expected that the fetal injury risk for the driver situation should be higher than for the passenger situation where the steering wheel is absent. It is interesting that, while the crash-injury database for pregnant occupants is not yet large enough to make distinctions between fetal injury risk levels for pregnant passengers and drivers in frontal crashes, the measures of peak anterior bladder pressure from the MAMA-2B sled-impact tests show the expected trends. Peak anterior pressure values for the two driver tests are 99.2 and 102.4 kPa, but only 69 kPa for the passenger test. Using the injury criterion relationship in Equation 1, these correspond to approximately 70% and 30% risks of adverse fetal outcome, respectively. Additional in-depth investigations of real-world crashes involving pregnant occupants are needed to develop fetal risk curves for specific combinations of occupant position and restraint type.

A potential limitation to using peak anterior internal bladder pressure as the sole predictor of fetal injury is that it is related to the volume of fluid displaced rather than the deflection of the abdomen. For example, an object with a small surface area penetrating deep into the bladder can produce the same peak pressure as an object with a large surface area penetrating a small distance into the bladder. These two loading scenarios are not likely to have the same effect on fetal injury, but would predict the same probability of the risk for adverse fetal outcome based on peak anterior pressure alone. However, this limitation may not be a significant disadvantage in today's crash-test environments where the likelihood of small objects impacting the abdomen is low. In addition, the pressure histories would probably be significantly different for abdominal loading by objects with small and large surfaces. Thus, it may be possible to use the shape of the pressure curve (e.g., the slope of the curve to peak pressure) to enhance the prediction of adverse fetal outcome using only anterior bladder pressure measurements.

Assessing the Likelihood of Placental Abruption

The peak-anterior-pressure-based injury criterion used in the MAMA-2B predicts the likelihood of adverse fetal outcome without regard for specific injuries or specific injury mechanisms. However, crash-injury data indicate that placental abruption (i.e., separation of the placenta from the uterus) is the most likely cause of adverse fetal outcome in motor-vehicle crashes, accounting for up to 70% of fetal losses, and is frequently observed alone or in conjunction with other common crash-induced injuries to the pregnant abdomen, such as direct fetal or uterine injury (Pearlman 1990). Therefore, the tolerance to placental abruption is believed to be substantially lower than for other injuries to the pregnant abdomen.

During the course of developing the MAMA-2B, consideration was given to designing and instrumenting the abdomen to assess directly for the probability of tissue failure at the uterine-placental interface. In the absence of definitive data on the mechanisms of UPI failure, computer modeling and engineering principles were used to establish two scenarios by which placental abruption is believed to occur in motor-vehicle crashes (Rupp et al. 2001). These include failure of the UPI due to shear strains caused by changes in curvature of the uterus resulting from direct abdominal loading, and failure of the UPI due to inertially induced tensile strains generated by negative pressure in the posterior uterus. UPI failure due to direct loading is hypothesized to occur when an object deforms the pregnant abdomen in the area of the uterus where the placenta is attached. UPI failure due to tensile strain is hypothesized to occur from the pressure gradient that develops across the uterus when it is decelerated through its attachment to the pelvis and its viscous coupling to the lumbar spine. This hypothesized injury mechanism is similar to contre-coup head injury where a deceleration of the head results in positive pressure in the anterior surfaces of the skull and negative pressures at the posterior surface. Provided that the placenta is located in the superior or fundal region of the uterus, which is true in 80% of pregnancies (Fried 1978), these negative pressures can result in tensile strains at the UPI.

Based on these two hypothesized mechanisms for placental abruption, efforts were made to measure negative pressure in the posterior of the fluid-filled MAMA-2B abdomen and deformation of the midline contour of the bladder, with the intent that injury criteria would be developed to relate these measurements to comparable quantities measured in a finite element model of the human uterus under similar loading conditions. These efforts were eventually abandoned due to project time constraints, limitations of the available biomechanical tissue data, and limitations of currently available instrumentation (Rupp et al. 2001). However, it is recommended that exploration of placental-abruption based injury criteria, and instrumentation to measure the necessary response parameters, be continued.

SUMMARY AND CONCLUSIONS

The goal of this project was to develop and evaluate a new and improved pregnant abdomen for the small-female Hybrid III ATD. The project was conducted in conjunction with several related studies, including a study of seated anthropometry during pregnancy (Klinich et al. 1999b, Pearlman et al. 2000), in-depth investigations of motor-vehicle crashes involving pregnant occupants (Klinich et al.1999a, Pearlman et al. 2000), a study of the impact response of the human abdomen (Hardy and Schneider 2001), and the development of a reusable, rate-sensitive abdomen for the midsize-male Hybrid III (Rouhana 2001).

Based on data from these other studies, previous medical research on fetal injury and loss in motor-vehicle crashes (Pearlman et al. 1990), and recognized limitations and deficiencies of the first pregnant dummy (Viano and Pearlman 1996), three primary objectives for the design and performance of the MAMA-2B were established early in the project. These include improving the size and shape of the pregnant abdomen, enhancing the mechanical response of the pregnant abdomen, and implementing instrumentation and an injury criterion for assessing the risk of adverse fetal outcome in frontal crashes.

The prototype dummy, referred to as the MAMA-2B, uses a water-filled silicone rubber bladder that closely approximates the size and shape of the human uterus at 30-weeks gestation. Achieving the desired size and position of the pregnant uterus required several other design changes to the standard small-female Hybrid III. These include trimming the 4th through 6th ribs, using the modified Hybrid III pelvis from the original pregnant dummy, and adding mass to the pelvis and spine box to obtain the appropriate total body mass and seated center of gravity.

The MAMA-2B abdomen demonstrates humanlike response for a range of loading conditions, and includes instrumentation to measure anterior internal pressure in a uterine-like bladder. Peak bladder pressure has been related to the risk of adverse fetal outcome using risk curves developed from real-world crash-injury data. Despite these improvements, the MAMA-2B requires further development and testing and should only be used in its current form as a research tool to study the interaction of restraint systems and vehicle interior components with the pregnant occupant in a frontal crash environment.

Future work on completing the development of the MAMA-2B should focus on improving the manufacturability and reproducibility of the abdomen components, and on conducting additional testing to validate durability and improve the ability to predict adverse fetal outcome in a wider range of crash scenarios. A second MAMA-2B should also be developed and tested to verify that the MAMA-2B abdomen design is reproducible. Additional investigation of crashes involving pregnant occupants is needed to improve the pressure-based MAMA-2B injury criterion by providing additional crash-injury data that can be used to develop relationships between the likelihood of adverse fetal outcome and crash severity for different restraint types and occupant locations. It is also recommended that additional efforts be directed toward developing and implementing instrumentation that can be used for the likelihood of adverse fetal outcomes based on injury criteria that are directly linked to the hypothesized mechanisms of placental abruption.

ACKNOWLEDGMENTS

The research described in this report was financed by General Motors (GM) pursuant to an agreement between GM and the U.S. Department of Transportation.

The authors gratefully acknowledge the contributions of Steve Rouhana, Ali Elhagediab, and Tony Walbridge whose pioneering work in the development of a new fluid-filled silicon abdomen for the midsize male Hybrid III ATD provided an invaluable resource for the current effort. Appreciation is also expressed to Dave Smith of First Technology Safety Systems, Inc. for the fabrication of prototype dummy components. Finally, the authors would like to thank several UMTRI staff members for their engineering and technical assistance throughout the project. These include Warren Hardy, Miriam Manary, Charlie Bradley, Brian Eby, Stewart Simonett, and Monica Prasad.

REFERENCES

Beckman, D. L., McElhaney, J. H., Roberts, V. L., Stalnaker, R. L. (1971). Impact Tolerance - Abdominal Injury. Final Report No. HSRI-71-102/ DOT HS 800 549. Highway Safety Research Institute, Ann Arbor, MI.

Cavanaugh, J.M., Nyquist, G.W, Goldberg, S.J., and King, A.I. (1986). Lower abdominal tolerance and response. Proceedings of the 30th Annual Stapp Car Crash Conference. pp. 41-63. Society of Automotive Engineers, Warrendale, PA.

Crosby, W. M., Snyder, R. G., Snow, C. C., Hanson, P. G. (1968). Impact injuries in pregnancy I: experimental studies. American Journal of Obstetrics and Gynecology, 101(1): 100-110.

Culver, C. C., Viano, D. C. (1990). Anthropometry of seated women during pregnancy: Defining a fetal region for crash protection research. Human Factors, 32(6):625-636.

Dummy Testing Equipment Subcommittee. (1998). User's Manual For the Hybrid III Small Adult Female Dummy. SAE Engineering Aid 25. Society of Automotive Engineers Warrendale, PA.

Eppinger, R.H. (1978). Prediction of thoracic injury using measurable parameters. Proceedings of the 6th Internatiotional Techincal Conference on Experimental Safety Vehicles. pp. 770-780. National Highway Traffic Safety Association, Washington DC.

Freid, A.M. (1978). Distribution of the bulk of the normal placenta: Review and classification of 800 cases by ultrasonography. American Journal of Obstetrics & Gynecology, 132(6): 675-80.

Hardy, W.N and Schneider, L.W. (2001). Development and Refinement of Abdominal-Response Corridors, Project D2a - Development of a Reusable, Rate-Sensitive Abdomen. Final Report No. UMTRI-2001-10. University of Michigan Transportation Research Institute, Ann Arbor, MI

Hardy, W.N and Schneider, L.W (2001). Development and refinement of abdominal-response corridors. Proceedings of the 45th Stapp Car Crash Conference. Society of Automotive Engineers, Warrendale PA.

Institute of Medicine (1990). Nutrition During Pregnancy, Weight Gain, and Nutrient Supplements. Report of the Subcommittee on Nutritional Status and Weight Gain during Pregnancy, Subcommittee on Dietary Intake and Nutrient Supplements during Pregnancy. Committee on Nutritional Status during Pregnancy and Lactation, Food and Nutrition Board. National Academy Press, Washington, DC

Jensen, R.K., Doucet, S., Treitz, T. (1996). Changes in segment mass and mass distribution during pregnancy. Journal of Biomechanics, 29(2): 251-256.

King, A. I., Crosby, W. M., Stout, L. C., Eppinger, R. H. (1972). Effects of lap belt and three-point restraints on pregnant baboons bubjected to deceleration. Proceedings of the Fifteenth Stapp Car Crash Conference, pp. 68-83. Society of Automotive Engineers, Warrendale PA.

Klinich, K.D., Schneider, L.W., Moore, J.A., and Pearlman, M.D. (1999a). Investigatons of Crashes Involving Pregnant Occupants. Final Report NO. UMTRI 99-29. University of Michigan Transportation Research Institute, Ann Arbor, MI.

Klinich, K.D., Schneider, L.W., Eby, B., Rupp, J.D., and Pearlman, M.D. (1999b). Seated Anthropometry During Pregnancy. Final Report No. UMTRI 99-16. University of Michigan Transportation Research Institute, Ann Arbor, MI.

Klinich, K.D., Schneider, L.W., Rupp, J.D., Eby, B., and Pearlman, M.D. (1999c). Challenges in frontal crash protection of pregnant drivers based on anthropometric considerations. Technical Paper 1999-01-0711. Society of Automotive Engineers, Warrendale, PA.

Melvin, J.W., King, A.I., and Alem, N.M. (1988). AATD System Technical Characteristics, Design Concepts, and Trauma Assessment Criteria. AATD Task E-F Final Report (1985) in DOT-HS-807-224. U.S. Department of Transportation, National Highway Traffic Safety Administration, Washington, DC.

National Vital Statistics Reports. (October 5, 1999). 47(25):35-36.

Pearlman, M. D., Tintinalli, J. E., Lorenz, R. P. (1990). Blunt trauma during pregnancy. New England Journal of Medicine, 323(23):1609-1613.

Pearlman, M. D., Viano, D. (1996). Automobile crash simulation with the first pregnant crash test dummy, American Journal of Obstetrics and Gynecology, 175: 977-981.

Pearlman, M.D., Kinich, K.D., Schneider, L.W., Rupp, J.D., Moss, S., and Ashton-Miller, J. (2000). A comprehensive program to improve safety for pregnant women and fetuses in motor vehicle crashes: A preliminary report. American Journal of Obstetrics and Gynecology, 182(6):1554-1564.

Pritchard, J.A., MacDonald, P.C., and Gant, N.F. (1985). Williams Obstetrics. Seventeenth Edition. Appeltic-Century-Crofts, Norwalk, CT.

Rouhana, S.W., Elhagediab, A.M. Chapp, J.J. (1998) A high-speed sensor for measuring chest deflection in crash test dummies. Proceedings of the Sixteenth International Technical Conference on Enhanced Safety Vehicles. Paper No. 98-S9-O-15. National Highway Traffic Safety Adminstration, Washington DC.

Rouhana, S.W., Elhagediab, A.M., Walbridge, T. (2001) Development of a reusable rate-sensitive abdominal insert for the Hybrid III family of dummies. Proceedings of the 45th Stapp Car Crash Conference. Society of Automotive Engineers, Warrendale PA.

Rupp, J.D., Schneider, L.W., Klinich, K.D., Moss, S., Zhou, J., and Pearlman, M.D. (2001). Design, Development, and Testing of a New Pregnant Abdomen for the Small Female Hybrid III Crash Test Dummy. Final Report No. UMTRI 2001-07-11. University of Michgan Transportation Research Institute, Ann Arbor, MI.

Schneider, L.W., Ricci, L.L, King, A.I., Neathery, R.F., and Beebe, M.S. (1994). Design and Development of an Advanced ATD Thorax System for Frontal Crash Environments, Volume 2: Exploration of Alternative Design Approaches. Report No DOT-HS-808-139. National Highway Traffic Safety Administration, Washington DC.

Snyder, R.G., Snow, C.C., Crosby, W.M., Hanson, P., Fineg, J., and Chandler, R. (1966). Impact injury to pregnant female and fetus in lap belt restraint. Proceedings of the 10th International Stapp Car Crash Conference. pp. 249-259. Society of Automotive Engineers, Warrendale, P.A.

Viano, D., Smrcka, J., Jedrzejczak, E., Deng, B., Kempf, P, and Pearlman, M.D. (1996). Belt and airbag testing with a pregnant Hybrid III female dummy. Proceedings of the 15th Experimental Safety Vehicles Conference, Paper No. 96-S1-O-03. National Highway Traffic Safety Adminstration, Washington DC.

CHAPTER 5

MATERIAL PROPERTIES OF PREGNANT TISSUES

CHAPTER 5: MATERIAL PROPERTIES OF PREGNANT TISSUES

Sarah J. Manoogian [1,2] and Stefan M. Duma [2]
1. Biodynamic Research Corporation, San Antonio, TX
2. Virginia Tech – Wake Forest University, Center for Injury Biomechanics

Adverse fetal outcome can be the result of premature rupture of the fetal membranes, placental abruption, direct fetal injury, or uterine laceration. These complications all result from mechanical loading on the fetus or maternal pregnant tissues. To better understand why these complications occur, research has focused on determining the material properties of the fetus, umbilical cord, amnion, chorion, placenta, and uterus. However, testing these materials is difficult for several reasons. The tissues can change dramatically in size and structure throughout the gestational period, at the time of delivery, and in the time following delivery. Even for a specific time during the gestational period, it is a complex task to obtain the tissues for testing. Moreover, once the tissues are obtained, they are complicated in structure and function and require sophisticated testing to appropriately characterize the full material properties. Imaging techniques to monitor the tissue response in vivo is not feasible for extensive research efforts. Regardless of these obstacles, researchers have been able to obtain tissues and quantify the strength of the fetal and maternal tissues to a reasonable degree. A review of the literature regarding the strength of these tissues is presented in this chapter, as well as a summary of material properties of uterine tissue (Table 1). Included for reference are a few of the more recent research papers in this field of study.

FETUS

Material property data are limited for the fetus because of the complications in obtaining tissues. However, there have been two primary efforts to obtain the material properties of the skull and brain tissue in infants and fetuses. Research in the early 1980s by Kriewall and McPherson measured the stiffness of 20 parietal bones from fetal cadavers (McPherson and Kriewall 1980, Kriewall et al. 1981, Kriewall 1982). This work was followed by research published by Margulies who matched the response of porcine and fetal skull responses so that a porcine surrogate could be used for a more extensive analysis (Margulies and Thibault 2000, Prange and Margulies 2002, Coats and Margulies 2006). Through those research efforts, the differences between infant and adult skull and suture strength were quantified. The research by Margulies included tissue testing of brain tissue to complete a finite element model of a one month old infant skull and brain.

Using three-point bending of samples from an infant parietal bone, the rupture stresses range from 3.1 MPa to 17.0 MPa for the subjects age 25 weeks gestation, 30 weeks gestation, and 1 week old (Margulies and Thibault 2000). The rupture stress for a porcine infant bone at a quasi-static loading rate had an average rupture stress of 17.4 ± 2.1 MPa. Using a tensile test, the porcine infant cranial bone had an average rupture stress of 10.6 ± 1.6 MPa. The corresponding ultimate strain was 0.034 ± 0.007 mm/mm. The results from the work by Margulies et al. compared favorably to the results published by McPherson when the fibers were perpendicular as they were in the Margulies study.

Additionally, very early research in the field by Okamoto reported the strength of human nerves and included the sciatic nerve of 38 fetuses and median nerve of 38 fetuses (Yamada 1970). Through uniaxial tension tests, the ultimate elongation of the fiber bundles was measured to be 18.8 ± 0.5% for the fetus at term. The corresponding ultimate strength of these bundles was 9.42 ± 0.25 MPa.

UMBILICAL CORD

Researchers have evaluated the tensile strength of the human umbilical cord to gain an understanding of the forces required to elongate the tissue and restrict the umbilical vessels. In the early 1960s, Katake examined the umbilical cord strength of 62 fetuses (Yamada 1970). While this research examined the strength and elongation of the umbilical cord throughout gestation, the strength and elongation for the fetus at term, which is the greatest, is restated here. The force and elongation at failure for the tissue at term was 1471 ± 49 kPa and 0.59 ± 0.02 strain. An approximate linear modulus in the failure region of the stress versus strain diagram is 14 MPa. Moreover, Katake also indicated that ruptures most frequently occur in the section of the umbilical cord closest to the fetus and failure initiates as a slit in the long axis of the cord. More recently in 2001, Pennati published work using fresh umbilical cords at term (Pennati 2001). Through uniaxial tension and stress relaxation tests, he established the nonlinear and anisotropic response of the tissue. The elastic modulus in the circumferential direction was 2.4 MPa, while the elastic modulus in the longitudinal direction was 10 MPa, on average. The strength in the longitudinal direction is on the same order of magnitude as the work done 40 years prior by Katake. There are additional researchers who have published work on the strength of the umbilical cord (Morris and Hunt 1966, Crichton 1973, Ghosh et al. 1984).

CHORIOAMNIOTIC MEMBRANE

The motivation for researching the strength of the chorioamnion as well as these two layers independently is the high incidence of premature rupture of membranes, which can lead to fetal complications. These tissues have been tested more extensively than other pregnant tissues. One reason for this is because they are easily obtainable after delivery. Other reasons for the extensive testing are that many samples may be obtained from one donor and the tissue is thin, which makes it easier to grip in simple material testing machines. Research on the strength of amniotic membranes has been dated back to as early as 1886 (Benson-Martin et al. 2006). Since that time several researchers have obtained data on the strength of fetal membranes in tension (Danforth et al. 1953, Maclachlan 1965, Meudt and Meudt 1967). Early researchers determined there were differences in the thickness and elasticity of membranes that ruptured prematurely (Artal et al. 1976, Parry-Jones and Priya 1976). The viscoelastic response of the chorioamniotic membrane was established using biaxial stress relaxation and creep tests (Lavery and Miller 1977). More testing which assessed the effect of labor on the chorioamniotic membrane properties hypothesized that uterine activity could contribute to the weakening of the membrane, causing premature rupture (Lavery et al. 1982).

Using a materials testing machine, Oxlund et al. provided a more detailed investigation of the differences between the chorion and amnion layers. From this research, the chorionic component is more elastic and only contributes 10-15% of the strength of the chorioamniotic membrane. The orientation of the samples did not have a significant effect on the material properties measured (Oxlund et al. 1990). Research has shown conflicting results regarding the change in the material properties of the membrane before and after delivery. Research in Japan showed no differences (Manabe et al. 1991), while research in Denmark indicated that, after delivery, the amnion and chorion separate and also have a dramatic decrease in their mechanical strength (Helmig et al. 1993). Other researchers considered the effect of human ejaculate on the chorioamniotic membrane. The results were conflicting, with one study showing a weakening effect and one showing a strengthening effect (Romem et al. 1989, Harmanli et al. 1998). Puncture testing was performed to determine the membrane strength (Schober et al. 1994). In a more recent study, the effect of gestation on the rupture strength of the membrane was tested. Although these results were not reported as stress and strain, a trend of increase in rupture force was reported until 20 weeks, and then it maintained approximately the same rupture force until 39 weeks (Pressman et al. 2002).

The most recent research that has determined the response of the chorioamniotic membrane includes uniaxial stress-relaxation tests by Oyen et al. and uniaxial failure tests of preterm and term membranes by Benson-Martin et al. (Oyen et al. 2004, Benson-Martin et al. 2006). Previous researchers determined that the amnion is the primary strength-bearing part of the chorioamniotic membrane. Therefore, Oyen et al. took six specimens of human fetal chorioamniotic tissue and separated the chorion from the amnion (Oyen et al. 2004). Tissue samples 25 mm in length and approximately 6 mm in width were pulled in uniaxial tension to failure. The strain rate was 2.3%/s with an 11 mm grip-to-grip distance. All of the samples were submerged in saline for testing, and the thickness was assumed to be 50 μm for all tests. Results from this research indicated a failure force of 1.57 ± 0.62 N with a failure displacement of 4.1 ± 0.4 mm. The corresponding stiffness of the amnion layer was 0.62 ± 0.20 N/mm. When a linear modulus was calculated from the data, the average value was 23.1 ± 6.3 MPa, which corresponds to previous work by Helmig, who obtained a value of 29.5 MPa for the amnion layer alone.

In 2006, Benson-Martin published research that compared eight preterm and 15 term amnion membranes that had been separated from the chorion layer (Benson-Martin et al. 2006). Using a sample size of 50 mm by 20 mm, he loaded the samples in uniaxial tension at 10 mm/min until failure occurred. The results were reported in engineering stress and strain values. The average elastic modulus for the preterm membranes was significantly different (p=0.03) from the term membranes. The corresponding values were 3.6 MPa and 2.3 MPa for the two tissue groups. The average thicknesses of the tissues were 0.20 mm and 0.25 mm for the preterm and term membranes, respectively. Although these modulus values are much lower than the previous literature, the rate of loading is almost four orders of magnitude lower than the work by Oyen. Moreover, Benson-Martin reports that the differences are attributed to the hydration level of the tissue, which therefore affects the cross-sectional area.

PLACENTA

Although placental tissue is easily obtained for testing, determining the material properties of placental tissue is difficult because of its size. In 1999 Pearlman indicated in a report the results of his efforts to quantify the loading properties of placental tissue at a quasi-static rate (Pearlman 1999). He did uniaxial tension tests at subfailure levels. The gage lengths of the specimens were 30 mm, and they were pulled at a rate of 1.6 mm/s. For all five specimens, the average peak strain was 0.43 ± 0.16 and the average peak stress was 15.6 ± 7.6 kPa.

More recently, research has quantified the loading and the failure properties of chorion layer and the placenta without the chorion attached in uniaxial tension at dynamic loading rates (Manoogian et al. 2008a). A total of 20 dynamic uniaxial tensile tests on the maternal side of the placenta and ten dynamic uniaxial tensile tests on the chorion layer of the placenta were performed. These tests were completed from six human placentas to determine material properties at a strain rate of 7.0 strains/s. The data are reported in Green Lagrangian strain and true stress values. The results show that the average peak strain at failure for both the maternal portion and the chorion layer of the placenta were similar, with a value of 0.56 and 0.61, respectively. However, the average failure stress for the chorion layer, 167.8 kPa, was much higher than the average failure stress for the placenta with the chorionic plate removed, 18.6 kPa. The large difference in strength is due to differences in the structure and function of these layers in the placenta.

While the quasistatic tests are subfailure and the tests in the Manoogian study are failure, the dynamic testing data from the maternal side of the placenta are compared to quasistatic placenta tension research. The data from Pearlman's tests compare favorably to the maternal side tension tests but have significantly lower peak stresses than the chorion layer tension tests ($p<0.001$).

Dynamic uniaxial placental tissue testing was also performed by Hu et al. (2009). In the Hu study, 21 placental specimens were pulled to failure while measuring stress and strain. Five whole human placentas from cesarean sections were used to obtain 100 mm by 20 mm individual specimens without the chorion layer attached. The strain rate chosen for the testing was 12/s. The mean failure strain was 0.47 ± 0.1 with a corresponding failure stress of 34.8 ± 12.6 kPa.

The data collected by Manoogian (2008) and Hu (2009) are similar. However the method of cutting specimens, grip types, and initial conditions vary between the two studies. Manoogian et al. used a stamp to obtain uniform specimens. Hu et al. cut the samples by hand. Because of the thickness of the samples in the Hu study, cryogenic grips were necessary. In the Manoogian study, the samples were sliced thin with a tissue slicer and therefore were gripped at room temperature. The variance in the initial conditions likely affected the prestress and prestrain in the tissues.

Additional work was completed on placental tissue to address the issue of rate dependence in the tissue. Manoogian et al. (2009) performed uniaxial failure tests on placental specimens at 0.07/s, 0.7/s, and 7.0/s. This data set included 64 uniaxial tests on six placentas. Each placenta had three to five tests at each of the three strain rates. The methodology was the same as previous research, with a tissue slicer used to obtain thin slices of the tissue and then a stamp to cut uniform samples. The failure strains were 0.34 ± 0.07, 0.36 ± 0.06, and 0.37 ± 0.06, respectively, with no significant differences between the loading rates. The failure stresses were 10.8 ± 5.9 kPa, 11.4 ± 3.7 kPa, and 18.6 ± 5.4 kPa, with a corresponding increase in loading rates. The only significant difference in failure stress was between the highest and lowest strain rate. These experiments were then utilized to determine corresponding properties that are best suited for modeling (Yu 2009).

Secondary to obtaining the placental properties in dynamic tension, the researchers also developed material models and performed simulations to model the experiments. These results are discussed further in the modeling section.

UTERUS

Previous researchers have conducted material tests on nonpregnant, pregnant, human, and animal uterine tissues at quasi-static loading rates (Ohara 1953, Csapo and Goodall 1954, Conrad and Kuhn 1967, Yamada 1970, Pearsall and Roberts 1978, Mizrahi et al. 1980, Pearlman 1999, Deyer et al. 2000, Rupp et al. 2001). Research on uterine tissue material properties began in the early 1950s with a basic uniaxial tensile failure test at a quasistatic rate on animal uterus (Yamada 1970). Ohara found the tensile strength of a nonpregnant rabbit uterus in the longitudinal direction to be 177 kPa with a corresponding ultimate elongation of 1.5 (Ohara 1953). In addition to failure properties, Csapo and Goodall used electrical stimulation on rabbit uteri to quantify active muscle tensile strength. The result of their research was normalized force vs. normalized length activation curves for myometrium under estrogen and progesterone domination (Csapo and Goodall 1954). Myometrium is the smooth muscle layer of the uterus.

To follow this work, research progressed to study pregnant vs. nonpregnant myometrium. Wood pioneered human uterine tissue testing by using strips of human uterus that were extracted during a cesarean section at the incision location (Wood 1964). Uniaxial quasistatic tensile tests of these uterine samples of the lower uterine region had a failure true stress of 483 kPa and the elastic moduli of the samples at 69 kPa stress averaged to 1207 kPa (Pearsall and Roberts 1978). No failure strain was reported.

In 1967, Conrad used pregnant and nonpregnant human uterine samples in uniaxial quasistatic tensile tests with a step increase in load (Conrad and Kuhn 1967). Results of his study indicated the pregnant uterine tissue, from a cesarean section sample, had a lower elastic modulus at 69 kPa than the nonpregnant tissue from a uterus extracted during a hysterectomy. The respective elastic moduli from engineering stress-strain data were 586 kPa and 965 kPa (Pearsall and Roberts 1978).

By the late 1970s, Pearsall tested more than 20 nonpregnant human uterine specimens in uniaxial tension (Pearsall and Roberts 1978). Human uterine specimens were acquired from the fundal region of three nonpregnant uteri, which had been removed in hysterectomy operations. The tests were at a quasistatic rate of less than 0.01 strains/s and had failure stresses that ranged from 550-2069 kPa. The maximum failure strain from those tests was in the range of 0.30-0.95. Additional tests were completed to evaluate the tissue response to compressive loads. Results of these tests showed that the tissue has a stiffer response in tension than in compression.

Mizrahi's research goal was to measure strains during labor by placing strain rosettes on the skin and the uterus (Mizrahi et al. 1980). He recorded strains on 36 women while in their 36th to 42nd weeks of pregnancy. The peak strain recorded during labor was less than 0.03, which is inconsistent with other in vivo studies that report a much higher strain. Although Mizrahi's study has several limitations, the main result was that the uterus has an anisotropic response.

More recently, four subfailure uniaxial tensile tests were completed on human uterine samples taken from the location of a cesarean incision and frozen before testing. At a quasistatic rate, the uterine tissue had an average peak stress of 45 kPa with a maximum strain equal to 0.65 (Pearlman 1999). Deyer and Ashton-Miller researched the uterine tissue response during labor by quantifying the uterine wall thickness from ultrasound measurements (Deyer et al. 2000). From this study it was determined the uterus reaches maximum contractile strain before the placenta separates. The results reported contractile strains consistent with smooth muscle data. At the time of placental separation, the peak contractile uterine strains were 4.5 radial and 0.75 circumferential. A single failure test by Ashton-Miller performed on a uterus sample that included the uteroplacental interface (UPI) indicated 0.60 strain in the uterus when the UPI failed (Rupp et al. 2001).

In 2008, Manoogian presented stress-strain data for pregnant porcine uterine tissue when loaded biaxially in tension at a dynamic rate (Manoogian et al. 2008b). A total of 23 tests were completed and analyzed to develop stress-strain curves and peak strain data for pregnant porcine uterine tissue. At a loading rate of 0.7 strains/s, the circumferential direction peak stress component was 500 ± 219 kPa with a corresponding peak strain of 0.30 ± 0.09. The peak stress for the longitudinal direction was 320 ± 176 kPa with a peak strain of 0.30 ± 0.09.

Table 1: Summary of material properties of uterine tissue.

Source (Year)	Uterine Tissue Type	Pregnant	Test	Peak Stress Range (kPa)	Max True Strain	Linear Modulus Range (kPa)	Failure
Ohara (1953)	Rabbit	No	Quasistatic Uniaxial tension	177	0.92*		Yes
Wood (1964)	Human - cesarean section	Yes	Quasistatic Uniaxial tension	483		1207	
Conrad (1966)	Human – cesarean section	Yes	Quasistatic Uniaxial tension			586	
Conrad (1966)	Human - hysterectomy	No	Quasistatic Uniaxial tension			965	
Pearsall (1978)	Human-hysterectomy	No	Quasistatic Uniaxial tension (parallel to long axis)	550-2069	0.30-0.95	500-1378	20+ Failure
Pearlman (1999)	Human - cesarean section then frozen	Yes	Quasistatic Uniaxial tension	9-105	0.36-0.55*	20-279	4 Sub-Failure
Ashton-Miller (2001)	Human –with UPI	Yes	Quasistatic Uniaxial tension		0.47*		Yes
Manoogian (2008)	Porcine Circumferential Direction	Yes	Dynamic Biaxial	500±219 182-1094	0.30±0.09 0.15-0.49	695-3453	23 Sub-Failure
Manoogian (2008)	Porcine Longitudinal Direction	Yes	Dynamic Biaxial	320±175 76-726	0.30±0.09 0.16-0.48	507-2327	23 Sub-Failure

* These values have been converted to true strain from their published values.

REFERENCES

Artal, R., R. J. Sokol, M. Neuman, A. H. Burstein, and J. Stojkov. 1976 "The mechanical properties of prematurely and non-prematurely ruptured membranes. Methods and preliminary results." *American Journal of Obstetrics and Gynecology* 125(5): 655-9.

Benson-Martin, J., P. Zammaretti, G. Bilic, T. Schweizer, B. Portmann-Lanz, T. Burkhardt, R. Zimmermann, and N. Ochsenbein-Kolble. 2006. "The Young's modulus of fetal preterm and term amniotic membranes." *European Journal of Obstetrics & Gynecology and Reproductive Biology* 128(1-2): 103-7.

Coats, B. and S. S. Margulies. 2006. "Material properties of human infant skull and suture at high rates." *Journal of Neurotrauma* 23(8): 1222-32.

Conrad, J. T. and W. Kuhn. 1967. "The active length-tension relationship in human uterine muscle." *American Journal of Obstetrics and Gynecology* 97(2): 154-60.

Crichton, J. L. 1973. "Tensile strength of the umbilical cord." *American Journal of Obstetrics and Gynecology* 115(1): 77-80.

Csapo, A. and M. Goodall. 1954. "Excitability, length tension relation and kinetics of uterine muscle contraction in relation to hormonal status." *The Journal of Physiology* 126(2): 384-95.

Danforth, D. N., T. W. McElin, and M. N. States. 1953. "Studies on fetal membranes. I. Bursting tension." *American Journal of Obstetrics and Gynecology* 65(3): 480-90.

Deyer, T. W., J. A. Ashton-Miller, P. M. Van Baren, and M. D. Pearlman. 2000. "Myometrial contractile strain at uteroplacental separation during parturition." *American Journal of Obstetrics and Gynecology* 183(1): 156-9.

Ghosh, K. G., S. N. Ghosh, and A. B. Gupta. 1984. "Tensile properties of human umbilical cord." *Indian Journal of Medical Research* 79: 538-41.

Harmanli, O. H., R. J. Wapner, and J. F. Lontz. 1998. "Human ejaculate. Effects on the biomechanical properties of the human chorioamniotic membranes." *Journal of Reproductive Medicine* 43(9): 779-82.

Helmig, R., H. Oxlund, L. K. Petersen, and N. Uldbjerg. 1993. "Different biomechanical properties of human fetal membranes obtained before and after delivery." *European Journal of Obstetrics & Gynecology and Reproductive Biology* 48(3): 183-9.

Hu, J., K. Klinich, C. Miller, G. Nazmi, M. Pearlman, L. Schneider, and J. Rupp. 2009. "Quantifying dynamic mechanical properties of human placenta tissue using optimization techniques with specimen-specific finite-element models." *Journal of Biomechanics* 42(15) 2528-2534.

Kriewall, T. J. 1982. "Structural, mechanical, and material properties of fetal cranial bone." *American Journal of Obstetrics and Gynecology* 143(6): 707-14.

Kriewall, T. J., G. K. McPherson, and A. C. Tsai. 1981. "Bending properties and ash content of fetal cranial bone." *Journal of Biomechanics* 14(2): 73-9.

Lavery, J. P. and C. E. Miller. 1977. "The viscoelastic nature of chorioamniotic membranes." *Obstetrics and Gynecology* 50(4): 467-72.

Lavery, J. P., C. E. Miller, and R. D. Knight. 1982. "The effect of labor on the rheologic response of chorioamniotic membranes." *Obstetrics and Gynecology* 60(1): 87-92.

Maclachlan, T. B. 1965. "A Method for the Investigation of the Strength of the Fetal Membranes." *American Journal of Obstetrics and Gynecology* 91: 309-13.

Manabe, Y., N. Himeno, and M. Fukumoto. 1991. Tensile strength and collagen content of amniotic membrane do not change after the second trimester or during delivery." *Obstetrics and Gynecology* 78(1): 24-7.

Manoogian, S. J., J. A. Bisplinghoff, C. McNally, A. R. Kemper, A. C. Santago, and S. M. Duma. 2008a. "Dynamic Tensile Properties of Human Placenta." *Journal of Biomechanics* 41:3436-3440.

Manoogian, S. J., C. McNally, J. D. Stitzel, and S. M. Duma. 2008. "Dynamic Biaxial Tissue Properties of Pregnant Porcine Uterine Tissue." *Stapp Car Crash Journal* 52:167-185.

Manoogian, S., J. Bisplinghoff, C. McNally, A. Kemper, A. Santago, S. M. Duma. 2009. "Effect of Strain Rate on the Tensile Material Properties of Human Placenta." *Journal of Biomechanical Engineering* 131(9).

Margulies, S. S. and K. L. Thibault. 2000. "Infant skull and suture properties: measurements and implications for mechanisms of pediatric brain injury." *Journal of Biomechanical Engineering* 122(4): 364-71.

McPherson, G. K. and T. J. Kriewall. 1980. "The elastic modulus of fetal cranial bone: a first step towards an understanding of the biomechanics of fetal head molding." *Journal of Biomechanics* 13(1): 9-16.

Meudt, R. and E. Meudt. 1967. "Rupture of the fetal membranes. An experimental, clinical, and histologic study." *American Journal of Obstetrics and Gynecology* 99(4): 562-8.

Mizrahi, J., Z. Karni, and W. Z. Polishuk. 1980. "Isotropy and anisotropy of uterine muscle during labor contraction." *Journal of Biomechanics* 13(3): 211-8.

Morris, J. F. and A. C. Hunt. 1966. "Breaking strength of the umbilical cord." *Journal of Forensic Science* 11(1): 43-9.

Ohara, T. 1953. "On the comparison of strengths of the various organ-tissues." *Journal of Kyoto Preferred Medical University* 53: 577-597.

Oxlund, H., R. Helmig, J. T. Halaburt, and N. Uldbjerg. 1990. "Biomechanical analysis of human chorioamniotic membranes." *European Journal of Obstetrics & Gynecology and Reproductive Biology* 34(3): 247-55.

Oyen, M. L., S. E. Calvin, and R. F. Cook. 2004. "Uniaxial stress-relaxation and stress-strain responses of human amnion." *Journal of Materials Science: Materials in Medicine* 15(5): 619-24.

Parry-Jones, E. and S. Priya. 1976. "A study of the elasticity and tension of fetal membranes and of the relation of the area of the gestational sac to the area of the uterine cavity." *British Journal of Obstetrics and Gynaecology* 83(3): 205-12.

Pearlman, M. D. 1999. "Data Acquisition for Development of a Uteroplacental Interface for the Second-Generation Pregnant Abdomen." University of Michigan Health Systems, D.8. Final Report.

Pearsall, G. W. and V. L. Roberts. 1978. "Passive mechanical properties of uterine muscle (myometrium) tested in vitro." *Journal of Biomechanics* 11(4): 167-76.

Pennati, G. 2001. "Biomechanical properties of the human umbilical cord." *Biorheology* 38(5-6): 355-66.

Prange, M. T. and S. S. Margulies. 2002. "Regional, directional, and age-dependent properties of the brain undergoing large deformation." *Journal of Biomechanical Engineering* 124(2): 244-52.

Pressman, E. K., J. L. Cavanaugh, and J. R. Woods. 2002. "Physical properties of the chorioamniotic throughout gestation." *American Journal of Obstetrics and Gynecology* 187(3): 672-5.

Romem, Y., C. Sires, and R. Artal. 1989. "Effects of seminal ejaculate on the biomechanical properties of chorioamniotic membranes." *Journal of Reproductive Medicine* 34(3): 221-4.

Rupp, J. D., L. W. Schneider, K. D. Klinich, S. Moss, J. Zhou, and M. D. Pearlman. 2001. "Design, Development, and Testing of a New Pregnant Abdomen for the Hybrid III Small Female Crash Test Dummy." UMTRI-2001-07. University of Michigan Transportation Research Institute.

Schober, E. A., R. P. Kusy, and D. A. Savitz. 1994. "Resistance of fetal membranes to concentrated force applications and reconciliation of puncture and burst testing." *Annals of Biomedical Engineering* 22(5): 540-8.

Wood, C. 1964. "Physiology of Uterine Contractions." *Journal of Obstetrics and Gynaecology of the British Commonwealth* 71: 360-373.

Yamada, H. 1970. *Strength of Biological Materials*. F. G. Evans, ed. Williams and Wilkins, Baltimore.

Yu, M., S. Manoogian, S. M. Duma, and J. D. Stitzel. 2009. "Finite element modeling of human placenta tissue." *Annals of the Association of Automotive Medicine* Vol. 53, 257-270.

The following papers are included in their entirety to augment this chapter:

Hu, J., K. D. Klinich, C. S. Miller, G. Nazmi, M. D. Pearlman, L. W. Schneider, and J. D. Rupp. 2009. "Quantifying dynamic mechanical properties of human placenta tissue using optimization techniques with specimen-specific finite-element models." *Journal of Biomechanics* 42(15):2528-2534.

Manoogian, S. J., J. A. Bisplinghoff, C. McNally, A. R. Kemper, A. C. Santago, and S. M. Duma. 2008. "Dynamic tensile properties of human placenta." *Journal of Biomechanics* 41:3436-3440.

Manoogian, S. J., C. McNally, J. D. Stitzel, and S. M. Duma. 2008. "Dynamic Biaxial Tissue Properties of Pregnant Porcine Uterine Tissue." *Stapp Car Crash Journal* 52:167-185.

Manoogian, S. J., J. A. Bisplinghoff, C. McNally, A. R. Kemper, A. C. Santago, and S. M. Duma. 2009. "Effect of strain rate on the tensile material properties of human placenta." Paper No. 091008. *Journal of Biomechanical Engineering* 131(9).

Yu, M., S. J. Manoogian, S. M. Duma, and J. D. Stitzel. 2009. "Finite element modeling of human placenta tissue." *Annals of Advances in Automotive Medicine* 53:257-270.

Reprinted from Journal of Biomechanics, Vol. 42(15), 2009, Hu et al, "Quantifying dynamic mechanical properties of human placenta tissue using optimization techniques with specimen-specific finite-element models," pp 2528-2534, Copyright 2009, with permission from Elsevier.
Journal of Biomechanics 42 (2009) 2528–2534

Contents lists available at ScienceDirect

Journal of Biomechanics

journal homepage: www.elsevier.com/locate/jbiomech
www.JBiomech.com

ELSEVIER

Quantifying dynamic mechanical properties of human placenta tissue using optimization techniques with specimen-specific finite-element models

Jingwen Hu [a,*], Kathleen D. Klinich [a], Carl S. Miller [a], Giseli Nazmi [b], Mark D. Pearlman [b], Lawrence W. Schneider [a,c], Jonathan D. Rupp [a,d]

[a] University of Michigan Transportation Research Institute, 2901 Baxter Road, Ann Arbor, MI 48109, USA
[b] Obstetrics and Gynecology and Surgery, University of Michigan Health Systems, Ann Arbor, MI, USA
[c] Department of Biomedical Engineering, University of Michigan, Ann Arbor, MI, USA
[d] Department of Emergency Medicine, University of Michigan, Ann Arbor, MI, USA

ARTICLE INFO

Article history:
Accepted 8 July 2009

Keywords:
Placental abruption
Fetal loss
Dynamic tissue testing
Specimen-specific finite-element model
Optimization

ABSTRACT

Motor-vehicle crashes are the leading cause of fetal deaths resulting from maternal trauma in the United States, and placental abruption is the most common cause of these deaths. To minimize this injury, new assessment tools, such as crash-test dummies and computational models of pregnant women, are needed to evaluate vehicle restraint systems with respect to reducing the risk of placental abruption. Developing these models requires accurate material properties for tissues in the pregnant abdomen under dynamic loading conditions that can occur in crashes. A method has been developed for determining dynamic material properties of human soft tissues that combines results from uniaxial tensile tests, specimen-specific finite-element models based on laser scans that accurately capture non-uniform tissue-specimen geometry, and optimization techniques. The current study applies this method to characterizing material properties of placental tissue. For 21 placenta specimens tested at a strain rate of 12/s, the mean failure strain is 0.472 ± 0.097 and the mean failure stress is 34.80 ± 12.62 kPa. A first-order Ogden material model with ground-state shear modulus (μ) of 23.97 ± 5.52 kPa and exponent (α_1) of 3.66 ± 1.90 best fits the test results. The new method provides a nearly 40% error reduction ($p < 0.001$) compared to traditional curve-fitting methods by considering detailed specimen geometry, loading conditions, and dynamic effects from high-speed loading. The proposed method can be applied to determine mechanical properties of other soft biological tissues.

1. Introduction

Motor-vehicle crashes (MVC) are the leading cause of fetal deaths from maternal trauma in the United States (Weiss, 2001; Weiss and Strotmeyer, 2002), outnumbering the annual number of infant deaths from MVC (Pearlman et al., 2000). Placental abruption (separation of the placenta from the uterus) is the most common cause of traumatic fetal death (Weiss et al., 2001).

The placenta is a vascular organ that exchanges oxygen, nutrients, and waste between the mother and fetus. Microvilli attach the maternal side of the placenta to the decidual layer of the uterine wall. The umbilical cord attaches to the chorion covering the fetal side of the placenta. Placental abruption occurs at the utero-placental interface (UPI). To understand the mechanisms of placental abruption, the tensile mechanical properties of the uterus and placenta and the failure characteristics of the UPI in tension and shear are needed. The current study addresses placenta mechanical properties.

Computational models have already improved understanding of placental abruption mechanisms in MVC (Moorcroft et al., 2003; Pearlman et al., 2000). However, these models are limited because the mechanical properties of human placenta under the loading conditions that can occur in MVC have not been determined. Manoogian et al. (2007, 2008) report experimental data from uniaxial tensile testing of human placenta tissue at one dynamic loading rate. However, they did not develop a constitutive material model that characterizes placenta properties.

The dynamic mechanical properties of soft tissues are challenging to quantify. Previous studies on mechanical properties of biological tissues, such as brain (Miller and Chinzei, 2002), liver (Tamura et al., 2002), and kidney (Snedeker et al., 2005), have demonstrated that material models can be developed by applying curve-fitting techniques to experimental data. However, this approach requires testing with a uniform tissue specimen of known geometry, which is difficult to achieve with placenta tissue. Assuming uniform specimen geometry may cause errors in

* Corresponding author. Tel.: +1734 763 6398; fax: +1734 647 3330.
E-mail address: jwhu@umich.edu (J. Hu).

0021-9290/$ - see front matter © 2009 Elsevier Ltd. All rights reserved.
doi:10.1016/j.jbiomech.2009.07.003

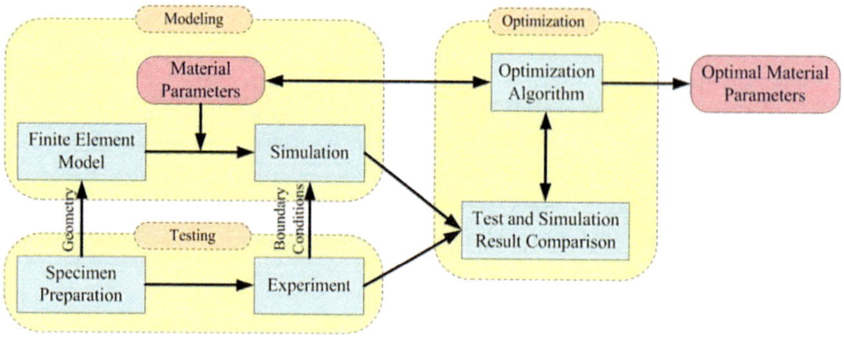

Fig. 1. Flowchart of steps to determine mechanical properties of soft biological tissues, such as placenta, using experiments, specimen-specific FE models, and optimization techniques.

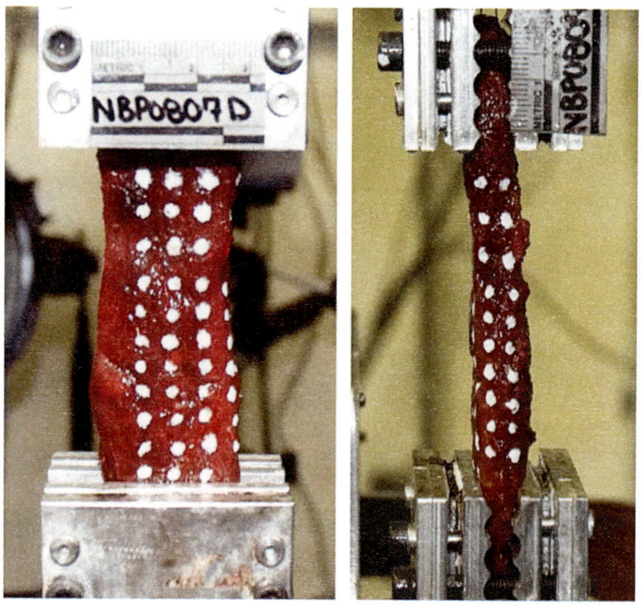

Fig. 2. Placental specimen in thermoelectrically cooled tissue-clamping device ready for uniaxial tensile testing in the high-speed MTS machine.

calculating the mechanical properties due to variation in cross-sectional area along the specimen. Further, the traditional curve-fitting method requires maintaining constant loading conditions during testing, which can be difficult at high loading rates.

In this study, a method to quantify the dynamic mechanical properties of human placenta tissue was developed that employs a combination of dynamic uniaxial tensile tests, specimen-specific finite-element (FE) models, and material model-fitting using optimization techniques. This method can also be applied to other very soft, non-uniform biological tissues.

2. Methods

Fig. 1 outlines how the mechanical properties of placenta tissue were estimated. After conducting high-speed uniaxial tensile tests, each test was reconstructed using a specimen-specific FE model to account for geometric and loading differences. A material model was selected for the FE model and optimization was used to vary simulation parameters to reduce the differences between test and simulation results. Optimal model parameters were typically achieved after several hundred iterations. To evaluate this method, material parameters for each specimen were compared using three methods: traditional curve-fitting techniques, specimen-specific FE model with optimization techniques, and simplified (constant cross-sectional area) FE model with optimization techniques.

2.1. Uniaxial tensile tests

Twenty-one placenta tissue specimens excised from five whole human placentas were tested. Donor tissues were collected after caesarian deliveries following informed-consent procedures approved by the University of Michigan institutional review board. Individual specimens measuring approximately $100 \times 20\,\text{mm}^2$, chosen to minimize visible discontinuity, were cut by a scalpel. The chorion layer was removed and the remaining full thickness of the placenta was tested. During preparation, the specimens were hydrated with a saline-saturated towel. As shown in Fig. 2, each end of the specimen was gripped in a high-speed MTS machine using a thermoelectrically cooled clamp. To secure the sample, sutures were applied to one end of the tissue, and the tissue suspended vertically from the top grip. After aligning the specimen, the jaws of the top grip were slightly tightened so that approximately 20 mm of the specimen was within the clamp, while limiting the tissue deformation. The sample was then lowered into the lower jaw and the process was repeated. While the cryogenic clamps were activated, the temperature in the active region of the specimen was monitored at five locations using a thermocouple and maintained at 20–26 °C by a radiant heater. The specimen was continually hydrated with saline spray. Once the ends were frozen, the clamps were tightened to prevent slipping within the grips. A load cell below the lower clamp measured the reaction-force (accuracy of 0.01 N), and the MTS displacement transducer measured global displacement (accuracy of 0.1 mm). Testing used a loading rate of 12/s based on estimated maximum strain rate to the pregnant abdomen from simulated frontal crashes using the pregnant dummy (Rupp et al., 2001). Ram acceleration and time of initial loading were also measured. A pattern of 2-mm diameter markers was manually applied to each specimen before testing and a high-speed video camera recorded marker motion at 1000 frames/s. Strain at the failure layer was calculated using four triangular elements constructed through the markers. Green-lagrange strain tensors were calculated for each element at the failure time, and the calculated strains in the loading direction were averaged to estimate failure strain. The failure stress was calculated using Eq. (1) by assuming tissue incompressibility.

$$Stress_{fail} = \frac{F_{fail}}{A_0} \times \left(1 + \frac{\Delta L_{fail}}{L_0}\right) \tag{1}$$

where, ΔL_{fail} is total specimen displacement at failure time, L_0 is original specimen length, F_{fail} is reaction-force at failure time, and A_0 is original cross-sectional area at the failure location.

After mounting the specimen, a FARO Laser ScanArm (accuracy of 50 μm) was used to scan the 3-dimensional geometry of the test specimen. Every region of the specimen was scanned from different angles to avoid possible missing points for a moist surface. All specimens were tested within 48 h after delivery. If the test cannot be conducted on the same day of delivery, the specimens were stored at 3 °C after being wrapped in a saline-saturated towel.

2.2. Specimen-specific finite-element model

A semi-automatic template-based algorithm (Fig. 3), previously reported by Hu et al. (2007), was used to develop specimen-specific FE models consisting of only hexahedral elements. The grid dataset collected with the FARO Laser ScanArm was processed by Geomagic Studio 10 (Geomagic, Inc.) into a smooth 3D surface. A $60 \times 20 \times 15\,\text{mm}^3$ 3D mesh composed of only ideal hexahedral elements (Jacobian = 1) was used as the baseline model, which was then rigidly registered with the geometric surface. After the registration, the baseline model was morphed using several control points; the FE exterior surface nodes of the baseline model were then projected to the geometric surface of the specimen to achieve a specimen-specific FE model. This procedure can be applied to develop subject-specific FE models with much more complicated geometries, such as a patient-specific brain model for navigational system in neurosurgery (Hu et al., 2007).

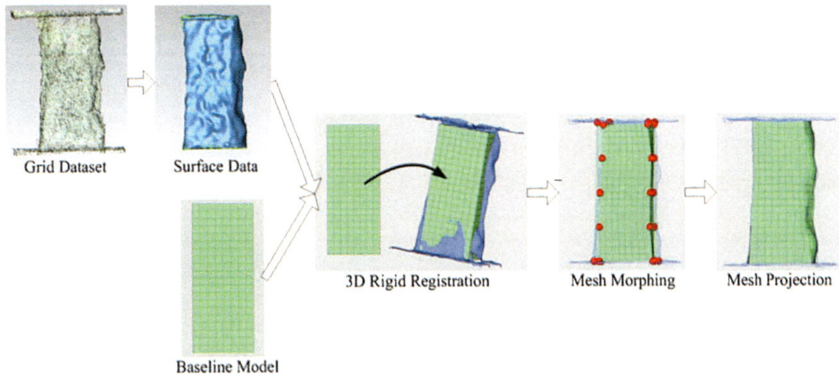

Fig. 3. A method to develop specimen-specific FE models for placenta tissue.

All simulations were performed using LS-DYNA3D 971 (LSTC), an explicit nonlinear dynamic FE code capable of simulating large deformations under high-speed loading. A mesh convergence analysis was initially conducted by comparing the model-predicted reaction-force history for successive mesh refinements. Models with 960 hexahedral elements (nominal mesh size of 1.5 mm) were sufficient (error < 0.1%) to achieve accurate predictions.

After establishing the specimen FE mesh, the measured velocity history was applied as a prescribed vertical velocity to the top node layer in the FE model, with the other two nodal displacement directions fixed. At the bottom node layers, displacements in all three directions were fixed.

2.3. Material property model

The models incorporated a hyperelastic material model (Ogden, 1972, 1997). This model considers the material to be fully incompressible, and the strain energy function expressed as:

$$W(\lambda_1, \lambda_2, \lambda_3) = \sum_{p=1}^{N} \mu_p(\lambda_1^{\alpha_p} + \lambda_2^{\alpha_p} + \lambda_3^{\alpha_p} - 3)/\alpha_p \tag{2}$$

where λ_1, λ_2, and λ_3 are the principal stretches, N is the order of Ogden model, and μ_p and α_p are the pth shear modulus and exponent, respectively. The ground-state shear modulus μ is defined as:

$$\mu = \sum_{p=1}^{N} \frac{\mu_p \alpha_p}{2} \tag{3}$$

In a uniaxial tensile test, let λ_1 be the stretch ratio in the direction of elongation. Given the incompressibility constraint, the other two principal strains $\lambda_2 = \lambda_3 = \lambda_1^{-1/2}$. Using a first-order Ogden model, the strain energy function from Eq. (2) simplifies to:

$$W(\lambda_1, \lambda_2, \lambda_3) = \mu_1(\lambda_1^{\alpha_1} + 2\lambda_1^{-\alpha_1/2} - 3)/\alpha_1 \tag{4}$$

Tensile stress is generally defined as force per unit unstrained area of cross-section. This stress, denoted by S_1 is given by:

$$S_1 = \frac{\partial W}{\partial \lambda_1} = \mu_1[\lambda_1^{\alpha_1 - 1} - \lambda_1^{-(1+\alpha_1/2)}] \tag{5}$$

Therefore, if the original test sample cross-sectional area is known and the reaction-force and global tissue displacement are measured during tensile tests, material parameters (μ_1 and α_1) for the Ogden model can be calculated by a simple curve-fitting technique. However, since the cross-sectional areas of the placenta tissue samples were not uniform longitudinally throughout the specimen, using a single value for cross-sectional area could potentially cause errors in the material model parameters. Furthermore, the above curve-fitting method neglects some dynamic effects resulting from tissue mass and acceleration.

2.4. Optimization method

This study employed a genetic algorithm (GA) to optimize material parameters for each specimen-specific FE placenta specimen model. GAs are search algorithms based on the mechanics of natural selection and genetics (Holland, 1975) that have recently been applied to injury biomechanics (Stitzel et al., 2005; Fu et al., 2004). The present study used the sum-of-squares reaction-force error shown in Eq. (6) as the objective function while considering two material parameters (μ and α_1) as input variables. An initial value of each material parameter was determined by simple curve-fitting described in Section 2.3, with lower (50% of initial value) and upper (200% of initial value) bounds used to restrain the scope of searching. If the final optimal value was near the searching bound, a new searching scope was selected for another optimization. A damping coefficient ranging from 0 to 0.3 was also varied during the optimizations to allow appropriate simulation of the dynamic test data. A Poisson ratio of 0.4999 and density of 1.0 g/mm^3 were used for all specimens. ModeFRONTIER 4.0 (ESTECO) was used for all optimizations.

$$F_{error} = \sqrt{\sum_{i=0}^{n}(Fsim_i - Ftest_i)^2} \tag{6}$$

where, $Fsim_i$ and $Ftest_i$ represent the simulated and experimentally measured reaction-force at each time point, respectively, and n is the number of points available from the start of the test until tissue failure.

2.5. Error calculation and comparison

To compare accuracies between FE-optimization and curve-fitting methods and demonstrate the importance of specimen-specific geometry, the accumulated reaction-force errors (Eq. (6)) were calculated using specimen-specific FE models with material parameters predicted by three different methods: (1) CurveFit: traditional curve-fitting method, (2) Specimen-FE-Optimal: specimen-specific FE models with optimizations, and (3) Simplified-FE-Optimal: constant cross-sectional area FE models with optimizations. In methods 1 and 3, the cross-sectional area measured at 5–8 locations along the specimen length (using the scanned geometric data) was averaged and assigned as the cross-sectional area of the specimen model. Paired t-tests were performed to compare the error from Specimen-FE-Optimal method to that from other two methods. The sampling frequencies of the reaction-forces in the test and simulation were both 10 kHz.

3. Results

3.1. Test results

The stress–stretch curves (Eq. (5)) until tissue failure for all 21 specimens and the average curve are shown in Fig. 4. The average cross-sectional area for each specimen was used for stress calculation. The failure point was defined as the time when the reaction-force reached its maximum value, representing the beginning of tissue total failure, although sub-failure damage of the tissue might occur before the total failure. Review of high-speed video data showed that the time of maximum force corresponded well with the start of tissue total failure. The stress–stretch curves demonstrate a very clear "toe" region, followed by a linear or convex upward region until failure. In the 21 specimens, 6 failed near the bottom, 4 failed near the top, and 11 failed in the middle. No specimen failed near the attachments to the clamps. The calculated failure strain and stress for each specimen based on marker digitization are shown in Table 1. The mean failure strain was 0.472 ± 0.097, and the mean failure stress was 34.80 ± 12.62 kPa. For comparison, Manoogian et al. (2008) reports a failure strain of 0.56 and failure stress of 18.6 kPa at a strain rate of 7/s. The average

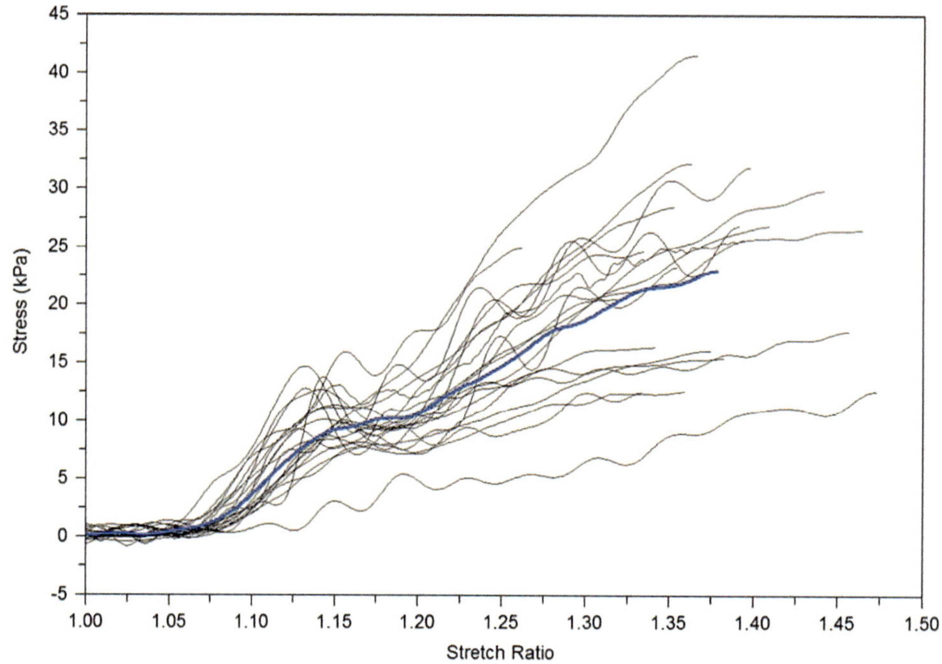

Fig. 4. Stress–stretch curves in uniaxial placenta-specimen tensile tests. The thin lines are the stress–stretch curves from each specimen, and the thick line is the average curve.

Table 1
Optimal material parameters and failure strain and stress by different methods.

Test no.	Material parameters						Failure stress and strain			
	CurveFit		Specimen-FE-Optimal		Simplified-FE-Optimal		Test and marker digitization		Specimen- FE	Simplified-FE
	α_1	μ (kpa)	α_1	μ (kpa)	α_1	μ (kpa)	σ_{Fail} (kpa)	ε_{Fail}	ε_{Fail}	ε_{Fail}
1	8.18	15.93	6.33	23.95	6.56	21.55	42.28	0.469	0.409	0.349
2	6.38	12.86	3.76	18.87	3.89	18.19	21.60	0.432	0.421	0.338
3	6.28	19.47	5.22	23.14	5.44	22.25	52.79	0.485	0.427	0.373
4	6.50	15.11	4.31	24.08	4.92	20.66	45.65	0.475	0.514	0.397
5	2.34	30.20	2.65	30.32	2.06	30.90	38.50	0.489	0.460	0.374
6	5.57	18.58	3.45	25.61	3.41	25.23	31.33	0.428	0.409	0.343
7	8.53	20.72	5.41	32.40	5.64	30.79	33.46	0.391	0.333	0.267
8	7.21	14.05	5.84	17.40	6.12	16.89	34.97	0.408	0.401	0.341
9	2.81	15.02	0.96	19.20	0.76	19.00	18.63	0.432	0.418	0.361
10	1.26	17.37	3.21	16.05	2.38	16.78	19.30	0.434	0.489	0.320
11	6.29	5.44	3.68	10.72	3.29	11.09	17.53	0.588	0.584	0.484
12	2.78	18.09	0.75	26.32	0.56	24.42	24.78	0.591	0.668	0.407
13	7.86	16.33	5.84	24.41	6.19	22.66	48.18	0.407	0.400	0.347
14	5.86	17.12	4.36	23.98	4.35	23.06	46.26	0.472	0.480	0.420
15	1.85	18.99	0.30	26.46	0.28	24.76	31.55	0.743	0.730	0.510
16	4.39	22.80	2.96	31.67	3.06	28.80	55.73	0.577	0.610	0.440
17	8.01	21.08	6.44	29.37	6.53	27.54	55.66	0.494	0.451	0.345
18	5.52	12.39	3.34	18.87	3.04	18.54	20.39	0.502	0.512	0.349
19	6.44	19.68	4.97	24.38	5.04	23.99	32.52	0.323	0.414	0.325
20	4.46	17.38	0.50	26.56	0.45	25.99	21.78	0.302	0.353	0.307
21	4.04	20.72	2.58	29.51	2.32	28.54	38.01	0.478	0.474	0.454
Mean	**5.36**	**17.59**	**3.66**	**23.97**	**3.63**	**22.93**	**34.80**	**0.472**	**0.474**	**0.374**
SD	**2.18**	**4.78**	**1.90**	**5.52**	**2.08**	**5.07**	**12.62**	**0.097**	**0.101**	**0.060**

Note: μ and α_1 are the ground-state shear modulus and the first exponent in the Ogden hyperelastic model, respectively. The first shear modulus (μ_1) used in Ls-dyna can be calculated by Eq. (3) using μ and α_1. ε_{Fail} is the failure green-lagrange strain in loading direction, and σ_{Fail} is the failure stress calculated by Eq. (1).

cross-sectional area of all specimens is $312.01 \pm 71.20\,\text{mm}^2$, and the average standard deviation among individual specimens is $35.56\,\text{mm}^2$. These results indicate a substantial variation in cross-sectional area within and between specimens.

3.2. *Specimen-specific FE models*

Since the FE models were developed using a template-based method, every model has the same number of nodes (1323) and elements (960). Moreover, use of the same node and element

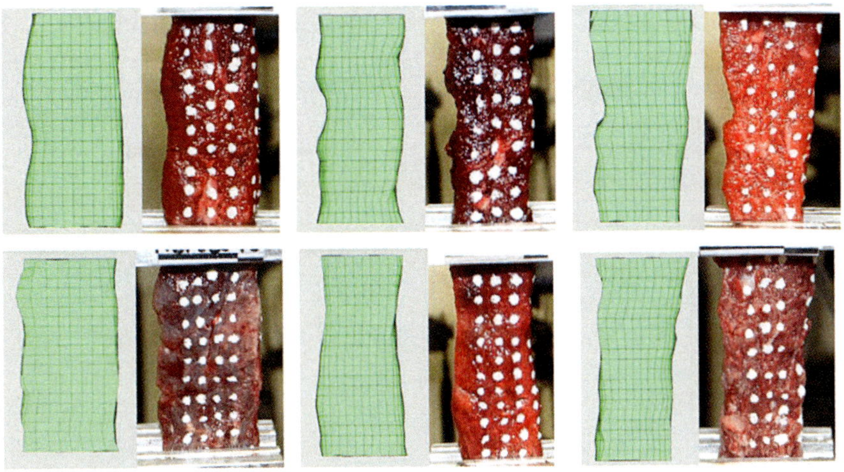

Fig. 5. Comparisons between FE models and photos of actual placenta specimens.

orders provided an automated method that allowed use of the same material model and boundary conditions for each simulation. Fig. 5 shows comparisons between the specimen-specific FE models and actual specimens for six placenta specimens. The FE models successfully capture the major geometric features of each specimen. In all the models, the smallest element size is about 1 mm, the smallest Jacobian is set to 0.3, and generally 90% of the elements have a Jacobian greater than 0.7. Approximately 10 min or less was needed to develop a specimen-specific placenta FE model from a geometric surface.

3.3. Optimal material parameters and error comparison

The material parameters (μ and α_1) for each specimen calculated by three methods in Section 2.5, namely CurveFit, Specimen-FE-Optimal, and Simplified-FE-Optimal, are shown in Table 1. Based on the results from Specimen-FE-Optimal method, placenta tissue is a hyperelastic material with $\mu = 23.97 \pm 5.52$ kPa and $\alpha_1 = 3.66 \pm 1.90$. The mechanical properties from the Specimen-FE-Optimal method are significantly different ($p < .001$) to that from CurveFit or Simplified-FE-Optimal methods. As seen in Table 2, the Specimen-FE-Optimal method generated smaller errors for almost every specimen than the CurveFit and Simplified-FE-Optimal methods and the differences are statistically significant ($p < 0.001$). In addition, the failure strains using marker data, specimen-specific and simplified FE models were also reported in Table 1. The failure strains calculated by specimen-specific models matched well (paired t-test: $p = 0.85$, correlation: $R = 0.91$) with those from marker digitization, while simplified FE models constantly underestimated the failure strain by 21% (paired t-test: $p < 0.001$).

Table 2
Reaction force sum-of-squares errors using material parameters from different methods (unit: N).

Test no.	CurveFit	Specimen-FE-Optimal	Simplified-FE-Optimal
1	16.33	4.90	9.46
2	9.28	5.63	5.85
3	12.55	11.59	11.61
4	19.70	5.32	7.87
5	11.21	11.17	11.19
6	10.05	5.81	5.96
7	11.73	3.52	4.02
8	6.82	6.43	6.37
9	9.25	5.68	7.45
10	4.68	3.72	3.72
11	14.96	4.79	5.68
12	14.74	4.45	6.90
13	12.18	3.26	4.16
14	21.17	6.49	9.35
15	21.32	6.99	10.57
16	13.62	5.37	7.26
17	22.54	8.36	12.21
18	13.33	1.95	4.40
19	5.95	4.81	4.99
20	8.99	4.71	4.88
21	15.42	5.06	8.86
Mean	**13.13**	**5.72**	**7.27**
SD	**5.06**	**2.33**	**2.65**
t*-Test: *p	**$<$0.001**	**–**	**$<$0.001**

Note: all reaction forces were calculated using specimen-specific FE models, and the errors were calculated using Eq. (6). Detailed explanations for each method were presented in Section 2.5.

4. Discussion

In this study, a method combining high-speed uniaxial tensile testing, specimen-specific FE modeling, and optimization techniques was used to quantify the dynamic mechanical properties of human placental tissue. The proposed method generated significantly more accurate material properties than those produced by a traditional curve-fitting method because of more detailed geometric representations and boundary condition definitions, as well as consideration of dynamic effects. Specimen-specific FE models can provide more accurate material predictions than using a simplified FE model (Table 2), due to the complicated geometry considered for each placenta specimen. The material parameters for placenta tissue estimated in this study can be applied to FE models on a larger scale, such as an FE abdomen model of a pregnant female used in vehicle crash simulations.

Fig. 6 shows reaction-force histories from specimen tests, curve-fitting method and specimen-specific model simulation results with material parameters from three methods described in Section 2.5. The specimen-specific FE model with optimal material parameters provided the most accurate estimations among all the methods used.

The FE method is better at capturing dynamic effects in high-speed loading conditions. Fig. 6 shows that the curve-fitting method alone cannot predict the "toe" regions that occur in the tests because Eq. (5) was derived under quasi-static loading conditions, and thus does not consider the dynamic effects of high-speed loading. In the current study, the load cell was installed below the fixed lower clamp, so it cannot detect the

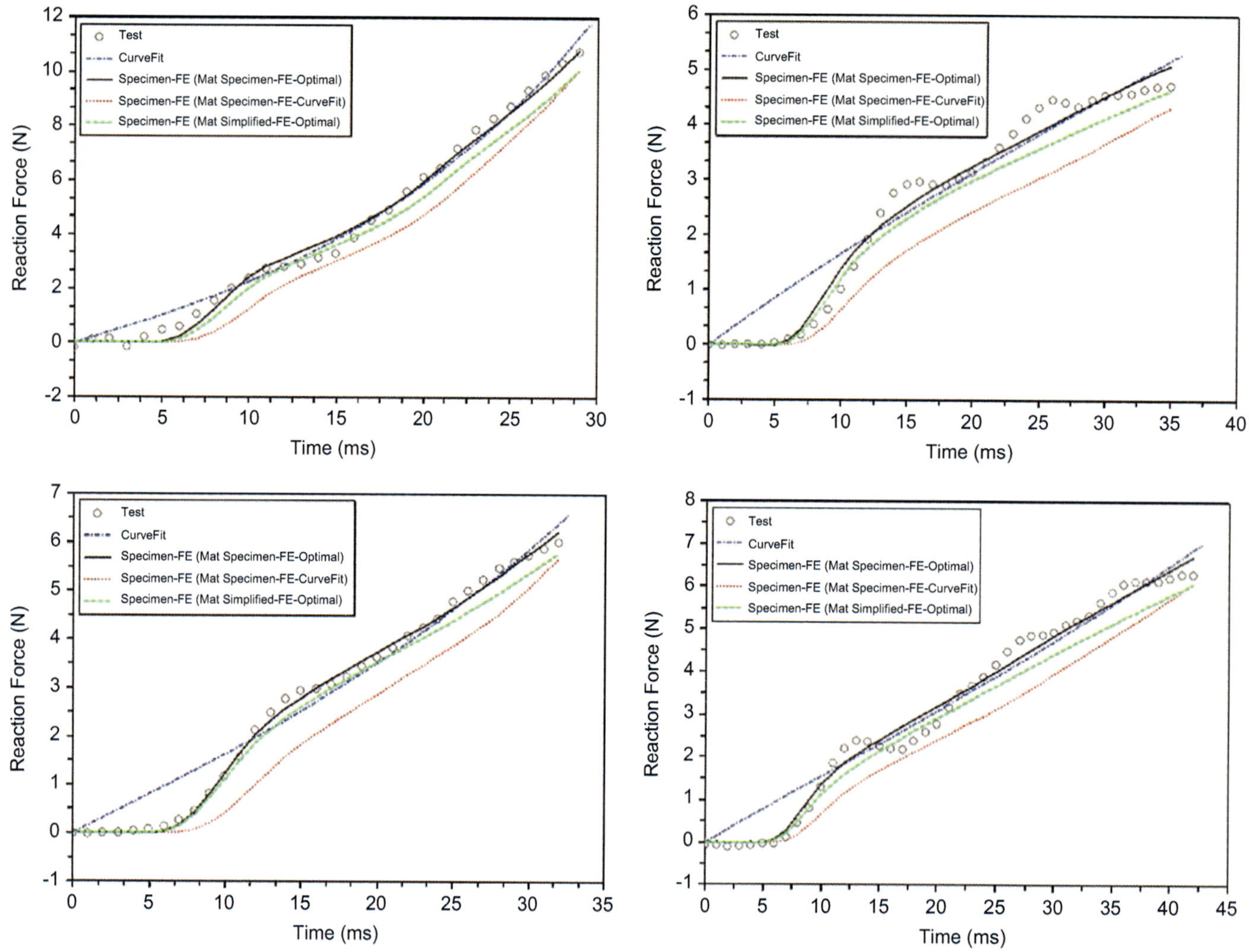

Fig. 6. Comparison of reaction-forces from test, curve-fitting method, and specimen-specific FE model with three sets of material parameters by three methods introduced in Section 2.5.

initial force during the test when the tissue was accelerated up to speed by moving the upper clamp. In Manoogian's study (2008), the "toe" region is not as distinct because load cells were installed at both the lower and upper clamps that could load the tissue simultaneously. If curve-fitting-predicted material parameters are input into the specimen-specific FE models, the model results clearly show a "toe" region, indicating the weakness of the curve-fitting method for high-speed loading conditions. Although a "toe" region may result solely from tissue properties, such as the uncrimping of collagen fibrils when tendon is stretched, the "toe" region was mainly caused by the dynamic effects during the current study.

The new method is also better at modeling geometric differences between samples. Fig. 6 illustrates clear differences between using specimen-specific geometry and a simplified uniform geometry based on average cross-sectional area. Specimen-specific models also provided significantly ($p<0.001$) better predictions of the reaction-force than simplified FE model-predicted material properties (Table 2), demonstrating the importance of using specimen-specific models to estimate material parameters. In the current study, average cross-sectional area was calculated based on the geometric scan data. Without these accurate scan data, a rough estimation of cross-sectional area and/or simply assuming the same cross-sectional area for every specimen would introduce even larger errors. Moreover, Table 1 demonstrates that specimen-specific models predicted more accurate failure strains than simplified FE models. Generally, higher strain occurred at the layer with the smallest cross-sectional area, and strain concentration may also occur at corners with irregular shapes. The simplified FE model neglected all these factors that may contribute to a larger failure strain, and thus underestimated the failure strain constantly. This result indicated that the strain distribution in a specimen-specific model is more reasonable than a more uniformly distributed strain field in a simplified FE model.

The study has several limitations. First, placenta tissues are visco-elastic and transversely isotropic. In the present study, uniaxial tensile tests were performed at one loading rate and the specimens were cut along the plane of the placenta. Therefore, visco-elastic and transversely isotropic characteristics of placenta tissue were not quantified. However, uniaxial tests using different loading rates are currently being conducted. New results along with a more complicated material model can be implemented with the current method, and will lead to a more comprehensive material characterization of human placenta tissue suitable for use in crash simulations of pregnant occupants.

Second, this study is only a first step toward understanding injury to pregnant occupants because it does not address the failure characteristics of UPI. Experiments to measure the UPI failure characteristics are needed to further understand the placental abruption mechanisms. Such data, along with a pregnant abdomen FE model with placenta-UPI-uterus structure, should be used to assess the potential of new restraint technologies to reduce the incidence of placental abruption in MVC.

Conflict of interest statement

There are no conflicts of interest in the work presented in the manuscript.

Acknowledgement

This research was funded by the National Highway Traffic Safety Administration under Contract DTNH22-05-H-01020. The authors thank Julia Samorezov, Elliot Hwang, Rajen Kumar, Brian Eby and Tom Jeffreys for help in performing and analyzing tests, as well as Dr. Sumeet Parashar and Dr. Chang Qi for modeFRONTIER technical support.

References

Fu, Y., Yang, R.J., Yeh, I., 2004. A genetic algorithm for optimal design of an inflatable knee bolster. Struct. Multidiscip. Optim. 26, 264–271.

Holland, J.H., 1975. Adaptation in Natural and Artificial Systems. The University of Michigan Press, Ann Arbor, MI.

Hu, J., Jin, X., Lee, J.B., Zhang, L., Chaudhary, V., Guthikonda, M., Yang, K.H., King, A.I., 2007. Intraoperative brain shift prediction using a 3D inhomogeneous patient-specific finite element model. J. Neurosurg. 106, 164–169.

Manoogian, S., McNally, C., Calloway, B., Duma, S., Mertz, H., 2007. Utilizing cryogenic grips for dynamic tension testing of human placenta tissue. Biomed. Sci. Instrum. 43, 354–359.

Manoogian, S.J., Bisplinghoff, J.A., McNally, C., Kemper, A.R., Santago, A.C., Duma, S.M., 2008. Dynamic tensile properties of human placenta. J. Biomech. 41, 3436–3440.

Miller, K., Chinzei, K., 2002. Mechanical properties of brain tissue in tension. J. Biomech. 35, 483–490.

Moorcroft, D.M., Stitzel, J.D., Duma, G.G., Duma, S.M., 2003. Computational model of the pregnant occupant: predicting the risk of injury in automobile crashes. Am. J. Obstet. Gynecol. 189, 540–544.

Ogden, R.W., 1972. Large deformation isotropic elasticity—on the correlation of theory and experiment for incompressible rubberlike solids. Proc. R. Soc. London A 326, 565–584.

Ogden, R.W., 1997. Non-linear Elastic Deformations. Dover Publications, INC., Mineola, New York.

Pearlman, M.D., Klinich, K.D., Schneider, L.W., Rupp, J., Moss, S., Ashton-Miller, J., 2000. A comprehensive program to improve safety for pregnant women and fetuses in motor vehicle crashes: a preliminary report. Am. J. Obstet. Gynecol. 182, 1554–1564.

Rupp, J.D., Klinich, K.D., Moss, S., Zhou, J., Pearlman, M.D., Schneider, L.W., 2001. Development and testing of a prototype pregnant abdomen for the small-female hybrid III ATD. Stapp Car Crash J. 45, 61–78.

Snedeker, J.G., Niederer, P., Schmidlin, F.R., Farshad, M., Demetropoulos, C.K., Lee, J.B., Yang, K.H., 2005. Strain-rate dependent material properties of the porcine and human kidney capsule. J. Biomech. 38, 1011–1021.

Stitzel, J.D., Gayzik, F.S., Hoth, J.J., Mercier, J., Gage, H.D., Morton, K.A., Duma, S.M., Payne, R.M., 2005. Development of a finite element-based injury metric for pulmonary contusion part I: model development and validation. Stapp Car Crash J. 49, 271–289.

Tamura, A., Omori, K., Miki, K., Lee, J.B., Yang, K.H., King, A.I., 2002. Mechanical characterization of porcine abdominal organs. Stapp Car Crash J. 46, 55–69.

Weiss, H.B., 2001. The epidemiology of traumatic injury-related fetal mortality in Pennsylvania, 1995–1997: the role of motor vehicle crashes. Accid. Anal. Prev. 33, 449–454.

Weiss, H.B., Songer, T.J., Fabio, A., 2001. Fetal deaths related to maternal injury. JAMA 286, 1863–1868.

Weiss, H.B., Strotmeyer, S., 2002. Characteristics of pregnant women in motor vehicle crashes. Inj. Prev. 8, 207–210.

Journal of Biomechanics 41 (2008) 3436–3440

Contents lists available at ScienceDirect

Journal of Biomechanics

journal homepage: www.elsevier.com/locate/jbiomech
www.JBiomech.com

Dynamic tensile properties of human placenta

Sarah J. Manoogian *, Jill A. Bisplinghoff, Craig McNally, Andrew R. Kemper, Anthony C. Santago, Stefan M. Duma

Virginia Tech-Wake Forest, Center for Injury Biomechanics, Randolph 100F, MC 0238, Blacksburg, VA 24061, USA

ARTICLE INFO

Article history:
Accepted 12 September 2008

Keywords:
Placenta
Pregnant
Tension
Chorion
Placental abruption

ABSTRACT

Automobile crashes are the largest cause of injury death for pregnant females and the leading cause of traumatic fetal injury mortality in the United States. Computational models, useful tools to evaluate the risk of fetal loss in motor vehicle crashes, are based on a limited number of quasi-static material tests of the placenta. This study presents a total of 20 dynamic uniaxial tensile tests on the maternal side of the placenta and 10 dynamic uniaxial tensile tests on the chorion layer of the placenta. These tests were completed from 6 human placentas to determine material properties at a strain rate of 7.0 strains/s. The results show that the average peak strain at failure for both the maternal portion and the chorion layer of the placenta are similar with a value of 0.56 and 0.61, respectively. However, the average failure stress for the chorion layer, 167.8 kPa, is much higher than the average failure stress for the placenta with the chorionic plate removed, 18.6 kPa. This is due to differences in the structure and function of these layers in the placenta. In summary, dynamic loading data for the placenta have been determined for use in computational modeling of pregnant occupant kinematics in motor vehicle crashes. Moreover the computational model should utilize the material properties for the placenta without the chorion layer.

1. Introduction

Automobile crashes are the largest cause of injury death for pregnant females and the leading cause of traumatic fetal injury mortality in the United States (US) (Attico et al., 1986; Weiss and Strotmeyer, 2002). An average estimate from several sources indicates each year in the US, approximately 160 pregnant women are killed in motor vehicle crashes and an additional 600 to 2600 fetuses are killed when the mother survives (Klinich et al., 1999a b; Pearlman, 1997; Weiss, 2001). The best way to protect the fetus is to protect the mother considering that maternal death has a near 100% fetal loss rate (Pearlman et al., 1990a). If the mother survives, protection of the fetus may be best accomplished by preventing placental abruption. Placental abruption, which is the premature separation of the placenta from the uterus, has been shown to account for 50–70% of fetal losses in motor vehicle crashes (Pearlman et al., 1990b). Computational modeling of pregnant occupant kinematics provides the widest basis for studying factors that could prevent adverse fetal outcome from a motor vehicle crash (Moorcroft et al., 2003a b c, 2004). However, material parameters for computational modeling come from early research on placenta tissue at quasi-static loading rates (Moorcroft et al., 2003a). Because abdominal loading in an automobile crash is at a dynamic rate, the placenta tissue material properties should be evaluated with a dynamic loading condition.

The microstructure of the human placenta is important in modeling the material response to dynamic loading in tension. The human placenta connects to the uterine wall, usually in the fundus of the uterus, and provides a means for the maternal–fetal exchange of gases and nutrients (Ramsey, 1975). However, because of the different roles it serves for the mother and fetus, there are microstructure differences on the maternal and fetal sides of the placenta (Cunningham and Williams, 2005). The fetal side of the placenta includes the chorionic plate. This layer of the placenta is filled with fetal blood vessels that originate at the umbilical cord and radiate outward to the edge of the placenta in a dense network. The maternal side of the placenta is a network of villi and intervillus spaces. Between the maternal side of the placenta and the uterus is the uteroplacental interface. This decidua basalis layer is where the placenta separates from the uterus during placental abruption. The purpose of this study is to quantify the loading and the failure properties of chorion layer and the placenta without the chorion attached in uniaxial tension at dynamic loading rates. While this does not address the strength of the uteroplacental interface, it does provide material data for computational models to accurately represent the response of the placenta to dynamic loading conditions.

* Corresponding author. Tel.: +540 231 1617; fax: +540 231 2953.
E-mail address: manoogsj@vt.edu (S.J. Manoogian).

0021-9290/$ - see front matter © 2008 Elsevier Ltd. All rights reserved.
doi:10.1016/j.jbiomech.2008.09.020

2. Methods

A total of 30 uniaxial tensile tests were performed on 6 whole placentas. Donor tissue followed the Wake Forest University Baptist Medical Center Institutional Review Board informed consent procedures. Following delivery, the tissue was stored in a saline solution and tested within 36 h. From each placenta there were 3 to 4 tests of the maternal side of the placenta at a dynamic strain rate of 7.00 strains/s. This strain rate represents loading expected in a motor vehicle crash as determined from previous computational modeling (Moorcroft et al., 2003c). For two of the placentas, additional tests were performed on the chorion layer to obtain material properties at the same strain rate. There were a total of 10 tests on the chorion layer and 20 tests on the non-chorion structure of the placenta.

All of the uniaxial tensile tests were performed on coupon specimens from either the maternal surface of the placenta or the chorion layer from the fetal side of the placenta. Thin slices of placenta were acquired from the whole placenta using a tissue slicer with a wet rotating blade. This allowed the maternal surface of the placenta to be sectioned into 5 mm slices and separated from the chorion layer of the placenta (Fig. 1). Because the placenta function requires it to have a lobe structure, the slices of the placenta had discontinuities. Sections with a continuous structure were chosen for acquiring a smaller coupon section. Each piece of tissue was aligned before stamping to ensure uniformity as much as possible with each sample. Guide rods provided a means for aligning both a stencil and ultimately the stamp. A coupon was then stamped out of the tissue using a steel bent stamp. A sharpened edge to the steel stamp minimized shear stress in the tissue during the stamping procedure. The result of the tissue stamping procedure was several coupons from the same donor tissue with a uniform shape (Fig. 2). Each coupon was kept hydrated throughout the preparation and prior to testing. Due to the process of obtaining the chorion layer of the placenta, the chorion slice had a larger thickness than 5 mm. Therefore, those specimens had an average cross-sectional area that was larger than the maternal side specimens. The average cross-sectional area of the chorion specimens is $42.2 \pm 13.2\,mm^2$. The maternal side specimens have an average cross sectional-area of $27.1 \pm 7.9\,mm^2$.

A custom designed system of linear motors was used for uniaxial dynamic tensile tests of placental tissue coupons (Fig. 3). The system included one multi-axis controller (Parker ACR9000, Irwin, PA) that served as a power supply for a motor driver (Parker ViX, Irwin, PA) which controlled the two linear motors with stages (Parker Daedal MX80S, Irwin, PA). The two individual stages were each instrumented with a load cell (OMEGA LC10, 10 lbf, Stamford, CT), potentiometer (Space Age Control, 160–1705, 540 mm, Palmdale, CA), and accelerometer (Endevco 7264B, 2000 G, San Juan Capistrano, CA). On each side of the test specimen, the direct path from the tissue clamp went through the load cell and was rigidly connected to the linear motor stage. The inertial load of the mass in front of the load cell was multiplied by the acceleration to compensate the measured reaction load. Synchronous motion of the motors allowed both of the stages to pull the tissue in tension at the same time for an overall velocity which corresponded to the desired strain rate.

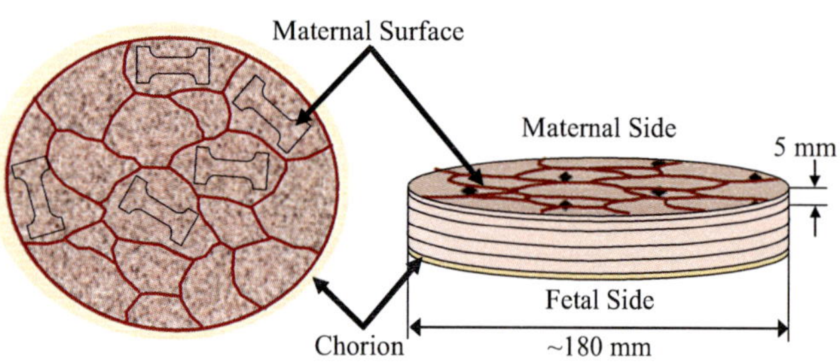

Fig. 1. Each placenta is sectioned into 5 mm slices.

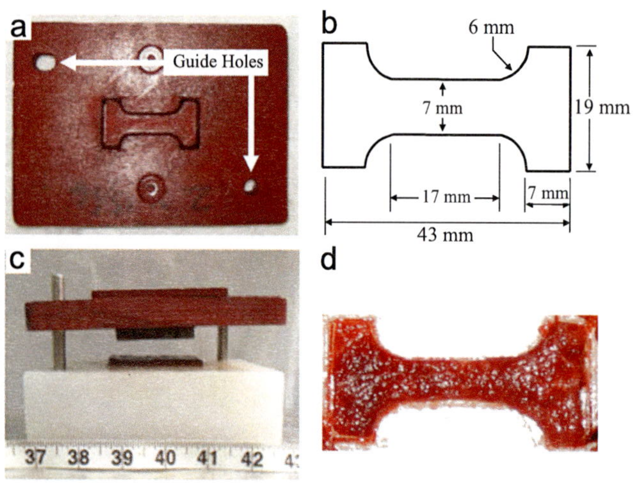

Fig. 2. (a) A steel bent stamp is used to cut the tissue into a dog bone shape. (b) The stamp provides uniform tissue samples. (c) Guide rods are used to align the stamp with the specimen. (d) The coupon shape is cut where the tissue has a uniform consistency.

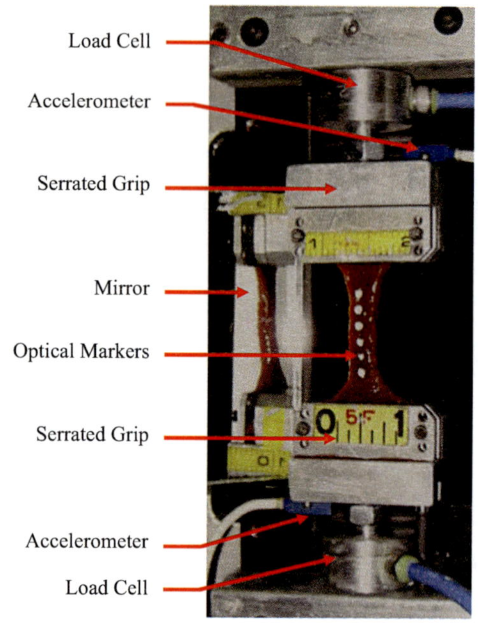

Fig. 3. The specimen is mounted between two serrated grips which are instrumented with both a load cell and an accelerometer while high speed video records the test event.

Each specimen was mounted in the testing setup with the same initial conditions. The top grip assembly was removed and laid flat to align the specimen in the grip along the centerline of the loading axis. After the grip was secured, the top grip assembly was then reattached to allow the specimen to rest hanging in 1 g tension. By supporting only its own weight, the tissue had a standard but minimal preload condition. With the tissue hanging, the specimen was clamped in the bottom grip with no additional preload. Optical markers were applied to the tissue after the specimen was mounted.

Local stress and strain were calculated for each specimen. Local deformation was recorded using high-speed video (Phantom V4, Wayne, NJ) at 500 frames per second with 512 by 512 resolution. Optical marker tracking was performed with TEMA® Advanced Motion Analysis Software (Linköping, Sweden). Using the configuration for these tests, the error in the motion analysis is ± 0.030 mm. The optical markers on either side of the location where the specimen tore were used to calculate the local strain. In addition, a mirror aligned at a 45° angle was mounted next to the specimen so that the side of the specimen could be viewed in the high-speed video. The measured displacement data were fit with a 5th order polynomial. The average R^2 value for the displacement curve fit is 0.999 ± 0.001. Similarly, the measured force data were fit with a 5th order polynomial. The average R^2 value for the force curve fit is 0.994 ± 0.009. Local stress was acquired by determining the location on the specimen where failure occurred and then taking the width and thickness of the specimen in that location from the pretest pictures as the initial cross-sectional area. The error in the pretest pictures is ± 0.12 mm. The cross-sectional area at the time of failure is calculated using the local strain and a Poisson's ratio value of 0.49. This value has been determined in previous placental tissue testing (Manoogian et al., 2007). The stress and strain data are reported for each specimen until failure was evident in the video. Although, the tissue could have been still supporting the load, the optical markers were not accurately representing the strain in the tissue once there was sub-failure.

A two sample *t*-test procedure was used to determine statistically significant differences between the chorion and non-chorion test series for both failure strain and failure stress. The failure stresses and strains at failure were also compared between donor groups to tests for any significant differences. Additionally, a standard method of determining the characteristic average was used for each group of tests at one strain rate (Lessley et al., 2004). This process began with normalizing each strain curve to its respective maximum. Interpolation routines determined the stress for each test at set increments of strain. Stress values for all tests at that rate were averaged to determine one value at each strain. The strain

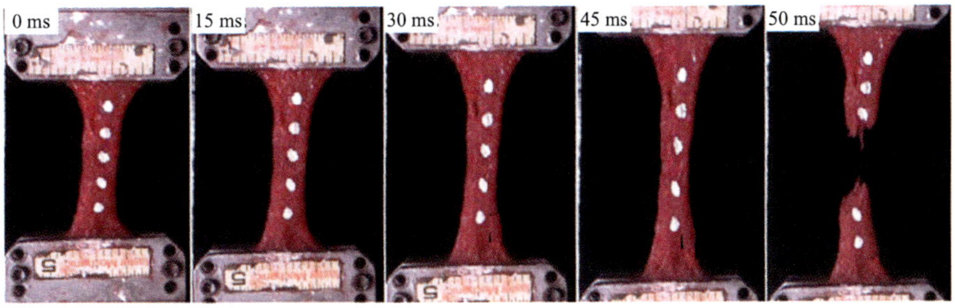

Fig. 4. Video captures of one test show a typical failure of a placental specimen pulled in uniaxial tension at 7.0 strains/s.

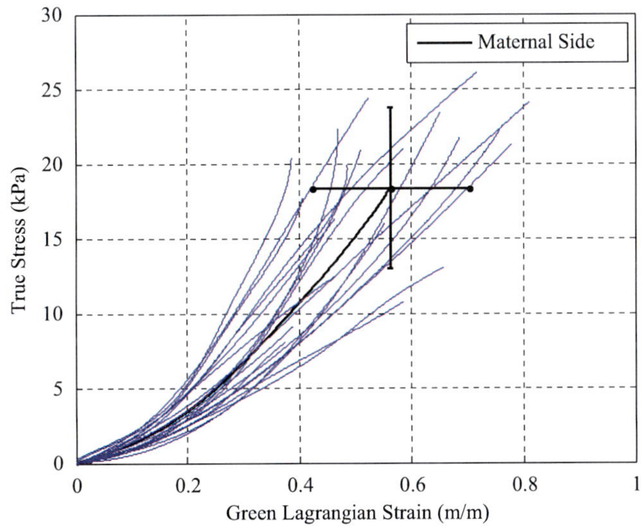

Fig. 5. All of the tests for the maternal side of the placenta have similar stress versus strain curves.

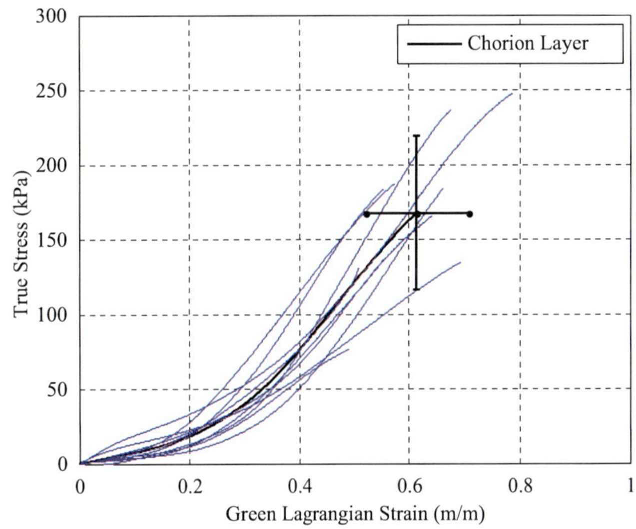

Fig. 6. All of the tests for the chorion layer of the placenta have similar stress versus strain curves.

data that corresponds to these stress values were obtained by multiplying the normalized strain by the average maximum strain of all tests. The result of this process was an average stress versus strain curve for all of the chorion tests and one for all of the non-chorion tests. Standard deviations of the stress and strain values were also calculated for each group of tests. Elastic modulus values were obtained from the characteristic average curves in the linear region before failure.

3. Results

Stress versus strain curves for the chorion layer and the maternal side of the placenta are presented for dynamic tensile tests. The data are reported in Green Lagrangian strain and true stress values. Frames from the high-speed video of a maternal side test show the tissue specimen fails in the gage length of the dog bone shape coupon (Fig. 4). This is also true of the chorion tests.

All of the stress versus strain data are plotted together for the maternal side of the placenta with the characteristic average and the standard deviations for the failure stress and strain (Fig. 5). The average failure stress for the maternal side of the placenta was 18.6 ± 5.4 kPa. The corresponding failure strain was 0.56 ± 0.14. The elastic modulus in the linear region of the characteristic average for this group of tests was 45.0 kPa. For the test series that included the chorion layer of the placenta, the material data are plotted together with the average and standard deviations for the failure stress and strain (Fig. 6). The failure strain, 0.61 ± 0.09, was similar to the maternal side only tests. However, the failure stress for the chorion layer was 167.8 ± 51.3 kPa. This value is higher than the failure stress for the placental tissue that did not include the chorion layer. As a result, the elastic modulus for this group of tests was much higher with a value of 445.8 kPa.

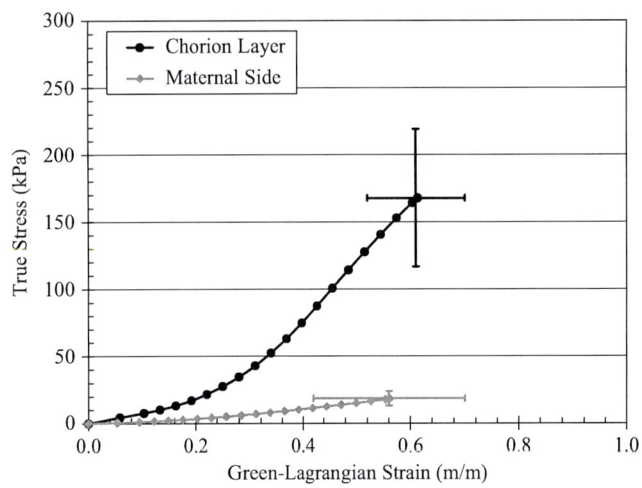

Fig. 7. Characteristic averages are shown with bars indicating the standard deviations of the failure values.

For the two donors of chorion tissue, there were no significant differences in failure stress or strain ($p = 0.50$, 0.85). None of the six donors for the maternal side of the placenta had a statistically significant different failure stress or strain than the others ($p > 0.2$). When comparing all of the chorion tests to the all of the tests on the maternal side of the placenta, the failure stresses are significantly different ($p < 0.001$) and the failure strains are not significantly different ($p = 0.24$). The characteristic averages show the chorion layer of the placenta provides a stiffer response than the maternal portion alone (Fig. 7).

4. Discussion

The results from this study give loading and failure information for the maternal side of the placenta and the chorion layer of the placenta. The maternal portion of the placenta, including only the villus structure, has a compliant stress versus strain response and therefore has a lower failure stress and elastic modulus than the chorion specimens. Due to the presence of the dense layer of blood vessels, the failure stresses are much higher for the chorion tissue. Both layers of placental tissue have similar failure strains.

The standard deviations in the failure values are assumed to be dependent on specimen variability since the testing procedure was the same for each specimen. The microstructure of the placental villi on the maternal side do not have a directional bias of their structure in the plane of the tissue specimens (Cunningham and Williams, 2005). It is assumed because of the transversely isotropic structure, the material properties would not be directionally dependent. While there are discontinuities, careful consideration was used when stamping the specimens to obtain a uniform gage length. However, the porous structure of the placenta could still contribute to specimen variability. Because of the fetal vessels that originate at the umbilical cord and disperse to the edge of the placenta, the chorion layer is not transversely isotropic in its structure (Cunningham and Williams, 2005). Although an effort was made to obtain samples in a consistent area of structure, the full complexity of the chorion layer is not addressed with these material tests. However, the average material response is represented. Moreover, the values are not significantly different between the two donors and the standard deviations are low.

These dynamic testing data from the maternal side of the placenta are compared to previous quasi-static placenta tension research (Pearlman, 1999). While the quasi-static tests are sub-failure and the tests in this study are failure, this is the most suitable previous research for comparison. The gage lengths of the specimens were 30 mm and they were pulled at a rate of 1.6 mm/s. For all five specimens, the average failure strain was 0.43 ± 0.16 and the average failure stress was 15.6 ± 7.6 kPa. However, it is not clearly stated in the literature if the reported strain values are Green Lagrangian. The data from Pearlman's tests compare favorably to the maternal side tension tests but have significantly lower peak stresses than the chorion layer tension tests ($p < 0.001$).

For dynamic testing of placental tissue, sectioning the tissue in thin coupon specimens provided necessary advantages. One advantage of this method was the ability to grip the tissue throughout the tissue thickness. Gripping the placenta without sectioning would be difficult because the internal layers need to strain the same amount as the external layers. Another advantage of the test configuration was the ability to measure local stress and strain of the tissue. In addition to these advantages, there are limitations with this study. It is assumed that since the placenta tissue maintains its size and shape after delivery, the tests postpartum provide a close model to the performance of the tissue *in vivo*. Further, the body temperature and blood flow of the tissue *in vivo* were not modeled in the experiments. In order to model the blood filled organ, the tissue was kept hydrated and saturated prior to testing. The blood flow *in vivo* would likely increase both the viscoelasticity and incompressibility of the placental tissue. Because the tissue is tested hydrated and the failure mode is tension in the fibers, it is expected that the failure properties *in vivo* would be similar those measured in this study. Previous researchers have evaluated the effect of temperature on material properties of ligaments (Hasberry and Pearcy, 1986; Lam et al., 1990). From those studies they have found that there is not a significant difference in tensile failure properties at room temperature versus body temperature. Because the viscosity of blood increases with a decrease in temperature, it is likely the placenta would have a less viscous response at body temperature than was measured in the experiments. However, it is assumed the strength of the fibers that compose the placental tissues would not be greatly affected by the change in temperature. Overall, the effect of blood flow and temperature are expected to have a minimal effect on the material properties of the placenta.

Pregnant occupant computational models should utilize the material properties for the placenta without the chorion layer when predicting placental abruption. As future pregnant occupant models become more detailed, they can incorporate the chorion layer as a transition from the umbilical cord to the placenta. For that application, it will be important to distinguish between the different materials and their respective material properties.

5. Conclusions

This study presents a total of 20 dynamic uniaxial tensile tests on the maternal side of the placenta and 10 dynamic uniaxial tensile tests on the chorion layer of the placenta. These tests were completed from 6 human placentas to determine material properties at a strain rate of 7.0 strains/s. The results show that the average peak strain at failure for both the maternal portion and the chorion layer of the placenta are similar with a value of 0.56 and 0.61, respectively. However, the average failure stress for the chorion layer, 167.8 kPa, is much higher than the average failure stress for the placenta with the chorionic plate removed, 18.6 kPa. This is due to differences in the structure and function of these layers in the placenta. In summary, dynamic loading data for the placenta have been determined for use in computational modeling of pregnant occupant kinematics in motor vehicle crashes.

Conflicts of interest

None .

Acknowledgment

The authors thank the Toyota Motor Corporation for sponsoring this research.

References

Attico, N.B., Smith, R.J., FitzPatrick, M.B., Keneally, M., 1986. Automobile safety restraints for pregnant women and children. The Journal of Reproductive Medicine 31, 187–192.

Cunningham, F.G., Williams, J.W., 2005. Williams Obstetrics. McGraw-Hill Professional, New York.

Hasberry, S., Pearcy, M.J., 1986. Temperature dependence of the tensile properties of interspinous ligaments of sheep. Journal of Biomedical Engineering 8, 62–66.

Klinich, K.D., Schneider, L.W., Eby, B., Rupp, J.D., Pearlman, M.D., 1999a. Seated Anthropometry During Pregnancy. University of Michigan Transportation Research Institute.

Klinich, K.D., Schneider, L.W., Moore, J.L., Pearlman, M.D., 1999b. Investigations of Crashes Involving Pregnant Occupants. University of Michigan Transportation Research Institute.

Lam, T.C., Thomas, C.G., Shrive, N.G., Frank, C.B., Sabiston, C.P., 1990. The effects of temperature on the viscoelastic properties of the rabbit medial collateral ligament. Journal of Biomechanical Engineering 112, 147–152.

Lessley, D., Crandall, J., Shaw, G., Kent, R., Funk, J., 2004. A Normalization Technique for Developing Corridors from Individual Subject Responses. SAE Technical Paper Series 2004-01-0288, 1–11.

Manoogian, S.J., McNally, C., Calloway, B.R., Duma, S.M., Mertz, H.L., 2007. Utilizing cryogenic grips for dynamic tension testing of human placenta tissue. Biomedical Sciences Instrumentation 43, 354–359.

Moorcroft, D.M., Duma, S.M., Stitzel, J.D., Duma, G.G., 2003a. Computational model of the pregnant occupant: predicting the risk of injury in automobile crashes. American Journal of Obstetrics and Gynecology 189, 540–544.

Moorcroft, D.M., Duma, S.M., Stitzel, J.D., Duma, G.G., 2003b. The effects of uterine ligaments on the fetal injury risk in frontal automobile crashes. Journal of Automobile Engineering Part D 217, 1049–1055.

Moorcroft, D.M., Duma, S.M., Stitzel, J.D., Duma, G.G., 2003c. A finite element and multi-body model of the pregnant occupant for the analysis of restraint effectiveness. SAE Technical Paper Series 2003-01-0157, 1–11.

Moorcroft, D.M., Duma, S.M., Stitzel, J.D., Duma, G.G., 2004. The effect of pregnant occupant position and belt placement on the risk of fetal injury. SAE Technical Paper Series 2004-01-0324, 1–8.

Pearlman, M.D., 1997. Motor vehicle crashes, pregnancy loss and preterm labor. International Journal of Gynaecology and Obstetrics 57, 127–132.

Pearlman, M.D., 1999. Data Acquisition for Development of a Uteroplacental Interface for the Second-Generation Pregnant Abdomen. University of Michigan Health Systems, Ann Arbor, MI.

Pearlman, M.D., Tintinalli, J.E., Lorenz, R.P., 1990a. Blunt trauma during pregnancy. The New England Journal of Medicine 323, 1609–1613.

Pearlman, M.D., Tintinallli, J.E., Lorenz, R.P., 1990b. A prospective controlled study of outcome after trauma during pregnancy. American Journal of Obstetrics and Gynecology 162, 1502–1510.

Ramsey, E.M., 1975. The Placenta of Laboratory Animals and Man. Holt, Rinehart and Winston, New York.

Weiss, H.B., 2001. The epidemiology of traumatic injury-related fetal mortality in Pennsylvania, 1995–1997: the role of motor vehicle crashes. Accident Analysis and Prevention 33, 449–454.

Weiss, H.B., Strotmeyer, S., 2002. Characteristics of pregnant women in motor vehicle crashes. Injury Prevention 8, 207–211.

Reprinted by the permission of the Stapp Association.

Stapp Car Crash Journal, Vol. 52 (November 2008), pp. 167-185

Stapp Car Crash Journal

2008-22-0007

Dynamic Biaxial Tissue Properties of Pregnant Porcine Uterine Tissue

Sarah J. Manoogian, Craig McNally, Joel D. Stitzel, Stefan M. Duma
Virginia Tech-Wake Forest Center for Injury Biomechanics

ABSTRACT – Automobile crashes are the largest single cause of death for pregnant females and the leading cause of traumatic fetal injury mortality in the United States. Current research for pregnant occupant safety utilizing computational models is limited by available pregnant tissue data. The purpose of this study is to collect experimental data from biaxial tissue tests on pregnant uterine tissue at a dynamic rate. Experimental tests were completed on pregnant porcine uterus which was chosen as a surrogate for the human pregnant uterus given its similarity and availability. Biaxial dynamic tensile tests were performed using a custom designed system of linear motors to pull a cruciform shaped specimen in tension simultaneously with four tissue clamps. The test series included 23 tests with corresponding peak stress and strain measurements of the central region of the specimen where optical markers tracked local displacements. The specimen was loaded at a rate of 0.7 strains per second to match the uterine strain rate in a motor vehicle crash. Experimental results include peak stresses and peak strains for the pregnant uterine tissue in tension. When loaded biaxially, the circumferential peak stress is 500 ± 219 kPa with a corresponding peak true strain 0.30 ± 0.09 and the longitudinal peak stress is 320 ± 176 kPa with a corresponding peak true strain 0.30 ± 0.09. This material information can be implemented in pregnant occupant models to evaluate the uterine tissue response to impact loading scenarios.

KEYWORDS – Pregnant, Uterus, Biaxial, Material Properties, Dynamic

INTRODUCTION

Injury Incidence

Automobile crashes are the largest cause of death for pregnant females (Attico et al. 1986) and the leading cause of traumatic fetal injury mortality in the United States (US) (Weiss and Strotmeyer 2002). In a three year study of fetal traumatic injury deaths in select states, 82% were a result of maternal injury in a motor vehicle crash (MVC) (Weiss 1999). Each year in the US, 160 pregnant women are killed in motor vehicle crashes and an additional 600 to 2600 fetuses are killed when the mother survives (Pearlman 1997, Klinich et al. 1999a, Klinich et al. 1999b, Weiss 1999). For comparison, Fatality Analysis Reporting System (FARS) data on child fatalities in the year 2005 indicate pedestrian, bike, and infant MVC fatality numbers are less than half the number of fetal deaths (Figure 1) (FARS 2005).

The first way to protect the fetus is to protect the mother considering that maternal death has a near 100% fetal loss rate (Pearlman et al. 1990a). If the mother survives, protection of the fetus may best be accomplished by preventing placental abruption. Placental abruption, which is the premature separation of the placenta from the uterus, has been shown to account for 50% to 70% of fetal losses in motor vehicle crashes (Pearlman et al. 1990b). Information gathered from crash investigations shows that placental abruption can occur without other, more severe injuries, such as uterine rupture or direct fetal injury. However, when these more severe injuries do occur they are often accompanied by placental abruption (Rupp et al. 2001a). The different material properties of the uterus and the placenta are one reason they separate naturally after birth when the uterus contracts and the placenta maintains its shape (Pearlman 1997). A similar mechanism can occur prematurely in a motor vehicle crash causing a local failure of the uteroplacental interface (Pearlman 1997).

Address correspondence to Dr. Stefan Duma, 114 Randolph Hall Virginia Tech, Blacksburg, VA 24061. Electronic mail: duma@vt.edu

Since real world crash data for pregnant occupants is limited to a retrospective analysis and pregnant cadaver studies are not feasible, crash test dummies and computational modeling have been utilized to evaluate the risk of adverse fetal outcome. The most current version of a pregnant female crash test dummy is called the Maternal Anthropomorphic Measurement Apparatus version 2B (MAMA-2B) (Rupp et al. 2001a). This pregnant surrogate is a 30-week gestation 5th percentile female crash test dummy. Validation tests have related the peak pressure inside the simulated uterus in the MAMA-2B with real world fetal loss risk data. The pregnant surrogate has been used in further analysis to determine the risk of fetal loss for different restraint conditions in a frontal impact.

Previous studies have utilized a validated Mathematical Dynamic Modeler (MADYMO) computer model of a 30-week pregnant occupant to assess the risk of fetal loss based on occupant position, belt placement, impact direction, advanced restraint systems, and standard frontal barrier impacts (Moorcroft et al. 2003a, Moorcroft et al. 2003b, Moorcroft et al. 2003c, Moorcroft et al. 2004, Duma et al. 2006, Manoogian et al. 2007a). Although pregnant occupant research has progressed with these tools, they are based on a limited amount of data on material properties of maternal tissues (Rupp et al. 2001a, Moorcroft et al. 2003a). In order to have more accurate research tools, better pregnant tissue material data are needed.

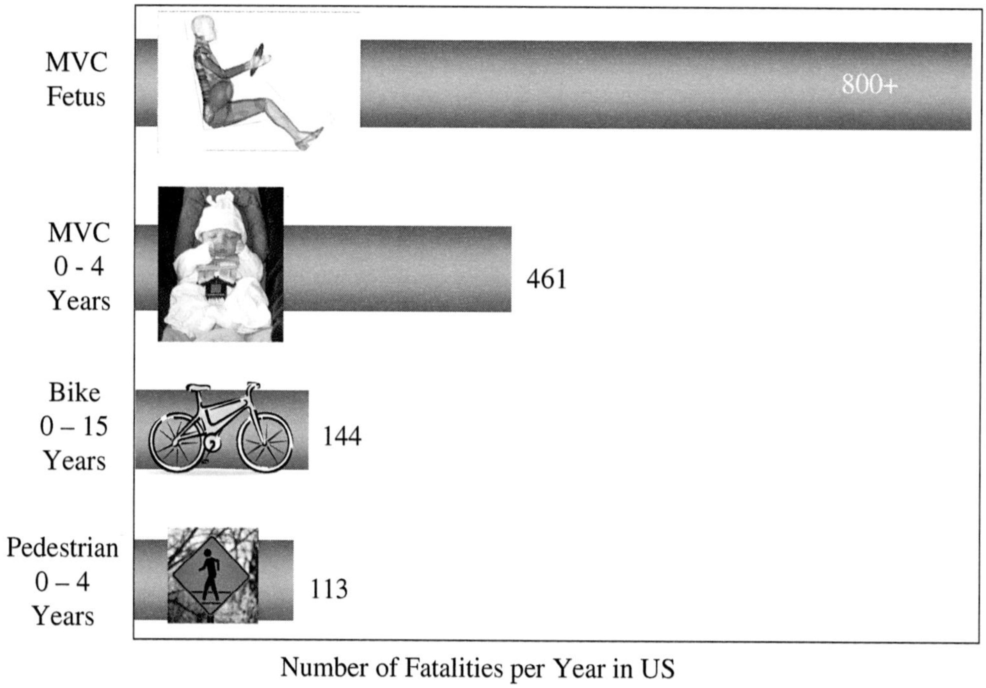

Figure 1: A comparison of fetal deaths to child deaths each year in the US (Klinich et al. 1999a, FARS 2005).

Previous Uterine Tissue Research

Previous researchers have conducted material tests on non-pregnant, pregnant, human, and animal uterine tissues at quasi-static loading rates (Ohara 1953, Csapo and Goodall 1954, Conrad and Kuhn 1967, Yamada 1970, Pearsall and Roberts 1978, Mizrahi et al. 1980, Pearlman 1999, Deyer et al. 2000, Rupp et al. 2001b). Research on uterine tissue material properties began in the early 1950's with a basic uniaxial tensile failure test at a quasi-static rate on animal uterus (Yamada 1970). Ohara found the tensile strength of a non-pregnant rabbit uterus in the longitudinal direction to be 177 kPa with a corresponding ultimate elongation of 1.5 (Ohara 1953). In addition to failure properties, Csapo and Goodall used electrical stimulation on rabbit uteri to quantify active muscle tensile strength. The result of their research was normalized force versus normalized length activation curves for myometrium under estrogen and progesterone domination (Csapo and Goodall 1954). Myometrium is the smooth muscle layer of the uterus.

To follow this work, research progressed to study pregnant versus non-pregnant myometrium. Wood pioneered human uterine tissue testing by using strips of human uterus that were extracted during a cesarean section at the incision location (Wood 1964). Uniaxial quasi-static tensile tests of these uterine samples of the lower uterine region had a failure true stress of 483 kPa and the elastic moduli of the samples at 69 kPa stress averaged to 1207 kPa (Pearsall and Roberts 1978). No failure strain was reported.

In 1967, Conrad used pregnant and non-pregnant human uterine samples in uniaxial quasi-static tensile tests with a step increase in load (Conrad and Kuhn 1967). Results of his study indicated the pregnant uterine tissue, from a cesarean section sample, had a lower elastic modulus at 69 kPa than the non-pregnant tissue from a uterus extracted during a hysterectomy. The respective elastic moduli from engineering stress-strain data were 586 kPa and 965 kPa (Pearsall and Roberts 1978).

By the late 1970's, Pearsall tested more than 20 non-pregnant human uterine specimens in uniaxial tension (Pearsall and Roberts 1978). Human uterine specimens were acquired from the fundal region of three non-pregnant uteri which had been removed in hysterectomy operations. The tests were at a quasi-static rate of less than 0.01 strains/s and had failure stresses that ranged from 550-2069 kPa. The maximum failure strain from those tests was in the range 0.30-0.95. Additional tests were completed to evaluate the tissue response to compressive loads. Results of these tests showed that the tissue has a stiffer response in tension than in compression.

Mizrahi's research goal was to measure strains during labor by placing strain rosettes on the skin and the uterus (Mizrahi et al. 1980). He recorded strains on 36 women while in their 36th to 42nd weeks of pregnancy. The peak strain recorded during labor was less than 0.03, which is inconsistent with other in vivo studies that report a much higher strain. Although Mizrahi's study has several limitations, the main result was that the uterus has an anisotropic response.

More recently, four sub-failure uniaxial tensile tests were completed on human uterine samples taken from the location of a cesarean incision and frozen before testing. At a quasi-static rate, the uterine tissue had an average peak stress of 45 kPa with a maximum strain equal to 0.65 (Pearlman 1999). Deyer and Ashton-Miller researched the uterine tissue response during labor by quantifying the uterine wall thickness from ultrasound measurements (Deyer et al. 2000). From this study it was determined the uterus reaches maximum contractile strain before the placenta separates. The results reported contractile strains consistent with smooth muscle data. At the time of placental separation, the peak contractile uterine strains were 4.5 radial and 0.75 circumferential. A single failure test by Ashton-Miller performed on a uterus sample that included the uteroplacental interface (UPI) indicated 0.60 strain in the uterus when the UPI failed (Rupp et al. 2001b).

There are limitations with non-pregnant, cesarean, and hysterectomy uterus samples as material models for the fundal region of a pregnant uterus. A pregnant uterus undergoes several changes necessary to the care and delivery of the fetus. Hypertrophy, increase in elastic tissue, accumulation of fibrous tissue, and uterine stretching are all physiological changes to strengthen the uterus for care and delivery of the fetus (Cunningham and Williams 2005). A non-pregnant uterus will not have the same shape or structure as a pregnant one and therefore has different material properties. Although some researchers tested pregnant uterine tissue from the cesarean incision location, they were limited in number and the tests were performed at quasi-static loading rates. Moreover, whole uteri available from a hysterectomy are not accurate models because they are non-pregnant and there is additional deterioration of the myometrial and endometrial layers if the uterus is from a postmenopausal donor (Cunningham and Williams 2005). Additional considerations make these previous data sets unsuitable for use in modeling high rate impact events for pregnant occupants in motor vehicle crashes. All of the previous tests are at quasi-static rates but loading in a motor vehicle crash is at a dynamic rate. Soft tissue such as the uterus is viscoelastic; therefore, it should be loaded at the rate in which the material properties are needed (Weiss et al. 1996, Sacks and Sun 2003).

For an anisotropic material that is loaded in biaxial tension *in vivo*, a biaxial material test is more suitable than a uniaxial test to determine the material response (Humphrey et al. 1987, Lanir et al. 1996, Bass et al. 2001, Shah et al. 2006). Specifically biaxial tests load multiple fiber directions at one time. This is a better model for the pregnant uterus than uniaxial tension which loads only one fiber direction at a time. Inflation testing would provide failure stress and strain data and has been applied by previous researchers on thin membranes (Wineman et al. 1979). However, it is limited in its ability to correctly characterize the anisotropic nature of soft tissue such as the uterus (Sacks 2000). Previous pregnant occupant modeling studies have measured the uterine response to a dynamic input, and indicate uterine strain is a result of loading the uterine wall in equiaxial tension (Moorcroft et al. 2003a, Moorcroft et al. 2003b, Moorcroft et al. 2003c). Stress and strain values for a biaxially loaded anisotropic specimen are significantly different from stress and strain values for a uniaxially loaded anisotropic specimen (Bass et al. 2004, Shah et al. 2006). Therefore, to measure material properties of the uterine tissue biaxial tests were chosen.

Animal Surrogate.

This study utilizes pregnant porcine uterine tissue as a model of the pregnant human uterus because of its similarity and availability. The similarities between a porcine uterus and a human uterus are in the structure of the uterine layers and their development during gestation. In a porcine uterus, as in a human uterus, the inner lining of the uterine cavity is the endometrial layer. This layer is attached to the myometrium and contains connective tissue and epithelium. During pregnancy, the porcine myometrium grows in size due to hypertrophy and develops elastic tissues as in the human (Marrable 1971). The human uterus has an anisotropic response with two groups of smooth muscle fibers, circular and longitudinal, that originate at the fallopian tubes and fan outward (Marrable 1971, Mizrahi and Karni 1975, Young and Hession 1999, Weiss et al. 2006). The myometrium for a sow is also anisotropic with two groups of muscle fibers, circular and longitudinal, that cross perpendicularly (Marrable 1971). Externally the porcine uterus has two uterine horns supported by the mesometrium, which serves the same function as the supporting ligaments for the human uterus (Figure 2 and Figure 3).

There are differences between the porcine and human uterus. One difference is the shape, the human uterus has a simplex shape, and the porcine uterus is bicornuate, the closest stage to the simplex in evolution (Finn and Porter 1975). The simplex shape means that there is only one part that is the uterus (Figure 2). Bicornuate means the uterus is comprised of two parts referred to as horns (Figure 3). The function of the porcine uterus differs from the human uterus in that the number of fetuses and gestational period is different. A typical human birth has a single fetus for 40 weeks while a typical porcine birth has more than 6 fetuses for a 16 week gestation (Marrable 1971). Moreover, the placenta for the porcine is diffuse and fills the entire uterus while the human placenta is discoid and only connects to part of the human uterus, usually at the fundus of the uterus (Ramsey 1975). In spite of the different shape, size, and placenta, the porcine and human uterine walls have similar myometria with anisotropic composition. For this study, only the material properties of the uterine wall are considered and therefore the differences described above do not eliminate the pregnant porcine uterus as a useful model for the pregnant human uterus.

Although there have been several research efforts to determine the loading and failure properties of uterus and pregnant uterus, the data are limited in their ability to accurately represent the fundal uterine tissue response to loading in a motor vehicle crash. The purpose of this study is to augment the currently available data by collecting material data from pregnant porcine uterine tissue in biaxial tension at a dynamic loading rate. The peak data collected from sub-failure loading are compared to previous research on uterine tissue from non-pregnant and pregnant uterine tissue research. Future studies can incorporate these data into computational models for predicting the response of the pregnant occupant to dynamic impacts.

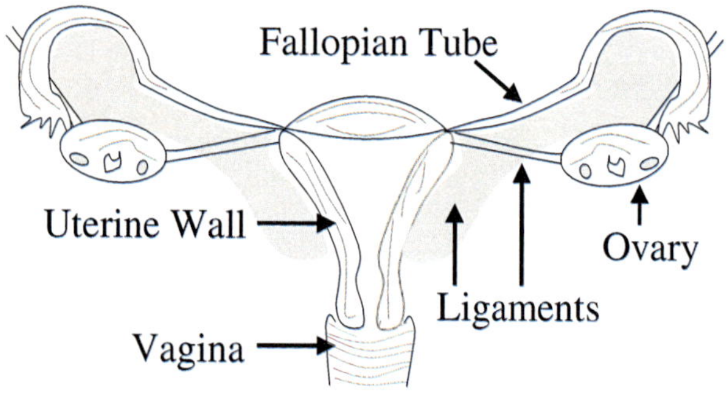

Figure 2: Diagram of a simplex human uterus.

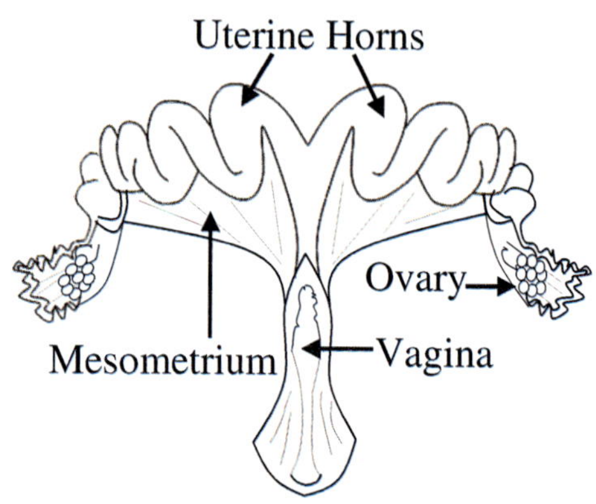

Figure 3: Diagram of a bicornuate porcine uterus.

METHODS

Tissue Sample Preparation. Whole porcine uterine tissues were collected from five pregnant sows at the time of death. All tissues were kept fresh in saline, which preserved the integrity of the tissue, and tested within 7 days of extraction. Individual test specimens were cut from the whole uterus. There were approximately 5 samples from each sow for a total of 23 specimens. Only one uterine horn was used from each donor. In a porcine uterus, the uterine wall meets at the mesometrium and forms a tubular shape. The samples for this study were taken on the side of the uterine wall opposite the mesometrium connection (Figure 4). This region of the uterine horn is most distant from the mesometrium connection and as a result is more uniform.

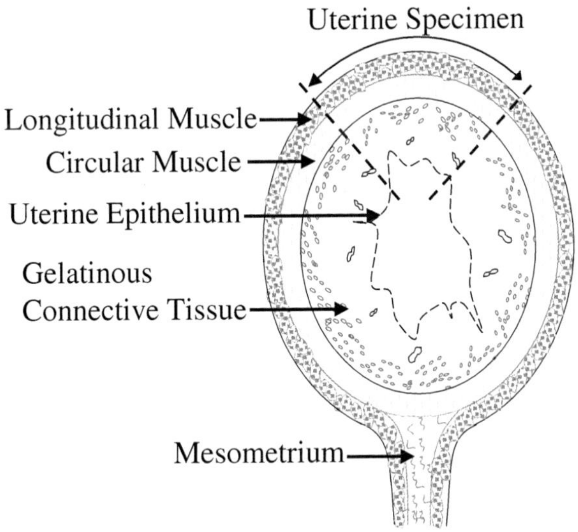

Figure 4: Cross sectional view of a porcine uterus.

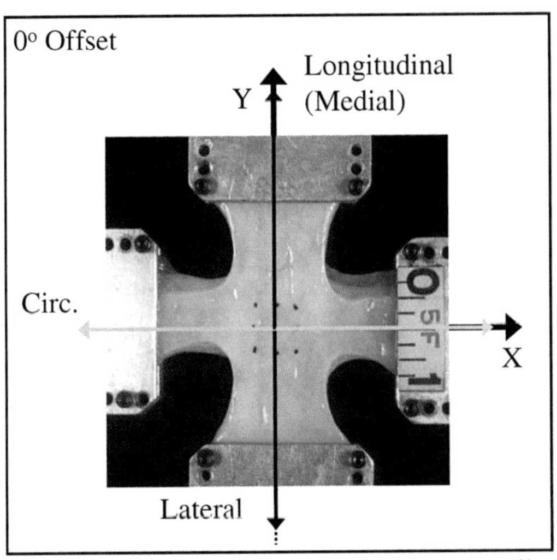

Figure 5: Specimens have the material axes aligned with the loading axes.

Each uterine horn was separated at the mesometrium connection along the length of the horn. This enabled the hollow uterine tube to be laid out flat and provided easy alignment of the longitudinal fibers. Each specimen was then cut using a steel rule die stamp. When the tissue was aligned to be stamped, the longitudinal fibers of the uterine specimen were aligned with the center axis of the clamp (Figure 5). This is possible because the longitudinal fibers are identifiable macroscopically for each specimen. The cruciform specimen is loaded along the circumferential and longitudinal fibers simultaneously.

Biaxial Test System. Biaxial dynamic tensile tests were performed using a custom designed system of linear motors to pull the cruciform specimen in tension simultaneously with four tissue clamps (Manoogian et al. 2007b) (Figure 6). The system included one multi axis controller (Parker ACR9000, Irwin, PA) that served as a power supply for two motor drivers (Parker ViX, Irwin, PA) which controlled four linear motors with stages (Parker Daedal MX80S, Irwin, PA). The four individual stages were each instrumented with a load cell (OMEGA LC201, 100 lbf, Stamford, CT), potentiometer (Space Age Control, 160-1705, 540 mm, Palmdale, CA), and accelerometer (Endevco 7264B, 2000 G, San Juan Capistrano, CA). The inertial load of the mass (0.0374 kg) in front of the load cell was multiplied by the acceleration to compensate the measured reaction load.

Specimens were kept moist prior to testing by storing them in saline, and during testing they were sprayed with saline solution. No preconditioning was applied before the test. Each test began with the same initial conditions. To ensure a proper initial geometry, the specimens were mounted in the grips so that the initial shape matched the original stamp shape (Fung 1967, Martin 2000, Duma et al. 2003). This resulted in a slight variation of the grip loads giving a standard deviation in the preload for all tests of ±0.13 N which corresponds to approximately 1% of the average peak load for each test. This technique ensured that each test began with no slack in the specimen and the initial condition for each stress versus strain curve was the resting state of the porcine uterus *ex vivo*.

The thickness for each specimen was acquired before the test once the specimen was installed in the test fixture. A FARO Laser ScanArm (FARO Technologies Inc., Platinum Laser ScanArm, Lake Mary, FL) was used to scan hundreds of points on both the top and bottom surfaces of the tissue specimen. This data provided two surface geometries with an accuracy of ±0.1 mm that represented the top and bottom of the tissue. From the surface geometries, an average thickness could be obtained by sectioning the scan in the region of interest for each individual specimen. The average thickness across the region of interest is used in the stress calculations.

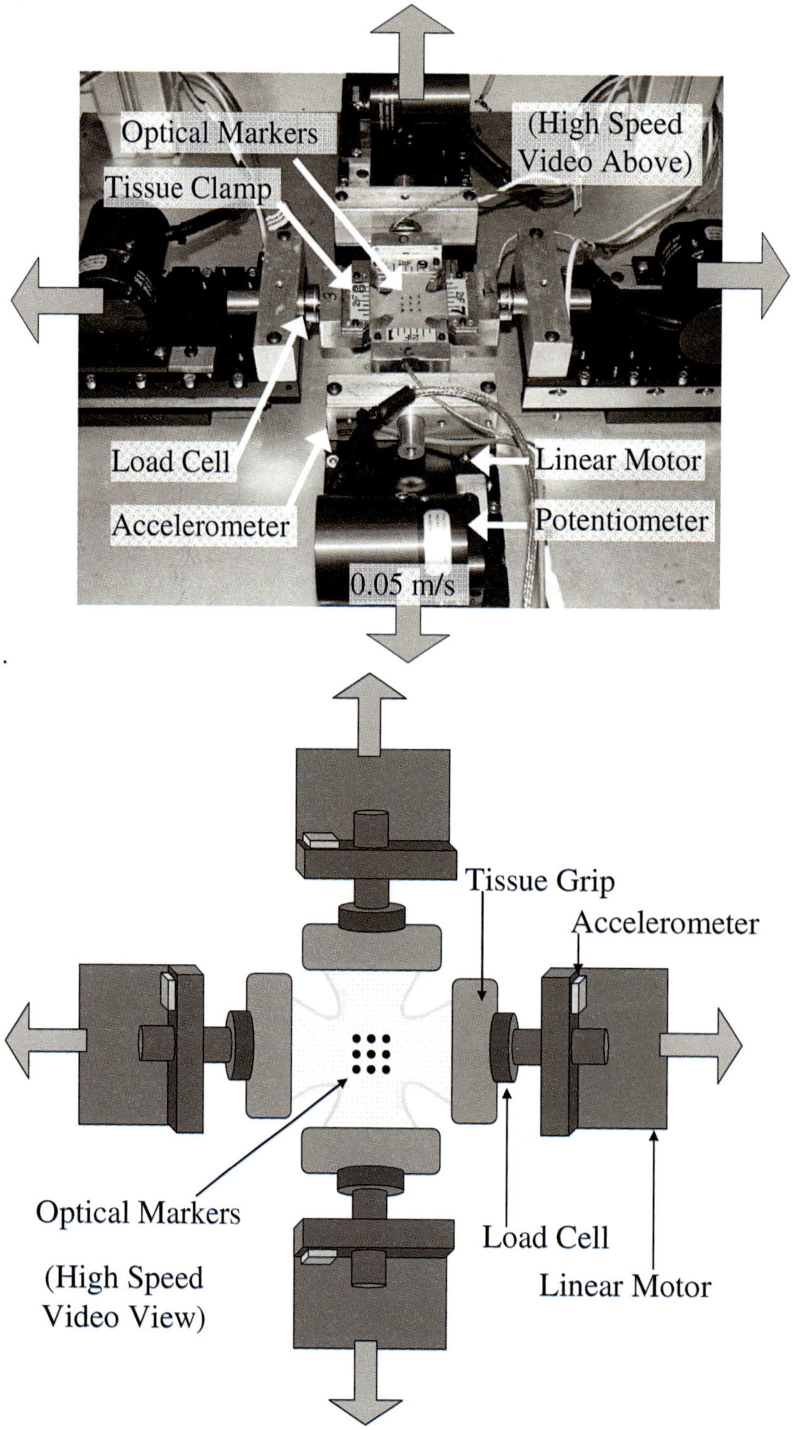

Figure 6: Four individual platforms pull the tissue in biaxial tension and are instrumented with an accelerometer, potentiometer, and load cell
(Photograph Top, Illustration Bottom).

On each of the 4 sides of the test specimen, the direct path from the tissue clamp went through the load cell and was rigidly connected to the linear motor stage (Figure 6). A multi axis controller provided synchronized motion of all four linear stages to move at 0.05 m/s for a displacement of 0.06 m. This distance assured failure occurred and corresponded to a strain rate of approximately 0.7 strains/s. This rate was approximated from previous research on modeling of a pregnant occupant, restrained properly by a three-point seat belt, in a frontal impact with a 35 kph change in velocity (Moorcroft et al. 2003a, Moorcroft et al. 2003c).

Local deformation was recorded using high-speed video (Phantom V4, Wayne, NJ) at 1000 frames per second with 512 by 512 resolution to track ink markers added to the tissue immediately before testing. Optical marker tracking was performed with TEMA® Advanced Motion Analysis Software (Linköping, Sweden). Using the configuration for these tests, the error in the motion analysis was 0.065 mm.

Data Processing and Analysis. The test series included 23 tests with corresponding circumferential and longitudinal stress and strain measurements from the central region of the specimen. Two stresses were calculated for these tests. One was an average in the circumferential direction and the other was the average in longitudinal direction. These calculations were established in previous biaxial testing research (Sacks and Sun 2003, Shah et al. 2006). The process was the same for both directions but the circumferential direction is discussed as an example. For the circumferential direction, this process took the measured loads at grips X_1 and X_2, inertially compensated and scaled them individually. Scaling of the loads was performed so that the stress and strain measurements were both for the area encompassed by the optical markers. Assuming an equal transfer of load from the grip to the center of the specimen, a length ratio could be used to scale each load (Figure 7, eq. 1). All of the load must be transferred through the smallest width of the cruciform arm, L_1. The load from this region was scaled down to the load that passed through the outer two optical markers. The distance between those markers is L_2 and is measured optically. This scaling process was completed for each side of the specimen. The average of the two scaled loads in the circumferential direction gave an average load at the region of interest for that direction. By assuming incompressibility, the tissue thickness at each time step was calculated. An initial thickness was measured for each specimen as previously discussed. The time history of the region of interest surface area was calculated from the optical markers. Given incompressibility, the thickness at each new time step could be calculated. Although the markers start in a square, they do not remain in that shape. Therefore, in order to determine the cross section for the stress calculation, an average length was found using two opposing sides of the region of interest. The true stress calculation for the horizontal direction used the average force and the average cross sectional area at each time step. The same procedure was completed for the vertical direction. The true stress in the circumferential and longitudinal directions as previously described are reported for each biaxial test.

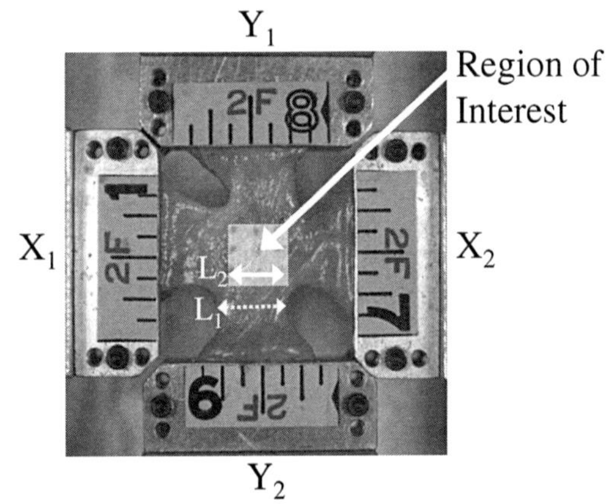

Figure 7: The four corner markers create a region of interest in which the strain and stress were calculated.

$$F_{AreaofInterest} = \frac{(tL)_{AreaofInterest} F_{grip}}{(tL)_{grip}} = \frac{L_2 F_{grip}}{L_1} \qquad \text{(eq. 1)}$$

The Green-Lagrange strains were calculated for the circumferential and longitudinal directions using the optical markers on the specimen (Humphrey et al. 1987, Sacks and Sun 2003, Shah et al. 2006). An average was calculated using the four triangles that connect the corners of the area of interest to the center point (Figure 8). The deformation gradient, F, was calculated from the matrix representing the deformed state of the markers times the inverse of the matrix with the undeformed coordinates of the markers (Figure 9, eq. 2). For each triangle, the deformation gradient tensor provided information to compute the Green-Lagrange strain tensor (eq. 3). The principal Green-Lagrange strains, E_{11} and E_{22}, were averaged separately to represent the respective X and Y directions. These were used to find the stretch ratio and calculate the true strain values which are presented in this paper. The eigenvalues of the Green-Lagrange strain tensor are the maximum and minimum principal strains (Ivarsson et al. 2000). The shear strain was determined from the change in angle between the two originally orthogonal vectors and the final state (Humphrey et al. 1987) (eq. 4).

A standard method of determining the characteristic average was used for both the circumferential and longitudinal sets of stress versus strain data (Lessley et al. 2004). This process began with normalizing each strain curve to its respective maximum. Interpolation routines determined the stress for each test at set increments of strain. These stress values were averaged for the average stress data. The strain data that corresponds to these stress values were obtained by multiplying the normalized strain by the average maximum strain of all tests. The result of this process was average stress-strain curves for both the circumferential and longitudinal directions for all 23 specimens. The linear elastic modulus is calculated in two regions for the characteristic average curve of each data set. One linear elastic modulus is a minimum value measured in the toe region of the curve and another value is a maximum slope of the stress versus strain curve before the peak value. Matched paired t-tests were used to test for significant differences between the circumferential and longitudinal directions for peak strain and peak stress. A one way ANOVA analysis tested all 5 donors for any significant difference from the others for peak stress and strain in the circumferential and longitudinal directions. A regression analysis was performed on the peak values to determine any regional significance based on proximity to the ovary or vagina on the uterine horn. Proximity of the tissue specimen was categorized as increasing in distance from the vagina as specimens were extracted from the whole uterine horn.

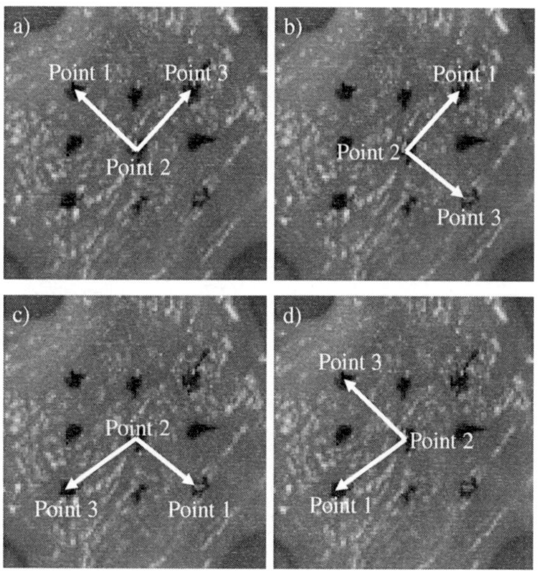

Figure 8: The four groups of vectors (a-d) connecting the center point to two of the corner points are used to calculate X and Y strains that are averaged to establish an X and Y strain value for the specimen.

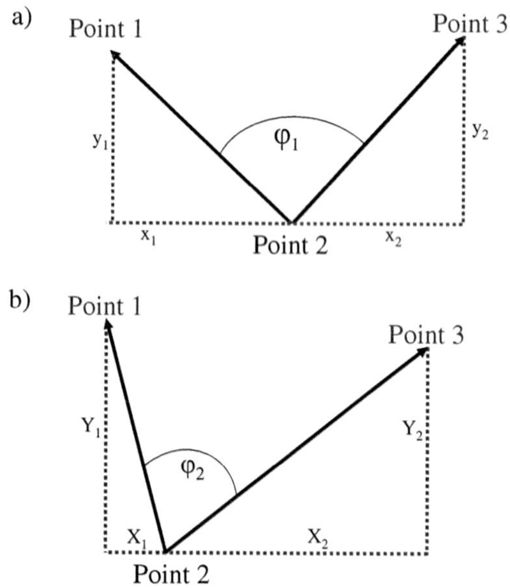

Figure 9: a) The initial state of two corner markers is described by vectors (x_1,y_1) and (x_2,y_2). b) The deformed state of two corner markers is described by vectors (X_1,Y_1) and (X_2,Y_2).

$$\mathbf{F} = \left(\begin{bmatrix} X_1 & X_2 \\ Y_1 & Y_2 \end{bmatrix}_{t=1} \right) \left(\begin{bmatrix} x_1 & x_2 \\ y_1 & y_2 \end{bmatrix}_{t=0} \right)^{-1} \quad \text{(eq. 2)}$$

$$\mathbf{E} = \begin{bmatrix} E_{11} & E_{12} \\ E_{21} & E_{22} \end{bmatrix} = \frac{1}{2}\left(F^T F - I\right) \quad \text{(eq. 3)}$$

$$\gamma = \frac{1}{2}\arcsin\left(\frac{2E_{12}}{\sqrt{1+2E_{11}}\sqrt{1+2E_{22}}} \right) \quad \text{(eq. 4)}$$

RESULTS

In this section, stress versus strain curves in the circumferential and longitudinal directions are presented for the biaxial tension loading of the porcine uterine specimens. These results are for 23 tests with the material axes inline with the loading axes. For the strain measurements, the tracking accuracy of the markers is 0.065 mm/pixel which corresponds to approximately 1.6% of the original distance between two corner markers that compose one side of the central region. The displacement measurement error of 1.6% also affects the load scaling and cross sectional area part of the stress calculation. The preload, thickness, and strain rate for each test is tabulated (Table 1).

Table 1: Test information is listed for each specimen.

Test	*Thickness (mm)*	*Preload (N)*	*True Strain Rate (Circ.)*	*True Strain Rate (Long.)*
1	2.06	0.09	0.84	0.50
2	2.65	0.22	0.39	0.86
3	1.86	0.07	0.70	0.47
4	2.41	0.02	0.77	0.81
5	2.94	0.13	0.64	0.63
6	2.17	0.12	0.70	0.85
7	3.65	0.18	0.44	0.66
8	1.92	0.17	0.58	0.67
9	2.28	0.30	0.71	0.54
10	2.74	0.32	0.57	0.70
11	2.64	0.08	0.61	0.85
12	2.23	0.10	0.59	0.80
13	1.79	0.10	0.55	0.70
14	1.81	0.04	0.63	0.87
15	1.06	0.17	0.81	0.60
16	1.62	0.08	0.90	0.74
17	1.70	0.12	0.76	0.76
18	1.87	0.08	0.83	0.69
19	1.02	0.07	0.82	0.74
20	2.21	0.11	1.00	0.73
21	2.64	0.14	0.54	0.71
22	1.93	0.14	0.73	0.64
23	2.28	0.21	1.00	0.73

Failure occurred across either of the circumferential cruciform arms for 22 of the 23 tests. This failure pattern is seen in video captures of a specimen with material axes aligned with the loading axes and pulled in biaxial tension at 0.7 strains/s (Figure 10).

All of the stress-strain data for the tests are plotted together with the average and standard deviation for the stress at each strain for the circumferential direction (Figure 11). The average peak stress in the circumferential direction is 500±219 kPa. The corresponding peak true strain is 0.30±0.09. All of the stress-strain data are plotted together with the average and standard deviation for the longitudinal direction (Figure 12). The peak stress for the longitudinal direction is 320±176 kPa. The peak true strain in the longitudinal direction is 0.30±0.09. Although the peak stresses are significantly different for the two directions ($p<0.001$), the peak strains are not significantly different ($p=0.82$). The characteristic average curves show the average shape of the stress-strain curves in both directions for the 23 tests of pregnant porcine uterine tissue with the material axes aligned with the loading axes (Figure 13). The linear elastic moduli for the biaxial porcine data are reported for the minimum and maximum slopes of the characteristic average curves. The minimum slopes are 695 kPa and 507 kPa for the circumferential and longitudinal directions. The maximum elastic moduli values are 3453 kPa and 2327 kPa for the circumferential and longitudinal directions respectively.

The five uteri used for the current study do not have significant differences in peak stress or peak strain values for the circumferential and longitudinal directions ($p>0.11$). The regression analysis indicated there were no significant differences in the peak stress and strain of the specimens based on their proximity to the ovary or vagina along the uterine horn ($p>0.30$).

DISCUSSION

Donors. Because the stress and strain peaks are not significantly different for the five uteri, the data from all uteri were averaged to present loading and peak stress-strain data. Since the information regarding sow age and number of fetuses for each uterus is not known, this study could not address the effects of those variables. Future research could perform more testing to address variability in the stress-strain curve as a function of several gestational variables. No regional variation was expected based on the uniformity of the uterine horn along its length and no significant regional dependence was measured.

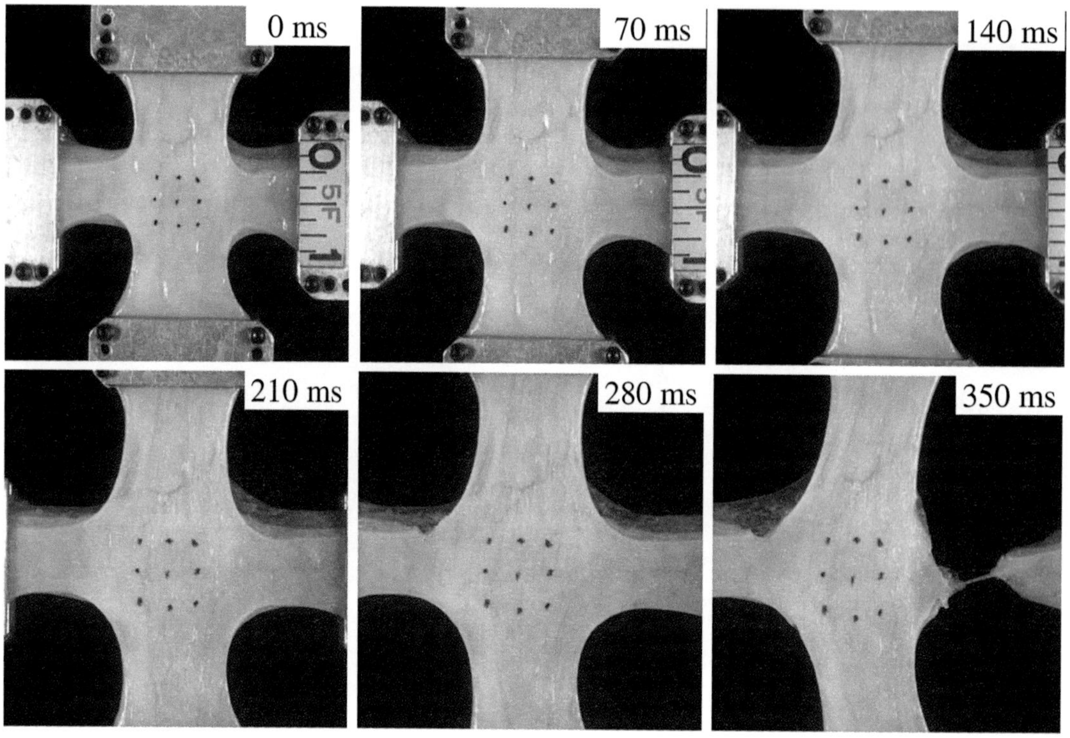

Figure 10: The specimens with the material axes aligned with the loading axes fail in the arm of the cruciform shape.

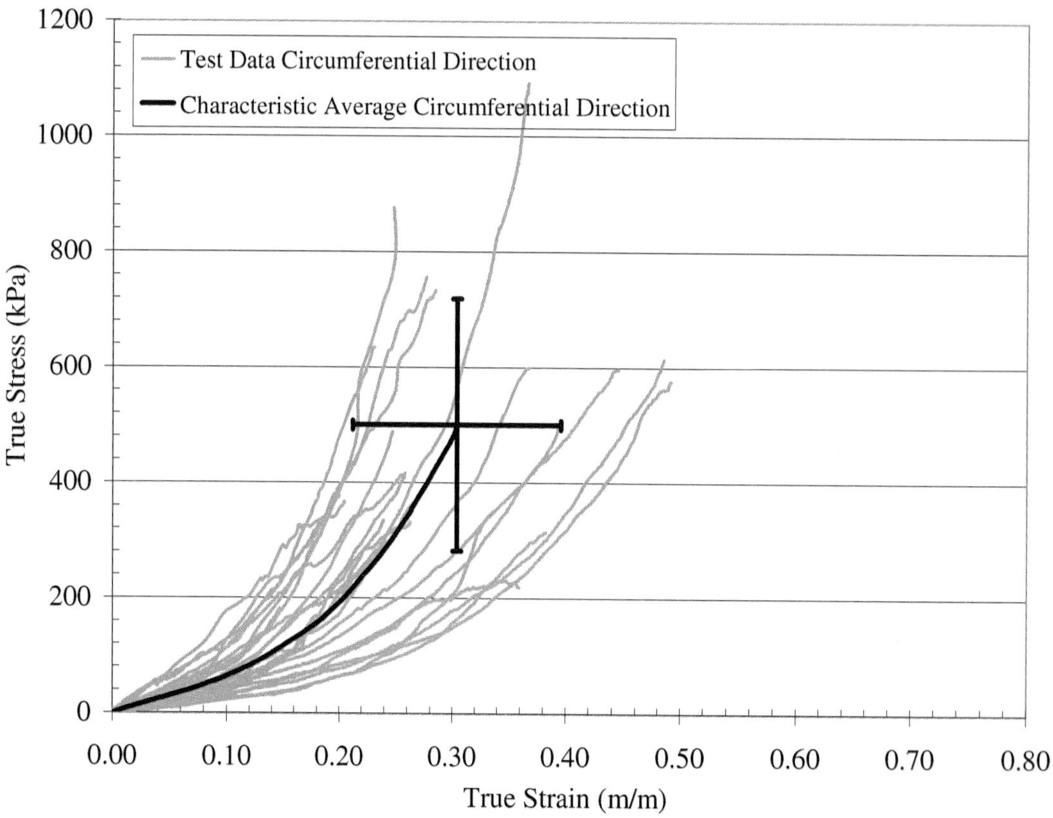

Figure 11: The average stress-strain curve in the circumferential direction has a peak of 500±219 kPa and 0.30±0.09 strain.

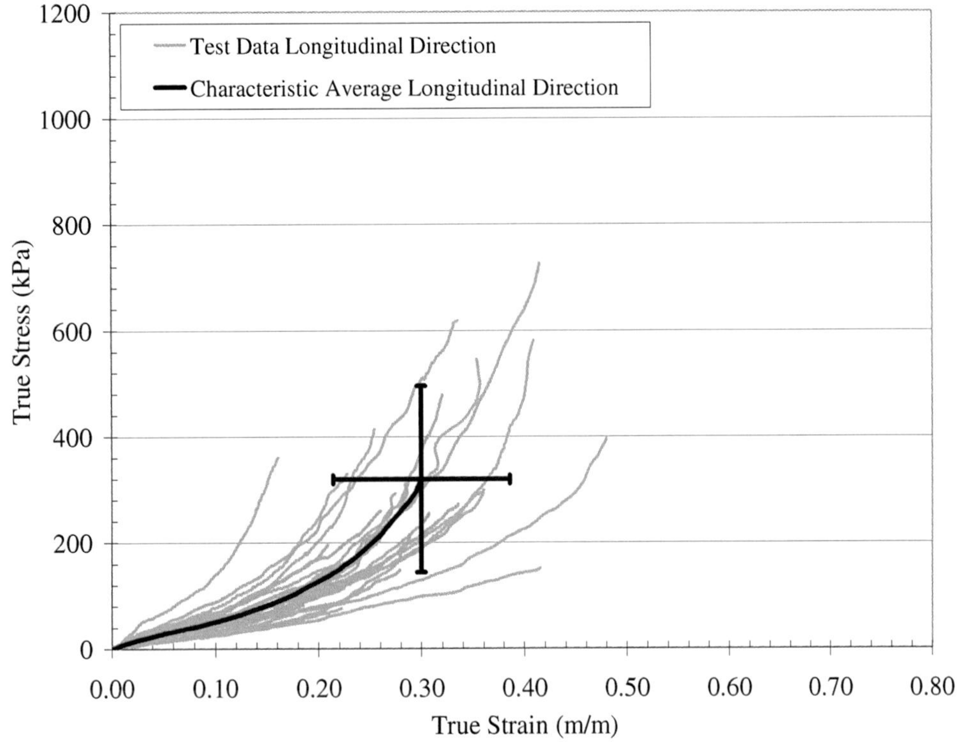

Figure 12: The average stress-strain curve in the longitudinal direction has a peak of 320±176 kPa and 0.30±0.09 strain.

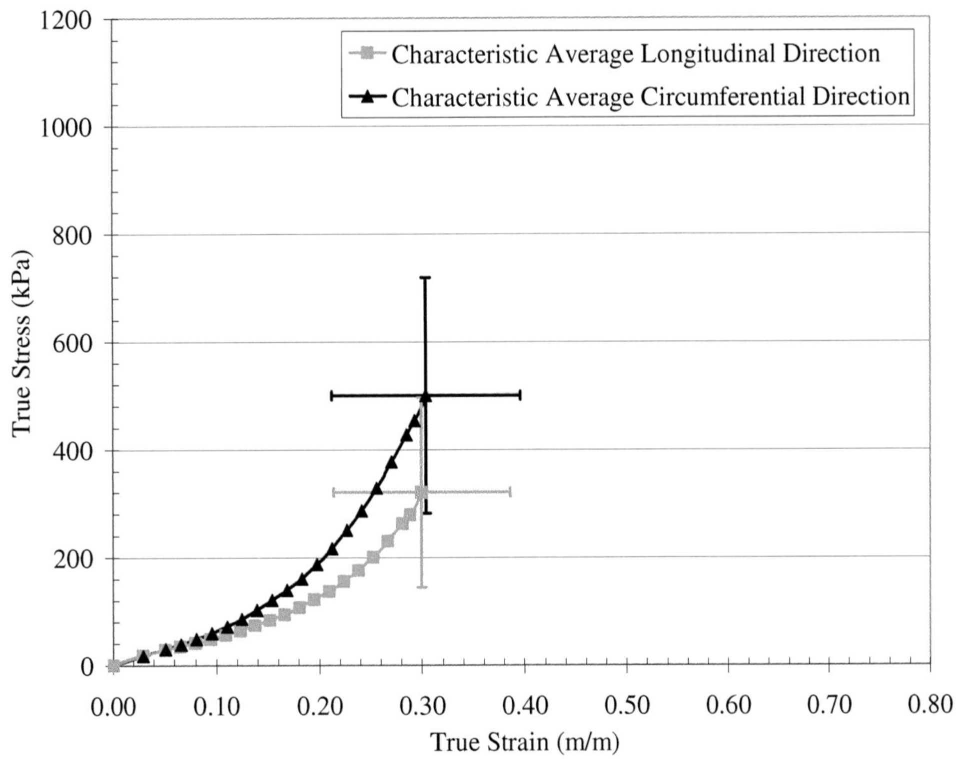

Figure 13: The average curve for the circumferential direction and the average curve for the longitudinal direction.

Comparison to Previous Studies

The stresses, strains, and elastic moduli from the current study are compared to previous research studies (Table 2). The previous published data in engineering strain values were converted to true strain values for comparison. Because there has not been a study that tested porcine uterus samples or uterus with a biaxial test setup, there is not a direct source for comparison. The data are compared primarily to the two most extensive previous studies by Pearlman and Pearsall (Pearsall and Roberts 1978, Pearlman 1999). The other studies are limited in their data available for comparison as well as the number of tests completed.

Peak Stress. The peak stress data in the current study are similar to the peak stress reported for unfrozen human pregnant uterine tissue tested in uniaxial tension at quasi-static rates (Wood 1964). The quasi-static peak stress of 483 kPa, measured by Wood (1964), is similar to the circumferential direction average peak stress for dynamic biaxial tests of the uterine tissue measured in this study. The longitudinal direction peak stress in the current study is slightly lower. Due to the viscoelastic behavior of muscle, it is expected a quasi-static loading rate such as Wood et al. tested would provide a lower peak stress than the current study using dynamic loading conditions. Peak strain is not reported for the human tissue tests performed by Wood.

When compared to non-pregnant human uterus material data, the peak stresses from biaxial porcine samples are lower (Pearsall and Roberts 1978). The range of peak stresses for Pearsall's failure tests on human non-pregnant uteri is 550-2069 kPa and the range of peak stresses for the current study is 76-1094 kPa. There are a few reasons for this difference. First, pregnant tissue is more elastic due to the changes in the myometrium and can strain more with a lower peak stress than non-pregnant tissue (Cunningham and Williams 2005). Moreover, the peak stress reported in the biaxial tests is only indicative of the loading until failure occurred at any location in the tissue. Therefore, this peak is a lower bound on failure stress, which should be greater than or equal to this value due to the stress concentration at the corner of the specimen. Finally, it is possible that this difference could be because the porcine uterus might have a lower peak stress than human uterus; however, there are no studies to show this.

Peak Strain. The peak strain for the porcine uterine tissue, reported as true strain, had a minimum value of 0.15 and a maximum value of 0.49. Previously published research on uniaxial human uterine tissue failure tests reported peak true strain values from 0.30 to 0.95 (Pearsall and Roberts 1978). This includes a higher range of peak strains than the current study using dynamic biaxial testing methods. The loading rate of the tissue could contribute to this difference. The biaxial tests are at a dynamic loading rate which is associated with a lower failure strain for viscoelastic tissue. In addition, the biaxial specimens are being strained equally in two dimensions, which causes the peak strain to be less than a peak uniaxial strain for the same tissue (Bass et al. 2004). Furthermore, for the biaxial tests the peak strain is recorded at the time the tissue fails in the corner of the specimen which is sub-failure strain at the region in the center of the specimen.

Pearlman's lower uterine segment samples had sub-failure strains from uniaxial tension in the range 0.36 to 0.55 when converted to true strain (Pearlman 1999). The only study to use pregnant uterine tissue from the fundal region of the uterus was one test by Aston-Miller in which the peak strain in the uterus was recorded as 0.60 when the UPI failed (Rupp et al. 2001b). Assuming this is reported as engineering strain, the true strain equivalent is 0.47. Both of the lower uterine segment samples as well as the single test from the fundal region have peak strain values are slightly above the range of the current biaxial pregnant porcine uterine tissue data set.

Moreover, when considering the long term goal of this research to provide information necessary in predicting placental abruption, it is noted that the material response of the uterus is very different than the material response previously reported for placental tissue (Figure 14) (Manoogian et al. 2007c). Placental abruption, failure of the uteroplacental interface, was discussed as the injury mechanism due to different material properties of the uterus and placenta (Pearlman 1997). Since it is not possible at this time to complete material tests on the human uteroplacental interface, knowing the material properties and peak strains of the uterus and placenta are important factors in addressing failure of the uteroplacental interface.

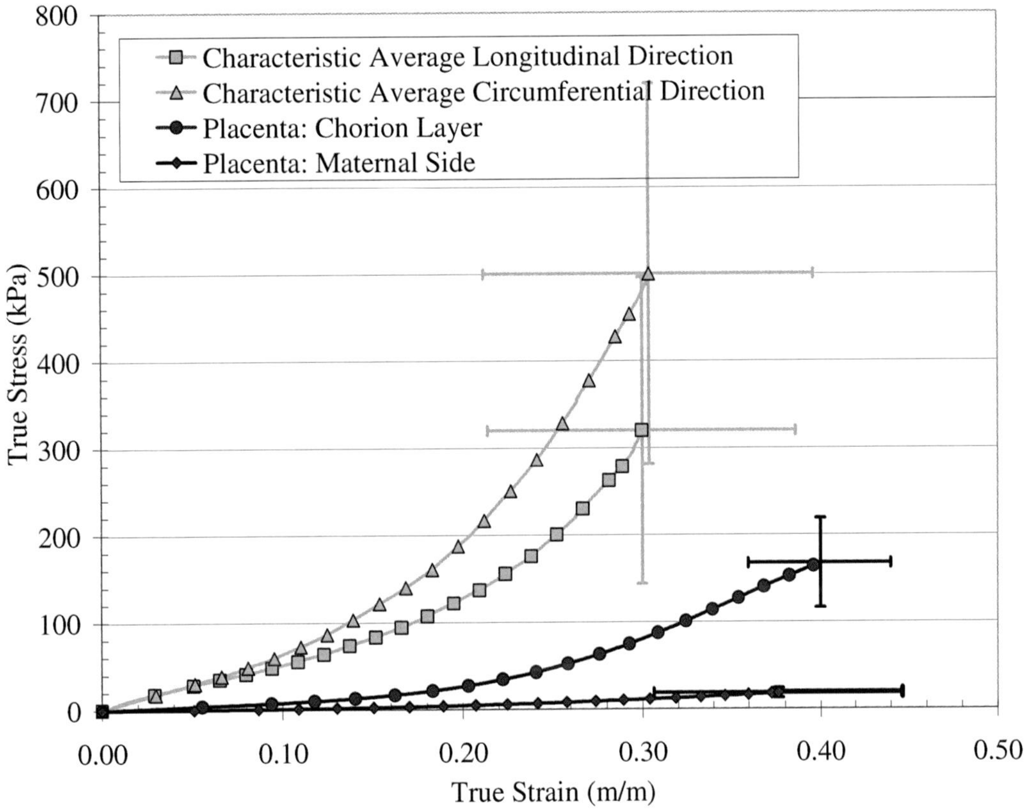

Figure 14: The pregnant porcine uterus characteristic curves are compared to human placental tissue data.

Elastic Modulus. Pearsall's human non-pregnant tissue had linear elastic modulus values from 500 kPa to 1378 kPa when tested in uniaxial tension at a quasi-static rate (Pearsall and Roberts 1978). The moduli for the toe region of the porcine stress-strain curves are at the lower range of moduli reported for Pearsall's tests. When the maximum slope is used for the modulus calculation, the biaxial porcine tests greatly exceed the range for non-pregnant quasi-static tissue tests. For a non-linear material response, the region of the curve in which the modulus is calculated can dramatically change the value. Pearsall calculated the tangent to the stress-strain curve at a stress value of 70 kPa for all of the specimens. This is approximately 3%-12% of the failure stress. As a result of taking the modulus in this region, the values reported by Pearsall are generally under representative of the slope of the curve during failure.

In addition to the possible inconsistency between studies in calculating a linear elastic modulus, there are other reasons the dynamic biaxial porcine moduli would be different than the non-pregnant quasi-static uniaxial testing moduli. For a viscoelastic tissue, such as the uterus, a dynamic loading condition should produce a higher elastic modulus. However, pregnant tissue is more elastic from the changes in the myometrium and therefore would have a lower elastic modulus (Cunningham and Williams 2005). Because the non-pregnant uterus is tested in uniaxial tension and the pregnant porcine tissue is tested in biaxial tension, the difference in elastic modulus can be a result of the different loading conditions (Bass et al. 2004). The reason for this is that with biaxial testing the tissue is resisting load in two dimensions rather than one. As a result, the higher peak stress at a given strain is associated with biaxial tests compared to uniaxial tests (Bass et al. 2004). This would cause a higher elastic modulus for a biaxial test of the same tissue tested in uniaxial tension.

Table 2: Summary of material properties of uterine tissue.

Source (Year)	Tissue Type	Pregnant	Test	Peak Stress Range (kPa)	Max True Strain	Linear Modulus Range (kPa)	Failure
Ohara (1953)	Rabbit	No	Quasi-static Uniaxial tension	177	0.92*		Yes
Wood (1964)	Human - cesarean section	Yes	Quasi-static Uniaxial tension	483		1207	
Conrad (1966)	Human –cesarean section	Yes	Quasi-static Uniaxial tension			586	
Conrad (1966)	Human - hysterectomy	No	Quasi-static Uniaxial tension			965	
Pearsall (1978)	Human-hysterectomy	No	Quasi-static Uniaxial tension (parallel to long axis)	550-2069	0.30-0.95	500-1378	20+ Failure
Pearlman (1999)	Human - cesarean section then frozen	Yes	Quasi-static Uniaxial tension	9-105	0.36-0.55*	20-279	4 Sub-Failure
Ashton-Miller (2001)	Human –with UPI	Yes	Quasi-static Uniaxial tension		0.47*		Yes
Manoogian (Current Study)	Porcine Circumferential Direction	Yes	Dynamic Biaxial	500±219 182-1094	0.30±0.09 0.15-0.49	695-3453	23 Sub-Failure
Manoogian (Current Study)	Porcine Longitudinal Direction	Yes	Dynamic Biaxial	320±175 76-726	0.30±0.09 0.16-0.48	507-2327	23 Sub-Failure

* These values have been converted to true strain from their published values.

Specimen Properties

Specimen Shape. In previous research, biaxial tensile tests have been conducted primarily on square or cruciform shape specimens. While both geometries have their limitations, the more common geometry is the cruciform shape (Waldman and Lee 2005). In biaxial research for composite, fabric, and planar connective tissues, the cruciform shape has been used extensively (Monch and Galster 1963, Flynn et al. 1998, Langdon et al. 1999, Tandon et al. 2002). Due to Saint-Venant's principle stating that the gripping of the specimen only affects the local stresses, the long sample arms provide the advantage of maximizing the distance from the grip to the regions where the strains are measured. For this study, the modified cruciform shape was chosen. The modified cruciform shape with the arm width at the grip being wider than the width at the innermost region of the arm provides the best loading profile for a dynamic test (Shah et al. 2005, Shah et al. 2006).

Tissue Degradation. All uterine tissues used for testing in this study were extracted from the animal immediately following sacrifice. They were stored submerged in saline and refrigerated until just prior to testing when they were allowed to reach room temperature. All tests were completed within 7 days of the hysterectomy. A previous study on muscle tissue analyzed the effects of postmortem time and storage on material properties (Van Ee et al. 2000). It was determined that muscle tissue would not break down over the course of a few days if kept hydrated and refrigerated (Van Ee et al. 2000). The uterine tissue for this study was never frozen. The *in vivo* state of the tissue thickness and hydration are not known. Therefore, the stresses using the thickness at the time of testing are presented in this study. Additional studies could be preformed to track the changes of the uterine thickness from the initial removal from the donor and as time progressed after removal. As a result of storing the tissue in saline, refrigerated, and never frozen, the integrity of the muscle was maintained prior to testing for the current study.

Biaxial Test Considerations

Peak Data. The cruciform shape creates a stress concentration in the corner of the central region when tension is applied. Stresses and strains were measured in the central region of the specimen where the optical markers were located. Because data were only collected until the tissue failure was observed in the video, the peak stresses and strains from the central region are sub-failure measurements. Failure in the video was used to mark the end of a test because strain data was collected with optical markers which no longer accurately represented strain once failure occurred in the specimen. Moreover, the load was no longer distributed evenly between the four grips. These sub-failure peak stresses and strains are considered a conservative estimate for the failure stresses and strains. Although the muscle structure of the uterus is uniform, the presence of blood vessels creates local material differences. It was not possible to control for vasculature when extracting samples. As a result, this contributes to the variability in peak values from the porcine tissue.

Testing muscle tissue in dynamic loading without preconditioning is more similar to the *in vivo* response of the tissue than it is with preconditioning (Van Ee et al. 2000). Because this tissue was not preconditioned prior to dynamic loading, it is assumed that the dynamic response measured from the specimen and reported in this study is most similar to the *in vivo* state of the porcine uterine tissue.

Tissue Grips. The two methods to grip biaxial specimens are clamps or sutures. While sutures are recommended for reducing the boundary condition effects, in a dynamic test clamps provide a more rigid coupling to the tissue (Sun et al. 2005). Sutures can expand and tear through the material. Additionally, clamps more closely model *in vivo* loading of the tissue by pulling uniformly across the specimen (Sacks and Sun 2003). Research has shown that gripping the tissue specimen with clamps provides a uniform loading of the material fibers versus discrete loads from sutures (Waldman and Michael Lee 2002, Waldman and Lee 2005). To prevent slipping of the tissue, the clamps in this study had a serrated grip.

Assumptions. The assumptions made to calculate the material properties are that the soft tissue is incompressible and has a uniform thickness. The assumption that soft tissue is incompressible has been established by previous researchers for uterine muscle tissue (Deyer et al. 2000). A Poisson's ratio of 0.5 was used for this study. If the actual Poison's ratio were less than this, it would result in an overestimate of the actual tissue thickness. Therefore, the actual stresses would be slightly higher than reported in this paper.

A necessary assumption to calculate stresses at the center of the specimen is that there is an even load transmission from the grip to the center of the specimen. It is not possible to quantify the edge effects from the grips and consider those when scaling the load to the central region where the optical markers are present. Thus, the load at the grip is simply scaled linearly using the length relationship between the central markers to the width where the specimen arm is connected. Previous biaxial tissue research using small angle light scattering mapped collagen fiber orientations in specimens during loading with both clamps and sutures as the gripping mechanism (Waldman et al. 2002). While the specimen shape was square and not crucible, both gripping mechanisms resulted in edge effects that made the stress distribution in the center of the specimen not uniform. Previous research with crucible shaped specimens demonstrates that not compensating for edge effects causes a small over prediction of stress (Shah et al. 2006).

Future Applications

Studies on uterine tissue properties prior to the current study focused on uniaxial tension response with quasi-static loading conditions. Uterine tissue response at dynamic rates for biaxial loading will advance the design of pregnant occupant models to represent more accurately *in vivo* pregnant uterine tissue response to high rate impacts. Uterine tissue loading information combined with research on placental material properties will augment the current methods of predicting placental abruption. Placental abruption, failure of the uteroplacental interface, occurs much more often than uterine rupture. The peak stress data for the uterine wall will give researchers insight on uteroplacental failure properties since that tissue has been too difficult to acquire for materials testing. Moreover, updating the material properties in the uterine wall for pregnant occupant models will provide a more accurate non-linear response to dynamic impacts. Updated pregnant computational models can also aid in the development of pregnant crash test dummies. Additionally, parametric studies with pregnant computational models can provide a method to define global injury measures that correspond to local tissue failure. These global injury measures such as deflection or pressure can then be used with pregnant crash test surrogates.

CONCLUSION

Currently pregnant occupant computational models and crash test dummies are designed from limited available uterine tissue material data. The current study presents stress-strain data for pregnant porcine uterine tissue when loaded biaxially in tension at a dynamic rate. A total of 23 tests were completed and analyzed to develop stress-strain curves and peak strain data for pregnant porcine uterine tissue. At a loading rate of 0.7 strains/s, the circumferential direction peak stress component is 500 ± 219 kPa with a corresponding peak strain 0.30 ± 0.09. The peak stress for the longitudinal direction is 320 ± 176 kPa with a peak strain 0.30 ± 0.09. The results presented in this paper will be useful in modeling pregnant uterine tissue response to dynamic loading and evaluating the risk of placental abruption due to strain in the uterus.

ACKNOWLEDGMENTS

The authors wish to acknowledge Toyota Motor Corporation for providing the funding for this research. The authors also wish to acknowledge Animal Technologies for their assistance in providing animal tissue for use in this study as well as Dr. Raffaella De Vita of Virginia Tech for her guidance and assistance with soft tissue mechanics.

REFERENCES

Attico, N.B., Smith, R.J., FitzPatrick, M.B., and Keneally, M. (1986) Automobile safety restraints for pregnant women and children. The Journal of Reproductive Medicine 31(3): 187-92.

Bass, C.R., Darvish, K., Bush, B., Crandall, J.R., Srinivasan, S.C., Tribble, C., Fiser, S., Tourret, L., Evans, J.C., Patrie, J., and Wang, C. (2001) Material properties for modeling traumatic aortic rupture. Stapp Car Crash J 45: 143-60.

Bass, E.C., Ashford, F.A., Segal, M.R., and Lotz, J.C. (2004) Biaxial testing of human annulus fibrosus and its implications for a constitutive formulation. Annals of Biomedical Engineering 32(9): 1231-42.

Conrad, J.T., Kuhn, W. (1967) The active length-tension relationship in human uterine muscle. American Journal of Obstetrics and Gynecology 97(2): 154-60.

Csapo, A., Goodall, M. (1954) Excitability, length tension relation and kinetics of uterine muscle contraction in relation to hormonal status. The Journal of Physiology 126(2): 384-95.

Cunningham, F.G., Williams, J.W. (2005) Williams obstetrics. McGraw-Hill Professional, New York.

Deyer, T.W., Ashton-Miller, J.A., Van Baren, P.M., and Pearlman, M.D. (2000) Myometrial contractile strain at uteroplacental separation during parturition. American Journal of Obstetrics and Gynecology 183(1): 156-9.

Duma, S.M., Cormier, J.M., Hurst, W.J., Stitzel, J.D., and Herring, I.P. (2003) The effects of Seam Design on Airbag Induced Skin Abrasions from High-Rate Shear Loading. The Proceedings of the IRCOBI Conference: 95-107.

Duma, S.M., Moorcroft, D.M., Gabler, H.C., Manoogian, S.M., Stitzel, J.D., and Duma, G.G. (2006) Analysis of pregnant occupant crash exposure and the potential effectiveness of four-point seatbelts in far side crashes. The Proceedings of the 50th Association for the Advancement of Automotive Medicine Conference 50: 177-88.

Fatality Analysis Reporting System. (2005) National Highway Traffic Safety Administration.

Finn, C.A., Porter, D.G. (1975) The uterus. Elek, London.

Flynn, D.M., Peura, G.D., Grigg, P., and Hoffman, A.H. (1998) A finite element based method to determine the properties of planar soft tissue. Journal of Biomechanical Engineering 120(2): 202-10.

Fung, Y.C. (1967) Elasticity of soft tissues in simple elongation. American Journal of Physiology 213(6): 1532-44.

Humphrey, J.D., Vawter, D.L., and Vito, R.P. (1987) Quantification of strains in biaxially tested soft tissues. Journal of Biomechanics 20(1): 59-65.

Ivarsson, J., Viano, D.C., Lovsund, P., and Aldman, B. (2000) Strain relief from the cerebral ventricles during head impact: experimental studies on natural protection of the brain. Journal of Biomechanics 33(2): 181-9.

Klinich, K.D., Schneider, L.W., Eby, B., Rupp, J.D., and Pearlman, M.D. (1999a) Seated Anthropometry During Pregnancy. University of Michigan Transportation Research Institute, UMTRI-99-16.

Klinich, K.D., Schneider, L.W., Moore, J.L., and Pearlman, M.D. (1999b) Investigations of Crashes Involving Pregnant Occupants. University of Michigan Transportation Research Institute, UMTRI-99-29.

Langdon, S.E., Chernecky, R., Pereira, C.A., Abdulla, D., and Lee, J.M. (1999) Biaxial mechanical/structural effects of equibiaxial strain during crosslinking of bovine pericardial xenograft materials. Biomaterials 20(2): 137-53.

Lanir, Y., Lichtenstein, O., and Imanuel, O. (1996) Optimal design of biaxial tests for structural material characterization of flat tissues. Journal of Biomechanical Engineering 118(1): 41-7.

Lessley, D., Crandall, J., Shaw, G., Kent, R., and Funk, J. (2004) A Normalization Technique for Developing Corridors from Individual Subject Responses. SAE Technical Paper Series 2004-01-0288: 1-11.

Manoogian, S.J., Duma, S.M., and Moorcroft, D.M. (2007a) Pregnant Occupant Injury Risk Using Computer Simulations with NCAP Vehicle Crash Test Data. The 20th International Technical Conference on the Enhanced Safety of Vehicles Conference, Lyon, France.

Manoogian, S.J., McNally, C., Calloway, B.R., and Duma, S.M. (2007b) Methodology for dynamic biaxial tension testing of pregnant uterine tissue. Biomedical Sciences Instrumentation 43: 230-5.

Manoogian, S.J., McNally, C., Calloway, B.R., Duma, S.M., and Mertz, H.L. (2007c) Utilizing cryogenic grips for dynamic tension testing of human placenta tissue. Biomedical Sciences Instrumentation 43: 354-9.

Marrable, A.W. (1971) The embryonic pig: a chronological account. Pitman Medical, London,.

Martin, P.G. (2000) Properties of Human Skin. PhD Dissertation, University of Virginia, Charlottesville.

Mizrahi, J., Karni, Z. (1975) A Mechanical Model for Uterine Muscle Activity during Labor and Delivery. Israel Journal of Technology 13: 185-191.

Mizrahi, J., Karni, Z., and Polishuk, W.Z. (1980) Isotropy and anisotropy of uterine muscle during labor contraction. Journal of Biomechanics 13(3): 211-8.

Monch, E., Galster, D. (1963) A method for producing a defined uniform biaxial tensile stress field. British Journal of Applied Physics 14: 810-812.

Moorcroft, D.M., Duma, S.M., Stitzel, J.D., and Duma, G.G. (2003a) Computational model of the pregnant occupant: predicting the risk of injury in automobile crashes. American Journal of Obstetrics and Gynecology 189(2): 540-4.

Moorcroft, D.M., Duma, S.M., Stitzel, J.D., and Duma, G.G. (2003b) The Effects of Uterine Ligaments on the Fetal Injury Risk in Frontal Automobile Crashes. Journal of Automobile Engineering Part D 217: 1049-1055.

Moorcroft, D.M., Duma, S.M., Stitzel, J.D., and Duma, G.G. (2003c) A finite element and multi-body model of the pregnant occupant for the analysis of restraint effectiveness. SAE Technical Paper Series 2003-01-0157: 1-11.

Moorcroft, D.M., Duma, S.M., Stitzel, J.D., and Duma, G.G. (2004) The effect of pregnant occupant position and belt placement on the risk of fetal injury. SAE Technical Paper Series 2004-01-0324: 1-8.

Ohara, T. (1953) On the comparison of strengths of the various organ-tissues. Journal of Kyoto Preferred Medical University 53: 577-597.

Pearlman, M.D. (1997) Motor vehicle crashes, pregnancy loss and preterm labor. International Journal of Gynaecology and Obstetrics 57(2): 127-32.

Pearlman, M.D. (1999) Data Acquisition for Development of a Uteroplacental Interface for the Second-Generation Pregnant Abdomen. University of Michigan Health Systems, D.8. Final Report.

Pearlman, M.D., Tintinalli, J.E., and Lorenz, R.P. (1990a) Blunt trauma during pregnancy. The New England Journal of Medicine 323(23): 1609-13.

Pearlman, M.D., Tintinallli, J.E., and Lorenz, R.P. (1990b) A prospective controlled study of outcome after trauma during pregnancy. American Journal of Obstetrics and Gynecology 162(6): 1502-10.

Pearsall, G.W., Roberts, V.L. (1978) Passive mechanical properties of uterine muscle (myometrium) tested in vitro. Journal of Biomechanics 11(4): 167-76.

Ramsey, E.M. (1975) The placenta of laboratory animals and man. Holt, Rinehart and Winston, New York.

Rupp, J.D., Klinich, K.D., Moss, S., Zhou, J., Pearlman, M.D., and Schneider, L.W. (2001a) Development and testing of a prototype pregnant abdomen for the small-female hybrid III ATD. Stapp Car Crash Journal 45: 61-78.

Rupp, J.D., Schneider, L.W., Klinich, K.D., Moss, S., Zhou, J., and Pearlman, M.D. (2001b) Design, Development, and Testing of a New Pregnant Abdomen for the Hybrid III Small Female Crash Test Dummy. University of Michigan Transportation Research Institute, UMTRI-2001-07.

Sacks, M.S. (2000) Biaxial Mechanical Evaluation of Planar Biological Materials. Journal of Elasticity 61: 199-246.

Sacks, M.S., Sun, W. (2003) Multiaxial mechanical behavior of biological materials. Annual Review of Biomedical Engineering 5: 251-84.

Shah, C.S., Hardy, W.N., Mason, M.J., Yang, K.H., Van Ee, C.A., Morgan, R., and Digges, K. (2006) Dynamic biaxial tissue properties of the human cadaver aorta. Stapp Car Crash Journal 50: 217-46.

Shah, C.S., Mason, M.J., Yang, K.H., Hardy, W.N., Van Ee, C.A., Morgan, R., and Digges, K. (2005) High-Speed Biaxial Tissue Properties of the Human Cadaver Aorta. ASME International Mechanical Engineering Congress 2005-82085: 1-9.

Sun, W., Sacks, M.S., and Scott, M.J. (2005) Effects of boundary conditions on the estimation of the planar biaxial mechanical properties of soft tissues. Journal of Biomechanical Engineering 127(4): 709-15.

Tandon, G.P., Kim, R.Y., and Bechel, V.T. (2002) Fiber-matrix interfacial failure characterization using a cruciform-shaped specimen. Journal of Computational Materials Science 36: 2667-2691.

Van Ee, C.A., Chasse, A.L., and Myers, B.S. (2000) Quantifying skeletal muscle properties in cadaveric test specimens: effects of mechanical loading, postmortem time, and freezer storage. Journal of Biomechanical Engineering 122(1): 9-14.

Waldman, S.D., Lee, J.M. (2005) Effect of sample geometry on the apparent biaxial mechanical behaviour of planar connective tissues. Biomaterials 26(35): 7504-13.

Waldman, S.D., Michael Lee, J. (2002) Boundary conditions during biaxial testing of planar connective tissues. Part 1: dynamic behavior. Journal of Materials Science: Materials in Medicine 13(10): 933-8.

Waldman, S.D., Sacks, M.S., and Michael Lee, J. (2002) Boundary conditions during biaxial testing of planar connective tissues. Part 2: Fiber orientation. Journal of Materials Science Letters 21: 1215-1221.

Weiss, H.B. (1999) The Epidemiology of Perinatal Traumatic Fetal Injury Mortality. PhD Dissertation, University of Pittsburgh, Pittsburgh.

Weiss, H.B., Strotmeyer, S. (2002) Characteristics of pregnant women in motor vehicle crashes. Injury Prevention 8(3): 207-211.

Weiss, J.A., Maker, B.N., and Govindjee, S. (1996) Finite element implementation of incompressible, transversely isotropic hyperelasticity. Computer Methods in Applied Mechanics and Engineering 135: 107-128.

Weiss, S., Jaermann, T., Schmid, P., Staempfli, P., Boesiger, P., Niederer, P., Caduff, R., and Bajka, M. (2006) Three-dimensional fiber architecture of the nonpregnant human uterus determined ex vivo using magnetic resonance diffusion tensor imaging. The Anatomical Record Part A 288(1): 84-90.

Wineman, A., Wilson, D., and Melvin, J.W. (1979) Material identification of soft tissue using membrane inflation. J Biomech 12(11): 841-50.

Wood, C. (1964) Physiology of Uterine Contractions. Journal of Obstetrics and Gynaecology of the British Commonwealth 71: 360-373.

Yamada, H. (1970) Strength of Biological Materials. F. G. Evans. Williams and Wilkins, Baltimore.

Young, R.C., Hession, R.O. (1999) Three-dimensional structure of the smooth muscle in the term-pregnant human uterus. Obstetrics and Gynecology 93(1): 94-9.

Effect of Strain Rate on the Tensile Material Properties of Human Placenta

Sarah J. Manoogian
e-mail: manoogsj@vt.edu

Jill A. Bisplinghoff

Craig McNally

Andrew R. Kemper

Anthony C. Santago

Stefan M. Duma[1]
e-mail: duma@vt.edu

Center for Injury Biomechanics,
Virginia Tech-Wake Forest,
Blacksburg, VA 24061

Automobile crashes are the largest cause of injury death for pregnant females and the leading cause of traumatic fetal injury mortality in the United States. Computational models, useful tools to evaluate the risk of fetal loss in motor vehicle crashes, are based on a limited number of quasistatic material tests of the placenta. This study presents a total of 64 uniaxial tensile tests on coupon specimens from six human placentas at three strain rates. Material properties of the placental tissue were evaluated at strain rates of 0.07/s, 0.70/s, and 7.00/s. The test data have average failure strains of 0.34, 0.36, and 0.37, respectively. Failure stresses of 10.8 kPa, 11.4 kPa, and 18.6 kPa correspond to an increase in strain rate from 0.07/s to 7.0/s. The results indicate rate dependence only when comparing the highest strain rate of 7.0/s to either of the lower rates. There is no significant rate dependence between 0.07/s and 0.70/s. When compared with previous testing of placental tissue, the current study addresses the material response to more strain rates as well as provides a much larger set of available data. In summary, tensile material properties for the placenta have been determined for use in computational modeling of pregnant occupant kinematics in events ranging from low impact activities to severe impacts such as in motor vehicle crashes. [DOI: 10.1115/1.3194694]

1 Introduction

Automobile crashes are the largest cause of injury death for pregnant females [1] and the leading cause of traumatic fetal injury mortality in the United States (U.S.) [2]. In a 3 year study of fetal traumatic injury deaths in select states, 82% were a result of maternal injury in a motor vehicle crash (MVC) [3]. An average estimate from several sources indicates that each year in the U.S., approximately 160 pregnant women are killed in motor vehicle crashes and an additional 600–2600 fetuses are killed when the mother survives [3–6]. The first way to protect the fetus is to protect the mother considering that maternal death has a near 100% fetal loss rate [7]. If the mother survives, protection of the fetus may be best accomplished by preventing placental abruption. Placental abruption, which is the premature separation of the placenta from the uterus, was shown to account for 50–70% of fetal losses in motor vehicle crashes [8]. The information gathered from crash investigations shows that placental abruption can occur without other, more severe injuries, such as uterine rupture or direct fetal injury. However, when these more severe injuries do occur they are often accompanied by placental abruption [9]. The different material properties of the uterus and the placenta are one reason they separate naturally after birth when the uterus contracts and the placenta maintains its shape [6]. A similar mechanism can occur prematurely in a motor vehicle crash causing a local failure of the uteroplacental interface [6].

Since real world crash data for pregnant occupants are limited to a retrospective analysis and pregnant cadaver studies are not feasible, crash test dummies and computational modeling have been utilized to evaluate the risk of adverse fetal outcome. The most current version of a pregnant female crash test dummy is called the maternal anthropomorphic measurement apparatus version 2B (MAMA-2B) [9]. This pregnant surrogate is a 30 week gestation fifth percentile female crash test dummy. Validation tests have related the peak pressure inside the simulated uterus in the MAMA-2B with real world fetal loss risk data. The pregnant surrogate has been used in further analysis to determine the risk of fetal loss for different restraint conditions in a frontal impact. Previous studies have utilized a validated mathematical dynamic models (MADYMO) computer model of a 30 week pregnant occupant to assess the risk of fetal loss based on occupant position, belt placement, impact direction, advanced restraint systems, and standard frontal barrier impacts [10–15]. Although pregnant occupant research has progressed with these tools, they are based on tissue data with limited availability [9,10]. In order to have more accurate research tools, better pregnant tissue material data are needed.

The microstructure of the human placenta is important in modeling the material response to dynamic tensile loading. The human placenta connects to the uterine wall, usually in the fundus of the uterus, and provides a means for the maternal-fetal exchange of gases and nutrients [16]. However, because of the different roles it serves for the mother and fetus, there are microstructure differences in the maternal and fetal sides of the placenta [17]. The fetal side of the placenta includes the chorionic plate. This layer of the placenta is filled with fetal blood vessels that originate at the umbilical cord and radiate outward the edge of the placenta in a dense network. The maternal side of the placenta is a network of villi and intervillus spaces.

Recent research has analyzed the dynamic response of placenta tissue. From this study it was determined that the maternal and fetal sides of the placenta have different tensile material properties at a dynamic loading rate [18]. However, placental abruption occurs between the maternal side of the placenta and the uterine wall at the uteroplacental interface. Therefore, this study will focus on the material properties of the maternal side of the placenta. Although previous research has addressed the material response of placental tissue at dynamic and quasistatic loading rates, a larger sample size of material tests, which addresses the rate dependence of the tissue, is necessary. The purpose of this study is to obtain material properties for the maternal side of the placenta at three strain rates. This information will advance the ability of pregnant occupant models to accurately model the response of the placenta in a variety of impact conditions.

[1]Corresponding author.

Contributed by the Bioengineering Division of ASME for publication in the Journal of Biomechanical Engineering. Manuscript received June 25, 2008; final manuscript received April 14, 2009; published online August 7, 2009. Review conducted by Susan Margulies.

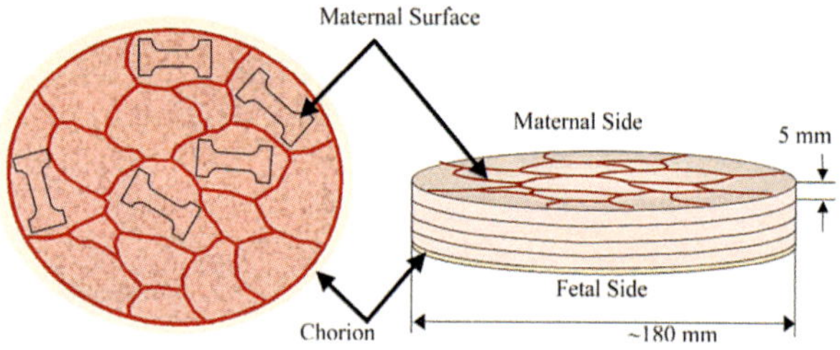

Fig. 1 Each placenta is sectioned into 5 mm slices

2 Methods

A total of 64 uniaxial tensile tests was performed on six whole placentas. Donor tissue followed the Wake Forest University Baptist Medical Center Institutional Review Board informed consent procedures. Following delivery, the tissue was stored in a saline solution and tested within 36 h. Each placenta had three to five tests at each of the three strain rates. The three strain rates chosen for this study were 0.07/s, 0.70/s, and 7.00/s. These rates represent loading from a low level impact up to the level of strain rate expected in a motor vehicle crash as determined from previous computational modeling [12].

All of the uniaxial tensile tests were performed on coupon specimens from the maternal surface of the placenta. Thin slices of placenta were acquired from the whole placenta using a tissue slicer with a wet rotating blade. This allowed the maternal surface of the placenta to be sectioned into 5 mm slices and separated from the chorion layer of the placenta (Fig. 1). Tissue specimens were acquired in the plane of the tissue parallel to the maternal and fetal surfaces. Because the placenta function requires it to have a lobe structure, the slices of the placenta had discontinuities. Sections with a continuous structure were chosen for acquiring a smaller coupon section. Each piece of tissue was aligned before stamping to ensure uniformity as much as possible with each sample. Guide rods provided a means for aligning both a stencil and ultimately the stamp. A coupon was then stamped out of the tissue using a steel bent stamp. A sharpened edge to the steel stamp minimized sheer stress in the tissue during the stamping procedure. The result of the tissue stamping procedure was several coupons from the same donor tissue with a uniform shape (Fig. 2). Each coupon was kept hydrated with a saline solution throughout the preparation and prior to testing. The tests were performed at room temperature. The coupon specimens have an average cross-sectional area of 28.4 ± 7.1 mm.

A custom designed system of linear motors was used for uniaxial dynamic tensile tests of placental tissue coupons (Fig. 3). The system included one multiaxis controller (Parker ACR9000, Irwin, PA) that served as a power supply for a motor driver (Parker ViX, Irwin, PA), which controlled the two linear motors

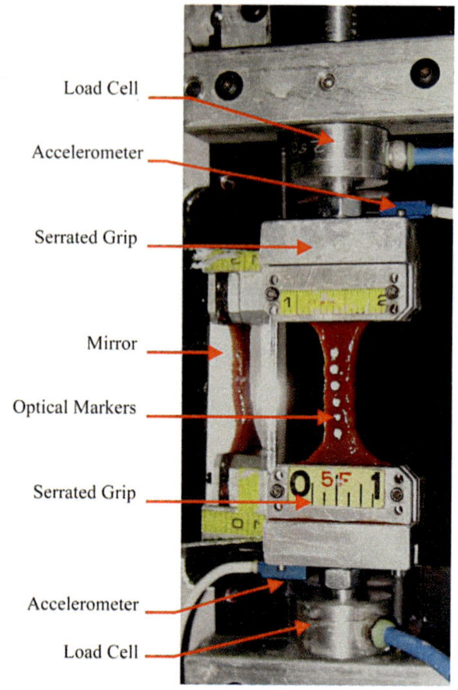

Fig. 3 The specimen is mounted between two serrated grips, which are instrumented with both a load cell and an accelerometer while high-speed video records the test event

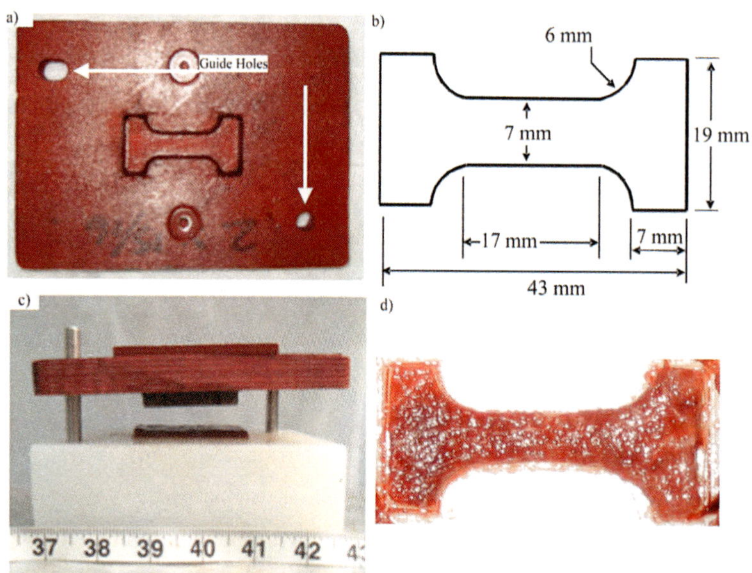

Fig. 2 (*a*) A steel bent stamp is used to cut the tissue into a dog bone shape, (*b*) the stamp provides the uniform tissue samples, (*c*) guide rods are used to align the stamp with the specimen, and (*d*) the coupon shape is cut where the tissue has a uniform consistency

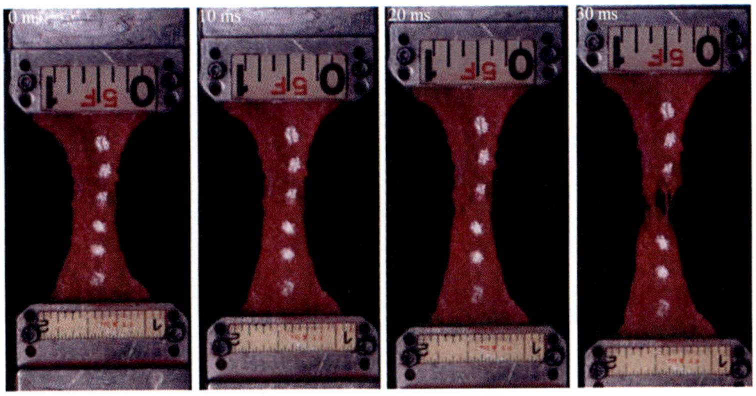

Fig. 4 The video captures of one test show a typical failure of a placental specimen pulled in uniaxial tension at 7.0/s

with stages (Parker Daedal MX80S, Irwin, PA). The two individual stages were each instrumented with a load cell (OMEGA LC10, 10 lbf, Stamford, CT), potentiometer (Space Age Control, 160-1705, 540 mm, Palmdale, CA), and accelerometer (Endevco 7264B, 2000 G, San Juan Capistrano, CA). On each side of the test specimen, the direct path from the tissue clamp went through the load cell and was rigidly connected to the linear motor stage. The inertial load of the mass in front of the load cell was multiplied by the acceleration to compensate the measured reaction load. Synchronous motion of the motors allowed both of the stages to pull the tissue in tension at the same time for an overall velocity, which corresponded to the desired strain rate.

Each specimen was mounted in the testing setup with the same initial conditions. The top grip assembly was removed and laid flat to align the specimen in the grip along the centerline of the loading axis. The grip assemblies consisted of two plates with sandpaper on the inner surfaces. The specimen was placed on the fixed side of the grip and aligned and then the second plate was placed over the tissue. After the grip was secured, the top grip assembly was then reattached to allow the specimen to rest hanging in 1 g tension. By supporting only its own weight, the tissue had a standard but minimal preload condition. With the tissue hanging, the specimen was clamped in the bottom grip with no additional preload. Again, the grip was lined with sandpaper to hold the tissue in place. Optical markers were applied to the tissue after the specimen was mounted.

Local stress and strain were calculated for each specimen. Local deformation was recorded using a high-speed video (Phantom V4, Wayne, NJ) at 500 frames per second with 512 by 512 resolution for the 0.7/s and 7.0/s tests and 70 frames per second with 512 by 512 resolution for the 0.07/s tests. Optical marker tracking was performed with TEMA® advanced motion analysis software (Linköping, Sweden). Using the configuration for these tests, the error in the motion analysis is ± 0.030 mm. The optical markers on either side of the location where the specimen tore were used to calculate the local strain. In addition, a mirror aligned at a 45 deg angle was mounted next to the specimen so that the side of the specimen could be viewed in the high-speed video. The measured displacement data were fit with a fifth order polynomial. The average R^2 value for the displacement curve fit is 0.999 ± 0.001. Similarly, the measured force data were fit with a fifth order polynomial. The average R^2 value for the force curve fit is 0.991 ± 0.013. The local stress was acquired by determining the location on the specimen where failure occurred and then taking the width and thickness of the specimen in that location from the pretest pictures as the initial cross-sectional area. The error in the pretest pictures is ± 0.12 mm. The cross-sectional area at the time of failure is calculated using the local strain and a Poisson's ratio value of 0.49. This value was determined in previous placental tissue testing [19]. The stress and strain data are reported for each specimen until failure was evident in the video. Although, the tissue could have been still supporting the load, the optical markers were not accurately representing the strain in the tissue once there was subfailure.

Statistical variations were examined between specimens from the three rates. To account for the dependence of rate on the corresponding donor, a linear regression was performed on each response variable. The regression model was chosen within statistical analysis software (SAS) (SAS Institute Inc., Cary, NC) to predict the response variable based on each rate, which was nested within each donor. This accounts for the interaction between rate and donor and allows for F tests to be used to assess the significance of region effects. The analysis of variance (ANOVA) statistics produced using the same models were used to test the null hypothesis that there is no effect of rate for the given response. The ANOVA p-values represent the probability that the effect of a given parameter is not zero.

A standard method of determining the characteristic average was used for each group of tests at one strain rate [20]. This process began with normalizing each strain curve to its respective maximum. Interpolation routines determined the stress for each test at set increments of strain. Stress values for all tests at that rate were averaged to determine one value at each strain. The strain data that correspond to these stress values were obtained by multiplying the normalized strain by the average maximum strain of all tests. The result of this process was an average stress versus strain curve for all tests at 0.07/s, 0.70/s, and 7.0/s. Standard deviations of the stress and strain values were also calculated for each group of tests.

3 Results

The data for dynamic tensile tests at three different strain rates are reported in true strain and true stress values. Frames from the high-speed video of a 7.0/s test show the tissue specimen fails in the gauge length of the dog bone shape coupon (Fig. 4). The stress versus strain data for all tests at 0.07/s have an average failure strain of 0.34 ± 0.07 and a corresponding failure stress of 10.8 ± 5.9 kPa. Similarly, the stress versus strain data for all of the tests at 0.7/s have an average failure strain of 0.36 ± 0.06 and a corresponding failure stress of 11.4 ± 3.7 kPa. The most dynamic strain rate, 7.0/s, has a corresponding average failure stress of 18.6 ± 5.4 kPa and 0.37 ± 0.06 strain.

ANOVA analyses were used to evaluate the failure stresses and strains among the three strain rate groups. There was no significant difference between the failure stresses at 0.07/s and 0.70/s ($p=0.81$). However, the tests at 7.0/s had a significantly higher failure stress than either the test series at 0.07/s or 0.70/s ($p < 0.01$). The average failure stresses for each strain rate group are plotted on a log scale with the standard deviations to show the

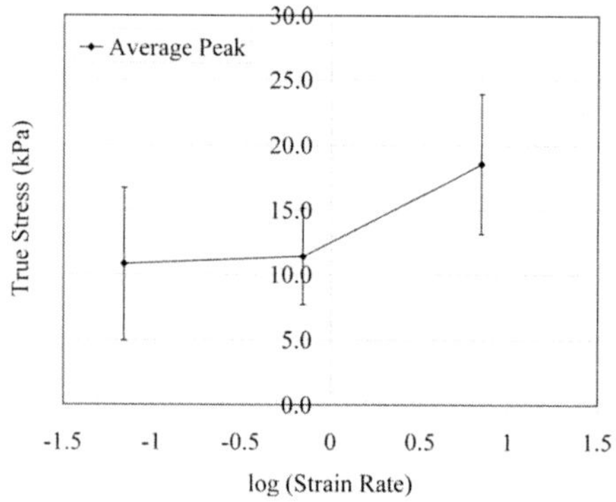

Fig. 5 The placenta failure stress by strain rate

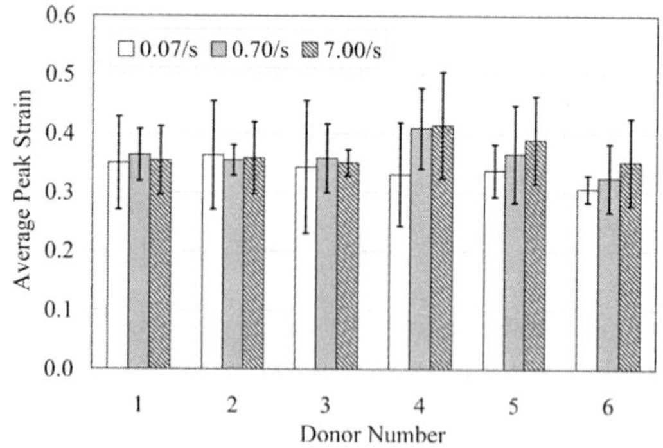

Fig. 8 The average failure strain at each rate is shown for each donor

relationship between strain rate and failure stress (Fig. 5). There were no statistically significant differences among the failure strains for the three groups ($p=0.26$, $p=0.15$, and $p=0.74$). When plotted together on a log scale with the standard deviations, the lack of change in failure strain over the three rates is evident (Fig. 6). The relationship between strain rate and corresponding stress versus strain data is also shown by comparing the characteristic averages for all of the strain rates tested (Fig. 7).

These results can also be categorized by donor. The average failure strains with the corresponding standard deviations are plotted for each donor group separated by strain rate (Fig. 8). For the failure strain values there are no statistically significant differences between any of the rates for any of the donors ($p>0.2$). The average failure stresses with the corresponding standard deviations are also plotted for each donor group separated by strain rate (Fig. 9). These data indicated significant differences between the lowest and highest strain rate for donors 3, 4, and 6 ($p<0.05$). Additionally, there are significant differences between the failure stresses for the intermediate and highest strain rate groups for donors 5 and 6 ($p=0.03$). None of the other relationships had statistical significance ($p>0.05$).

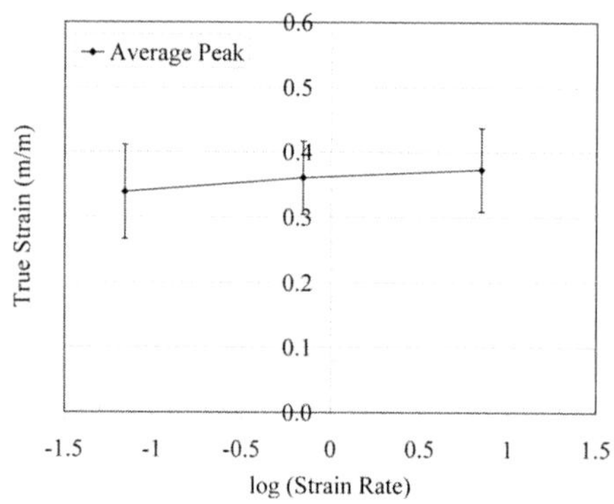

Fig. 6 The placenta failure strain by strain rate

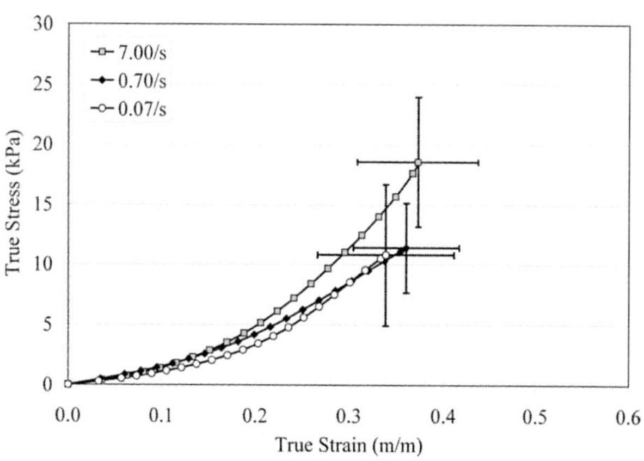

Fig. 7 The characteristic averages for each group of tests are shown together with the standard deviations of the failure values

4 Discussion

The results from this study give stress versus strain loading and failure information for the villous portion of the maternal side of the placenta for three different loading rates. The standard deviations in the failure values are assumed to be dependent on specimen variability since the testing procedure was the same for each specimen. Each test was carefully examined through both video and data analyses. By tracking the optical markers and the load time history, it was ensured that the tissue did not fail at the grip locations.

The microstructure of the placental villi on the maternal side does not have a directional bias of their structure in the plane of the tissue specimens [17]. It is assumed because of the transversely isotropic structure, the material properties would not be directionally dependent. While there are discontinuities, careful consideration was used when stamping the specimens to obtain a uniform gauge length. However, the fibrous structure of the placenta could still contribute to specimen variability. Variability can also occur due to several gestational variables.

Although previous material testing on human placenta is limited, Pearlman [21] tested full thickness placenta in quasistatic tension. These tests were subfailure and only consisted of five specimens. The gauge lengths of the specimens were 30 mm and they were pulled at a rate of 1.6 mm/s. For all five specimens, the average peak strain was 0.43 ± 0.16 and the average peak stress was 15.6 ± 7.6 kPa. The peak stresses from Pearlman's tests are similar to the failure stresses measured in the current study.

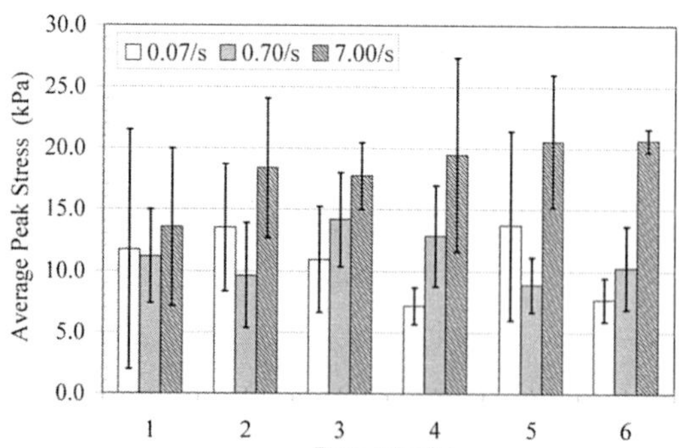

Fig. 9 The average failure stress at each rate is shown for each donor

Additionally, Manoogian et al. [18] performed failure testing on ten placental tissue specimens in dynamic uniaxial tension. These tests determined the material properties of placental tissue with the chorion layer attached at a strain rate of 7.0/s. Both the maternal portion and the placenta with the chorion attached have similar failure true strains of 0.38 and 0.49, respectively. However, the failure stress for the placenta with chorion, 167.8 kPa, was much higher than the failure stress for the placenta with the chorion layer removed, 18.6 kPa. The maternal portion of the placenta, including only the villus structure, has a compliant stress versus strain response and therefore has a lower failure stress and elastic modulus than the chorion specimens. Due to the presence of the dense layer of blood vessels, the failure stresses are much higher for the chorion tissue. Both layers of placental tissue have similar failure strains. There are no data available in literature to quantify the stress relaxation or rate-dependent properties of placental tissues.

For an increase in the dynamic rate of testing of the same viscoelastic tissue it is expected that the failure strain is reduced and the failure stress and elastic modulus are both increased [22]. Although the placenta is a viscoelastic tissue, there was not a large amount of rate dependence over the range tested. This is similar to published results for other tissues. Jin et al. [23] evaluated the bovine pia-arachnoid complex at 0.05/s, 0.50/s, and 5.0/s. Similar to the results in this study, the ultimate stress was significantly different ($p < 0.05$) at 5.0/s than the two slower rates, but over that range there was no significant difference in the ultimate strain recorded ($p > 0.05$). Likewise, previous quasilinear viscoelastic studies on ligaments showed a minimal change in the stress-strain response from 0.01/s to 1.0/s [24,25]. A significant rate-dependent response was reported for tissues when the range of strain rates tested is increased to 100 or 1000/s [23,26]. Strain rates of this magnitude were not predicted in the computational model and therefore were not considered in characterizing the response of the placental tissue to a dynamic impact.

There are limitations with this study. It is assumed that since the placenta tissue maintains its size and shape after delivery, the tests postpartum provide a close model to the performance of the tissue in vivo. Further, the body temperature and blood flow of the tissue in vivo were not modeled in the experiments. In order to model the blood filled organ, the tissue was kept hydrated and saturated prior to testing. The blood flow in vivo would likely increase both the viscoelasticity and incompressibility of the placental tissue. Because the tissue tested is hydrated and the failure mode is tension in the fibers, it is expected that the failure properties in vivo would be similar those measured in this study. The hydrating fluid was a simple saline solution that is used commonly in hydrating human tissues prior to impact testing. Previous researchers evaluated the effect of temperature on material properties of ligaments [27,28]. From those studies they have found that there is not a significant difference in tensile failure properties at room temperature versus body temperature. Because the viscosity of blood increases with a decrease in temperature, it is likely that the placenta would have a less viscous response at body temperature than was measured in the experiments. However, it is assumed that the strength of the fibers that compose the placental tissues would not be greatly affected by the change in temperature. Overall, the effect of blood flow and temperature are expected to have minimal effect on the material properties of the placenta.

5 Conclusion

This study presents a total of 64 uniaxial tensile tests on the villous structure of the placenta. Six human placentas were used to acquire coupon specimens for tensile testing at three strain rates. Material properties of the placental tissue were evaluated at strain rates of 0.07/s, 0.70/s, and 7.00/s. The test data have average failure strains of 0.34, 0.36, and 0.37, respectively. Failure stresses of 10.8 kPa, 11.4 kPa, and 18.6 kPa correspond to an increase in strain rate from 0.07/s to 7.0/s. The results indicate rate dependence only when comparing the highest strain rate of 7.0/s to either of the lower rates. There is no significant rate dependence between 0.07/s and 0.70/s. When compared with previous testing of placental tissue, the current study addresses the material response to more strain rates and provides a much larger set of available data. In summary, tensile material properties for the placenta have been determined for use in computational modeling of pregnant occupant kinematics in events ranging from low impact activities to severe impacts such as in motor vehicle crashes.

Acknowledgment

The authors thank the Ford Motor Co. for sponsoring this research.

References

[1] Attico, N. B., Smith, R. J., FitzPatrick, M. B., and Keneally, M., 1986, "Automobile Safety Restraints for Pregnant Women and Children," J. Reprod. Med., **31**(3), pp. 187–192.

[2] Weiss, H. B., and Strotmeyer, S., 2002, "Characteristics of Pregnant Women in Motor Vehicle Crashes," Inj. Prev., **8**(3), pp. 207–211.

[3] Weiss, H. B., 1999, *The Epidemiology of Perinatal Traumatic Fetal Injury Mortality*, University of Pittsburgh, Pittsburgh, PA.

[4] Klinich, K. D., Schneider, L. W., Eby, B., Rupp, J. D., and Pearlman, M. D., 1999, "Seated Anthropometry During Pregnancy," University of Michigan Transportation Research Institute, Report No. UMTRI-99-16.

[5] Klinich, K. D., Schneider, L. W., Moore, J. L., and Pearlman, M. D., 1999, "Investigations of Crashes Involving Pregnant Occupants," University of Michigan Transportation Research Institute, Report No. UMTRI-99-29.

[6] Pearlman, M. D., 1997, "Motor Vehicle Crashes, Pregnancy Loss and Preterm Labor," Int. J. Gynaecol. Obstet., **57**(2), pp. 127–132.

[7] Pearlman, M. D., Tintinalli, J. E., and Lorenz, R. P., 1990, "Blunt Trauma During Pregnancy," N. Engl. J. Med., **323**(23), pp. 1609–1613.

[8] Pearlman, M. D., Tintinallli, J. E., and Lorenz, R. P., 1990, "A Prospective Controlled Study of Outcome After Trauma During Pregnancy," Am. J. Obstet. Gynecol., **162**(6), pp. 1502–1510.

[9] Rupp, J. D., Klinich, K. D., Moss, S., Zhou, J., Pearlman, M. D., and Schneider, L. W., 2001, "Development and Testing of a Prototype Pregnant Abdomen for the Small-Female Hybrid III ATD," Stapp Car Crash J., **45**, pp. 61–78.

[10] Moorcroft, D. M., Duma, S. M., Stitzel, J. D., and Duma, G. G., 2003, "Computational Model of the Pregnant Occupant: Predicting the Risk of Injury in Automobile Crashes," Am. J. Obstet. Gynecol., **189**(2), pp. 540–544.

[11] Moorcroft, D. M., Duma, S. M., Stitzel, J. D., and Duma, G. G., 2003, "The Effects of Uterine Ligaments on the Fetal Injury Risk in Frontal Automobile Crashes," Proc. Inst. Mech. Eng., Part D (J. Automob. Eng.), **217**, pp. 1049–1055.

[12] Moorcroft, D. M., Duma, S. M., Stitzel, J. D., and Duma, G. G., 2003, "A Finite Element and Multi-Body Model of the Pregnant Occupant for the Analysis of Restraint Effectiveness," SAE Technical Paper Series No. 2003-01-0157.

[13] Moorcroft, D. M., Duma, S. M., Stitzel, J. D., and Duma, G. G., 2004, "The Effect of Pregnant Occupant Position and Belt Placement on the Risk of Fetal Injury," SAE Technical Paper Series No. 2004-01-0324.

[14] Duma, S. M., Moorcroft, D. M., Gabler, H. C., Manoogian, S. M., Stitzel, J. D., and Duma, G. G., 2006, "Analysis of Pregnant Occupant Crash Exposure and the Potential Effectiveness of Four-Point Seatbelts in Far Side Crashes," *The Proceedings of the 50th Association for the Advancement of Automotive Medicine Conference*, Vol. 50, pp. 177–188.

[15] Manoogian, S. J., Duma, S. M., and Moorcroft, D. M., 2007, "Pregnant Occupant Injury Risk Using Computer Simulations With NCAP Vehicle Crash Test Data," The 20th International Technical Conference on the Enhanced Safety of Vehicles Conference, Lyon, France.

[16] Ramsey, E. M., 1975, *The Placenta of Laboratory Animals and Man*, Holt, Rinehart and Winston, New York.

[17] Cunningham, F. G., and Williams, J. W., 2005, *Williams Obstetrics*, McGraw-Hill Professional, New York.

[18] Manoogian, S. J., Bisplinghoff, J., McNally, C., Kemper, A. R., Santago, A. C., and Duma, S. M., 2008, "Dynamic Tensile Properties of Human Placenta," J. Biomech. **41**(16), pp. 3436–3440.

[19] Manoogian, S. J., McNally, C., Calloway, B. R., Duma, S. M., and Mertz, H. L., 2007, "Utilizing Cryogenic Grips for Dynamic Tension Testing of Human Placenta Tissue," Biomed. Sci. Instrum., **43**, pp. 354–359.

[20] Lessley, D., Crandall, J., Shaw, G., Kent, R., and Funk, J., 2004, "A Normalization Technique for Developing Corridors From Individual Subject Responses," SAE Technical Paper Series No. 2004-01-0288.

[21] Pearlman, M. D., 1999, *Data Acquisition for Development of a Uteroplacental Interface for the Second-Generation Pregnant Abdomen*, University of Michigan Health Systems, Ann Arbor, MI.

[22] Yamada, H., 1970, *Strength of Biological Materials*, Williams and Wilkins, Baltimore, MD.

[23] Jin, X., Lee, J. B., Leung, L. Y., Zhang, L., Yang, K. H., and King, A. I., 2006,

"Biomechanical Response of the Bovine Pia-Arachnoid Complex to Tensile Loading at Varying Strain-Rates," Stapp Car Crash J., **50**, pp. 637–649.

[24] Funk, J. R., Hall, G. W., Crandall, J. R., and Pilkey, W. D., 2000, "Linear and Quasi-Linear Viscoelastic Characterization of Ankle Ligaments," ASME J. Biomech. Eng., **122**(1), pp. 15–22.

[25] Woo, S. L., Gomez, M. A., and Akeson, W. H., 1981, "The Time and History-Dependent Viscoelastic Properties of the Canine Medical Collateral Ligament," ASME J. Biomech. Eng., **103**(4), pp. 293–298.

[26] van Dommelen, J. A. W., Ivarsson, J., Jolandan, M. M., Millington, S. A., Raut, M., Kerrigan, J. R., Crandall, J. R., and Diduch, D. R., 2005, "Characterization of the Rate-Dependent Mechanical Properties and Failure of Human Knee Ligaments," SAE Technical Paper Series No. 2005-01-0293.

[27] Hasberry, S., and Pearcy, M. J., 1986, "Temperature Dependence of the Tensile Properties of Interspinous Ligaments of Sheep," ASME J. Biomed. Eng., **8**(1), pp. 62–66.

[28] Lam, T. C., Thomas, C. G., Shrive, N. G., Frank, C. B., and Sabiston, C. P., 1990, "The Effects of Temperature on the Viscoelastic Properties of the Rabbit Medial Collateral Ligament," ASME J. Biomech. Eng., **112**(2), pp. 147–152.

Finite Element Modeling of Human Placental Tissue

Mao Yu[1,2], Sarah Manoogian[3], Stefan M. Duma[2], Joel D. Stitzel[1,2]
[1]Wake Forest University School of Medicine
[2]Virginia Tech – Wake Forest University Center for Injury Biomechanics
[3]Biodynamic Research Corporation

ABSTRACT – Motor vehicle crashes account for a large portion of placental abruption and fetal losses. To better understand the material properties of the human placenta, a Finite Element (FE) model of human placenta tissue was created and verified using data from uniaxial tension tests. Sixty-four tensile tests at three different strain rates of 7% strain/s, 70% strain/s, and 700% strain/s from six whole human placentas were used for model development. Nominal stresses were calculated by dividing forces at the grips by the original cross-sectional area. Nominal strains were calculated by dividing cross-head displacement by the original gauge length. A detailed methodology for interpreting experimental data for application to material model development is presented. A model of the tension coupon was created in LS-DYNA and stretched in the same manner as the uniaxial tension tests. The behavior of the material was optimized to the uniaxial tension test using a multi-island genetic algorithm. The results demonstrate good correlation between experiments and the model, with an average difference of 2% between the optimized FE and experimental first principal stress at the termination state. The material parameters found in this study can be utilized in FE models of placental tissues for behavior under dynamic loading.

INTRODUCTION

Motor vehicle crashes are the largest cause of death for pregnant females and the leading cause of traumatic fetal injury mortality in the United States (Weiss, 2001a, Weiss, 2001b, Agran et al., 1987). Specifically, placental abruption accounts for up to 70% of these fetal losses. Therefore, the purpose of this study is to characterize human placental tissues in order to ultimately develop FE models of pregnant occupants.

While there have been many computational efforts on the pregnant occupant (Moorcroft et al., 2003, Duma et al., 2004, Duma et al., 2006, Manoogian et al., 2008b, Delotte et al., 2006), computer modeling of the placenta has not had much precedent. Moorcroft's computer model of the pregnant occupant had a simple linear elastic placenta material, but the placenta material was not the focus of the study (Moorcroft, 2003).

There have been some experimental tests of placental tissue. Pearlman (Pearlman, 2000) performed tensile testing of the placenta at a 100 mm/min displacement rate and found the average Young's modulus to be 32 kPa. Benson-Martin (Benson-Martin et al., 2006) performed tensile testing of the placenta at a 10 mm/min displacement rate for 5 cm x 2 cm samples and found the Young's modulus to be 2.29 MPa. The discrepancy in the Young's moduli between these two studies shows that further investigation into the material is needed, as well as exploration of the effects of differing strain rates, since these two studies did not use comparable strain rates.

In any FE simulation, one of the most significant inputs which affects the realism of the results is the material model. While linear elastic models have fairly robust material formulations and are often used, they represent an ideal that is not physically realistic for placental tissue. The material model created should not be overly simple and it should be validated with experimental tests.

Correlating experimental data with FE material behavior becomes especially important when simulating soft tissues such as the placenta. While most FE material models are widely used for structural analysis of inorganic objects and have been well validated, the behaviors of biological soft tissues are not understood as well. As a result, many of the biological soft tissue material models used in simulations are based upon hyperelastic models that have been fitted to experimental data. The usage of these hyperelastic materials is justified often with the fact that the material parameters are fitted to the real experimental stress vs. strain curve. However, the material parameters for the stress vs. strain curves are only one aspect of the behavior of the model. Effects of strain rate, stress relaxation, and compressibility must also be accounted for. Therefore, an iterative approach that optimizes the error between the fitted model behavior and the actual experimental behavior

Reprinted from Annuals of the Association of Automotive Medicine, Vol. 53, pp 257-270, 2009.

has the potential to generate more accurate material parameters.

A material model for the placenta was created by attempting to match the nominal stress and nominal strain in a coupon tension test in the FE simulation with the same measurements in the actual experiment. In addition to optimizing the model parameters to reduce error between experimental tests and the FE model, this study also used a material model that employs a direct table lookup method to calculate the stress vs. strain curve.

METHODS

The study is composed of three parts: experimental material testing, FE mesh creation, and optimization of material parameters. The methods described below detail each of these steps.

1. Experimental material testing

The experimental data is based on the work of Manoogian (Manoogian et al., 2008a) where the placenta was tested without the chorion attached. The experimental methods given here represent a brief overview of that work. The focus of this study is the interpretation and utilization of that data in the development of an FE material model, not the actual experimental procedures.

A total of 64 uniaxial tensile tests were performed on six whole human placentas (Manoogian et al., 2008a). Following delivery, the tissue was stored in saline solution and tested within 36 hours. The placenta tissue was prepared into tensile coupons and pulled to failure using a custom designed system of linear motors at one of three strain rates: 7% strain/s, 70% strain/s, and 700% strains/s. These values represent a range of loading rates applicable to a motor vehicle crash. Local strain was calculated for each specimen using dot tracking with high speed video, and nominal stress was determined using original cross-sectional area measurements combined with the load sensor data from each grip. The cross-sectional area was calculated by multiplying the original width by the thickness of the specimen at the center, using the assumption of a rectangular cross-section. The number of specimens for each strain rate is shown in Table 1.

Table 1: Experiment strain rates and specimen count

Strain Rate (% strain/s)	Specimen count
7% strain/s	22
70% strain/s	22
700% strain/s	20

Engineering strain rates were calculated from a constant displacement velocity divided by the gauge length of the specimen.

The specimens were cut into coupon shapes and stretched to failure from both ends at the specified strain rates. The strain rate was multiplied by the specimen gauge length to get the displacement rate of the cross-heads. The forces in the grips and the displacement of the grips were measured over time. The experimental material testing part of this study was performed by Manoogian et al (2008).

Experimental data processing

After data was collected, the data was processed with MATLAB to prepare it for validation with FE simulations.

First, any data before t=0 was truncated. This was performed to remove data where displacement was not being applied. The data for the displacement of the specimen between the grips was then normalized by the gauge length into the total stretch of the specimen. Strain was calculated from the formula $\varepsilon=\lambda-1$, where ε is the strain and λ is the stretch. Engineering stress was calculated by dividing the force at the end point by the original cross-sectional area. Engineering stress was used instead of the force in order to account for geometry. The stress vs. strain characteristic was used as the comparison curve.

To prepare experimental data for input into the FE simulation, the results of all the tests for a specific strain rate were averaged to get a characteristic curve for that strain rate. There were two sets of curves that were needed as input for the FE simulation: the strain vs. time response, and the stress vs. strain response.

Strain vs. time

The strain vs. time response for all the tests, separated into discrete strain rates, are shown in Figure 1. The strain is plotted instead of the displacement so that effects of differing specimen lengths are minimized. Strain rate is more easily controlled at lower strain rates, and was well controlled from 10-90% of the time for the two lower rates and from 40-90% of the time for the higher rate.

The strain rate vs. time responses for a given strain rate can have differing end points for different specimens, as the specimens break at different times. Each breakpoint is shown as a slight horizontal line that protrudes from the constant slope strain vs. time plot. This effect is magnified in Figure 2.

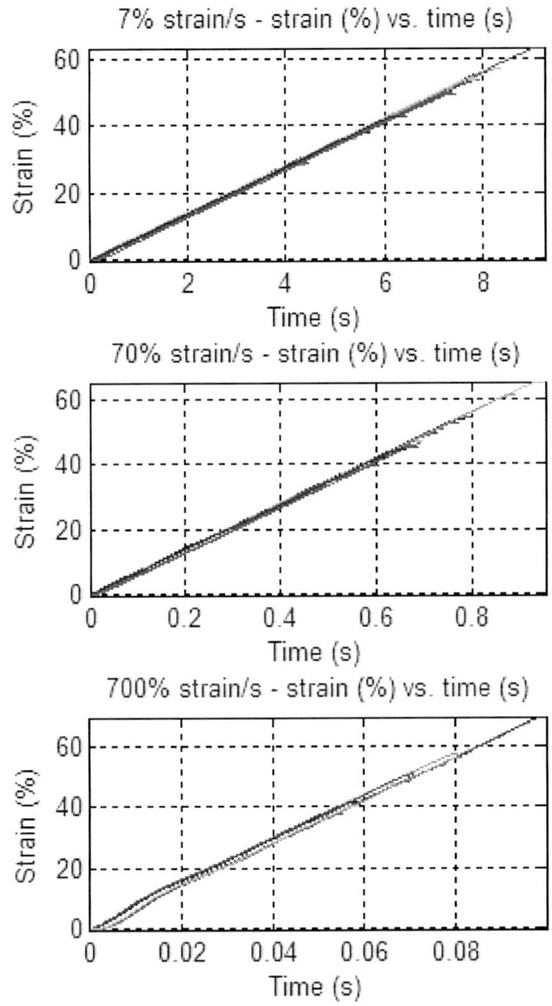

Figure 1: Specimen strain vs. time for all three strain rates

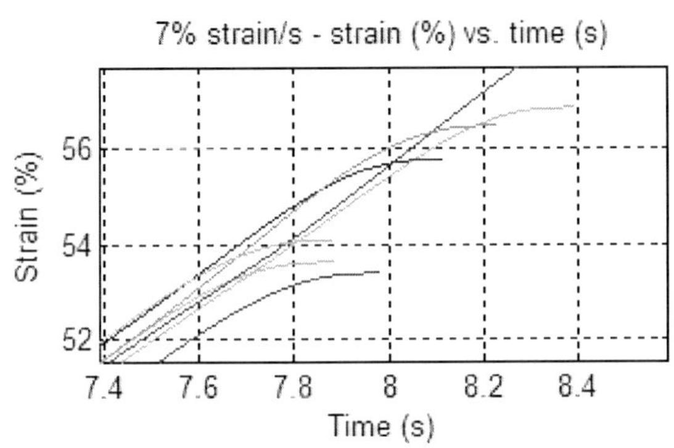

Figure 2: Effects of different termination points on strain

It can be observed that the different termination point creates changes in slope at the moment before the specimen breaks, which gives the strain vs. time response a unique shape that depends on the time of failure. A simple average of all the curves would remove this behavior and only show this change in slope for the test with the longest strain before failure. Furthermore, simple averaging would also create discontinuities in strain vs. time response due to the breaking of individual samples. A third problem with simple averaging is that the termination point of the characteristic average would always be the termination point of the test with the longest run time. This is not the best way to create a characteristic average of the strain vs. time curve. Extrapolating the strain vs. time curve would also create problems with the end point. If a linear extrapolation routine was used, the final average would have a linear segment as most of the curves appear linear before the curve ends.

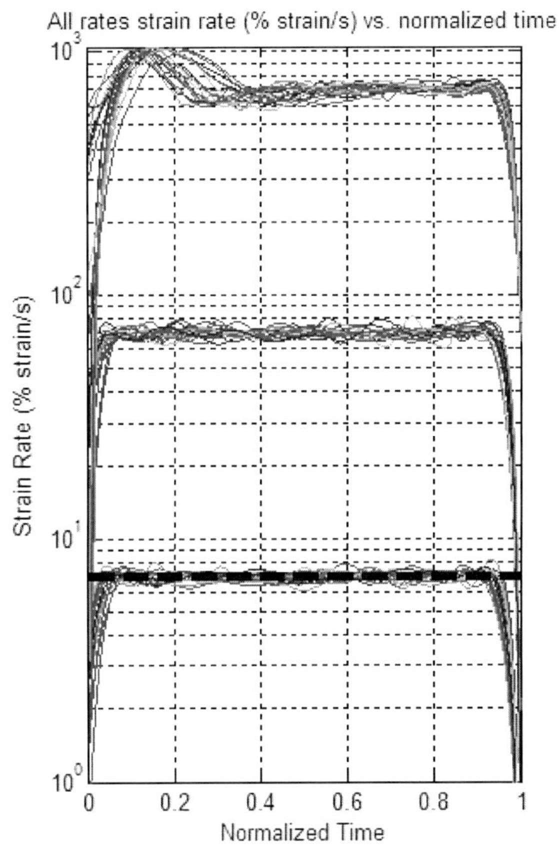

Figure 3: Strain rate (log scale) vs. time for all three strain rates. The dark black line represents the target rate.

The averaging algorithm developed in this study uses a normalized time scale to scale all the curves first by the break point, so that the final displacement value for all the curves become 1. Then averaging was performed. Not only does this preserve the general shape of the curves, but it also allows flexibility in rescaling the curve to an average termination time, instead of the longest time (Lessley, 2004). The strain rate vs. normalized time of the raw data is shown in Figure 3 (raw data filtered at 60 Hz for 700% strain/s, 6 Hz for 70% strain/s, and 0.6 Hz for 7% strain/s according SAE J211 algorithm). It is seen that the general shape of the strain vs. time curve has been preserved. There is an initial ramp up to the target strain rate, and a decrease in the strain rate before the specimen breaks.

To validate filtering levels were appropriate, the filtered curve was subtracted from the unfiltered curve to obtain the noise. The mean of the noise across all experimental tests is then observed to see if there is significant deviation from zero, indicating significant signal content. This is shown in Figure 4. Note that the filtering is performed on the raw data, not the normalized data, due to the fact that time information is used in the filtering algorithm. It can be seen that there is no significant drift in the mean noise response.

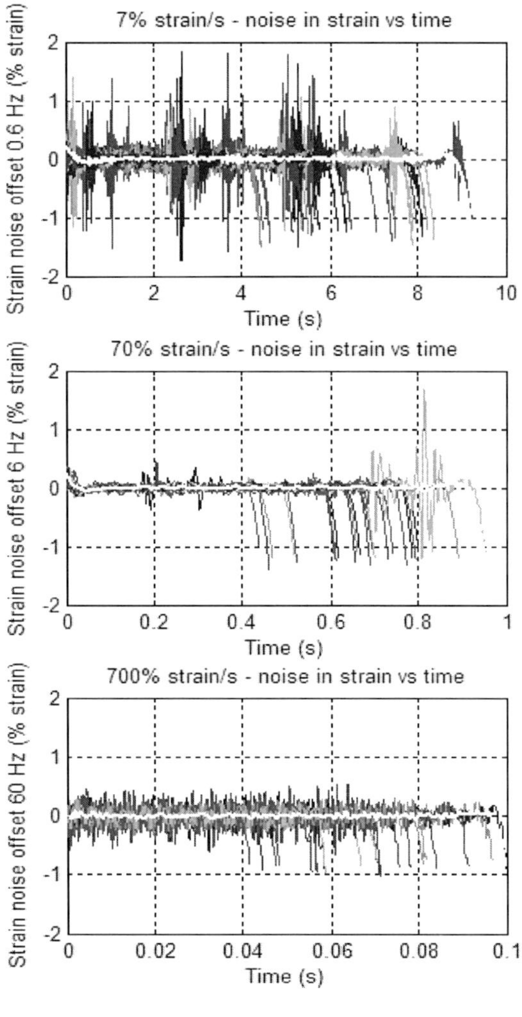

Figure 4: Noise subtracted from the strain vs. time curve

Stress vs. strain

The stress vs. strain curves were processed in a similar manner to the strain vs. time curve. The curve was first scaled in the x-axis by the maximum strain for the individual specimens, and then it was also scaled in the y-axis by the maximum stress. The end results are shown in Figure 5. Note that normalizing stress or force would produce identical results.

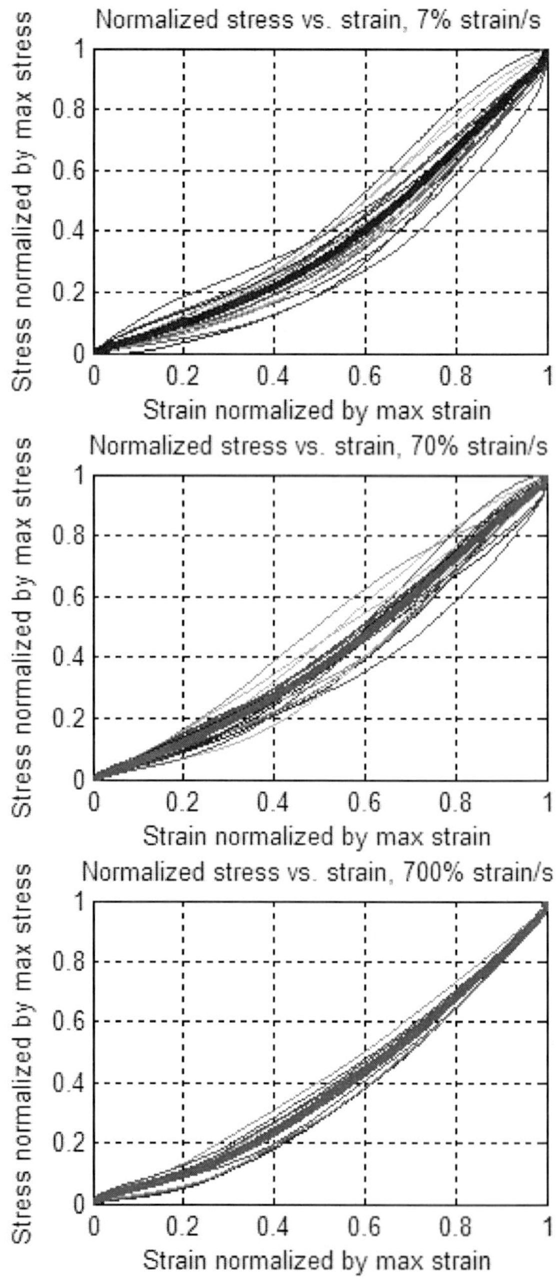

Figure 5: Normalized stress vs. strain for all strain rates

The resulting curves were also fitted to a fourth order polynomial in order to create a function. This function allows interpolation between points without loss of resolution and also ensured that the curve passes through (0, 0) by not allowing a constant in the equation. The function also offers limited smoothing of the data. After fitting the fourth order polynomial, scaling factors were needed to re-scale the functions of the normalized curves back to stress vs. strain curves. This was done by taking a simple 2D average of all the specimen end points for each strain rate, as shown in Figure 6.

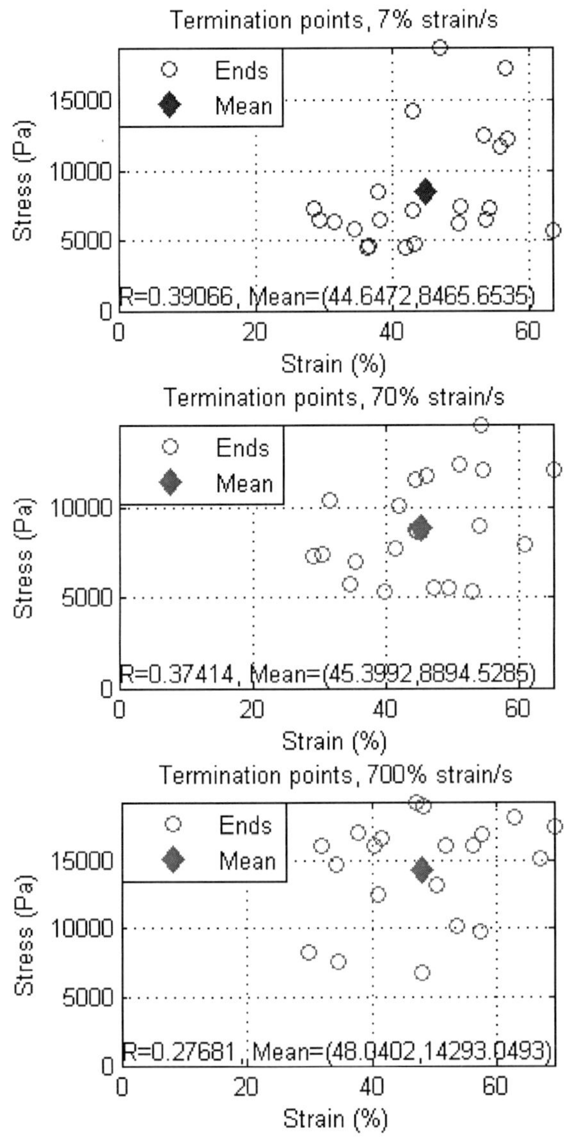

Figure 6: Individual and average end points for stress vs. strain

2. FE model

Specimen dimensions
The experimental specimens ranged in lengths from 24 mm to 48 mm, thicknesses from 2.68 mm to 7.7 mm, and gauge widths from 4.6 mm to 19 mm. The FE model created used the average of these values. These values are shown in Table 2.

Table 2: FE model specimen dimensions

Type	Value
Gauge Length	**31.9 mm**
Gauge Width	**7.0 mm**
Gauge Thickness	**4.2 mm**
Grip Width	**14.9 mm**

The FE model for the placental tissue is shown in Figure 7. The model was hand meshed in an attempt to contour the specimen to the expected deformation. The element sizes were chosen to ensure accuracy and stability (approximately 1 mm element side length). The final dimensions of the FE coupon are shown in Figure 7.

Boundary conditions
The applied boundary conditions consisted of applied velocities at the top and bottom layer of nodes. The applied velocity at each point was half of the total velocity, so that the specimen was stretched at both ends according to the target strain rate. Forces were tracked at the top face of the middle layer of elements, while displacement was controlled and based on specimen end (grip) location.

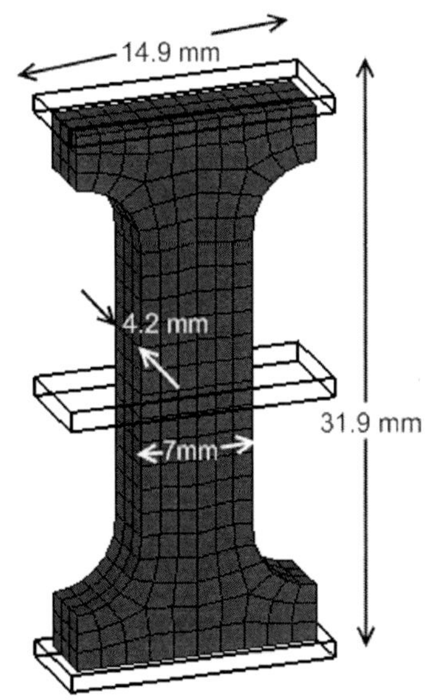

Figure 7: Specimen dimensions and boundary conditions

The strain rate applied is shown in Figure 8. Three simulations were set up, each representing a unique strain rate. The strain rate applied was smoothed using a spline-based interpolation to remove irregularities in the curves. This was most important for the highest strain rate, wherein the initial overshoot was to corrected to have a more constant strain rate. This was a necessity to satisfy the assumption of the material model that input curves represent distinct strain rates.

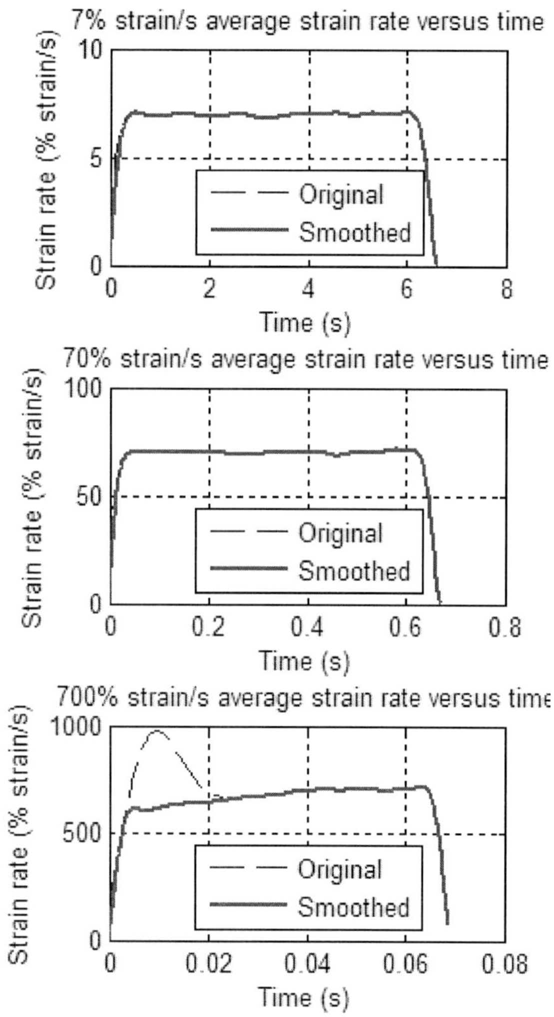

Figure 8: Strain rate applied

Material properties

The material model used in this study is *MAT_SIMPLIFIED_RUBBER (DuBois, 2003). The material model was chosen after evaluating 14 rubber/foam material models in LS-DYNA. Simulations were run using each of the models and the stress vs. strain response for each strain rate was evaluated against the target (average) stress vs. strain for that strain rate. It was found that the simplified rubber material offered the closest potential match to the target curve and also offered flexibility in the strain rate input.

A list of material parameter inputs into the *MAT_SIMPLIFIED_RUBBER deck is detailed in Table 3. Note that there were some additional properties including simulation parameters, which are not relevant to the current study, and stress vs. strain curves, which are explained separately.

Table 3: Simplified rubber material properties

Parameter	Value
Bulk Modulus	1.91 MPa
Shear Modulus for Damping	2.65 kPa
Stress Limit for Damping	26.5 Pa
Tension Type	Rate effects in loading only
Poisson's ratio	0.3
Stress vs strain	(set of curves)

The density chosen was approximately that of water. The bulk modulus was calculated from the average slope of stress vs. strain relationship by using linear elastic relationships.

No damping was necessary for these simulations. The tension type chosen allowed the runs to be more stable as unloading rate effects caused numerical noise in the solution to the simulation. The Poisson's ratio chosen allowed the most realistic behavior of the model.

The material properties also allowed direct input of a set of curves at discrete strain rates that define the force vs. change in gauge length. During element stress update, the stress calculation is based on an interpolation from this set of curves, or a table lookup. Because the load curves were already defined in terms of engineering stress vs. strain, the specimen gauge length, gauge width, and gauge thickness were all set to one. This allowed the stress vs. strain curve to be inputted into the material deck without scaling.

Stress vs. strain properties utilized for the material were modified to add compression behavior. Since only tension tests were available, compression data were generated from a negated and scaled tangent function so that as the compression reaches 80% strain, the forces approach infinity. This helps to ensure stability in cases of extreme compression of the material.

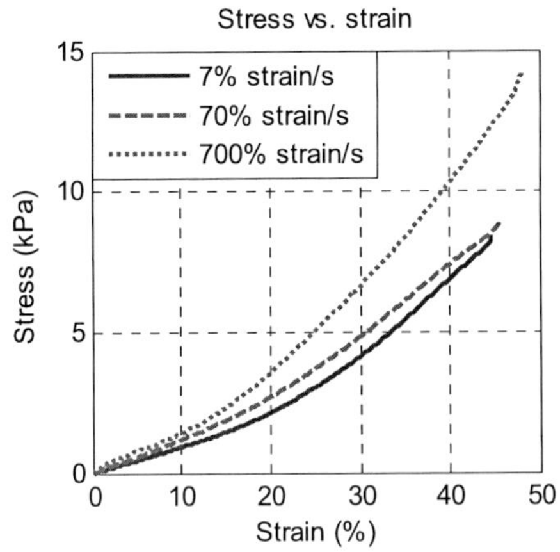

Figure 9: Average experimental stress vs. strain (unmodified)

In addition, due to numerical stability concerns, the curves cannot cross each other except at the origin. They also must start and end at the same level of strain. To achieve curves conforming to these rules, the normalized curves for each strain rate could not be scaled independently. Instead, a global characteristic average for all tests regardless of strain rate must be first created, and this curve scaled to the end termination strain for each strain rate. This ensures that the input curve for each strain rate does not intersect. After the curves were scaled back, the curves were truncated to 30% strain. The 30% strain threshold was used because the two lower strain rate curves began converging after this strain level, a phenomenon that cannot be accounted for by the material model. After the truncation was performed, a tangent function extrapolation was added after the threshold to define behavior past 30% strain.

The results of these operations are shown in Figure 10. While some variation in the stress-strain relationship for each strain rate is lost with these operations, it is necessary to ensure stability of the simulation.

Simulation parameters

In addition to the material specific parameters mentioned in the previous section, some simulation-specific parameters are also required. No hourglass control was used in this simulation as it was stable without hourglass control.

The simulations are run using the SMP version of LS-DYNA (Livermore Software Technology Corporation) explicit code (971 R4.2) on an Intel Pentium 4 (3.4 GHz, 3GB of RAM). The simulation end times for each strain rate are 6.6 s, 0.66 s, and 0.066 s for strain rates 7% strain/s, 70% strain/s, and 700% strain/s, respectively.

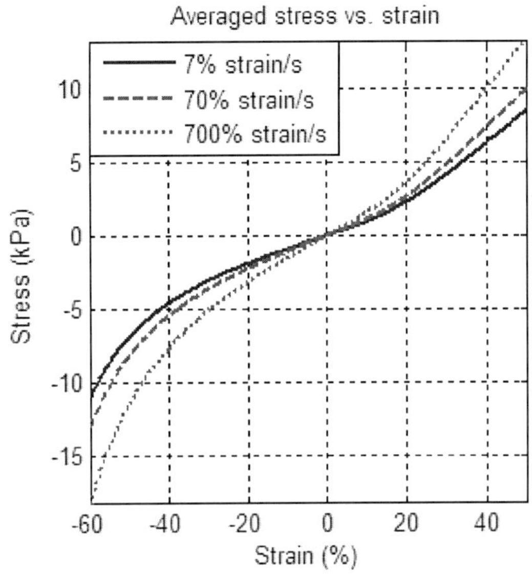

Figure 10: Stress vs. strain curves input to the material deck

The forces were measured with the LS-DYNA section force keyword and the displacements between the crossheads were measured with the node out keyword.

3. Optimization

Three simulations were created, one for each strain rate. An optimization routine was performed on all three simulations in which the difference between the FE output and the experimental stress vs. strain was optimized by varying the material input curves.

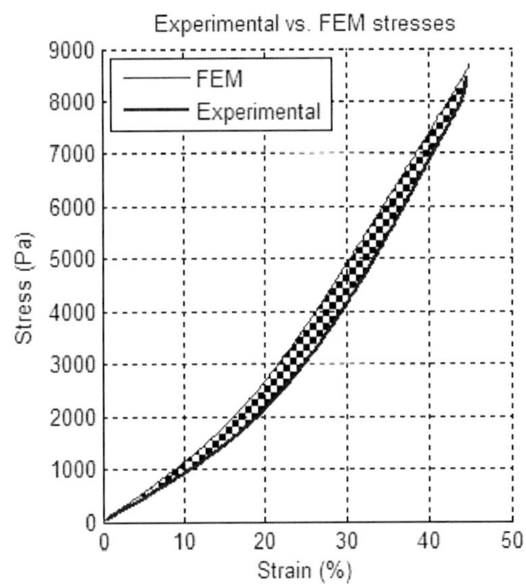

Figure 11: Minimization of the error (between experimental and FEM)

The stress in the FE model was calculated by dividing the force in the cross-section in the middle by the initial cross-section area as part of the optimization procedure. Engineering stress was used instead of true stress to ensure the FEM was comparable to the experimental results. The strain of the specimen was calculated by taking the difference between the crossheads and dividing it by the initial gauge length. This was performed also to ensure that the strain was comparable between the FEM and experimental strain.

The optimization algorithm used was a Multi-Island Genetic Algorithm (MIGA). The specifications for the MIGA algorithm are shown below in Table 4.

Table 4: MIGA parameters

Parameter	Value
Size of subpopulation	10
Number of islands	5
Number of generations	10
Total number of runs	500
Gene size	32
Rate of crossover	1
Rate of mutation	0.01
Rate of migration	0.5
Interval of migration	5
Elite size	1
Relative tournament size	0.5

A total of 500 runs were performed, which took an estimated 10 hours. There were three objective functions that were minimized: one objective function for each strain rate. The objective function for each strain rate was defined as the average absolute difference between the FE stress vs. strain response and the experimental response (stress difference).

Equation 1: Objective function for each strain rate

$$Obj = \frac{\sum_{i=1}^{N} | F(i)_{\exp} - F(i)_{FEM} |}{N}$$

In Equation 1, Obj is the objective function, $F(i)_{exp}$ is the experimental value at point i, $F(i)_{FEM}$ is the FE model's response at point i, and N is the number of points the curves were discretized to. In this study, curves were discretized to 200 points. The curves were truncated at 30% strain to ensure uniformity between experimental results and the FE model at differing strain rates. The absolute value minimization was chosen to minimize the effects of numerical outliers. The global objective function is used in the optimization to minimize error for each strain rate and to penalize variation between the stress-strain characteristics at different strain rates. This creates sensible stress-strain relationships, by ensuring that small variation between curves generated for two strain rates will not be tolerated at the expense of large variation between curves between another combination of strain rates. A global objective function is created from two sources. The first is sum squared error from each strain rate. The second is a penalty function which minimizes large variation in objective function values for different strain rates. The goal of the first set of error terms is to create material parameters that fit the experimental data well at each strain rate. The goal of the second set of error terms is to ensure material parameters do not fit exceptionally well at one strain rate, and poorly at other strain rates. The global objective function for all the strain rates is shown in Equation 2. Obj_{Global} is the global objective function value, $Obj_{7\%}$, $Obj_{70\%}$, and $Obj_{700\%}$ are objective function values for the individual strain rates. The values are squared to penalize large errors.

Equation 2: Global objective function

$$\begin{aligned} Obj_{Global} &= (Obj_{7\%})^2 + (Obj_{70\%})^2 + (Obj_{700\%})^2 \\ &+ (Obj_{7\%} - Obj_{70\%})^2 \\ &+ (Obj_{70\%} - Obj_{700\%})^2 \\ &+ (Obj_{700\%} - Obj_{7\%})^2 \end{aligned}$$

The independent variables that were optimized were scaling factors for the stress vs. strain curves. Since the simplified rubber material uses a table lookup, where the stress is interpolated from the material model input curves, to calculate the stress vs. strain property at any given point, there were no coefficients of fit that could be used as independent variables. The material properties given in Table 3 are computational parameters that did not have as much of an effect on the stress vs. strain behavior of the material. Therefore, it was decided that by scaling the stress vs. strain curve of the input deck, the material properties could be matched to the experimental stress vs. strain curves since the shape should be similar.

The scale factors were initially set to 0.9 for all strain rates. This was used to give a good initial estimate in terms of the stress vs. strain behavior. The scale factors were constrained between 0.6 and 1.4 for the optimization, as this allowed a 0.3 lower range and a 0.5 upper range from the initial guess. If any run terminated due to instability, the objective function was assigned an extremely high value of 10^{20}.

RESULTS

This section will first describe the results of the final optimized run, which was run 429 (hereafter called the 'optimal run') out of 500. The strain rates and stresses, with the behavior of the specimen are compared between the FE model and the experimental tests. Then the optimization results will be described, comparing the errors between the FEM and the experimental results, the progression of the runs, and the stability of the runs.

Strain rate characteristics – Optimal Run

Figure 12, Figure 13, and Figure 14 show the force of the model at the center cross-sectional area and relative velocity of the model between the top and bottom grips. The force vs. time curve has the same shape as the stress vs. time curve of the experimental data. The velocities for the lower two strain rates were within 5% of the target velocity of 2.23 mm/s for the 7% strain/s model and 22.3 mm/s for the 70% strain/s model. Initially, the velocity for the highest strain rate was less than the target velocity of 223 mm/s by more than 10%, but gradually increased to the target velocity around 0.05 seconds into the simulation. The final force for all the strain rates were less than 0.3 N for the time simulated.

The strain rates at mid and top elements in the FE model are recorded by taking the derivative of the mid-plane maximum principal strain.

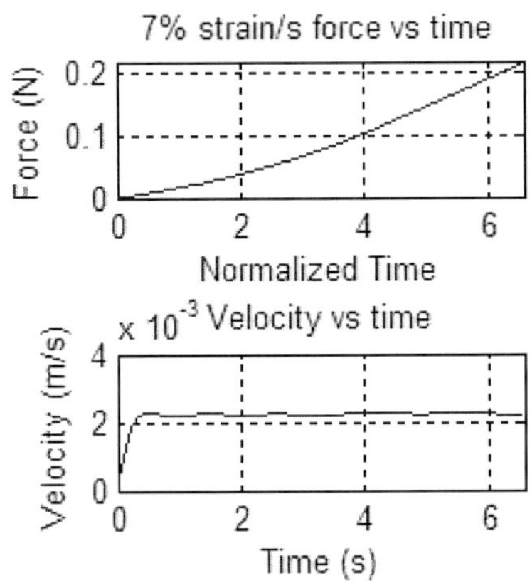

Figure 12: Force and velocity between the crossheads for 7% strain/s

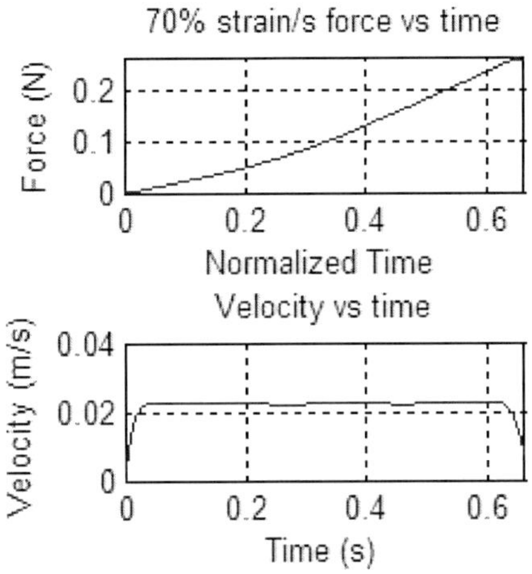

Figure 13: Force and velocity between the crossheads for 70% strain/s

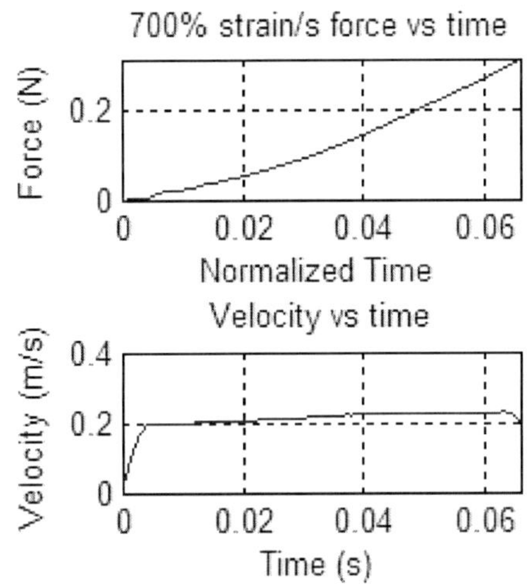

Figure 14: Force and velocity between the crossheads for 700% strain/s

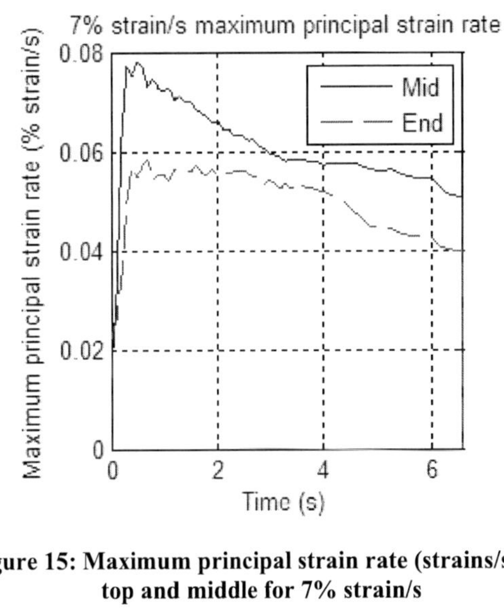

Figure 15: Maximum principal strain rate (strains/s) for top and middle for 7% strain/s

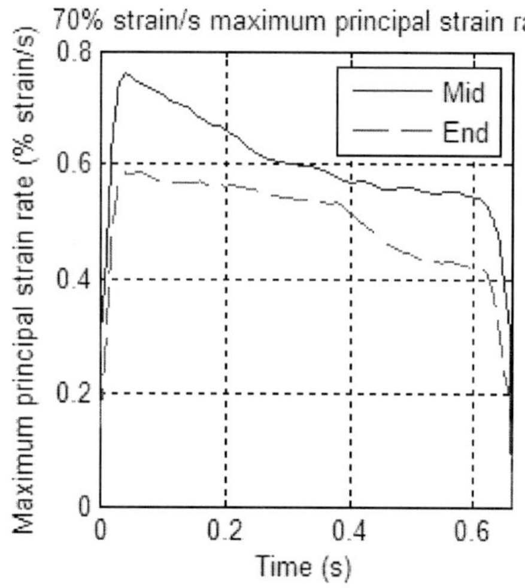

Figure 16: Maximum principal strain rate (strains/s) for top and middle for 70% strain/s

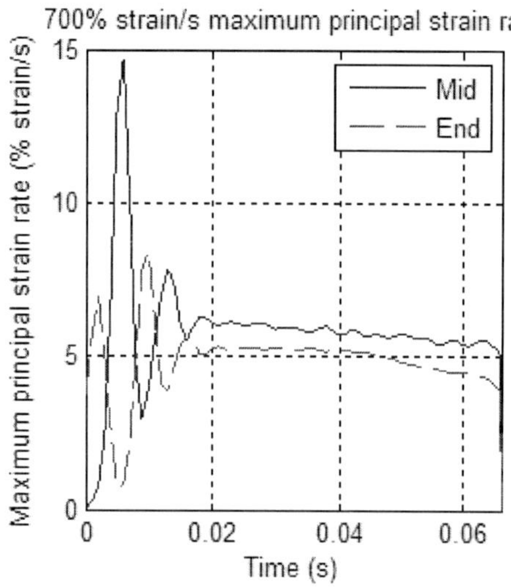

Figure 17: Maximum principal strain rate (strains/s) for top and middle for 700% strain/s

It can be seen that the elements near the grip (end) consistently undershoot the target strain rate while the elements near the middle consistently overshoot the target strain rate. In addition, strain rate curves for the 700% strain/s simulation show more oscillation in the early portion of the simulation.

Stresses – Optimal Run

The maximum stresses at the end points are shown in Table 5. The longitudinal stresses along the specimen were within 1% of the first principal stresses.

Table 5: Maximum stress at final state

Strain Rate	First principal stress	Longitudinal stress
7% strain/s	10.14 kPa	10.08 kPa
70% strain/s	12.28 kPa	12.21 kPa
700% strain/s	14.43 kPa	14.34 kPa

The stress vs. strain comparison between the experimental and the FE simulation is shown in Figure 18 and Figure 19. The 7% strain/s and 700% strain/s FE simulations initially followed the behavior of the experimental stresses, but dropped below them at high strains of 31% strain and 20% strain, respectively. However, the 70% strain/s simulation increased above the experimental stress at 22% strain.

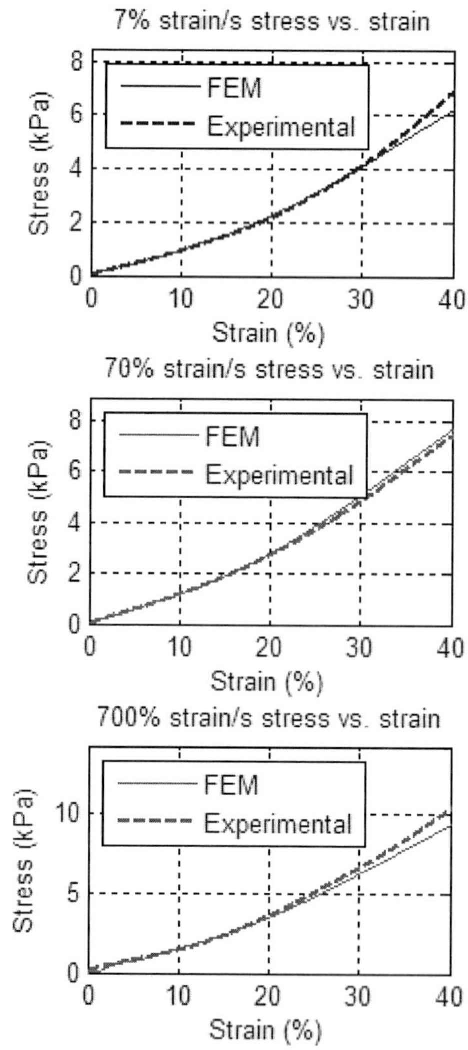

Figure 18: Experimental and FE stress vs. strain

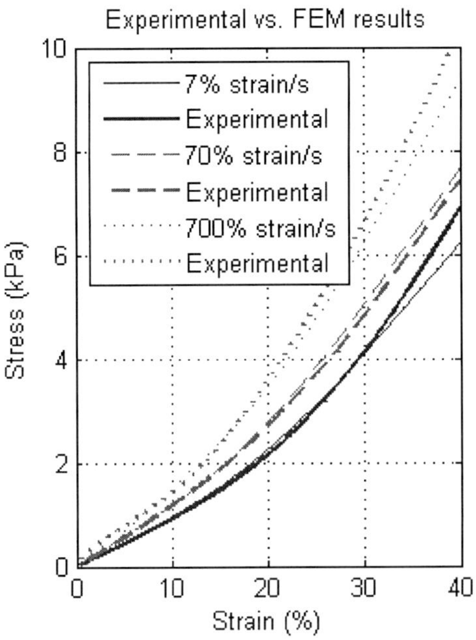

Figure 19: Experimental and FE stress vs. strain, all three strain rates

Optimization results

The most optimal run, the run with the least difference in stress between experimental results and FE model, is shown below along with the progression of the runs.

Optimal Run

The average difference in stresses between the experimental results and the FE model is shown in Table 6. The global average difference is higher than the 7% strain/s and the 70% strain/s difference, while the 700% strain/s difference is higher than the global average difference.

Table 6: Average difference in stress from 0-30% strain

Strain Rate	Value
7% strain/s	51.41 Pa
70% strain/s	65.15 Pa
700% strain/s	133.25 Pa
Average of all strain rates	83.27 Pa

The optimal scale factors determined are shown in Table 7. Note that the scale factors presented here are not the overall scale factors for the global average stress vs. strain curve, but additional scale factors for the stress vs. strain curve input for each strain rate.

The scaled stress vs. strain curves based on the optimized scale factors are shown in Figure 20. The scale factors did not change by more than 0.1 from the initial guess.

Table 7: Optimized scale factors

Strain rate	Scale factor
7% strain/s	0.867
70% strain/s	0.920
700% strain/s	0.835

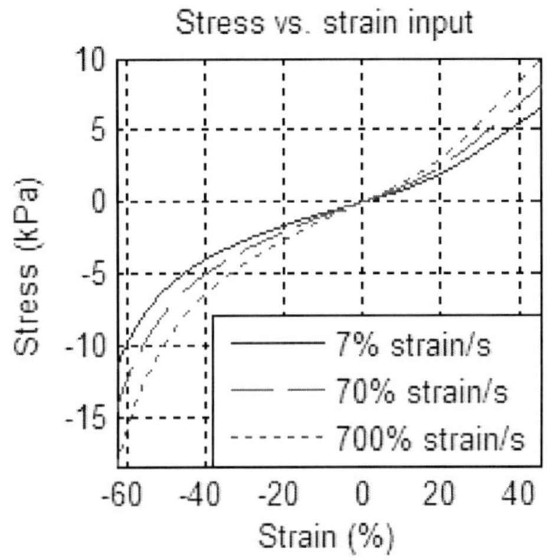

Figure 20: Optimized input stress vs. strain curves

Run progression

There were no failed runs in the optimization. This is shown in Figure 21. The top shows the exact globally averaged stress difference for each run, the bottom shows stress difference with a 10 point running average to show the trend better.

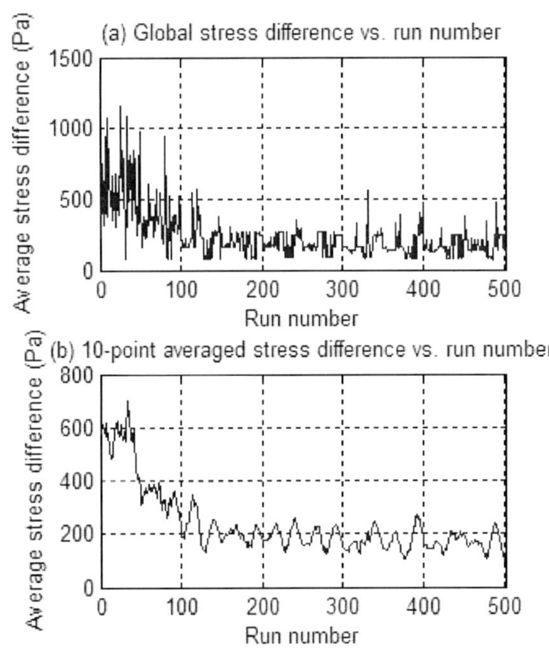

Figure 21: (a) Global stress difference vs. run number (b) 10-point running average of global stress difference vs. run number

The progressions of the stress difference, the objective function value, for each of the three strain rates are shown in Figure 22. The 700% strain/s stress difference has a higher difference on average than the other two strain rates.

The progressions of the scale factors are shown in Figure 23. The 7% strain/s scale factor had large oscillations in its magnitude. None of the scale factors reach the upper bound of 1.2 or lower bound of 0.6.

DISCUSSION

The experimental stress vs. strain curves show a stiffening of the response as the strain rate increased. This behavior is also seen in the FE model. The experimental stress vs. strain curves also showed that the curve of the two lower strain rates started to converge past 30% strain, which was not captured in the FE material model. The converging of the lower two strain rates suggests that there could be a lower bound to the stiffening behavior such that stiffening only occurs if a certain strain rate limit is exceeded.

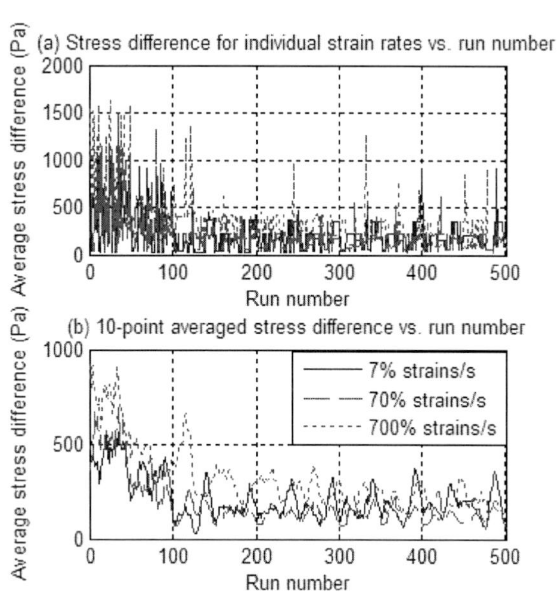

Figure 22: (a) Individual stress difference vs. time (b) 10-point running average of global stress difference vs. run number

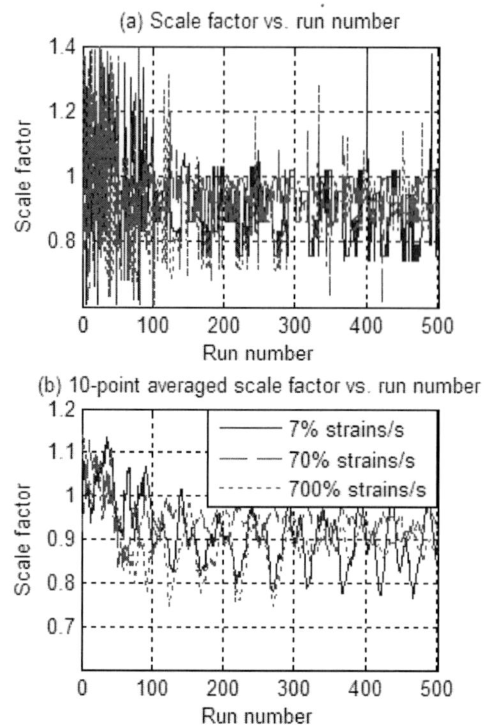

Figure 23: (a) Scale factor vs. run number (b) 10-point running average of scale factor vs. run number

The 30% threshold captured most of the initial stiffening behavior of the specimen, but did not include any behavior that could contribute to failure. Since failure was not defined for the material model, the threshold prevented inclusion of any behavior in the input curve that could be a result of failure. However, failure can be defined for the material model as a threshold, and the simulation run out to much higher strains to see the effect of failure. To ensure a stable model, sometimes it is desirable not to define failure.

Previous studies of the placenta utilized linear elastic material models or were experimental studies. Although a hyperelastic model was used in this study, an approximate Young's modulus can be calculated from the stress vs. strain curve by taking the maximum stress and dividing it by the maximum strain. The comparison results with this Young's modulus are shown in Table 8. This approximated modulus for this study is slightly softer than the Young's modulus from Moorcroft (2003), who chose a linear elastic model. Pearlman's modulus (2000) is approximately twice that of this study while Benson-Martin et al (2006) had a much stiffer modulus of 2.29 MPa for term placenta. Since this study's material model attempts to mimic the stress vs. strain behavior of the experimental test performed, the deviation from literature value is most likely due to variation in experimental procedures. There is high confidence in the material parameters presented here since they are derived from well-controlled experimental tests with a robust approach to material model parameter determination for a FE model.

Both Benson-Martin and Pearlman used a 10 N load cell. While Pearlman's specimens were well within the 10 N ranges, Benson-Martin had a much higher maximum force that exceeded 10 N if the force based on their reported peak stresses is calculated by multiplying the peak stress by the cross-sectional area. This seems to indicate that Benson-Martin may have used an inadequate load cell. While Pearlman's modulus is more than twice that of this study's approximated modulus, the standard deviation of Pearlman's modulus is 18.6 kPa, which means there is a high variation in the modulus. This study's modulus is within one-standard deviation of Pearlman's modulus.

Table 8: Values of placenta properties from the literature

Study	Material	Young's Modulus (kPa)	Poisson's ratio
Moorcroft (2003)	Linear Elastic	20	0.45
Pearlman (2000)	Experimental	32.7 +/- 18.6	not reported
Benson-Martin (2006)	Experimental	2290	not reported
Current Study	Simplified Rubber	~16.7	0.3

The average difference between the experimental stress vs. strain curve and the FE stress vs. strain curve was 83.27 Pa; this is less than 2% of the maximum stress experienced by the specimen. The accuracy of the FE model appears to be very good for the lower strains, but the behavior for all the strain rates show that the stress vs. strain curves started to diverge significantly from the experimental data past a strain of 30%. This phenomenon was expected as all the stress difference calculations were performed with the strain truncated at 30% for comparable averaging. Future studies could implement an averaging algorithm that can discretize to variable end strains.

The averaging algorithm used in this study seems to preserve the general shape of the curves. However, due to the requirement of having non-intersecting curves, a globally averaged (averaging all the strain rate curves) stress vs. strain curve had to be used to generate the table of curves. Different scaling factors were used to create the curves for different strain rates, but they all share the same shape. This means that any individual shape differences that were a result of the strain rates were lost. The stiffening behavior was still preserved, but this is one of the drawbacks of using this material model for implementing strain rate behavior. However, no other material model in LS-DYNA can account for this behavior directly, so it is not a limitation of this specific material model as that of the current FE implementations. The averaging algorithm developed was able to capture the general shape for multiple strain rates and preserve that shape for good comparison to the experimental tests.

Compression behavior was not tested in this study, but if experimental compression tests were available, they could be implemented into the material model.

The optimization technique gave a better result than could be achieved with curve fitting. With scale factors of 0.9 for each of the stress vs. strain curves, the stress difference was almost 5 times that of the optimized stress difference. The improvements to the scale factors tended to decrease after 150 runs. This suggests 200 runs would probably have been sufficient to achieve an optimal solution that is not much worse than the current solution. The optimization used in this study demonstrated a robust way of concurrently accounting for multiple objectives.

CONCLUSION

A FE model for placental tissue was created in LS-DYNA using geometry from experimental tension coupon tests. The velocity vs. time of the crossheads in the experimental tests were used as boundary conditions of the FE simulations. The stress vs. strain of the tension coupon was used as the input into the simplified rubber material model. The FE model stress vs. strain curves was optimized using a multi-island genetic algorithm to minimize the average difference in stress between the FE output and the experimental results for the same strain. This was performed by using scale factors for the curves as the independent variable.

It was found that the FE stress vs. strain output matched those of the experiments within 2% for strains up to 30%. This study demonstrated a sophisticated method of processing experimental data for multiple strain rates and multiple tests to create a characteristic average. It also demonstrated a method for concurrently optimizing simulation and experimental data to create a FE model that can be well-correlated to experimental data.

This material model can be used in FE models of the placenta as the run times are low for a complex material model, and the resulting model should approximate the behavior of placental tissues well at a wide variety of strain rates, from quasi-static to dynamic.

ACKNOWLEDGMENTS

The work was supported by Wake Forest University Graduate student stipend funding.

REFERENCES

Agran, P. F., Dunkle, D. E., Winn, D. G. & Kent, D. (1987) Fetal death in motor vehicle accidents. Annals of emergency medicine, 16, 1355-1358.

Benson-Martin, J., Zammaretti, P., Bilic, G., Schweizer, T., Portmann-Lanz, B., Burkhardt, T., Zimmermann, R. & Ochsenbein-Kölble, N. (2006) The Young's modulus of fetal preterm and term amniotic membranes. European Journal of Obstetrics & Gynecology and Reproductive Biology, 128, 103-107.

Delotte, J., Behr, M., Baque, P., Bourgeon, A., De Peretti, F. & Brunet, C. (2006) Modeling the pregnant woman in driving position. Surgical and radiologic anatomy : SRA, 28, 359-363.

Dubois, P. A. (2003) A simplified approach to the simulation of rubber-like materials under dynamic loading. 4th European LS-DYNA Users Conference.

Duma, S. M., Moorcroft, D. M., Stitzel, J. D. & Duma, G. G. (2004) Evaluating pregnant occupant restraints: the effect of local uterine compression on the risk of fetal injury. Annual proceedings / Association for the Advancement of Automotive Medicine. Association for the Advancement of Automotive Medicine, 48, 103-114.

Duma, S. M., Moorcroft, D. M., Stitzel, J. D. & Duma, G. G. (2006) Biomechanical modeling of pregnant occupants in far-side vehicle crashes. Biomedical sciences instrumentation, 42, 154-159.

Lessley, C., Shaw, Kent, and Funk (2004) Normalization Technique for Developing Corridors from Individual Subject Responses. Injury Biomechanics Research.

Manoogian, S. J., Bisplinghoff, J. A., Mcnally, C., Kemper, A. R., Santago, A. C. & Duma, S. M. (2008a) Dynamic tensile properties of human placenta. Journal of biomechanics, 41, 3436-3440.

Manoogian, S. J., Moorcroft, D. M. & Duma, S. M. (2008b) Pregnant occupant injury risk in severe frontal crashes using computer simulations. Biomedical sciences instrumentation, 44, 249-255.

Moorcroft, D., Stitzel (2003) A Finite Element and Multi-body Model of the Pregnant Female Occupant for the Analysis of Restraint Effectiveness. Society of Automotive Engineers.

Moorcroft, D. M., Stitzel, J. D., Duma, G. G. & Duma, S. M. (2003) Computational model of the pregnant occupant: predicting the risk of injury in automobile crashes. American journal of obstetrics and gynecology, 189, 540-544.

Pearlman, K., Schneider, Rupp, Moss, Ashton-Miller (2000) A comprehensive program to improve safety for pregnant women and fetuses in motor vehicle crashes: A preliminary report. Am J Obstet Gynecol, 182.

Weiss, H. (2001a) The epidemiology of traumatic injury-related fetal mortality in Pennsylvania, 1995-1997: the role of motor vehicle crashes. Accid Anal Prev, 449-454.

Weiss, S., and Fabio (2001b) Fetal deaths related to maternal injury. JAMA, 1863-1868.

CHAPTER 6

COMPUTATIONAL MODELING OF PREGNANT OCCUPANTS

CHAPTER 6: COMPUTATIONAL MODELING OF PREGNANT OCCUPANTS

Stefan M. Duma and Joel D. Stitzel
Virginia Tech – Wake Forest University, Center for Injury Biomechanics

Computational models of the human body are often used to investigate the injury biomechanics of automobile crashes. Given the complex anatomy and injury mechanisms of the pregnant occupant, researchers have used modeling as a tool to develop increased knowledge of pregnant interactions during impact loading conditions. This chapter presents a review of the published models as well as an investigation of low acceleration loading of the pregnant female. Although it is likely that more pregnant models exist than are discussed here, only models that have been published are presented.

PUBLISHED MODELS OF THE PREGNANT OCCUPANT

An overview of the pregnant models is presented (Table 1). The first and most extensively published pregnant occupant computational model is the one developed at Virginia Tech. It was first presented in 2002 (Moorcroft, 2002) and published in full-length peer-reviewed manuscripts in the spring of 2003 (Moorcroft, 2003a, 2003b). This model starts with the small female human multibody model developed by MADYMO. This model is a 5th percentile female (5 ft tall, 110 lb) and was modified to include a 30 week pregnant finite element abdomen so that the weight of the pregnant occupant model is 135 lb. The multibody human model provides biofidelic response of an occupant in a motor vehicle crash, while reducing the computational time compared to a full finite element human model. The finite element model of the pregnant abdomen offers more-detailed information and it includes representations of the uterus, placenta, and supporting ligaments. The exterior abdominal contour of the pregnant model closely matches the data used for the second-generation pregnant dummy. The fluid-filled pregnant uterus does not contain a fetus, as previous research illustrated that the fetal interaction with the uterus was not a significant mechanism in the risk of placental abruption.

Table 1: Summary of the published pregnant occupant computational models.

Name	Platform	Gestation	Description	Publications
Virginia Tech	MADYMO Multibody and Finite Element	30 Weeks	Small female human model as occupant with FE pregnant uterus and surrounding structures	Moorcroft 2002, Moorcroft 2003a, Moorcroft 2003b, Moorcroft 2003c, Moorcroft 2004, Duma 2004, Duma 2006, Manoogian 2008a, Manoogian 2008b
Volvo 'Linda'	RADIOSS Finite Element	36 Weeks	HIII dummy model as occupant with FE pregnant uterus and fetus inserted in dummy CAD	Thackray 2003, 2004 Volvo press releases
Loughborough	MADYMO Multibody and Finite Element	38 Weeks	Small Female Human Model as Occupant with FE pregnant uterus and rigid body fetus	Acar 2006, Acar 2009
INRETS	Radioss HUMOS Finite Element	39 Weeks	50th Percentile male model with pregnant uterus and fetus	Delotte 2006, Behr 2006, Delotte 2009

Four techniques were used to validate the Virginia Tech pregnant model. First, a global biofidelity response was evaluated by using a seatbelt to compress dynamically the pregnant abdomen (Moorcroft 2003b). The force vs. compression results were within the published corridors from scaled cadaver tests (Hardy 2001). Second, a similar validation procedure was performed with a rigid bar (Moorcroft 2003b), and these results were also consistent with previous data (Hardy 2001). The third technique involved validating the model against real-world crashes to investigate the model's ability to predict injury. Using fatal crashes from pregnant occupants (Klinich 1999), the model showed strong correlation (R2 = 0.85) between peak strain at the uteroplacental interface (UPI) as measured in the model compared to risk of fetal demise as reported in the real-world crashes over a range of impact velocities and restraint conditions (Moorcroft 2003a). The fourth method compared the physiological failure strain from placental tissue tests to the failure strain measured in the model. Tissue tests by Rupp et al. (2001) suggested approximately a 60% failure strain for UPI tissues, which is in agreement with the model's prediction of 75% risk of fetal loss at a 60% strain in the UPI. In summary, the global, injury, and tissue level validation techniques all indicate the model is reasonable for predicting injurious events for the pregnant occupant.

Following the initial development and validation papers, the Virginia Tech model was used to illustrate two important aspects of the pregnant occupant that should be incorporated into all models. First, the uterus has a series of supporting ligaments that should be included in the model as they affect the strain patterns in the uterus (Moorcroft 2003c). Second, the location of abdomen deformation relative to the placental attachment location is the critical measure when attempting to predict UPI strain (Duma 2004). In other words, where the abdomen compression is measured is very important, and it is most important in the region of the placental attachment.

In the fall of 2003, an abstract about a pregnant model was presented by Laura Thackray of Volvo, and the pregnant model was called Linda (Thackray 2003). Details about this model were obtained through conversations with the author and from reading Volvo press releases starting from 2004 and currently advertised on the Volvo web page. The Volvo-Linda model starts with a finite element model of the Hybrid III dummy. Next, a 36 week pregnant abdomen is inserted between the dummy pelvis and ribs. Unfortunately, there are no full-length publications about this model and it is not possible to describe the validation process or results.

The Loughborough model was published in 2006 and is similar to the Virginia Tech model (Acer 2006). This model also starts with the MADYMO small female human model and adds a finite element pregnant uterus and surrounding structures. The Loughborough model then adds a rigid-body fetus inside the uterus at the size of a 38 week gestational age. Little validation data are reported. Through simulations with unrestrained, belted, and belted-with-an-airbag pregnant occupants at impact velocities ranging from 15-45 km/h, the Loughborough model found similar results compared to the Virginia Tech pregnant model. Both models showed that the unrestrained case is the most severe and that strain approaching 60% in the area where the placenta attaches to the uterus is associated with high risk of fetal injury.

The most recent pregnant model was presented by INRETS and first published in 2006 (Delotte 2006). This model starts with the 50th percentile male finite element model called Radioss HUMOS. A finite element representation of a pregnant abdomen and fetus at 39 weeks is then inserted in the model to simulate the pregnant occupant. The manuscript presents very good detail regarding the anatomical considerations for the model; however, there are no data presented regarding validation of the model. In 2009 the INRETS model was used to investigate the effect of placental location on the risk of placental abruption during an automobile crash (Delotte 2009). The study found peak strains at the uteroplacental interface up to 20%, and it concluded that there was no significant effect of placenta location relative to the risk of placental abruption. For that study, only frontal impact speeds were tested at 20 km/h (12.4 mph), and it is possible that different results may be found at higher speeds or different directions.

In summary, researchers have utilized a variety of modeling techniques to investigate the injury biomechanics of the pregnant occupant. All studies generally agree that the seatbelt and airbag provide the best protection for the pregnant occupant.

LOW ACCELERATION OF THE PREGNANT FEMALE

While most research focuses on medium to high severity crashes, it is useful to examine low-level accelerations of the pregnant occupant. In 2008 a study was published using the Virginia Tech model to examine the effect of everyday accelerations on the pregnant female (Manoogian 2008a). Acceleration data measured on the pelvis of female volunteers during daily activities such as walking and sitting was collected and then applied to the pregnant model. Another low-impact study was conducted investigating low-speed side impacts in a far-side configuration (Duma 2006). For both of these studies, the resulting peak UPI strains were determined and the risk of fetal injury was calculated using the data from the Moorcroft et al. (2003a) study (Tables 1 and 2).

Table 2: Everyday acceleration data analyzed with the Virginia Tech pregnant female model and with updated fetal risk values (Manoogian 2008a).

Activity		Volunteer Pelvic Acceleration (g)	UPI Peak Strain (%)	Published Fetal Risk (%)	Updated Fetal Risk (%)
Walking	Minimum	0.92	2.3	3.1	0.0
	Maximum	0.67	5.1	6.9	0.0
Sitting	Minimum	0.45	3.9	5.3	0.0
	Maximum	1.70	5.0	6.8	0.0
Jumping Jacks	Minimum	5.67	7.6	10.3	3.3
	Maximum	10.56	8.6	11.7	4.9
Leap off Step	Minimum	2.92	7.2	9.8	2.7
	Maximum	9.75	9.2	12.5	5.8
Jump in Place	Minimum	7.15	8.4	11.4	4.5
	Maximum	3.49	9.5	12.9	6.2
Running	Minimum	2.13	8.2	11.1	4.2
	Maximum	2.59	13.9	18.8	12.9

Table 3: Far-side impact response of the Virginia Tech pregnant female model and with updated fetal risk values (Duma 2006).

Pregnant Female Occupant Restraint	Delta-V Speed (mph)	UPI Peak strain (%)	Published Fetal Risk (%)	Updated Fetal Risk (%)
No Restraint	5	5.7	7.8	0.4
Lap Belt	5	5.7	7.8	0.4
3-Point Belt	5	5.4	7.3	0.0
4-Point Belt	5	5.3	7.2	0.0
No Restraint	15	29.4	40.0	36.8
Lap Belt	15	27.1	36.9	33.3
3-Point Belt	15	27.5	37.4	33.9
4-Point Belt	15	24.6	33.5	29.4
No Restraint	25	32.8	44.6	42.0
Lap Belt	25	28.9	39.3	36.0
3-Point Belt	25	44.9	61.0	60.6
4-Point Belt	25	40.1	54.5	53.3

There are two possible ways to determine risk from the Moorcroft et al. (2003a) data (Fig. 1). First, the best correlation between UPI strain and risk of fetal injury uses a curve that is not forced through zero and gives an R^2 of 0.8461. This correlation was used originally in the Moorcroft et al. (2003a) study, and it is shown here as the best solution given the correlation as well as the more reasonable risk values at the lower end shown during everyday acceleration. These values are shown as the "Updated Fetal Risk" values in Tables 1 and 2. The second method is to force the correlation through zero, which gives a lower R^2 of 0.8327 and shows unreasonably high values of risk for everyday activities. This second method was used in the published versions of the Manoogian et al. (2008a) and the Duma et al. (2006) studies and is shown in Tables 1 and 2 as the "Published Fetal Risk." Based on the higher correlation and the more reasonable prediction at the lower-level acceleration tests, the updated fetal risk values are presented as the best data, and the first method for determining the correlation should be used. There is no meaningful difference between the two curves at the higher acceleration levels.

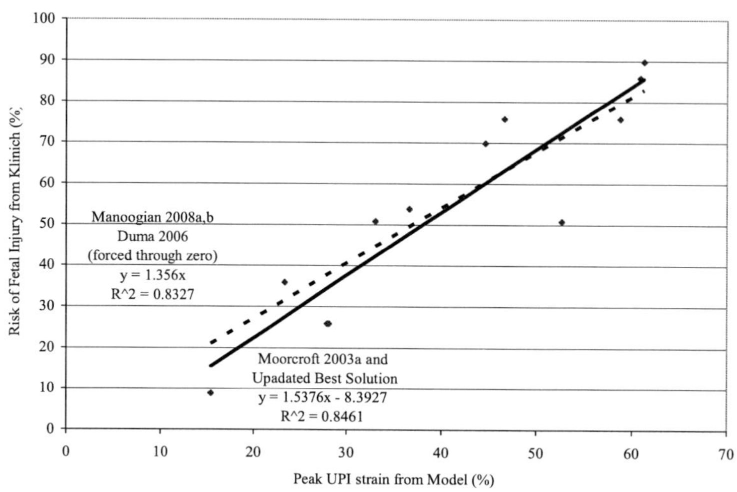

Figure 1: Correlations Between Fetal Injury Risk and the Peak UPI Strain from the Virginia Tech Model: Updated Best Solution Shown by Solid Line (Moorcroft 2003a).

FUTURE OF PREGNANT OCCUPANT MODELING

The future of pregnant occupant modeling will likely undergo large methodological changes. A paradigm shift in human occupant modeling for injury prediction is currently underway, and pregnant occupant models will likely follow this trend.

Some researchers consider the most advanced human model to be the Total Human Model for Safety (THUMS), a new version of which will be available in the Fall of 2010. The updated THUMS v4 will contain detailed internal organ anatomy. In addition, the Global Human Body Models Consortium (GHBMC) was formed and began model development efforts in 2008. The GHBMC 50th male geometry has been developed and is currently being meshed with delivery of the 50th percentile male model scheduled for 2011. The Global Human Body Models Consortium is a group of manufacturers that have come together to fund development of the world's standard human body models for injury prediction.

Greater modeling of anatomic detail of the uteroplacental interface, along with an improved understanding of the mechanism and pathogenesis of placental abruption are needed to improve FE models of the pregnant woman. Developments in morphing technology will allow for the creation of uterine models which are shaped and sized for different female anatomy and levels of gestation. Quantitative studies of variation in placental attachment are needed to make decisions regarding average anatomy for injury prediction. Opportunity exists to use Lagrangian-Eulerian coupling to model the fluid-filled nature of the uterus for application to injury prediction. It is likely that the mechanics of uterine deformation, and therefore the uterine wall strains with relation to the placental wall strains, are quite complex as a consequence of the unique anatomy and fluid filling. Future models should probably incorporate this to improve discriminative capability and accuracy of prediction.

REFERENCES

Acar, B. S. and D. van Lopik. 2006. "A Pregnant Occupant Model for Crash Impacts." Proceedings of International Crashworthiness Conference, Athens, Greece, May 2006.

Acar, B. S. and D. van Lopik. 2009. "Computational pregnant occupant model, 'Expecting', for crashsimulations." Proceedings of IMechE Part D: *Journal of Automobile Engineering* Vol. 223, 891-902.

Behr, M., L. Thollon, J. Delotte, P. J. Arnoux, and C. Brunet. 2006. "A Pregnant Version of the Radioss Humos Model." International Research Council on the Biomechanics of Impact, September, 393-396.

Delotte, J., M. Behr, P. Baque, A. Bourgeon, F. Peretti, and C. Brunet. 2006. "Modeling the pregnant woman in driving position." *Surgical Radiology Anatomy* 28: 359-363.

Delotte, J., M. Behr, L. Thollon, A. Bongain, and C. Brunet. 2009. "Does placenta position modify the risk of placental abruption in car crashes?" *Computer Methods in Biomechanics and Biomedical Engineering* Vol. 12 No. 4, August, 399-405.

Duma, S., D. Moorcroft, J. Stitzel, and G. Duma. 2004. "Evaluating Pregnant Occupant Restraints: The Effect of Local Uterine Compression on the Risk of Fetal Injury." *Association for the Advancement of Automotive Medicine,* 48.

Duma, S., D. Moorcroft, H. Gabler, S. Manoogian, J. Stitzel, and G. Duma. 2006. "Analysis of Pregnant Occupant Crash Exposure and the Potential Effectiveness of Four-Point Seatbelts in Far Side Crashes." *Association for the Advancement of Automotive Medicine,* 50.

Klinich, K. D, L. W. Schneider, J. L. Moore, and M. D. Pearlman. 1999. "Investigations of crashes involving pregnant occupants." UMTRI-99-29.

Hardy, W. N., L. W. Schneider, and S. W. Rouhana. 2001. "Abdominal impact response to rigid-bar, seatbelt, and airbag loading." *Stapp Car Crash Journal* 45, 1-32.

Manoogian, S. J., S. M. Duma, and D. M. Moorcroft. 2008a. "Evaluation of Pregnant Female Injury Risk During Everyday Activities." *Biomedical Sciences Instrumentation* 44: 183-188.

Manoogian, S. J., S. M. Duma, and D. M. Moorcroft. 2008b. "Pregnant Occupant Injury Risk in Severe Frontal Crashes Using Computer Simulations." *Biomedical Sciences Instrumentation* 44: 249-255.

Moorcroft, D., S. Duma, and G. Duma. 2002. "A Finite Element Model of the Pregnant Female Occupant: Analysis of Injury Mechanisms and Restraint Systems." Invited presentation and poster at 46th Association for the Advancement of Automotive Medicine Conference, Tempe, AZ, October.

Moorcroft, D. M., S. M. Duma, J. D. Stitzel, and G. G. Duma. 2003a. "Computational Model of the Pregnant Occupant: Predicting the Risk of Injury in Automobile Crashes." *American Journal of Obstetrics and Gynecology* Vol. 189 (2), 540-544.

Moorcroft, D. M., S. M. Duma, J. D. Stitzel, and G. G. Duma. 2003b. "A finite element and multi-body model of the pregnant occupant for the analysis of restraint effectiveness." SAE Paper No. 2003-01-0157. SAE International, Warrendale PA.

Moorcroft, D., J. Stitzel, S. Duma, and G. Duma. 2003c. " The Effects of Uterine Ligaments on the Fetal Injury Risk in Frontal Automobile Crashes." *Journal of Automobile Engineering* Vol. 217, Part D, 1049-1055.

Moorcroft, D., J. Stitzel, S. Duma, and G. Duma. 2004. "The Effect of Pregnant Occupant Position and Belt Placement on the Risk of Fetal Injury." SAE Paper No. 2004-01-0324. Presented at SAE International Congress and Exposition, Detroit, MI, March.

Rupp J. D., K. D. Klinich, S. Moss, J. Zhou, M. D. Pearlman, and L. W. Schneider. 2001. "Development and testing of a prototype pregnant abdomen for the small-female Hybrid III ATD." *Stapp Car Crash Journal* 45, 61-78.

Thackray, L. 2003. Pregnant model called Linda; International Traffic Medicine Conference, Budapest, September.

The following papers are included in their entirety to augment this chapter:

Duma, S. M., D. M. Moorcroft, J. D. Stitzel, and G. G. Duma. 2004. "Evaluating Pregnant Occupant Restraints: The Effect of Local Uterine Compression on the Risk of Fetal Injury." *Annals of Advances in Automotive Medicine.* 48:103-114.

Duma, S. M., H. C. Gabler, D. M. Moorcroft, S, J. Manoogian, J. D. Stitzel, and G. G. Duma. 2006. "Analysis of Pregnant Occupant Crash Exposure and the Potential Effectiveness of Four-Point Seatbelts in Far Side Crashes." *Annals of Advances in Automotive Medicine* 50:187-198.

Moorcroft, D. M., S. M. Duma, J. D. Stitzel, and G. G. Duma. 2003. "Computational Model of the Pregnant Occupant: Predicting the Risk of Injury in Automobile Crashes." *American Journal of Obstetrics and Gynecology* 189(2):540-544.

Moorcroft, D. M., J. D. Stitzel, S. M. Duma, and G. G. Duma. 2004. "The Effect of Pregnant Occupant Position and Belt Placement on the Risk of Fetal Injury." SAE Paper No. 2004-01-0324. Presented at SAE International Congress and Exposition, Detroit, MI, March.

EVALUATING PREGNANT OCCUPANT RESTRAINTS: THE EFFECT OF LOCAL UTERINE COMPRESSION ON THE RISK OF FETAL INJURY

Stefan M. Duma
David M. Moorcroft
Joel D. Stitzel
Virginia Tech – Wake Forest, Center for Injury Biomechanics
Blacksburg, Virginia

Greg G. Duma
University of Cincinnati, School of Medicine
Cincinnati, Ohio

ABSTRACT

In order to develop effective restraint systems for the pregnant occupant, injury criteria for determining fetal injury risk must be developed. This study presents computer simulations of a 30 week pregnant occupant that illustrate the importance of local uterine compression on the risk of fetal injury. Frontal impact simulations with a range of velocities and belt positions were used to identify the best correlation between local uterine compression and peak strain measured at the uterine-placental interface. It is suggested that future pregnant dummy development and specifically pregnant injury criteria should be based on local uterine compression relative to the placental attachment location.

Automobile crashes are the largest single cause of death for pregnant females (Attico et al., 1986) and the leading cause of traumatic fetal injury mortality in the United States (US) (Weiss, 2002). Each year, 160 pregnant women are killed in motor-vehicle crashes (MVCs) and an additional 800 to 3200 fetuses are killed when the mother survives (Klinich et al., 1999b) in the US. Unfortunately, fetal injury in motor vehicle crashes is difficult to predict due to the fact that real world crash data is limited and cadaver studies are not feasible.

48th Annual Proceedings
Association for the Advancement of Automotive Medicine
October, 2004, pp 165-179, Key Biscayne, Florida

In the non-pregnant female, the uterus is a muscular organ the size of a lemon located within the abdominal cavity. As the fetus grows during pregnancy, the abdomen stretches to the size of a watermelon. The internal volume increases from 0.005 L to 5 L and as much as 10 L (Rupp et al., 2001). The uterine wall is uniform prior to delivery, with a thickness of about 1 cm. The uterosacral and round ligaments extend from the uterus to the pelvis and act to support the uterus. The interior of the uterus contains the fetus, which is surrounded by amniotic fluid and the placenta (Figure 1). The placenta is a vascular organ that acts as a permeable membrane that exchanges oxygen, nutrients, and waste products between the mother and fetus via the umbilical cord. It is a flat, roughly circular structure 2 cm thick in the center and is typically located in the upper half of the uterus (Fried, 1978).

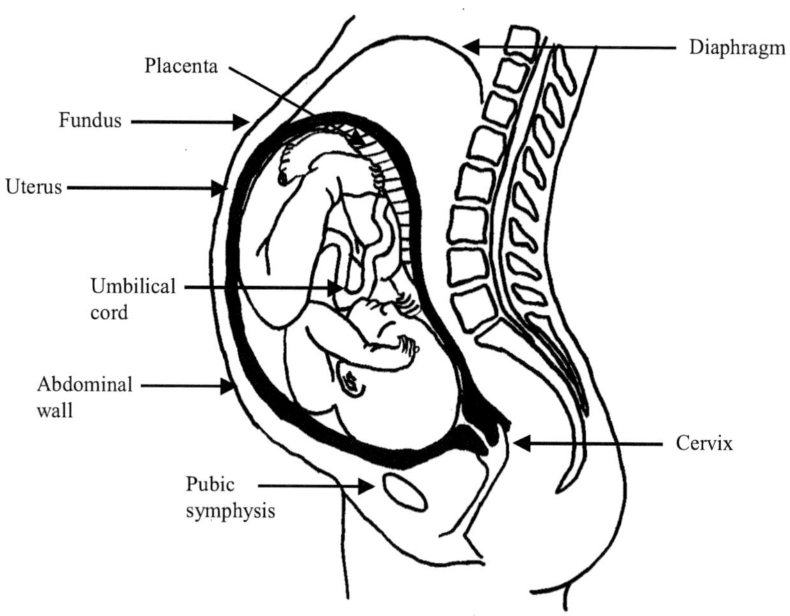

Figure 1: Anatomy of a 40-week pregnant woman (ligaments not shown).

In an effort to reduce the risk of injury to pregnant occupants in car crashes, a pregnant anthropometric test dummy (ATD) has been developed at the University of Michigan Transportation Research Institute (Rupp et al., 2001). The Maternal Anthropomorphic Measurement Apparatus Version 2B (MAMA-2B) is a second-generation prototype ATD that is a retrofit Hybrid III small female dummy. One of the primary limitations of the pregnant dummy is the lack of injury criteria for the fetus. The MAMA-2B was designed to measure anterior and posterior pressure in the fluid-filled abdomen insert as well as the strain on the perimeter of the insert. However, only the anterior pressure measurements were repeatable (Rupp et al., 2001). Therefore, it would be beneficial to have an injury criterion for the pregnant dummy that utilizes currently established ATD measurement methods. One leading example would be to measure overall abdominal compression in a similar manner that used to measure chest compression. For example, this could be done by using a string potentiometer as is done in the chest.

The most common cause of fetal death from motor vehicle accidents is placental abruption, which is the premature separation of the placenta from the uterus (Klinich et al., 1999b). Both the pregnant dummy and the pregnant model used in this study utilize this injury mechanism to predict fetal outcome (Moorcroft et al., 2003a). However, due to the difficulties in measuring this mechanism in the pregnant dummy, such as strain and pressure, another type of measurement is desired that can accurately predict fetal injury risk. The purpose of this paper is to identify the best correlation between uterine compression that could be measured by a dummy and the risk of fetal injury as predicted by the strain at the UPI.

METHODOLOGY

A total of 15 computer simulations of a 30 week pregnant occupant in a pure frontal impact were performed using a model previously developed and validated by Moorcroft et al. (2003a, 2003b, 2003c). The model is a modified MADYMO small female human model with a pregnant abdomen (Figure 2). The abdomen consists of the uterus, placenta, and amniotic fluid. The uterus is supported by two pairs of ligaments and surrounded by fat.

Four techniques were used to validate the pregnant model. First, a global biofidelity response was evaluated by using a seatbelt to compress dynamically the pregnant abdomen (Moorcroft et al., 2003b). The force versus compression results were within the published corridors from scaled cadaver tests (Hardy et al., 2001). Second, a similar validation procedure was performed with a rigid bar (Moorcroft et al., 2003b) and these results were also consistent with previous data (Hardy et al., 2001). The third technique involved validating the model against real-world crashes in order to investigate the model's ability to predict injury. Using fatal crashes from pregnant occupants (Klinich et al., 199b), the model showed strong correlation (R2 = 0.85) between peak strain at the uterine-placental interface (UPI) as measured in the model compared to risk of fetal demise as reported in the real-world crashes over a range of impact velocities and restraint conditions (Moorcroft et al., 2003a). The forth method compared the physiological failure strain from placental tissue tests to the failure strain measured in the model. Tissue tests by Rupp et al. (2001) suggested approximately a 60% failure strain for UPI tissues which is in agreement with the model's prediction of 75 % risk of fetal loss at a 60% strain in the UPI. In summary, the global, injury, and tissue level validation techniques all indicate the model is good at predicting injurious events for the pregnant occupant.

The right-front passenger seat was chosen for all simulations because the potentially confounding effects of the steering wheel are not present and therefore the occupant loading is dominated by the restraint system (Kent et al., 2001). In addition, the airbag was not deployed in any simulations in order to focus on the uterine compression. As with the omission of the steering wheel, the lack of airbag deployment allows for the lap belt to load the abdomen, thus giving a consistent loading pattern and allowing for the location of the abdominal measurement to be assessed. This is effectively a rigid bar impact to the abdomen, but by using lap belt, the load conforms more realistically around the abdomen and a better contact interaction is achieved between the two finite element objects.

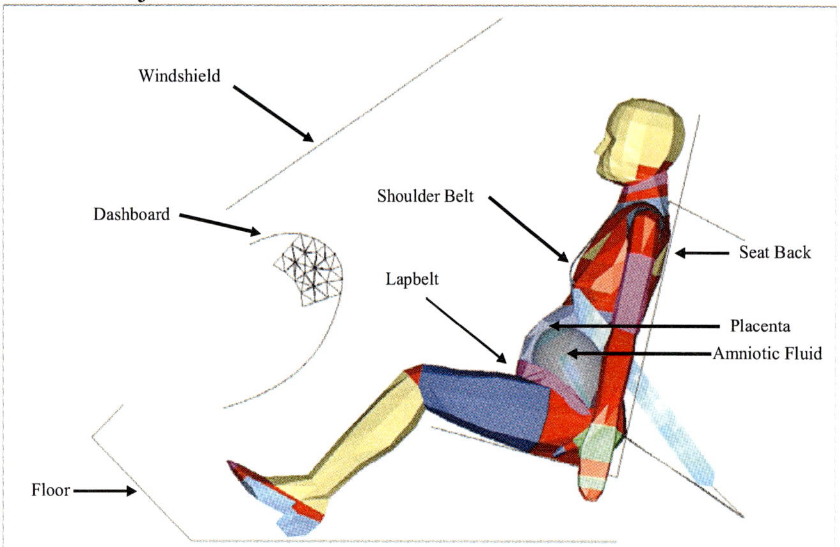

Figure 2: Simulation configuration of pregnant occupant in passenger position with three-point belt.

The passenger interior is a typical MADYMO passenger compartment with a seatback, seat cushion, floor, dashboard, and windshield. The setback distance of the passenger seat is consistent with that of a typical passenger compartment and representing a fully aft seat position with an abdominal clearance of 40 cm. This positioning of the pregnant occupant was based on the seated anthropometry of a pregnant woman in her 30th week of pregnancy as defined by Klinich et al. (1999a). The MADYMO finite element belt was constructed of nonlinear triangular membrane elements. The material model was isotropic with hysteresis and the loading curve had an approximate slope of 1.75 GPa and an unloading slope of approximately 200 MPa. The belt was 5 cm wide.

The test matrix consisted of simulations with five different lap belt locations (Figure 3). The initial lap belt location is representative of the recommended position. The top belt placement represents a lap belt that has slid over the top of the uterus. Three intermediate belt locations were also defined, with nominal locations, with respect to the initial belt location, of +0.03 m, +0.07 m, and +0.09 m. Positioning of the belts was accomplished by placing a straight belt at a specific vertical location and then pulling the belt across the abdomen until the belt was taut. The initial, or recommended location, is based on the recommended position for all occupant which has the belt as low as possible on the abdomen to facilitate loading of the pelvis (Rupp et al., 2001).

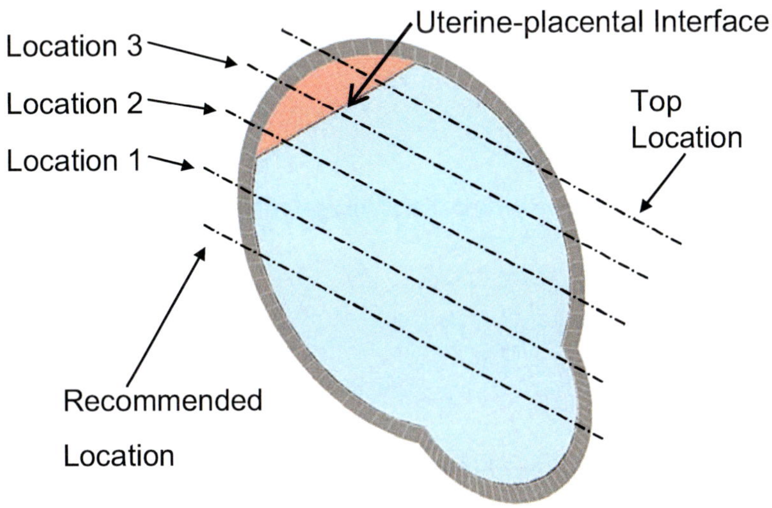

Figure 3: Five belt locations relative to pregnant uterine model.

Abdominal measurements were taken at five locations, named AB1, AB2, AB3, AB4 and TH1 to correspond to the names of the flexible layers in MADYMO (Figure 4). The first abdominal layer, AB1, is at the same height as the pelvis. The fourth abdominal layer, AB4, is directly in line with the placenta. The first thoracic layer, TH1, is above the uterus. For each layer, a central node on the skin, both on the anterior and posterior surfaces of the abdomen, was selected. The magnitude of the distance between the two points was outputted. The magnitude was chosen to account for the rotation of the body. The initial distance ranged from 217 mm to 248 mm. Peak UPI strain was also recorded for each simulation.

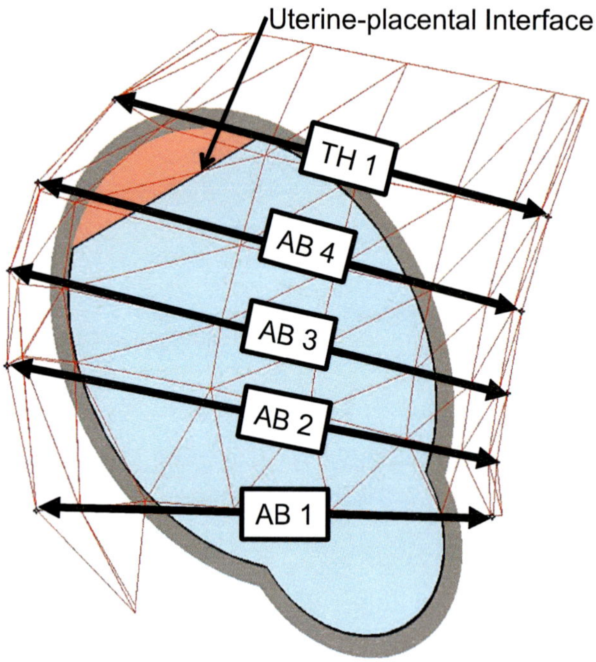

Figure 4: Local uterine compression measurement locations.

Each of the five lap belt positions was evaluated at three velocities (13 kph, 35 kph, 55 kph) for a total of 15 simulations. The applied sled pulse was a half-sine wave imposed for a duration of 100 ms. Linear egression analysis was used to correlate the peak uterine compression values to the peak UPI strain and subsequent risk of fetal injury.

RESULTS

The peak uterine strain increased with crash speed and lap belt height except for the top belt location (Table 1). However, overall compression, which is the peak compression value for all locations, was constant over the range of lap belt locations for a given speed; 20% for 13 kph, 35% - 40% for 35 kph, and 50% for 55 kph. In nearly all simulations, the peak compression occurred at the same time as the peak strain. Therefore, the same event is being recorded by strain and compression. Next, the compression at each location was evaluated (Table 2). The measurement locations each showed a bias towards localized loading. The bottom two locations and the top measurement location responded minimally to remote loading.

Table 1: Simulation matrix, peak results and timing.

Run Number. Belt Location	Speed	Peak Strain (%)	Time of peak strain (ms)	Overall Peak Compression (%)	Time of peak compression (ms)
1.Recommended	13	15.8	78	17.0	80
2.Location 1	13	23.7	76	20.2	89
3.Location 2	13	32.8	89	22.6	93
4.Location 3	13	29.7	91	18.0	93
5.Top Location	13	13.7	88	15.9	89
6.Recommended	35	33.7	81	35.1	90
7.Location 1	35	49.5	93	40.7	87
8.Location 2	35	96.9	80	45.1	80
9.Location 3	35	83.1	77	35.4	77
10.Top Location	35	35.2	88	31.4	75
11.Recommended	55	37.4	81	44.6	99
12.Location 1	55	51.5	90	49.8	97
13.Location 2	55	145.4	122	50.8	113
14.Location 3	55	97.5	77	51.4	117
15.Top Location	55	41.8	84	36.2	94

Table 2: Peak abdominal compression for each uterine location.

Run Number. Belt Location	AB 1 (%)	AB 2 (%)	AB 3 (%)	AB 4 (%)	TH 1 (%)
1.Recommended	3.6	8.8	17.0	5.6	2.8
2.Location 1	1.6	9.5	20.2	12.6	4.5
3.Location 2	0.0	2.7	16.8	22.6	11.7
4.Location 3	0.0	0.4	7.9	18.0	16.7
5.Top Location	0.0	0.4	1.3	5.5	16.0
6.Recommended	10.0	16.3	35.1	11.9	4.7
7.Location 1	0.0	18.1	40.7	17.3	6.3
8.Location 2	0.0	5.7	37.8	45.1	17.9
9.Location 3	0.0	0.4	16.9	35.2	35.4
10.Top Location	12.7	0.4	2.0	9.7	31.4
11.Recommended	12.6	18.4	44.7	14.9	5.2
12.Location 1	0.0	19.4	49.8	19.5	6.5
13.Location 2	0.0	5.9	46.2	50.8	20.8
14.Location 3	0.0	0.4	18.7	51.4	43.9
15.Top Location	0.0	0.4	5.7	10.6	36.2

The kinematics of the simulations follow the same general trend showing submarining of the occupant below the lap belt occurring for belt positions above the recommended position (Figure 5). Additionally, large abdominal and subsequently uterine compression resulted from this submarining. The danger of this loading is evident in strain values which exceeded the 60% tissue limit in many of the simulations at 35 kph and 55 kph. The localized loading is evident in examination of the uterus itself (Figure 6). Loading at belt location 1 maximally compressed the uterus at the AB 3 location; however, minimal compression is observed at AB 4 and minimal deformation of the upper half of the uterus as well, thereby resulting in a moderate value of peak UPI strain.

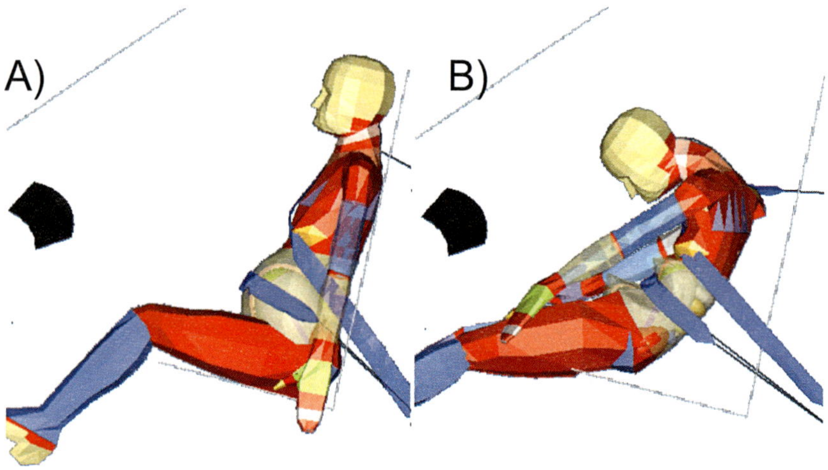

Figure 5: Simulation belt location 2 pre-test (A) and peak overall uterine compression (B).

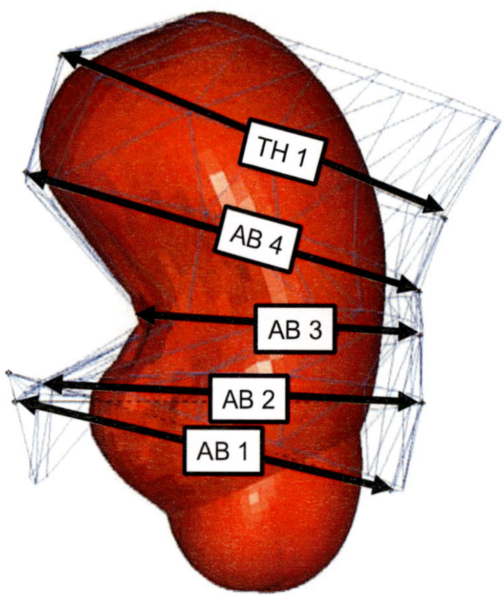

Figure 6: Example uterine compression peak at location AB 3.

A linear regression analysis of the data was performed for each measurement location (Table 3). The best correlation was between AB 4 and peak UPI strain for all three impact velocities (Table 3). Moreover, the AB 4 peak compression results in the best correlation coefficient (R=0.93) for all simulations compared when to the overall peak compression (R=0.75) (Figure 7).

Table 3: Correlation coefficient values for each uterine compression measurement versus peak UPI strain.

Measurement Location	13 kph	35 kph	55 kph	All speeds
AB 1	0.48	0.62	0.69	0.53
AB 2	0.50	0.69	0.77	0.65
AB 3	0.24	0.45	0.72	0.05
AB 4	0.92	0.97	0.92	0.93
TH 1	0.20	0.14	0.33	0.26

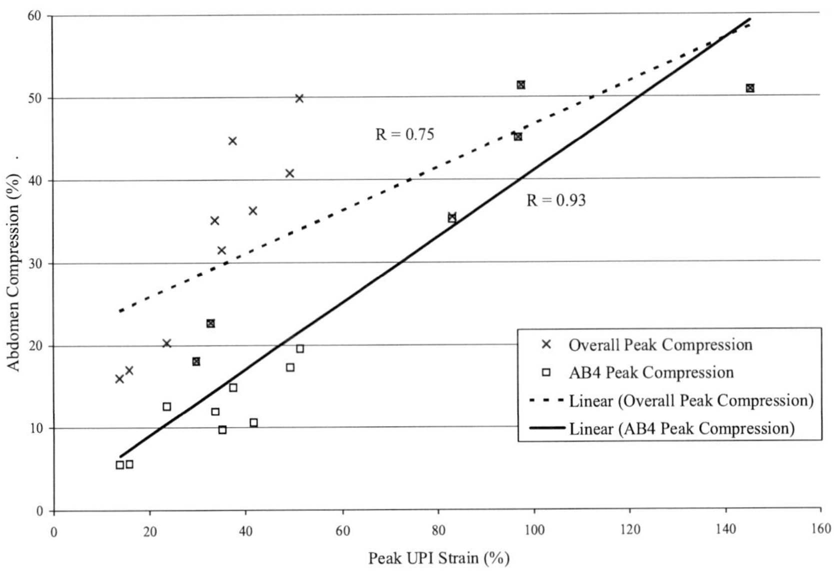

Figure 7: Overall peak compression and AB4 peak compression versus peak UPI strain.

DISCUSSION

The AB 4 measurement location correlated the best because the strain is measured in the uterus at the placental location, which is at approximately the same height as the AB4 measurement location. As noted above, the greatest risk for fetal injury results from placental abruption, which occurs when the strain in the uterus near the placenta exceeds the tissue limit. Therefore, in simulations where the abdominal loading was remote from the placenta, low strains were recorded and relatively low compression was recorded at the AB4 level as compared to the level of the loading. Although placental location varies from pregnancy to pregnancy, and can vary during pregnancy, as many as 95% of placenta's are located in the upper half of the uterus (Fried, 1978). Therefore, it is suggested as best practice to measure abdominal loading at the uppermost uterine location in an attempt to predict injury to the UPI and risk of fetal demise.

Predicting fetal injury from abdominal deflection is loosely analogous to using chest deflection to predict thoracic injury. As a simple comparison, chest deflection for the small female is limited to 52 mm by federal safety standards (Eppinger et al., 1999). A chest deflection of 52 mm is approximately 35% compression which corresponds to approximately 40% risk of an AIS 3 or greater injury (Mertz et al., 1991). Given the obvious anatomical differences between the thorax and pregnant uterus, it is interesting that 35% compression of the uterus at AB 4 is also the higher limit of injury. The abdominal deflection could be measured in the same manner as chest deflection, using a string potentiometer, chestband, or through processing of digital video. It is important to note that the measurements need to be taken from a pregnant dummy, which has the correct anthropometry and abdominal force-deflection response as a pregnant woman.

Although this study indicates a local compression measure is needed to predict injury to the fetus, it may be difficult to measure this in the MAMA-2B test dummy. However, the results of this study should be utilized to help design future versions of the pregnant dummy in order to measure local compression values. This is analogous to the progress that has been made in thoracic injury dummy measurements. A single chest compression measure is used currently in the HIII dummy to predict chest injuries; however, the THOR next generation male test dummy includes sensors to predict several local chest deflections that may provide more accurate thoracic injury information.

It is important to note that previous simulations indicate that for all frontal impacts it is safest for the pregnant occupant to ride in the passenger seat while wearing a three-point belt and utilizing the frontal airbag when appropriate (Moorcroft et al., 2004). However, for the purpose of this paper and examining uterine compression, the airbag was not utilized.

As with all computational models, this model is limited by the accuracy of input and simplifications made. The tissue data, from which the failure strain is derived, is sparse and simplifications are made to use that data as a material model. Additionally, the boundary conditions and geometry can and should be improved in future generations of the model. Furthermore, the model only looks at injury at the UPI. In cases with very large deflections, direct injury to the fetus may occur at injury rates different then those for placental abruption. It is recommended that the methods in this paper be applied to future generations of the pregnant occupant model to provide a continually improving understanding of pregnant occupant injury risk prediction.

CONCLUSIONS

Computer simulations of a 30 week pregnant small female were performed using a range of belt locations and impact velocities in order to evaluate a new injury predictor for the pregnant occupant. The results illustrate that local uterine compression correlates with peak uterine strain and is a better predictor then the overall global peak compression. It is critical for future pregnant dummy development and for overall protection of the pregnant occupant that injury criteria be evaluated. Abdominal compression is a simple measurement that should yield viable results when predicting the likelihood of injury to the fetus as a result of motor vehicle crashes,

ACKNOWLEDGEMENTS

The authors thank Altair Engineering Inc. for the use of HyperWorks as the pre and post processor, and TNO for the use of MADYMO. The authors would also like to thank Kathy Klinich at the University of Michigan Transportation Research Institute for her assistance in the development of the pregnant model.

REFERENCES

Attico NB, Smith III RJ, Fitzpatrick MB, Keneally. Automobile safety restraints for pregnant women and children. J Reprod Med, 1986; 31(3):187-92.

Eppinger R, Sun E, Bandak F, Haffner M, Khaewpong N, Maltese M, Kuppa S, Nguyen T, Takhounts E, Tannous R, Zhang A, Saul R (1999). Development of improved injury criteria for the assessment of advanced automotive restraint systems – II. NHTSA, Docket No. 99-6408-5.

Fried AM (1978). Distribution of the bulk of the normal placenta. Review and classification of 800 cases by ultrasonography. **American Journal of Obstetrics and Gynecology,** 132(6), 675-680.

Hardy WN, Schneider LW, Rouhana SW (2001). Abdominal impact response to rigid-bar, seatbelt, and airbag loading. **Stapp Car Crash Journal**, 45, 1-32.

Kent RW, Crandall JR, Bolton J, Prasad P, Nusholtz GS, Mertz HJ (2001). The influence of superficial soft tissues and restraint condition on thoracic skeletal injury prediction._**Stapp Car Crash Journal,** vol. 45, 2001-22-0008, 183-204.

Klinich KD, Schneider LW, Eby B, Rupp J, Pearlman MD (1999a). Seated anthropometry during pregnancy. UMTRI-99-16.

Klinich KD, Schneider LW, Moore JL, Pearlman (1999b). Investigations of crashes involving pregnant occupants. UMTRI-99-29.

Mertz HJ, Horsch JD, Horn G, Lowne RW (1991). Hybrid III sternal deflection associated with thoracic injury severities of occupants restrained with force-limiting shoulder belts. SAE Technical Paper 910812, 105-119.

Moorcroft DM, Duma SM, Stitzel JD, Duma GG (2003a) Computational Model of the Pregnant Occupant: Predicting the Risk of Injury in Automobile Crashes. **American Journal of Obstetrics and Gynecology** Vol. 189 (2), pp. 540-544, 2003.

Moorcroft DM, Duma SM, Stitzel JD, Duma GG (2003b). A finite element and multi-body model of the pregnant occupant for the analysis of restraint effectiveness. SAE Technical Paper. 2003-01-0157.

Moorcroft DM, Stitzel JD, Duma SM, Duma GG (2003c). The Effects of Uterine Ligaments on the Fetal Injury Risk in Frontal Automobile Crashes. **Journal of Automobile Engineering**, Vol. 217, Part D, pp. 1049-1055.

Moorcroft DM, Duma SM, Stitzel JD, Duma GG (2004). The effect of pregnant occupant position and belt placement on the risk of fetal injury. SAE Technical Paper. 2004-01-0324.

Rupp JD, Klinich KD, Moss S, Zhou J, Pearlman MD, Schneider LW (2001). Development and testing of a prototype pregnant abdomen for the small-female Hybrid III ATD. **Stapp Car Crash Journal**, 45, 61-78.

Pearlman MD, Ashton-Miller JA, Dyer T, Reis P (1999). Data acquisition for development to characterize the uteroplacental interface for the second-generation pregnant abdomen. Submitted to NHTSA.

Weiss HB, Strotmeyer S. Characteristics of pregnant women in motor vehicle crashes. Injury Prevention 2002:8(3):207-214.

Analysis of Pregnant Occupant Crash Exposure and the Potential Effectivenessof Four-Point Seatbelts in Far Side Crashes

Stefan M. Duma
David M. Moorcroft
Hampton C. Gabler
Sarah M. Manoogian
Joel D. Stitzel
Greg G. Duma
Virginia Tech – Wake Forest
Center for Injury Biomechanics
Blacksburg, VA

ABSTRACT

The purpose of this paper is to present the crash exposure patterns of pregnant occupants and to evaluate the effectiveness of restraint systems, including four-point seatbelts, in far side crashes. The NASS CDS database revealed that 53% of pregnant occupants are exposed to frontal crashes while 13% are exposed to far side impacts. Given that far side crashes were the second leading crash mode after frontal impacts, a previously validated MADYMO computer model of a 30 week pregnant occupant was utilized to investigate pregnant occupant biomechanics in far side crashes. Three impact speeds (5, 15, and 25 mph) were simulated with four restraint conditions: unbelted, lap-belt only, three-point belt, and a four-point belt. Direct abdominal contact from the shoulder strap of the three-point or four-point belt caused uterine-placental strain in contrast to the inertial loading induced strain in the lap-belt and unbelted cases. Overall, the three-point and four-point belt systems provide superior restraint effectiveness for the pregnant occupant compared to the lap-belt and no restraint cases. The four-point resulted in slightly better performance than the three-point belt by reducing the fetal injury risk and occupant excursion.

50th Annual Proceedings
Association for the Advancement of Automotive Medicine
October16-18, 2006, pp 187-198, Chicago, Illinois

Automobile crashes are the largest single cause of death for pregnant females and the leading cause of traumatic fetal injury mortality in the United States (Attico, 1986; Weiss, 2002). Each year, 160 pregnant women are killed in motor-vehicle crashes (MVCs) and an additional 800 to 3200 fetuses are killed when the mother survives in the United States (Klinich, 1999a; Klinich, 1999b). Unfortunately, fetal injury in motor vehicle crashes is difficult to predict due to the fact that real world crash data is limited and cadaver studies are not feasible.

A pregnant anthropometric test dummy (ATD) has been developed at the University of Michigan Transportation Research Institute (Rupp, 2001). From that research it was determined that the most common cause of fetal death from motor vehicle accidents is placental abruption, which is the premature separation of the placenta from the uterus. The pregnant dummy (MAMA-2B) utilizes this injury mechanism to predict fetal outcome. However, due to the difficulties in measuring this mechanism in the pregnant dummy, such as tissue strain and pressure, a computational model of the pregnant occupant was created (Moorcroft, 2003). This computer model has been used to evaluate frontal crashes and has shown that local uterine compression is a critical factor in predicting placental abruption (Duma, 2004).

Although several studies have investigated pregnant occupant biomechanics, there is much that remains unknown. In particular, this paper investigates two objectives. First, the overall crash distribution of pregnant occupants is analyzed. Second, computer simulations are presented to examine the effectiveness of current restraint systems as well as possible future four-point restraint systems in far side impacts.

METHODS

In order to determine the pregnant occupant crash exposure, the National Automobile Sampling System Crashworthiness Data System (NASS CDS) was searched for the 11 years between 1993 to 2003. All crashes involving pregnant occupants were collected and raw and weighted case data were analyzed for seating position and impact direction.

Given that far side crashes were the second leading crash mode, and no previous research had examined this event, the second part of this study investigated pregnant occupant biomechanics in far side impacts. For this study, a previously validated MADYMO computer model of a 30 week pregnant occupant was utilized to investigate pregnant occupant biomechanics in far side crashes (Moorcroft, 2003). The model is a modified MADYMO small female human model with a pregnant abdomen. The abdomen consists of the uterus, placenta, and amniotic fluid. The uterus is supported by two pairs of ligaments and surrounded by fat. Four techniques were used to validate the pregnant model. First, a global biofidelity response was evaluated by using a seatbelt to compress dynamically the pregnant abdomen (Moorcroft, 2003). The force versus compression results were within the published corridors from scaled cadaver tests (Hardy, 2001). Second, a similar validation procedure was performed with a rigid bar and these results were also consistent with previous data (Rupp, 2001; Hardy, 2001). The third technique involved validating the model against real-world crashes in order to investigate the model's ability to predict injury. Using fatal crashes from pregnant occupants, the model showed strong correlation ($R^2 = 0.85$) between peak strain at the utero-placental interface (UPI) as measured in the model compared to risk of fetal demise as reported in the real-world crashes over a range of impact velocities and restraint conditions (Klinich, 199b). The forth method compared the physiological failure strain from placental tissue tests to the failure strain measured in the model. Tissue tests suggested approximately a 60% failure strain for UPI tissues which is in agreement with the model's prediction of 75 % risk of fetal loss at a 60% strain in the UPI (Rupp, 2001). In summary, the global, injury, and tissue level validation techniques all indicate the model is good at predicting injurious events for the pregnant occupant.

A total of 12 simulations were performed using three impact speeds (5, 15, and 25 mph). The crash speeds, or change of velocity, of 5 mph, 15 mph, and 25 mph were chosen to represent mild, moderate, and very severe far-side impacts. To achieve the desired change in velocity, the occupant was accelerated with a triangular pulse for 100 ms. Maternal head and pelvis accelerations where recorded in addition to fetal injury risk as measured by peak strain at the uterine-placental interface as a predictor of placental abruption.

Four restraint conditions were simulated including unbelted, lap-belt only, three-point belt, and a four-point belt. The vehicle interior was comprised of a driver seat, passenger seat, center console, and passenger door with window (Figure 1). This interior was modeled with a combination of planes and ellipsoids. The driver's seat, modeled as two planes, was considered rigid to minimize the effect on the occupant's motion. The center console was modeled with an ellipsoid 2.5 inches wide, 6.5 inches tall, and 20 inches long. The top of the console is approximately 7.5 inches above the seat cushion-seat back intersection. Two contact properties were used for the center console, one for contact with the lower legs and one for contact with the pelvis and remainder of the occupant model. The pelvis contact was modeled as stiff to represent the actual car consol, while the leg contact was modeled as soft due to numeric instabilities in the human model from leg to leg contact that resulted in extreme local motion that caused the simulation to terminate. The passenger side of the vehicle was predominately modeled with planes for simplicity, with a few notable exceptions. The seat back was modeled as an ellipsoid because of the possibility of the head hitting the edge of the seatback and the window was modeled as an ellipsoid to allow for the observed angle. All contacts utilized the stiff console force-deflection property. The seat friction coefficient was set to 0.4.

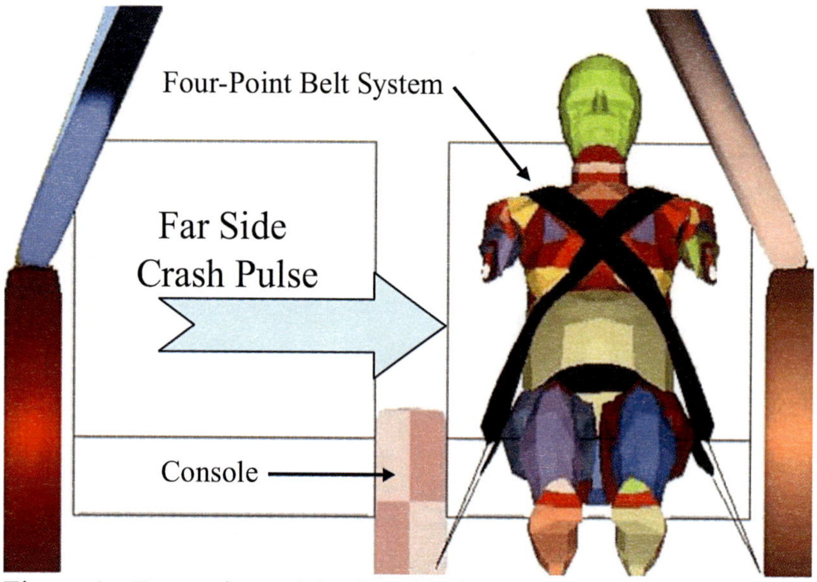

Figure 1 - Front view of the far-side impact pulse on the pregnant occupant with a four-point belt system

RESULTS

The NASS CDS search yielded a total of 220,672 weighted cases (876 raw cases) involvind pregnant occupants. For crash direction, 53% were frontal, 13% were far side, 13% were near side, 11% were rear, and 8% were rollover (Figure 2). For seat position, 75% were drivers, 22% were passengers, and 3% were in the rear seat. When accounting for seating position and crash direction, far side and near side impacts were tied for second at 13 % each (Figure 3).

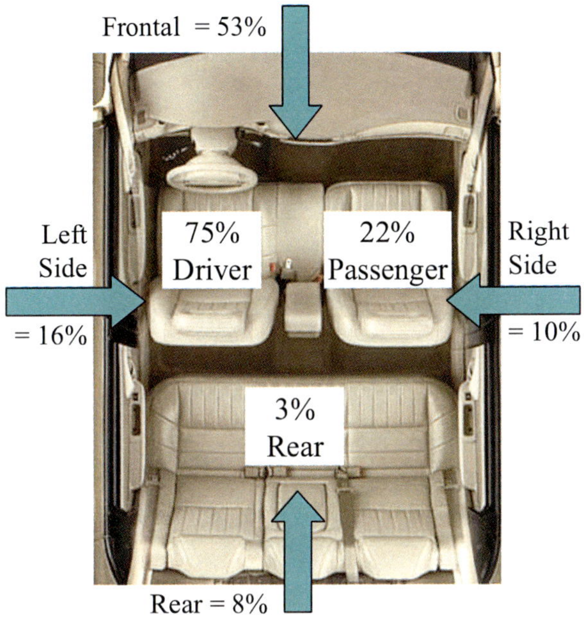

Figure 2 - Occupant seating position and impact direction distributions for pregnant occupants

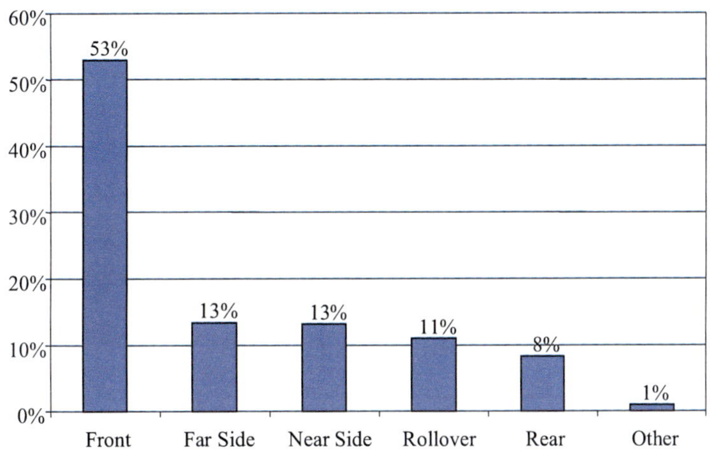

Figure 3 - Pregnant occupant crash exposure accounting for seating position and impact direction

Direct contact with the center console created the largest pelvic accelerations for all simulations (Figure 4). This interaction between the occupant and the interior causes the pelvis to absorb most of the energy to slow the occupant down. This occurred at 15 mph and 25 mph impact speeds. Although the peak pelvic acceleration was reduced with the presence of a restraint, it only varied slightly between the lap-belt only and lap-belt with a shoulder harness cases. Pelvic accelerations were as high as 591g and 539 g for the lap-belt only and four-point belt simulations respectively. For a slow speed impact of 5 mph, all cases had low pelvic accelerations because there was no contact with the console.

During the low speed crash all restraint configurations produce low Head injury criteria (HIC) values (Figure 5). With an increase in the crash severity, the three-point and four-point restraints have a greater affect in limiting the occupant excursion. For both the unbelted and lap-belt only cases, the mother's head strikes the opposite side door during a 25 mph crash. Thus, the result is a HIC value of 4000 which greatly exceeds the HIC injury threshold for skull fracture. The occupant's head also strikes the opposite side of the interior for the unbelted 15 mph crash resulting in a HIC value that is above the injury threshold. The restricted range of motion allowed by the restraints with a shoulder harness prevents contact with vehicle interior by the occupant's head which greatly reduces the HIC value.

The risk of serious fetal injury was calculated from the peak uterine strain so those results have the same trends. As with the other injury criteria, the low impact velocity cases had low values with negligible risk (Figure 6 and Figure 7). For 15 mph and 25 mph speed simulations, risk of fetal injury was between 33.5% and 61.0%. All of the computer simulations of pregnant occupants in far side impacts at 5 mph resulted in very low risk of maternal or fetal injury.

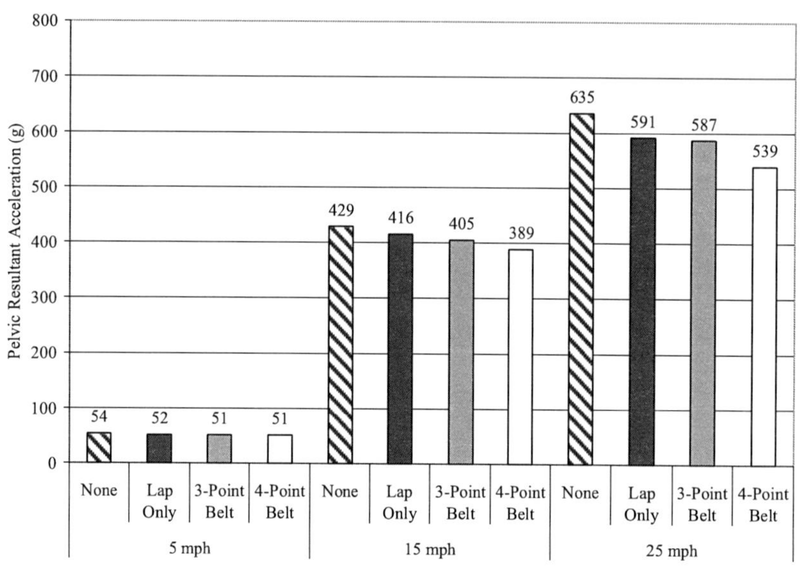

Figure 4 - Peak pelvic accelerations for all simulations

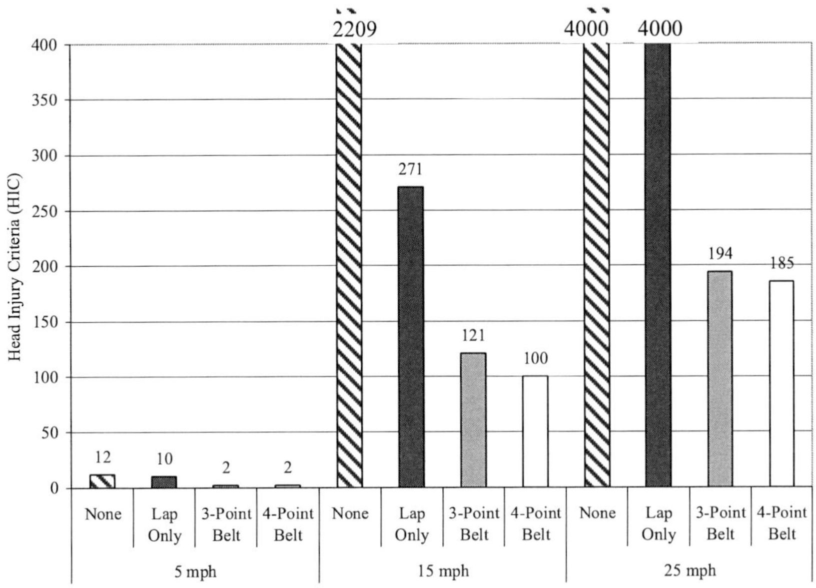

Figure 5 – Peak maternal HIC values for all simulations

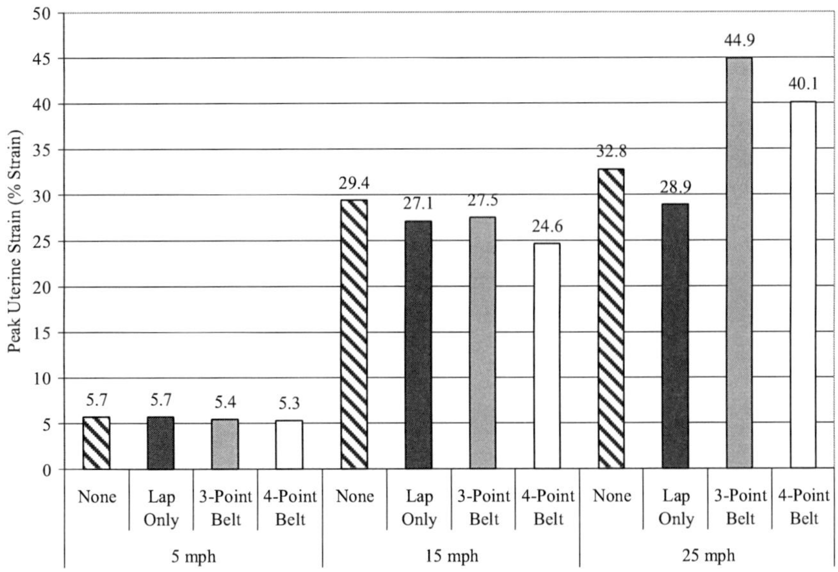

Figure 6 Peak UPI strain values for all simulations.

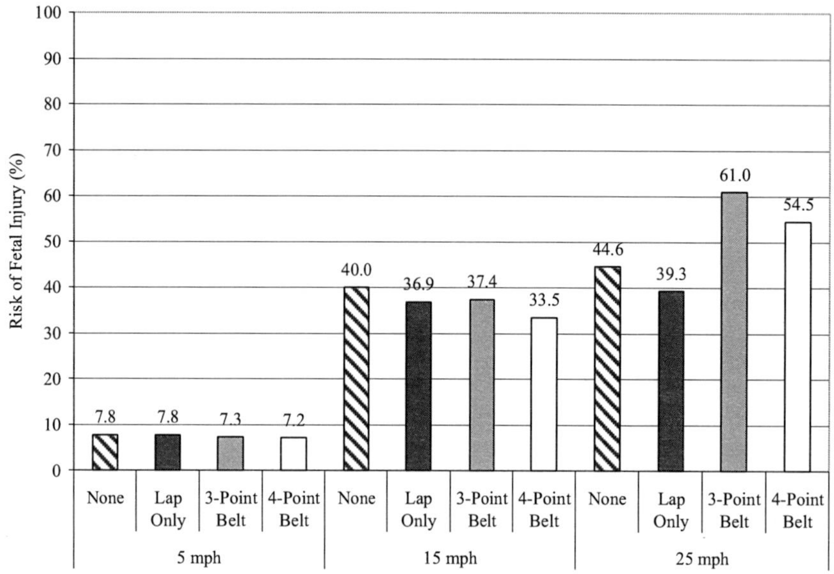

Figure 7 - Risk of serious fetal injury

The loading mechanism for the strain was dependent on the restraint configuration. With the shoulder belts in the three-point and four-point systems, the abdomen was loaded directly (Figure 8). The four-point belt distributed the load to the abdomen and had a reduced peak strain when compared to the three-point belt. However when there was no shoulder belt present the strain was induced by inertial loading. For all the simulations the center console was not a possible injury mechanism for the fetus. The occupant passed over the console in the unrestrained cases and rotated above it for the restrained cases.

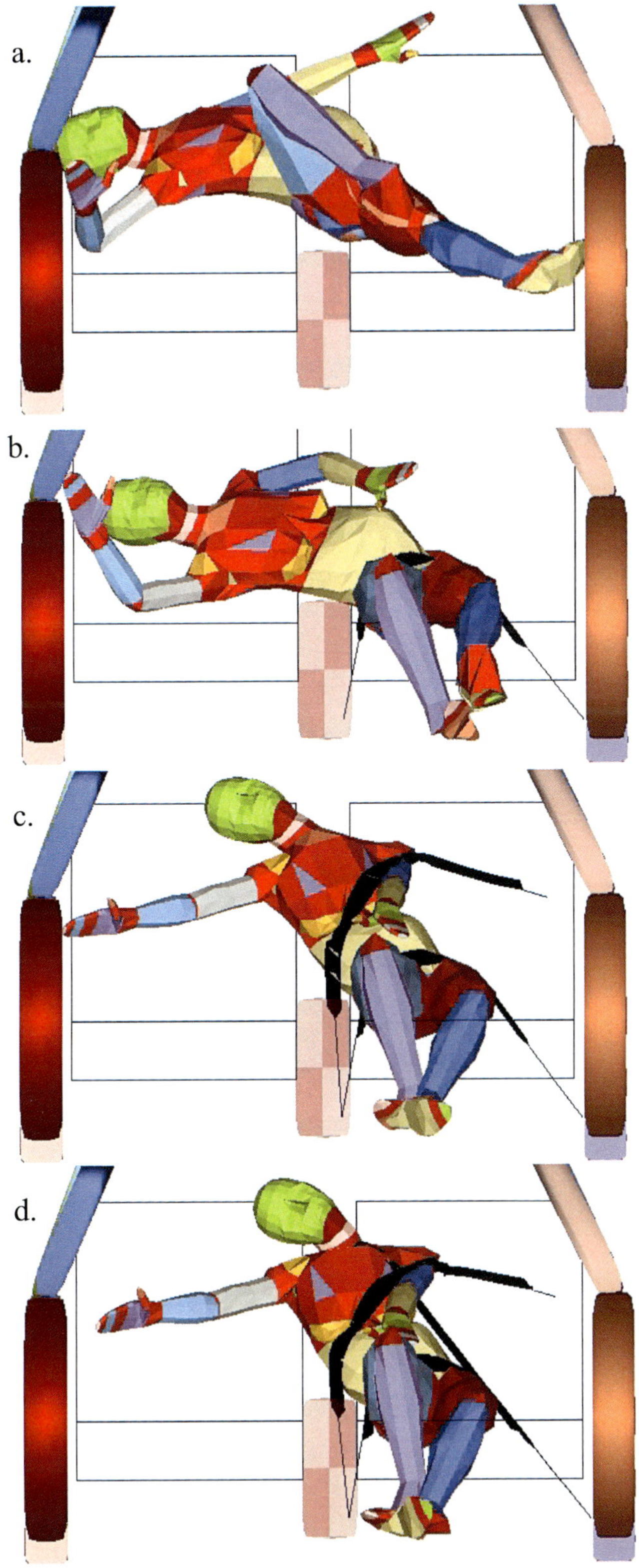

Figure 8 - The maximum excursion decreases with an increase in occupant restraint protection: a) no restraint, b) lap-belt only, c) three-point belt, d) four-point belt.

DISCUSSION

In order to restrain the pregnant occupant in any crash, forces must be applied to her in order to stop her momentum. Ideally, these forces are applied over the greatest duration in order to reduce their magnitude, and they are applied to the strong bones of the body. It was shown that the three-point belt and the four-point belt were superior in protecting the pregnant occupant by reducing the intrusion towards the far side door and therefore eliminating the head strike potential. However, this resulted in some force being applied through the abdomen and therefore it increased the risk of fetal injury. This is an acceptable trade-off given the most important factor in saving the fetus' life is keeping the mother alive. The four-point belt is better than the three-point belt in that some of the load is applied through the mother's neck and therefore less is applied to the abdomen in order to restraint her.

It is important to note several limitations of this study. The pregnant model approximates the pregnant female, and although it has been validated, there are still limitations. As with all computational models, this model is limited by the accuracy of input and simplifications made. The tissue data, from which the failure strain is derived, is sparse and simplifications are made to use that data as a material model. Additionally, the boundary conditions and geometry can and should be improved in future generations of the model. Furthermore, the model only looks at injury at the UPI. In cases with very large deflections, direct injury to the fetus may occur at injury rates different then those for placental abruption. It is recommended that the methods in this paper be applied to future generations of the pregnant occupant model to provide a continually improving understanding of pregnant occupant injury risk prediction.

Additional limitations of this study include the fact that only one type of four-point belt system was investigated. It is possible that other styles of four-point belts would have different and possibly better results. Another limitation is that vehicle crush and intrusion was not modeling in these simulations. By moving the far side door closer to the pregnant occupant, the intrusion into the passenger compartment becomes more important and the possibility of a head strike is increased.

Finally it is important to note that these results agree with previous simulations for frontal impacts. Overall, the results indicate for all frontal and side impacts that it is safest for the pregnant occupant to ride in the passenger seat while wearing a three-point belt, or four-point belt if possible, and utilizing the frontal airbag when appropriate (Moorcroft, 2004).

CONCLUSIONS

Overall, the crash exposure of pregnant occupants is nearly identical to the average driving population. Although the fetal injury risk is slightly increased in the three-point and four-point simulations, this negative trends is outweighed by the considerble benefit observed in the substantial reduction in maternal HIC when using the three or four-point belts. It has been demonstrated that the a critical factor in protecting the fetus is keeping the mother alive. While both the three-point and four-point belt systems provided good protection in far side simulated crashes, the four-point belt did a superior job of restraining the pregnant occupant and reducing overall translation. Given the magnitude and severity of pregnant occupant injuries, this study illustrates the need for manufacturers to consider four-point belt systems for this population.

ACKNOWLEDGEMENTS

The thank Kathy Klinich at the University of Michigan Transportation Research Institute for her assistance in the development of the pregnant model. The funding for this research has been provided by the Virginia Tech – Wake Forest Center for Injury Biomechanics Foundation. Also, the authors thank Altair Engineering Inc. for the use of HyperWorks as the pre and post processor, and TNO for the use of MADYMO.

REFERENCESS

Attico NB, Smith III RJ, Fitzpatrick MB. Automobile safety restraints for pregnant women and children. *J Reprod Med*; 31(3):187-92, 1986

Duma SM, Moorcroft DM, Stitzel JD, Duma GG. Evaluating pregnant occupant restraints: the effect of local uterine compression on the risk of fetal injury. Proceedings of the Association for the Advancement of Automotive Medicine, October 2004.

Hardy WN, Schneider LW, Rouhana SW. Abdominal impact response to rigid-bar, seatbelt, and airbag loading. *Stapp Car Crash Journal*, 45, 1-32, 2001.

Klinich KD, Schneider LW, Eby B, Rupp J, Pearlman MD (1999a). Seated anthropometry during pregnancy. UMTRI-99-16.

Klinich KD, Schneider LW, Moore JL, Pearlman (1999b). Investigations of crashes involving pregnant occupants. UMTRI-99-29.

Moorcroft DM, Duma SM, Stitzel JD, Duma GG (2003a) Computational Model of the Pregnant Occupant: Predicting the Risk of Injury in Automobile Crashes. *American Journal of Obstetrics and Gynecology* Vol. 189 (2), pp. 540-544, 2003.

Moorcroft DM, Duma SM, Stitzel JD, Duma GG (2004). The effect of pregnant occupant position and belt placement on the risk of fetal injury. SAE technical paper series. 2004-01-0324.

Rupp JD, Klinich KD, Moss S, Zhou J, Pearlman MD, Schneider LW. Development and testing of a prototype pregnant abdomen for the small-female Hybrid III ATD. *Stapp Car Crash Journal*, 45, 61-78, 2001.

Weiss HB, Strotmeyer S. Characteristics of pregnant women in motor vehicle crashes. *Injury Prevention*:8(3):207-214, 2002.

Reprinted from American Journal of Obstetrics & Gynecology, Vol. 189(2), 2003, Moorcroft et al, pp 540-544, "Computational model of the pregnant occupant: Predicting the risk of injury in automobile crashes."

Computational model of the pregnant occupant: Predicting the risk of injury in automobile crashes

David M. Moorcroft, MS,[a] Joel D. Stitzel, MS,[a] Greg G. Duma, MD,[b] and Stefan M. Duma, PhD[a]

Blacksburg, Va, and Cincinnati, Ohio

OBJECTIVE: The goal of this study was to create a computational model of the pregnant occupant of a motor vehicle to predict fetal outcome in crashes.

STUDY DESIGN: A finite element uterine model of a 7-month pregnant woman was created and integrated into a multibody human model. Unrestrained, three-point belt, and three-point belt plus airbag tests were simulated at speeds that ranged from 13 to 55 km per hour.

RESULTS: Peak uterine strain, as determined by the model, correlated well with the risk of fetal death, as determined by investigations of car crashes. The strain in the uterine wall exceeded the limits of the tissue in simulations of no restraint at 35 km per hour and three-point belt tests at 45 and 55 km per hour. The safest restraint for the pregnant driver is the combination three-point belt and airbag.

CONCLUSION: The model is a good first step toward the prediction of the risk of fetal death and verified experimental findings that note the importance of proper restraint use for the pregnant occupant. (Am J Obstet Gynecol 2003;189:540-4.)

Key words: Pregnant, model, fetal loss, injury introduction

For women of reproductive age, accidents and homicide are the leading causes of death worldwide.[1] It has been estimated that trauma affects between 6% and 7% of all pregnancies and that two thirds of all trauma during pregnancy is the result of motor vehicle crashes.[1] Abruptio placentae has been shown to account for 50% to 70% of fetal losses in motor vehicle crashes.[2] Uterine rupture and direct fetal injury account for <10% of fetal losses; the remaining 20% to 40% of fetal losses are the result of severe maternal injuries, including maternal death.[2]

The medical literature contains hundreds of individual case studies on injuries that were sustained by pregnant occupants in motor vehicle crashes.[2-13] These studies typically are reviews of hospital medical records, and although they present a good resource for incidence and types of injuries, the cases are limited in their ability to determine injury mechanisms and inherently are unable to predict the risk of injury in future crashes.

In the late 1960s and early1970s, several researchers investigated seatbelt effectiveness on pregnant baboons and monkeys. It was concluded that the three-point belt system provided superior protection to both mother and fetus compared with the lap belt only.[14-17] In the mid 1990s and early 2000, two prototype pregnant dummies were developed by Pearlman and Viano[18] and Rupp et al.[19] Both dummies were tested under several restraint conditions. The findings agreed with results from animal testing regarding the use of a three-point belt, and preliminary data suggest that an airbag coupled with a three-point belt system provides further protection to both mother and fetus. Testing with crash test dummies is costly and does not show the biofidelity of a human occupant.

Rupp et al[19] performed simplified computer modeling to determine the most likely causes of abruptio placentae from impact loading and found that pressure gradients generated by the inertia of the amniotic fluid and strains in the uteroplacental interface because of direct loading of the abdomen were the dominant mechanisms of abruptio placentae. The simplified model provided valuable information regarding the injury mechanisms of abruptio placentae; however, the model is not included in a human body model to determine fetal loss from motor vehicle crashes.

The purpose of this study was to create a computational model of a pregnant occupant and evaluate this occupant under a range of loading conditions.

From Virginia Tech, Impact Biomechanics Laboratory,[a] Blacksburg, Va, USA, and the Department of Obstetrics and Gynecology, University of Cincinnati School of Medicine[b] Cincinnati, Ohio, USA.

Received for publication September 16, 2002; revised December 2, 2002; accepted April 8, 2003.

Reprint requests: Stefan Duma, 114 Randolph Hall, Blacksburg, VA 24061. E-mail: Duma@vt.edu

0002-9378/2003 $30.00 + 0
doi:10.1067/S0002-9378(03)00519-2

Material and methods

Motor vehicle crashes were simulated with the use of MADYMO (TNO, Delft, Netherlands), a program designed specifically for occupant safety analysis. A finite element model of a pregnant uterus was developed and inserted into the abdomen of a multibody human model

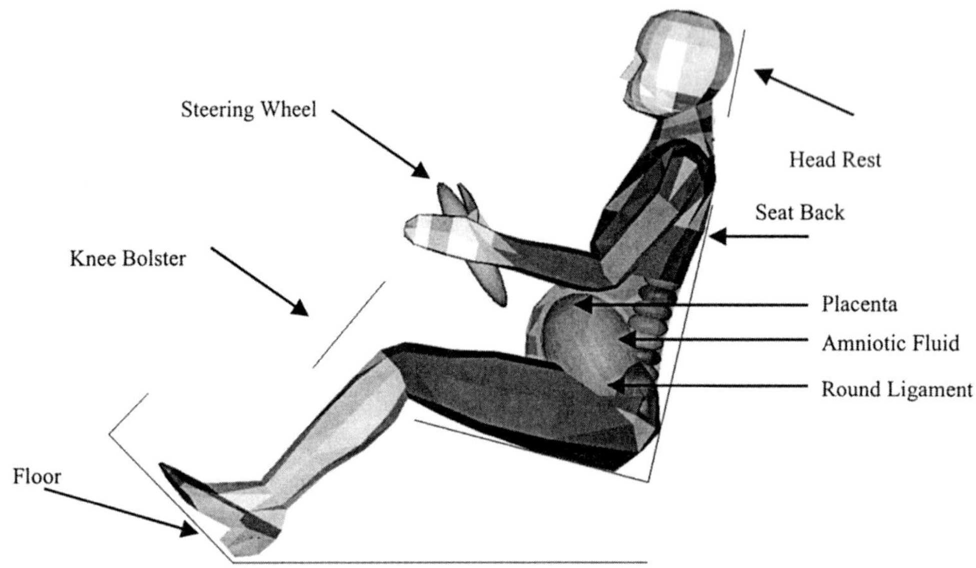

Fig 1. Pregnant occupant in vehicle interior.

(Fig 1). The human model is a 5th percentile human female model (5 feet tall, 110 pounds), and the uterus is designed to represent an occupant in her week 30 of gestation. The multibody human model provides the biofidelic response of an occupant in a motor vehicle crash, while reducing the computational time compared to a full finite element human model. The uterine model includes the uterus, placenta, and amniotic fluid and is attached within the human abdomen by the uterosacral and round ligaments and its connection in the cervical area (Fig 2). The uterine model is surrounded by fat to accurately represent the boundary between the uterus and spine, abdominal organs, and pelvis. A fetus was not included because the injury mechanisms that predominantly contribute to abruptio placentae, as described by Rupp et al,[19] are independent of the fetus. The uterus is 27 cm long, 18 cm wide, and 1 cm thick; the placenta is located at the fundus of the uterus and is 2 cm thick. The remainder of the interior of the uterus is filled with the amniotic fluid. The uterus, placenta, and amniotic fluid are modeled as linear elastic isotropic solids. The uterine model, including the ligaments and fat, is made up of 11,632 elements and 16,335 nodes. The weight of the pregnant model is 135 pounds.

Injury to the fetus can be determined through an injury risk curve developed by Klinich et al[20] from investigations of motor vehicle crashes that involved pregnant women. The curve predicts the risk of adverse fetal outcome based on crash speed for improper restraint and proper restraint. Proper restraint was defined as the correct use of a three-point belt or a three-point belt plus an airbag. Improper restraint included all other forms of restraint, from no restraint to shoulder belt only or airbag only. Injury can also be predicted from tissue failure. Abruptio placentae is believed to occur when the strain (change in length/original length) in the uteroplacental interface exceeds 60%, or when uterine strain exceeds 60%.[19]

The anthropometry of a pregnant woman was quantified also by Klinich et al.[22] The abdominal contour of the pregnant model closely matches the data of Klinich et al that were used for the second-generation pregnant dummy. Positioning of the pregnant occupant in the MADYMO vehicle was based on this data for the small female group at 30 weeks of gestation. The initial distance between the abdomen and the bottom of the steering wheel is 38 mm. The overlap of the uterus to the steering wheel is 12%. The overlap is defined as the ratio of the height of the uterus above the steering wheel to the total height of the uterus. The seatback angle, relative to vertical, is 13 degrees; and the steering wheel tilt is 29 degrees from vertical.

The pregnant model was validated by simulating belt loading and rigid bar impact tests. The abdominal response of the pregnant model is within the defined upper and lower bounds developed by Hardy et al[23] for the 50th percentile male and scaled to a 5th percentile pregnant women by Rupp et al.[19]

The test matrix for this study included 12 sled test simulations (Table). The simulations included occupants with no restraint, a three-point belt, and a three-point belt plus airbag under various crash speeds that ranged from 13 km per hour to 55 km per hour. The test matrix was designed to identify inertial, steering wheel, seatbelt, and airbag loading on the pregnant uterus, and to match testing of the second-generation pregnant dummy.[19] In each test, the pregnant occupant was a driver in a frontal crash. Simulations were run in excess of 100 milliseconds, to ensure that the forward motion of the occupant had

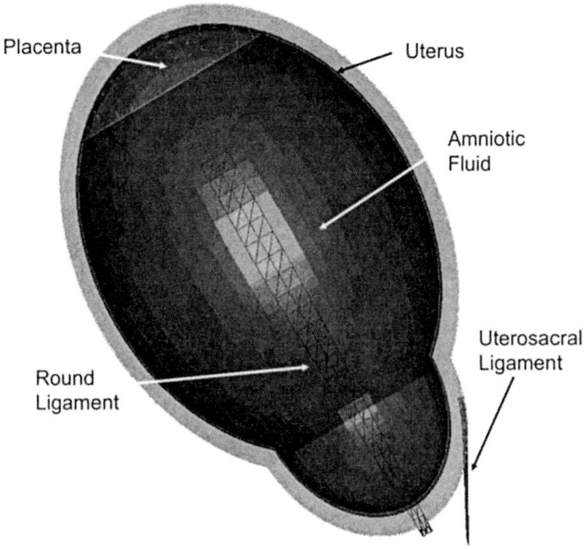

Fig 2. Uterine model shows placental location and supporting ligaments.

Table. Effect of restraint use and crash speed on the risk of fetal loss and peak uterine strain

Test No.	*Restraint configuration*	*Crash speed (kph)*	*Risk of adverse fetal outcome (%)**	*Maximum strain in uterus (%)*
1	None	13	36	23.2
2	None	20	54	36.6
3	None	25	70	44.6
4	None	35	86	60.8
5	3-pt belt	13	9	15.5
6	3-pt belt	25	26	27.9
7	3-pt belt	35	51	52.6
8	3-pt belt	45	76	58.7
9	3-pt belt	55	90	61.2
10	Airbag + 3-pt belt	25	26	28.1
11	Airbag + 3-pt belt	35	51	33.0
12	Airbag + 3-pt belt	45	76	46.6

*Based on crash investigations.[20]

stopped. The computational model uses peak Von Mises strain in the uterus near the placenta as the measure for the prediction of the risk of injury. This strain was compared to the risk of adverse fetal outcome, as defined by Klinich et al,[20] to determine whether strain could be used as a predictive tool.

Results

For all three restraint groups, the risk of adverse fetal outcome and the maximum uterus strain increased as the crash speed increased (Table). The risk of adverse fetal outcome, or fetal loss, is calculated on the basis of statistical of case report data analyses performed by Klinich et al.[20] The case report data included pregnant drivers in various crash and restraint conditions, where fetal loss was reported. The computational results are compared with these results for interpretation.

For tests that were run at the same speeds, the unrestrained pregnant female had the highest risk of adverse fetal outcome and the highest uterus strain compared with the tests with a three-point seatbelt or a three-point seatbelt and an airbag. In particular, the matched tests at 35 km per hour resulted in a 60.8% strain value for the unrestrained occupant compared with 52.6% strain for the three-point seatbelt and only 33.0% strain for the three-point seatbelt and airbag combination. The risks for identical speeds are identical because Klinich et al[20] considers restrained occupants equally, whether the occupant is belted, belted with airbag, or both, although the finite element model predicts different strains for the different speeds.

Three simulations at the same 35 km per hour crash speed were compared to reveal the kinematic and loading differences between the three restraint types tested (Fig 3). As with the three-point belt, the abdomen has minimal contact with the steering wheel; because the airbag absorbs some of the energy of the crash, the uterine strain is reduced by approximately one half compared with the no-restraint case. During the belted simulations, with or without an airbag, peak strain occurs when the forward momentum of the pregnant occupant is halted. In other words, the peak strain in the belted simulations is primarily a result of inertial loading.

In the course of model development, several parametric studies were performed to examine the location of stress in the uterine model and the effect of the modeled ligaments. A stress analysis with Von Mises stress shows that peak strains in the uterus for the unbelted condition (Fig 4) occur in the anterior portion of the uterus, whereas strains are lower in the same location for the belted condition.

The peak uterine strain was compared with the risk of adverse fetal outcome, as defined by Klinich et al,[20] to determine whether strain could be used as a predictive tool (Fig 5). A 2-variable linear regression of the data shows that the uterine strain from the computational model is a good predictive measure of the risk of fetal injury ($R^2 = 0.85$). This approach shows that the model can be validated empirically by comparison with the results from real world cases with pregnant occupants.

The main method of the protection of the fetus is to protect the mother. Head injury criteria (HIC) and viscous criterion were examined as a check of overall occupant protection. The small female head injury criterion was set at 1000.[24] The maximum head injury criterion in the simulations was 310 in the 55 km per hour, three-point belt test. In general, the head injury criterion was lowest for the three-point belt plus airbag tests and highest for the

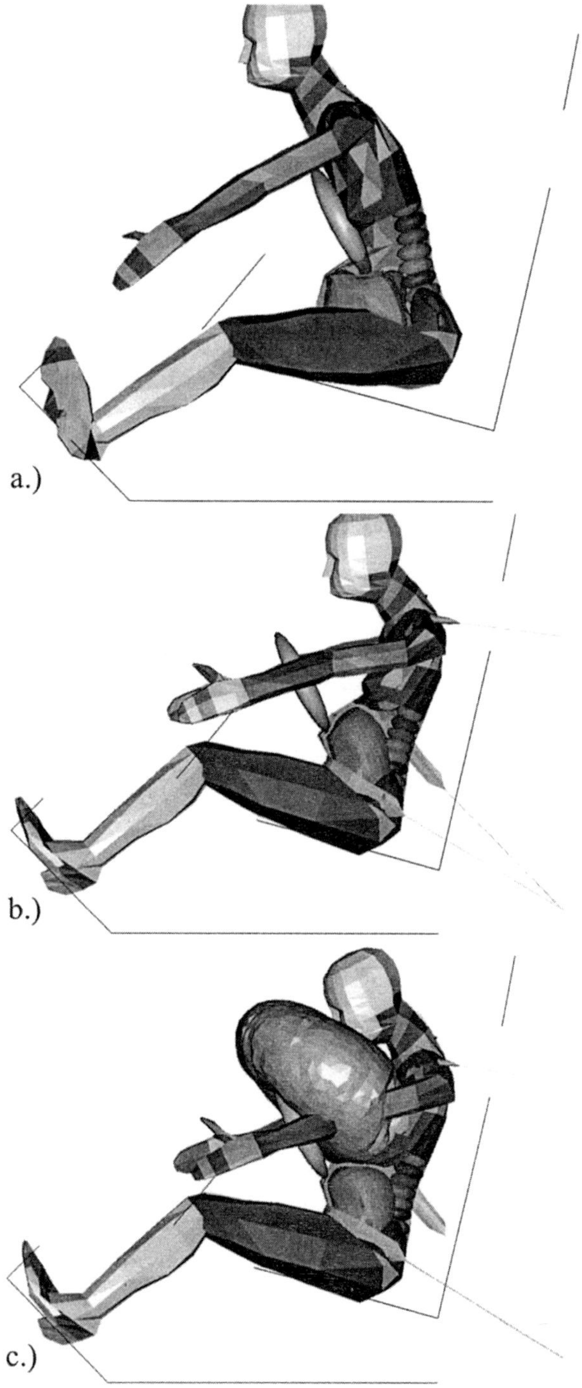

Fig 3. Pregnant occupant in a 35 km per hour crash at peak uterine strain. **a,** An unbelted pregnant occupant is shown contacting the steering wheel (no restraint at 101 milliseconds), which results in a large deformation of the uterus. The peak strain is the substantial steering wheel loading that results in large deformation of the uterus. **b,** A matched belted occupant simulation (three-point belt at 75 milliseconds) shows that steering wheel contact is minimal, which results in a lower strain in the uterus. However, there is considerable neck flexion, which could lead to maternal injury. **c,** An airbag deployment is shown that is combined with a three-point belt and airbag at 82 milliseconds.

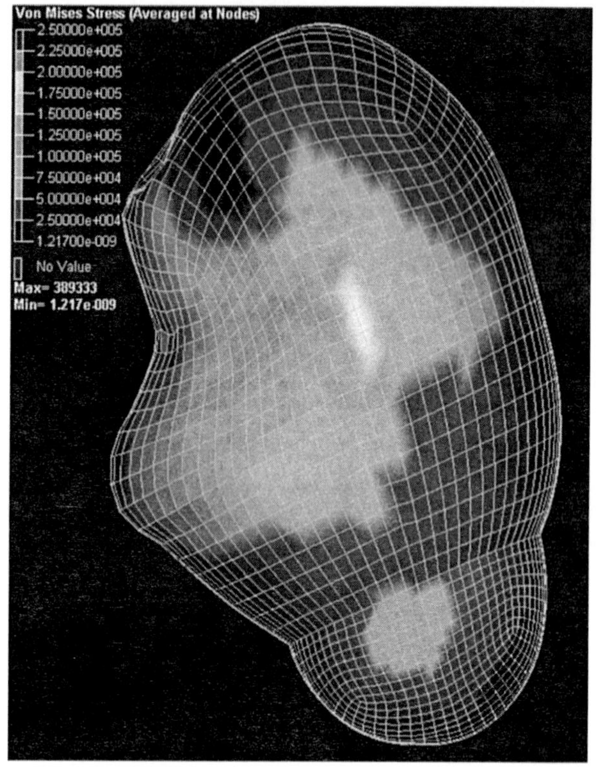

Fig 4. Von Mises stress in the uterus at 35 km per hour during a three-point belt test.

no-restraint tests. The viscous criterion limit is 1.0; the simulations recorded a high of 0.7 in the no restraint 35 km per hour test. The airbag tests showed a slight increase in viscous criterion compared with the three-point belt only. The three-point belt plus airbag restraint provided the greatest amount of protection to the mother.

Comment

The peak uterine strain was found to be a good predictor of fetal outcome. The three-point belted tests produced the greatest variation, typically predicting uterine strain greater than the airbag tests. This result is not unexpected. The three-point belt plus airbag restraint generally is considered safer than a three-point belt restraint, and the associated risk of adverse fetal outcome is based on proper restraint use, which includes three-point belt and three-point belt plus airbag restraint use. Therefore, three-point belt tests should report higher strains than the three-point belt plus airbag simulations.

The kinematic output of the model allows for a close-up view of the interior of the uterus. This reveals the motion and deformation of the uterine wall and placenta, which is an advantage over dummy testing. The location of tissue failure can be determined and used to design advanced restraint systems.

Several assumptions were made in the formulation of the finite element uterine model. The uterus, placenta,

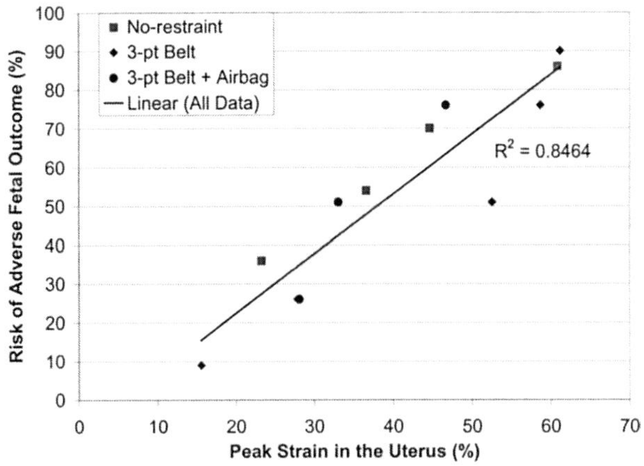

Fig 5. Peak strain in the uterus as a predictor of adverse fetal outcome.

and amniotic fluid were assumed to be linear, elastic, isotropic solid materials. The uterus is known to be anisotropic and viscoelastic; however, there is not currently enough data to apply these material properties to the model. The geometry, although an improvement over previous modeling efforts, is simplified. It is recommended that these improvements be included in future versions of the model.

A finite element model of the pregnant abdomen was created to predict fetal outcome after a motor vehicle crash. The model was incorporated into a human body model in a dynamic solver and was validated with data from previous studies. The model can distinguish between inertial, steering wheel, seatbelt, and airbag loading.

Uterine strain was found to be good predictor of fetal injury because of abruptio placentae. Injury to the mother, particularly head and chest trauma, must also be considered. The model has verified previous experimental findings regarding the importance of proper restraint use for the pregnant occupant. The combination of a three-point seatbelt and airbag was found to be the safest restraint system for the pregnant occupant. The model can be used to run numerous tests quickly and to design advanced restraint systems that will be designed specifically for pregnant occupants.

REFERENCES

1. Pearlman MD. Motor vehicle crashes, pregnancy loss and preterm labor. Int J Gynecol Obstet 1997;57:127-32.
2. Pearlman MD, Tintinalli JE, Lorenz RP. A prospective controlled study of outcome after trauma during pregnancy. Am J Obstet Gynecol 1990;162:1502-10.
3. Dahmus MA, Sibai BM. Blunt abdominal trauma: are there any predictive factors for abruptio placentae or maternal-fetal distress? Am J Obstet Gynecol 1993;169:1054-9.
4. Duma SM, Jernigan VM, Stitzel JD. The effect of frontal air bags on eye injury patterns in automobile crashes. Arch Ophthalmol 2002;120:1517-22.
5. Elliott M. Vehicular accidents and pregnancy. Aust N Z J Obstet Gynaecol 1966;6:279-86.
6. Esposito TJ, Gens DR, Smith LG, Scorpio R, Buchman T. Trauma during pregnancy. a review of 79 cases. Arch Surg 1991;126:1073-8.
7. Fort AT, Harlin RS. Pregnancy outcome after noncatastrophic maternal trauma during pregnancy. Obstet Gynecol 1970;35:912-5.
8. Goodwin TM, Breen MT. Pregnancy outcome and fetomaternal hemorrhage after noncatastrophic trauma. Am J Obstet Gynecol 1990;162:665-7.
9. Hoff WS, Lucke JF, Diamond DL, D'Amelio LF, Tinkoff GH. Maternal predictors of fetal demise in trauma during pregnancy. Surg Gynecol Obstet 1991;172:175-80.
10. Kissinger DP, Rozycki GS, Morris JA, Knudson MM, Copes WS, Bass SM, et al. Trauma in pregnancy: predicting pregnancy outcome. Arch Surg 1991;126:1079-86.
11. Rothenberger D, Quattlebaum FW, Perry JF Jr, Zabel J, Fischer RP. Blunt maternal trauma: a review of 103 cases. J Trauma 1978;18: 173-9.
12. Timberlake GA, McSwain NE Jr. Trauma in pregnancy: a 10-year perspective. Am Surg 1989;55:151-3.
13. Williams JK, McClain L, Rosemurgy AS, Colorado NM. Evaluation of blunt abdominal trauma in the third trimester of pregnancy: maternal and fetal considerations. Obstet Gynecol 1990;75:33-7.
14. Snyder RG, Snow CC, Crosby WM, Hanson P, Fineg J, Chandler R. Impact injury to the pregnant female and fetus in lap belt restraint. Proceedings of the 10th Stapp Car Crash Conference; 1966 Nov 8-9; Holloman AFB, New Mexico. Warrendale (PA): Society of Automotive Engineers; 1966. p. 249-59.
15. Crosby WM, Snyder RG, Snow CC, Hanson PG. Impact injuries in pregnancy. Am J Obstet Gynecol 1968;101:100.
16. van Kirk DJ, King AI. A preliminary study of an effective restraint system for pregnant women and children. Proceedings of the 13th Stapp Car Crash Conference; 1969 Dec 2-4; Boston, Massachusetts. Warrendale (PA): Society of Automotive Engineers; 1969. p. 353-64.
17. King AI, Crosby WM, Stout LC, Eppinger RH. Effects of lap belt and three-point restraints on pregnant baboons subjected to deceleration. Proceedings of the 15th Stapp Car Crash Conference; 1971 Nov 17-19; San Diego, California. Warrendale (PA): Society of Automotive Engineers; 1971. p. 83.
18. Pearlman MD, Viano D. Automobile crash simulation with the first pregnant crash test dummy. Am J Obstet Gynecol 1996;175:977-81.
19. Rupp JD, Klinich KD, Moss S, Zhou J, Pearlman MD, Schneider LW. Development and testing of a prototype pregnant abdomen for the small female Hybrid III small female ATD. Stapp Car Crash J 2001; 45:61-78.
20. Klinich KD, Schneider LW, Moore JL, Pearlman MD. Injuries to pregnant occupants in automotive crashes. Proceedings of the 42nd annual meeting of the Association for the Advancement of Automotive Medicine; 1998 Oct 5-7; Charlottesville, Virginia. Charlottesville: The Association; 1998. p. 57-91.
21. Reference deleted in press.
22. Klinich KD, Schneider LW, Rupp J, Eby B. Pearlman MD. Challenges in frontal crash protection of pregnant drivers based on anthropometric considerations. SAE technical paper.
23. Hardy WN, Schneider LW, Rouhana SW. Abdominal impact response to rigid-bar, seatbelt, and airbag loading. Stapp Car Crash J 2001;45: 1-32.
24. Hutchinson J, Kaiser MJ, Lankarani HM. The head injury criterion (HIC) functional. Appl Math Comput 1998;96:1-16.

2004-01-0324

The Effect of Pregnant Occupant Position and Belt Placement on the Risk of Fetal Injury

David M Moorcroft, Joel D Stitzel, Stefan M Duma
Virginia Tech – Wake Forest, Center for Injury Biomechanics

Greg G Duma
University of Cincinnati, School of Medicine

ABSTRACT

The goal of this project was to evaluate the effect of occupant seating and seatbelt placement on the risk of adverse fetal outcome from a motor vehicle crash. Unrestrained, 3-pt belt, and 3-pt belt plus airbag tests were simulated with the Virginia Tech pregnant occupant computational model in both a driver-side and passenger-side vehicle interior in frontal impacts at 35 kph. The pregnant occupant model is a small female human body model modified to include a finite element uterine model. The model was previously created and validated with abdominal force-deflection responses. Peak uterine strain was reduced by 30% to 50% in passenger-side simulations vs. driver-side simulations. However, in the unrestrained, passenger-side simulation, the pregnant occupant sustained a HIC score of 2820, suggesting immediate maternal death and a high likelihood of fetal death. Additional simulations were run in which the vertical position of the lap-belt was varied through three heights. It was found that the vertical position of the lap-belt can increase the risk of adverse fetal outcome by a factor of three. The worse case lap-belt height corresponds to the belt being placed directly over the placenta. This case produced a peak uterine strain of 97%, well above the reputed tissue limit of 60%. It is recommended that, when practical, the pregnant woman ride in the passenger seat with a 3-pt belt. The seat should be positioned as far rearward as possible and the lap-belt should be worn as low as possible

INTRODUCTION

For women of reproductive age, automobile crashes are the leading cause of death worldwide. It has been estimated that 40,000 women in the second half of pregnancy are involved in motor-vehicle crashes each year (Jernigan, 2002). It has also been estimated that between 300 and 3800 women will experience a fetal loss (Klinich, 1999b). Placental abruption, which is the premature separation of the placenta from the uterus, has been shown to account for 50% to 70% of fetal losses in motor vehicle crashes (Pearlman, 1990a).

The best way to protect the fetus is to protect the mother considering that maternal death has a near 100% fetal loss rate (Pearlman, 1990b). If the mother survives, protection of the fetus may best be accomplished by preventing placental abruption, which is thought to be the weakest injury mechanism. Information gathered from crash investigations shows that placental abruption can occur without other, more severe injuries, such as uterine rupture or direct fetal injury, and that when these injuries do occur, they are often accompanied by placental abruption (Rupp, 2001). Placental abruption is thought to occur when the strain in the uterine wall exceeds 60% (Rupp, 2001). This limit is based on experimental tests with uterine tissue and modeling of postpartum separation.

In motor vehicle crashes in the US involving a pregnant occupant, based on information from the NASS database for the years 1995 - 2000, the pregnant occupant has been the driver nearly 78% of the time and in the right front seat 20% of the time (Jernigan, 2002). The primary direction of impact was frontal in nearly half of the crashes involving a pregnant occupant (Jernigan, 2002). The American College of Obstetricians and Gynecologists (ACOG) recommends that pregnant women use the 3-pt belt system throughout pregnancy and that the airbag should not be disabled. The ACOG recommends that "the lap belt portion should be placed under the pregnant women's abdomen, over both anterior superior iliac spines and the pubic symphysis. The shoulder harness should be placed between the breasts". Excessive slack should be eliminated from both the lap and shoulder belts. However, it has been reported in the literature that the lap belt tends to ride up the abdomen during driving, resulting in discomfort of the women and potential hazard to the fetus due to abdominal compression (Klinich, 1999a). The purpose of this project was to explore the effect of seat position and lap-belt placement on the pregnant occupant.

METHODOLOGY

Motor vehicle crash tests of a pregnant occupant were simulated using a model previously developed and validated by Moorcroft (2003). The model is a modified Madymo small female human model with a finite element pregnant abdomen (Figure 1). The abdomen consists of the uterus, placenta, and amniotic fluid. The uterus is supported by two pairs of ligaments and surrounded by fat. The abdominal response of the complete model was validated with belt loading and rigid bar impacts (Moorcroft, 2003). The computational model uses peak von Mises strain in the uterus, near the placenta, as the measure for predicting risk of injury. In previous simulations with the pregnant model, the peak Von Mises strain correlated well with the risk of adverse fetal outcome defined by Klinich (Moorcroft, 2003). High risk was associated with the presumed 60% strain tissue limit allowing the prediction of fetal injury based on the strain.

The current simulations were chosen to determine the effect of occupant position on the response of the pregnant occupant. The test matrix consisted of 6 simulations, run at 35 kph (21.7 mph), with occupant position and occupant restraint variations (Table 1). The applied sled pulse is a half-sine wave imposed for a duration of 100 ms. Acceleration is defined with respect to time; therefore the area under the curve corresponds to the change of velocity of the crash. Two interiors were used in the simulations; a standard driver-side interior and a passenger-side interior. The driver interior is a typical Madymo interior made up of rigid planes to represent the seat, vehicle floor, and knee bolster. Positioning of the pregnant occupant was based on the seated anthropometry of a pregnant woman in her 30th week of pregnancy as defined by Klinich (1999a). Four parameters were chosen to define the position of the occupant, using the parameter values that correspond to the small female group in the Klinich study (average height: 5 ft., average weight: 134 lb). The abdominal clearance, defined as the distance between the abdomen and the bottom of the steering wheel, is 38 mm. The mean overlap of the uterus to the steering wheel is 12%, where the overlap is defined as the ratio of the vertical height of the uterus above the bottom of the steering wheel to the total vertical height of the uterus. The seatback angle, relative to vertical, is 13 degrees, and the steering wheel tilt is 29 degrees from vertical. Standard Madymo finite element belts are used for the three-point restraint condition. For the airbag tests, a Madymo 600 mm driver airbag (volume = 35 L) is used, with inflation triggered 15 ms into the simulation.

The passenger interior is a typical Madymo passenger compartment with a dashboard, and windshield (Figure 1). The seat used for the driver-side impacts is also used for the passenger side tests for consistency, however, the seat was moved backwards to match the distance seen in the typical passenger compartment and representing a fully aft seat position (abdominal clearance to dashboard = 40 cm). The same finite element belts were used for the passenger tests as for the driver tests, with the shoulder belt repositioned to extend from the right shoulder to the left hip. A standard Madymo passenger airbag (volume = 135 L) was used for tests with airbag deployment. Inflation was triggered 15 ms into the crash simulation.

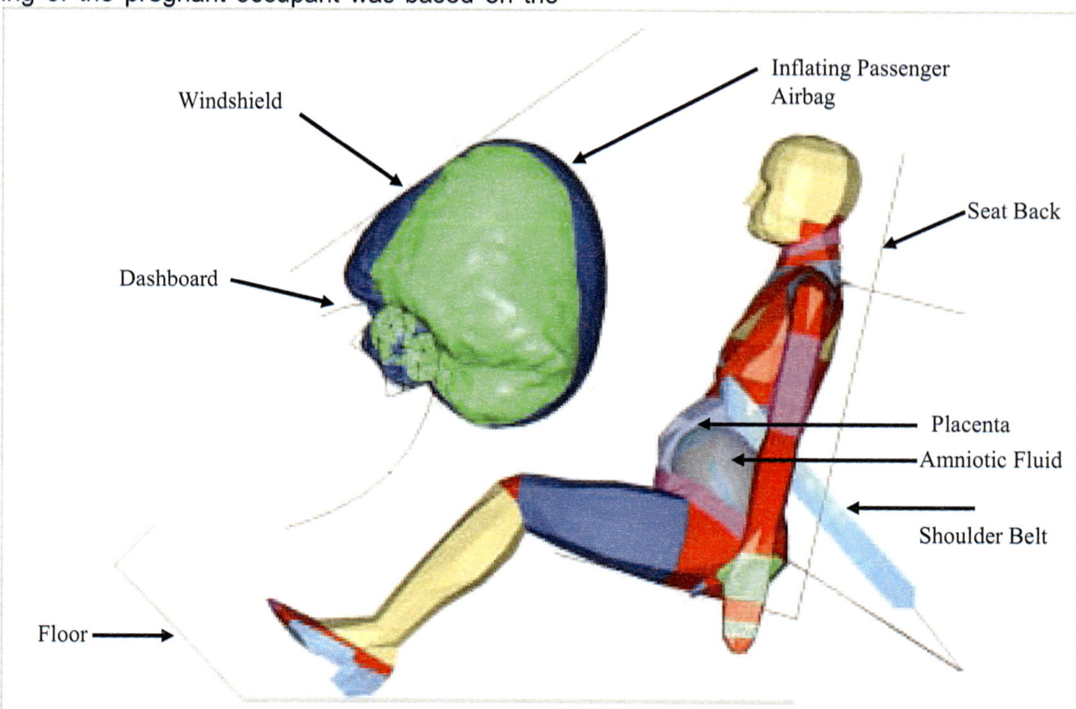

Figure 1: Pregnant occupant in passenger-side interior

To determine the effect of the vertical position of the lap-belt on the occupant and the uterus, two additional simulations were run (Table 2). The previously mentioned passenger compartment with finite element belts was used for frontal impact simulations. The right-front passenger seat was chosen because the effects of the steering wheel are not present and therefore the occupant loading is dominated by the restraint system (Kent, 2001). The lap-belt was varied through three locations (Figure 2); the recommended placement (position 1), directly over the placenta (position 2), and above the uterus (case 3). The initial lap-belt location is representative of the ACOG recommended position. The top belt placement represents a belt that has slid over the top of the uterus. Positioning of the belts was accomplished by placing a straight belt at a given vertical location and then pulling the belt across the abdomen until the belt was taut.

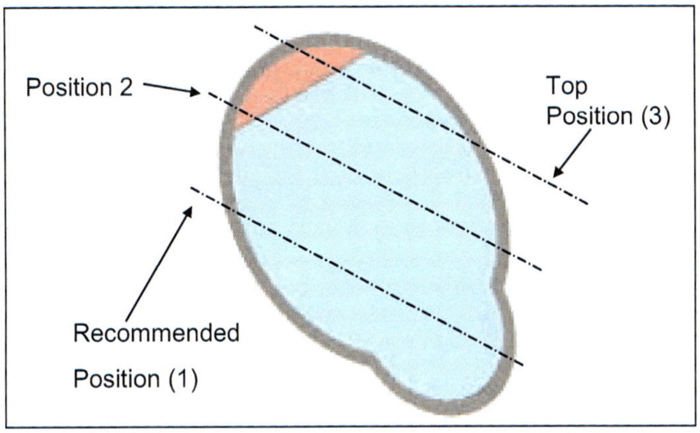

Figure 2: Lap-belt placement with respect to the uterine model (belt centerlines shown)

RESULTS

PART 1: OCCUPANT POSITION - Both strain in the uterus and maternal responses were considered with respect to fetal outcome (Table 1). Simulations in which the occupant was positioned in the passenger-side interior resulted in lower peak uterine strains then in the driver-side interior for all restraints tested (Figure 3). Substantial reductions were seen for the unrestrained and 3-pt belt cases. For belted simulations, the peak strain is 26% to 36% less in passenger-side simulations compared to driver-side simulations even though the forward motion of the occupant is roughly equal between simulations with the same restraint. The key difference in the tests is the presence or absence of the steering wheel. In the driver-side configuration, the occupant contacts the steering wheel to some degree in all the configurations tested. A lower peak strain is recorded in the unrestrained cases because the abdomen does not contact the steering wheel, due to the seatbelts in the belted cases and due to the contact between the head and the windshield in the unrestrained case (Figure 4). For the two belted cases, the occupant does not approach the dashboard and therefore, strain is primarily due to inertial effects. The main effect of varying the occupant position, therefore, appears to be to alter the abdominal loading pattern from contact in the driver-side cases to inertia in the passenger-side cases.

Table 1: Test matrix and results for Part 1: Occupant Position

Position	Restraint	Strain (%)	HIC	VC	Chest Deflection (mm)
Driver	None	60.8	156	0.72	88.7
Driver	3-pt Belt	52.6	185	0.12	35.6
Driver	Belt + Airbag	33.0	114	0.24	32.9
Passenger	None	28.2	2820	0.33	32.7
Passenger	3-pt Belt	33.7	181	0.30	51.5
Passenger	Belt + Airbag	24.4	140	0.27	47.8

The importance of examining the maternal response is highlighted in the unrestrained passenger-side case. Although this simulation produced a low peak strain, based on the HIC value of 2820, it is reasonable to predict maternal death. This elevated value is the result of severe contact between the occupant's head and the dashboard (Figure 4). HIC scores for the remaining simulations were generally low and consistent between seating position. Thorax response for the unrestrained occupant shows the same trend as for the strain, where the limited contact between the thorax and any vehicle surface reduces the passenger injury risk as compared to the driver response. For the restrained occupant, a slight increase is seen in thorax injury risk with the removal of the steering wheel. This is a result of the contact between the steering wheel and the pregnant abdomen in driver-side simulations reducing the load applied to the shoulder belt as compared to the passenger-side.

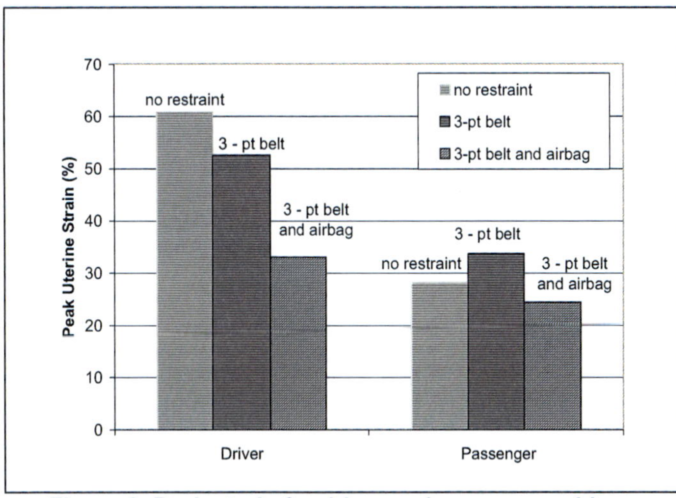

Figure 3: Peak strain for driver and passenger-side simulations

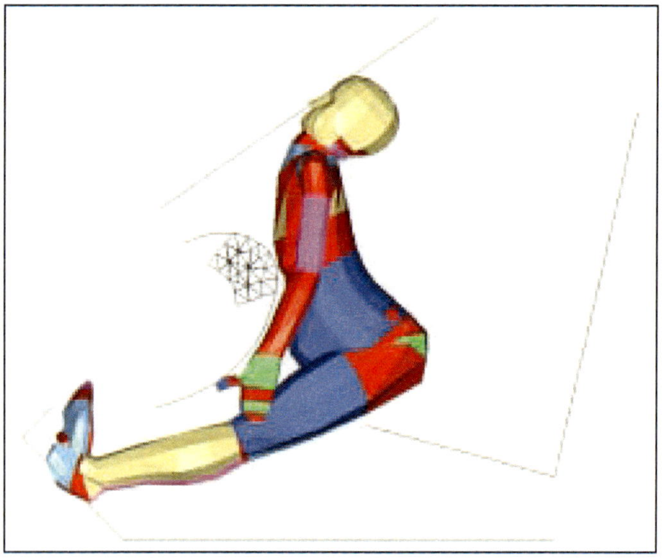

Figure 4: Unrestrained passenger simulation, HIC = 2820

PART 2: LAP-BELT LOCATION - The effect of lap-belt height is clearly seen in close-ups of the uterus, in particular when comparing the recommended belt position and location 2 (Figure 5). Maternal responses between the recommended lap-belt position (position 1) and position 2 are relatively unchanged (Table 2). However, the peak strain for position 2 is nearly three times that of position 1 and substantially higher then the 60% tissue limit (Figure 5). The peak uterine strain was approximately equal for the recommended and top belt locations, however the top belt location led to submarining, causing the maternal injury risk to increase (Figure 6). The height of the top belt placement positioned the lap-belt to partially compress the chest and therefore increased the VC and chest deflection values compared to the other belt placements.

Table 2: Test matrix and results for Part 2: Lap-belt Location

Lap-belt Location	Strain (%)	HIC	VC	Chest Deflection (mm)
Position 1	33.7	181	0.30	51.5
Position 2	96.9	172	0.24	44.6
Position 2	35.2	216	0.50	56.0

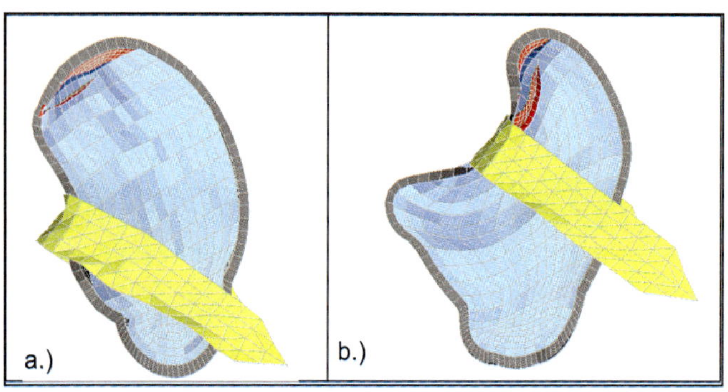

Figure 5: Uterus deformation at peak strain for a.) recommended belt placement and b.) location 2

DISCUSSION

Overall, the crashes simulated were generally safe for the pregnant occupant with unrestrained cases producing the highest risk. There is a high probability that placental abruption would occur in the driver-side, unrestrained, frontal impact simulation. In the passenger–side simulation, there is a near 100% risk of life-threatening maternal brain injury, and therefore a high risk for fetal death. The use of a 3-pt. belt, as well as an airbag, reduces the risk to the pregnant women and the fetus. The difference in abdominal clearance between the driver-side and passenger-side simulations played an important role in peak strain in the uterine wall. The strain was 26% to 54% less in passenger side simulations, primarily due to the presence or absence of the steering wheel. Based on the results of this study it is recommended that, when practical, the pregnant woman ride in the passenger seat with a 3-point seatbelt and airbag with the seat positioned as far rearward as possible.

Placental abruption is believed to occur when the strain in the uterine wall exceeds 60%. The risk for placental abruption is largest for high strains that occur near the placenta. Therefore, the peak strain was recorded in uterine elements near the placenta. As the lap-belt approaches the height of the placenta, which is located at the top of the uterus, the strain increases. Position 2, where the lap-belt is directly loading the uterus at the placental location, produced the highest recorded strain. Once the lap-belt height is above the placenta, the strain decreases, with the strain for the top belt position matching that seen for the recommended belt location. However, it is important to note that there is increased risk to the mother with incorrect lap-belt placement, including elevated head and chest injury response. This is important because the best way to protect the fetus is to protect the mother.

The kinematics of the pregnant occupant were similar for the various belt locations simulated. The main difference is the appearance of submarining seen in simulations with elevated lap-belt height. Because the lap-belt was at or above the umbilicus level, the forward motion of the occupant was reduced at chest level, while the forward motion of the pelvis and lower extremities was increased. Although lower extremity injuries are not typically life-threatening, trauma to the pregnant occupant can complicate the pregnancy. Coupled with the increased strain values recorded, the elevated lap-belt height scenario results in an increased risk of adverse fetal outcome and should be avoided. To accomplish this, the 3-point belt should initially be positioned in the ACOG recommended position, with the lap-belt placed under the abdomen, over both anterior superior iliac spines and the pubic symphysis, and the shoulder harness placed between the breasts. The occurrence of belt slippage under normal driving conditions complicates protection of both the mother and fetus. A belt system for the pregnant driver should be designed to prevent the lap-belt from sliding in the vertical direction.

Limitations - Several assumptions were made in the formulation of the finite element uterine model. The uterus, placenta, and amniotic fluid were assumed linear, elastic, isotropic solid materials. The uterus is known to be anisotropic and viscoelastic, however there is not currently enough data to apply these material properties to the model. The amniotic fluid was defined as a solid because Madymo does not allow fluids to be defined. The geometry, although an improvement over previous modeling efforts, is still simplified. It is recommended that these improvements be included in future versions of the FE model.

As a whole, the model was designed to be generic. The occupant is a standard fifth percentile size, with an average abdominal contour, weight gain, and seated posture for 30 weeks gestation. Due to the generic nature of the model, the vehicle model used, including belts and airbags, is also generic. A parameter sensitivity analysis has not been completed. As a result, the pregnant model is best suited to give overall trends, and is not suited to produce the exact outcome from any specific motor vehicle crash based on the current configuration. It is likely that many factors, including individual size differences, seat positioning, belt geometry and attachment locations, among others, will influence the loads applied to the occupant and therefore, the occupant response. It is the goal of the authors to show the general trends in occupant response to various crash scenarios, increase awareness, and provide a starting point for future modeling attempts, where these detailed analyses would be of more benefit.

CONCLUSION

The goal of this study was to elucidate the effect of occupant seating and seatbelt placement on the risk of adverse fetal outcome from a motor vehicle crash. Unrestrained, 3-pt belt, and 3-pt belt plus airbag tests were simulated with the pregnant occupant model in both a driver-side and passenger-side vehicle interior in frontal impacts. Peak uterine strain was reduced by 30% to 50% in passenger-side simulations vs. driver-side simulations, primarily due to increased distance between the abdomen and the nearest vehicle surface, namely the steering wheel for driver-side tests and the dashboard for passenger-side tests.

Additionally, simulations were performed to evaluate the effect of lap-belt vertical position on abdominal response. Simulations were run with the occupant in the passenger compartment to isolate the effect of the lap-belt from contact with internal vehicle surfaces. Simulations demonstrated that the vertical position of the lap-belt can increase fetal risk by a factor of three. Based on the results of this study it is recommended that, when possible, the pregnant woman ride in the passenger seat with a 3-pt belt. The seat should be positioned as far rearward as possible and the lap-belt should be worn as low as possible.

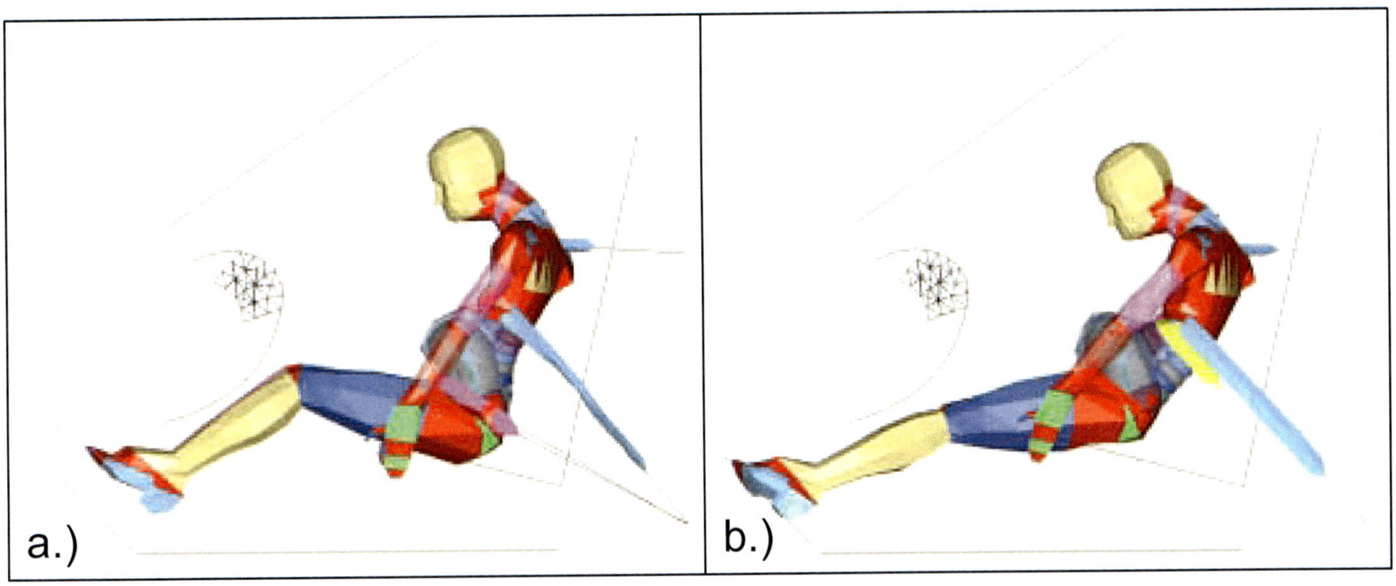

Figure 6: Pregnant occupant at peak strain, a.) recommended belt placement and b.) top belt location

REFERENCES

1. American College of Obstetricians and Gynecologists (1998). Obstetric aspects of trauma management. Educational Bulletin Number 251.

2. Jernigan MV (2002). Statistical analysis and computational modeling of injuries in automobile crashes. MS Thesis, Virginia Poltechnic Institute and State University.

3. Kent RW, Crandall JR, Bolton J, Prasad P, Nusholtz GS, Mertz HJ (2001). The influence of superficial soft tissues and restraint condition on thoracic skeletal injury prediction. Stapp Car Crash Journal, vol. 45, 2001-22-0008, 183-204.

4. Klinich KD, Schneider LW, Rupp JD, Eby B, Pearlman MD (1999a). Challenges in frontal crash protection of pregnant drivers based on anthropometric considerations. SAE technical paper series; 1999-01-0711: 1-23.

5. Klinich KD, Schneider LW, Moore JL, Pearlman (1999b). Investigations of crashes involving pregnant occupants. UMTRI-99-29.

6. Moorcroft DM, Duma SM, Stitzel JD, Duma GG (2003). A finite element and multi-body model of the pregnant occupant for the analysis of restraint effectiveness. SAE technical paper series. 2003-01-0157.

7. Pearlman MD, Tintinalli JE, Lorenz RP (1990a). A prospective controlled study of outcome after trauma during pregnancy. American Journal of Obstetrics and Gynecology, 162, 1502-1510.

8. Pearlman MD, Tintinalli JE, Lorenz RP (1990b). Blunt trauma during pregnancy. New England Journal of Medicine, 323, 23, 1690-1613.

9. Rupp JD, Klinich KD, Moss S, Zhou J, Pearlman MD, Schneider LW (2001). Development and testing of a prototype pregnant abdomen for the small-female Hybrid III ATD. Stapp Car Crash Journal, 45, 61-78.

CONTACT

Stefan Duma
Virginia Tech – Wake Forest, Center for Injury Biomechanics
114 Randolph Hall, Blacksburg, Virginia 24061
Phone: 540-231-3945,
Fax: 540-231-9100,
Email: duma@vt.edu

ABOUT THE AUTHORS

Stefan Duma, Ph.D.

Stefan Duma is a professor and department head of biomedical engineering at Virginia Tech, which is the joint program between Virginia Tech and Wake Forest University called the School of Biomedical Engineering and Sciences. Dr. Duma is responsible for 51 tenured and tenure track faculty with over $25 million in annual expenditures. In addition, Dr. Duma was the founding director of the Center for Injury Biomechanics (CIB) and has received external funding from the NIH, CDC, NSF, DOT-NHTSA, DOD, and industry partners including Toyota, Ford, and GM. He has published over 200 technical papers in the field of injury biomechanics, including 80 peer-reviewed journal papers. His research has been recognized with the AAAM Best Scientific Paper Award in 2004 and 2007, two AAAM Best Student Paper Awards, the Stapp Conference Best Paper Award, and four Best Student Awards at Stapp Conferences. Dr. Duma is a Fellow in AAAM. He is married to Dr. Chrissy Duma and they have two children, Lauren age 9 and Brock age 7.

Greg G. Duma, M.D.

Dr. Duma received his undergraduate degree in chemistry from the University of Tennessee in Knoxville, TN in 1987. Afterward, he received his M.D. from the University of Tennessee Center for Health Sciences in Memphis, TN. He completed his residency in the field of obstetrics and gynecology at the University of Cincinnati. Dr. Duma was an associate professor in clinical obstetrics and gynecology at The University of Cincinnati and was the Director of Gynecologic Laparoscopic Surgery at the medical center. Currently Dr. Duma is an adjunct professor in the Virginia Tech – Wake Forest University School of Biomedical Engineering and Sciences, and he is in private practice in Covington, KY.

Melissa C. Hulse, M.S.

Melissa Hulse received her M.S. degree from Virginia Tech in Industrial Systems Engineering/Human Factors in 1988. Having published over 20 publications and two book chapters, her main focus has been human factors as it pertains to automotive safety. She has worked at the University of Idaho, University of Iowa, and the Virginia Tech Transportation Institute.

In 2009, Ms. Hulse began working at the Center for Injury Biomechanics at Virginia Tech. Her research has included seatbelt safety, pregnancy trauma statistics, and child toy safety.

Kathleen D. Klinich, Ph.D.

Kathleen DeSantis Klinich is an assistant research scientist in UMTRI's Biosciences Group. She earned a Ph.D. in mechanical engineering from the University of Michigan and B.S. and M.S. degrees in mechanical engineering, as well as a B.A. in technical writing, from Case Western Reserve University. Prior to joining UMTRI in 1996, she worked as a research engineer for TRC, Inc. at NHTSA's Vehicle Research and Test Center.

Dr. Klinich's research interests focus on protecting occupants in motor-vehicle crashes. She has experience in the analysis of motor-vehicle crashes and crash databases, child passenger safety, pregnant occupant safety, crash dummy design, laboratory reconstruction of real-world loading events, injury criteria development, occupant-anthropometry and posture evaluation, child restraint usability, and finite element modeling. She has authored dozens of technical papers on these topics. She is active in the SAE Child Restraints and SAE Dummy Testing Equipment Committees.

Sarah J. Manoogian, Ph.D.

Sarah Manoogian joined Biodynamic Research Corporation in 2008 after completing her doctorate in mechanical engineering at Virginia Polytechnic Institute and State University. She is a licensed engineer in the state of Texas and consults in the fields of biomechanics, injury causation analysis, and accident reconstruction. She currently holds an ACTAR accreditation for accident reconstruction. Her primary areas of research are evaluating adverse fetal outcomes for pregnant women and analyzing head accelerations in sports. Additionally, she has published papers on everyday accelerations, neck injuries, bone strength, soft tissue properties, and various computational modeling tools.

Dr. Manoogian has researched several areas that contribute to the advancement of pregnant occupant protection. Using a computational model of a pregnant female, she has considered the effects of accelerations from everyday activities to loading in a 35 mile per hour barrier impact. She has quantified the material response of pregnant uterine tissue in dynamic biaxial loading. Dr. Manoogian has also tested placental tissues at various strain rates and determined the response of the two primary layers of the tissue. Her research studies on uterine and placental tissues have provided material properties for use in pregnant occupant models that experience dynamic loading.

Jonathan D. Rupp, Ph.D.

Dr. Rupp is an assistant research scientist at the University of Michigan Transportation Research Institute and a research assistant professor in the University of Michigan Department of Emergency Medicine. For the past 15 years, he has worked to improve vehicle safety by collecting and analyzing data on vehicle performance and occupant injury in crashes, characterizing human injury response and restraint system effectiveness, and developing tools that assess injury potential in crashes. Dr. Rupp currently directs research to study the effects of occupant characteristics (size, shape, gender, and age) on adult and pediatric impact response and injury outcome in crashes and the ability of new active and passive vehicle safety systems to mitigate the potential for injury in crashes. He is also actively developing new techniques for combined analysis of different crash-injury databases.

Lawrence W. Schneider, Ph.D.

Dr. Lawrence Schneider is a research professor, head of the Biosciences Group of the University of Michigan Transportation Research Institute (UMTRI), and an associate director of UMTRI. His research in biomechanics over 37 years spans a wide range of topics, including child and adult anthropometry for restraint system and vehicle interior design, studies of human impact response and injury tolerance, development of advanced anthropomorphic test devices and associated injury criteria, improved transportation safety for disabled travelers, and the application of in-depth crash investigations and crash/injury data to studying injury causation and assessing the effectiveness of occupant-protection systems.

Dr. Schneider is lead author or co-author on more than eighty peer-reviewed publications and more than fifty technical reports. He is a member of the Stapp Car Crash Conference Advisory Committee and has served as general chairman of five Stapp conferences. He is also an editor-in-chief of the Stapp Car Crash Journal, a member of the Stapp Student Awards Committee, a founding member of the Stapp Association, and member of the Stapp Board of Directors. He is also the director of the Rehabilitation Engineering Research Center on Wheelchair Transportation Safety funded by a five-year grant from the National Institute on Disability and Rehabilitation Research.

Joel D. Stitzel, Ph.D.

Dr. Joel D. Stitzel, Ph.D. is an associate professor of biomedical engineering at Wake Forest University School of Medicine. He is a core faculty member of the Virginia Tech – Wake Forest University School of Biomedical Engineering and Sciences. He is currently program leader and director of the Wake Forest Campus of the Virginia Tech – Wake Forest University Center for Injury Biomechanics. He received a B.S. in Engineering Science and Mechanics from Virginia Tech, an M.S. in Biomedical Engineering from the Medical College of Virginia, and Ph.D. in Mechanical Engineering from Virginia Tech in 2003.

His dissertation work involved modeling of the human rib cage and thorax. He is the engineering co-PI of Wake Forest University School of Medicine's Crash Injury Research and Engineering Network center. He is the PI of the Global Human Body Models Consortium Integration Center, the lead center on a global endeavor to create the industry-standard finite element model of the human body for trauma prediction. His interests include child and elderly safety, finite element modeling techniques, soft tissue biomechanics generally and lung injury, eye injury, brain injury, and carotid injury in particular. He has authored papers on biomechanics of the eye, pregnant occupants, anthropometry and its relationship to injury, and lung trauma. His future goals include enhancing the trauma research interface between clinicians and engineers, and improving human safety. His efforts focus on the prediction, mitigation, and prevention of trauma in a variety of scenarios, including automotive, sports, and military applications.